Serpentine Geoecology of Western North America

Serpentine Geoecology of Western North America

Geology, Soils, and Vegetation

E. B. Alexander
R. G. Coleman
T. Keeler-Wolf
S. Harrison

2007

OXFORD
UNIVERSITY PRESS

Oxford University Press, Inc., publishes works that further
Oxford University's objective of excellence
in research, scholarship, and education.

Oxford New York
Auckland   Cape Town   Dar es Salaam   Hong Kong   Karachi
Kuala Lumpur   Madrid   Melbourne   Mexico City   Nairobi
New Delhi   Shanghai   Taipei   Toronto

With offices in
Argentina   Austria   Brazil   Chile   Czech Republic   France   Greece
Guatemala   Hungary   Italy   Japan   Poland   Portugal   Singapore
South Korea   Switzerland   Thailand   Turkey   Ukraine   Vietnam

QE
462
.U4
S47
2007

Copyright © 2007 by Oxford University Press, Inc.

Published by Oxford University Press, Inc.
198 Madison Avenue, New York, New York 10016

www.oup.com

Oxford is a registered trademark of Oxford University Press

All rights reserved. No part of this publication may be reproduced,
stored in a retrieval system, or transmitted, in any form or by any means,
electronic, mechanical, photocopying, recording, or otherwise,
without the prior permission of Oxford University Press.

Library of Congress Cataloging-in-Publication Data
Serpentine geoecology of western North America : geology, soils, and vegetation /
E.B. Alexander . . . [et al.].
p. cm.
Includes bibliographical references and index.
ISBN-13 978-0-19-516508-1
ISBN 0-19-516508-X
1. Rocks, Ultrabasic—West (U.S.)   2. Rocks, Ultrabasic—Northwest, Canadian.
3. Soils—Serpentine content—West (U.S.)   4. Soils—Serpentine content—Northwest,
Canadian.   5. Environmental geology—West (U.S.)   6. Environmental
geology—Northwest, Canadian.   I. Alexander, Earl B.
QE462.U4S47 2006
581.7'2—dc22   2005034397

9 8 7 6 5 4 3 2 1
Printed in the United States of America
on acid-free paper

PREFACE

Ultramafic, or "serpentine" rocks are unique in the continental crust of the earth in that their chemical composition is similar to that of the mantle, which is overlain by 30–70 km of continental crust, or much thinner ocean crust. There are several different kinds of ultramafic rocks that reach the surface of the earth in different ways. Most of those in western North America are derived from the ocean floor. This is a fascinating story, but the main focus of this book is the effects of ultramafic rocks on terrestrial landscapes and ecosystems.

The unique ultramafic rock chemistry is largely inherited by serpentine soils, which support unique suites of plants. Because serpentine soils and vegetation are so unique, they warrant this comprehensive book. This book is a major source of information about serpentine soils, plant communities, and the organisms inhabiting them.

Although ultramafic rocks occur around the world, on all continents, from polar to equatorial regions, we chose to limit our geographic coverage to western North America and devote a large portion of our book to the principles of serpentine soil formation and serpentine plant ecology. Ultramafic rocks are well distributed from latitudes 24°N to 68°N in western North America. Practically all principles of serpentine soil formation, vegetation distribution, and floristic development can be illustrated with examples from this region. These principles are applicable to serpentine soils on all continents, making the book pertinent for pedologists and ecologists around the world. Some human health problems, such as those related to asbestos in dust and chromium in drinking water, are addressed, and there is a section on revegetation of serpentine soils.

The topics of this book, which together make up geoecology, span many disciplines. These disciplines have become so specialized that it is hardly possible for an expert in any one discipline to address the entire spectrum of serpentine geoecology with authority.

Therefore, four coauthors from three disciplines have collaborated to make *Serpentine Geoecology of Western North America* a comprehensive volume on the geoecology of serpentine (ultramafic) landscapes.

Although our aim was to make the presentation as interdisciplinary as possible, it was necessary to make some assignments of labor. Robert Coleman, a geologist, authored part I, Geology and Hydrology (chapters 2–4), with help from Earl Alexander on chapter 4. Alexander, a pedologist, wrote part II, Soils and Life in Them (chapters 5–8), with help from Susan Harrison on chapter 7. Todd Keeler-Wolf and Harrison, plant ecologists, wrote part III, Plant Life on Serpentines (chapters 9–12). Part IV, Serpentine Domains of Western North America (chapters 13–22), was jointly written by all of us. Coleman was responsible for the choices of domains and localities within domains and for the maps showing their locations. Part V, Social Issues and Epilogue (chapters 23 and 24), was a coordinated effort of all four coauthors.

This book was written for readers with minimal scientific backgrounds, yet with details for specialists in soils, plant ecology, and other environmental disciplines. Some parts will be more appealing to general readers, and other parts will be referred to by specialists.

A. K. Kruckeberg has written two books that address many of the topics in our book, but they are more directly oriented to lay readers: *Geology and Plant Life* and *Introduction to California Soils and Plants: Serpentine, Vernal Pools, and Other Geobotanical Wonders.* These books were written from a botanical perspective and do not have the depth of serpentine geology and soils coverage that is a unique feature of our book.

Acknowledgments

For providing valuable information or reviews of the book, we thank Kerry Arroues, Doris Baltzo, Cici Brooks, Rebecca Burt, Dennis Churchill, Mark Coleman, Andrew Conlin, Harry Edenborn, Steve Edwards, Susan Erwin, James Frampton, Lisa Hoover, Gary Lewis, Carol Kennedy, Jaime Kienzle, Patricia Krosse, Dig McGahan, Jolanta Mesjasz-Przyblowicz, Neil Munro, Ivana Noell, Carolyn Parker, Jon Rebman, Bill Reed, Judy Robertson, Brad Rust, Randy Southard, Darlene Southworth, Richard Strong, David Swanson, Steven Trimble, and Melanie Werdon.

CONTENTS

1. Introduction   3

I. Geology and Hydrology
    2. Nature of Ultramafics   13
    3. Mineralogy and Petrology of Serpentine   18
    4. Water in Serpentine Geoecosystems   27
  Part I References   33

II. Soils and Life in Them
    5. Nature of Soils and Soil Development   37
    6. Serpentine Soil Distributions and Environmental Influences   55
    7. Animals, Fungi, and Microorganisms   79
    8. Serpentine Soils as Media for Plant Growth   97
  Part II References   142

III. Plant Life on Serpentine
    9. Responses of Individual Plant Species to Serpentine Soils   159
    10. Serpentine Plant Assemblages: A Global Overview   175
    11. Serpentine Plant Life of Western North America   190
    12. Serpentine Vegetation of Western North America   204
  Part III References   245

IV. Serpentine Domains of Western North America
    13. Baja California, Domain 1   259
    14. Sierra Motherlode, Domain 2   271
    15. Southern California Coast Ranges, Domain 3   288

16. Northern California Coast Ranges, Domain 4   308
17. Klamath Mountains, Domain 5   329
18. Blue Mountains, Domain 6   350
19. Northern Cascade-Fraser River, Domain 7   358
20. Gulf of Alaska, Domain 8   372
21. Denali-Yukon, Domain 9   382
22. Northern Alaska–Kuskokwim Mountains, Domain 10   388
Part IV References   395

V. Social Issues and Epilogue
23. Serpentine Land Use and Health Concerns   413
24. Synthesis and Future Directions   417
Part V References   422

Appendices
A. Nature of Minerals in Serpentine Rocks and Soils   427
B. Characteristics of the Chemical Elements–Ionic Properties and Toxicities   436
C. Soil Classification   442
D. Kingdoms of Life   460
E. Water Balance by the Thornthwaite Method   463
F. Western North America Protected Areas with Serpentine   467
G. California Plant Taxa Endemic to Serpentine   473

Glossary   483
Plant Index   491
General Index   501

Serpentine Geoecology of Western North America

1   Introduction

Ultramafic, or colloquially "serpentine," rocks and soils have dramatic effects on the vegetation that grows on them. Many plants cannot grow in serpentine soils, leaving distinctive suites of plants to occupy serpentine habitats. Plants that do grow on serpentine soils may be stunted, and plant distributions are commonly sparse relative to other soils in an area (fig. 1-1). Plant communities on serpentine soils are usually distinctive, even if one does not recognize the plant species. Because of these distinctive features, ultramafic rocks and serpentine soils are of special interest to all observers of landscapes.

Geology underlies both conceptually and literally the distinctive vegetation on serpentine soils. The occurrence of special floras on particular substrates within particular regions makes rocks and soils of key significance to plant evolution and biogeography. Sophisticated interpretations of these interrelationships require a combined knowledge of geology, soils, and botany that few people possess. Even highly specialized professionals generally lack the requisite expertise in all three disciplines. The science of ecology, which in principle concerns interactions among all aspects of the environment, seldom incorporates a deep understanding of rocks and soils. Some scientists have attempted to bridge this gap through creating a discipline known as geoecology (Troll 1971, Huggett 1995), which forms the basis for our interdisciplinary exploration of serpentine rocks and soils in western North America.

The term "serpentine" is applied in a general sense to all ultramafic rocks, soils developed from them, and plants growing on them. Ultramafic rocks are those with very high magnesium and iron concentrations. The word serpentine is derived from the Latin word *serpentinus*, meaning "resembling a serpent, or a serpent's skin," because many serpentine rocks have smooth surfaces mottled in shades of green to black.

Figure 1-1 Contrasting vegetation on serpentine and adjacent nonserpentine soils–open forest and shrub on serpentine soil (S) compared to dense forest with rocky openings on nonserpentine soils (N) of a granitic pluton and surrounding accreted allochthonous rocks of North Star Mountain in the Rattlesnake Creek terrane, Tehama County, California.

The distinctive chemistry of ultramafic rocks and serpentine soils restricts the growth of many plants and makes them refuges for plants that thrive in serpentine habitats, including serpentine endemics (species that are restricted to these soils) and other species that have evolved means of tolerating these habitats. Often the means of tolerance include visible adaptations such as slow growth and relatively thick, spiny foliage. Compared with many other rocks and soils, then, serpentine soils typically support vegetation that appears strikingly different even to the casual observer. Beneath the surface appearance of this unique ecological condition, which has been called the "serpentine syndrome" (Jenny 1980), lies a world of geoecological complexity that we endeavor to explain here.

Ultramafic rocks occur in orogenic belts and on stable interior platforms on every continent. In western North America, ultramafic rocks of orogenic belts occur in a cordillera, or chain of mountains, that stretches from the Brooks Range in Alaska through Yukon Territory, British Columbia, Washington, Oregon, California, and Mexico, to Guatemala and Costa Rica. From Guatemala, traces of ultramafic rocks can be seen in a trend that follows an orogenic belt through Cuba and around the Antilles to reappear in the Andes Mountains. Our main consideration is the area from the Brooks Range to Baja California Sur, which extends from the Arctic to the subtropics and comprises a tremendous diversity of serpentine soils and plant communities.

Although the serpentine syndrome is most closely associated with plant communities, there are some accounts of the effects of serpentine soils and plants on animals and microorganisms. The best known examples of serpentine effects on animal distributions in western North America concern various butterflies whose larval host plants are confined to serpentine soils. There are so few accounts of the types of fungi, algae, and animals that live on ultramafic rocks and serpentine soils in western North America that it is necessary to refer to examples from other areas.

A book by Kruckeberg (1984) described the flora, vegetation, geology, soils, and management problems of serpentine in California from a primarily botanical viewpoint. Our book considers a broader geographic scope, and also provides more in-depth accounts of the geology (part I), soils (part II), and plant communities (part III) of serpentine soils area in western North America. In addition, there have been many studies since 1984 on the flora, plant ecology, and vegetation of serpentine in this region, and we explore this research in the context of a world view of serpentine flora and vegetation (part III). Part IV of this book provides a detailed region-by-region account of serpentine soils and vegetation in western North America. Part V addresses social and health concerns.

1.1 Geoecosystems

The concern in ecology is relationships of organisms to their environment and the interactions among organisms. A basic unit in ecology is the ecosystem, which is broadly defined as including all the living organisms in a given region, plus all the abiotic (non-living) elements that influence them. However, in practice, studies of ecosystems do not always explicitly consider the influences of rocks and soils. Troll (1971) first coined the term "geoecology" for a branch of science considering the relationships among rocks, soils, biotic communities, and atmosphere (climate). Geoecosystems, or interacting systems of the above components, are the basic units of geoecology (Huggett 1995). Although serpentine habitat presents one of the clearest cases of the need to adopt a geoecological perspective, there are many other examples of the strong effects of geology on plant life around the world (Kruckeberg 2004).

Geoecosystems have a practically unlimited number of components, but these can be considered to fall within four major compartments: rocks (the lithosphere), soils (the pedosphere), living communities (the biosphere), and the atmosphere. This book has separate sections on three of the major compartments of geoecosystems: rocks, soils, and plant communities. We follow many other geoecologists in using plant communities, or vegetation types, as surrogates for biotic communities as a whole. This is largely because communities of macroscopic animals are not as easily defined, measured, and mapped at large geographic scales as are plant communities, nor are they as clearly responsive to the influences of rocks and soils. Microorganisms are ubiquitous in all four compartments, but we consider them under the heading of the pedosphere because of the major role they play in the soil environment. The atmosphere is not known to be strongly affected by serpentine substrates because it is rapidly mixed to limit the residence time of any such effects.

Mass and energy flow in all directions in geoecosystems, particularly from soils and the atmosphere (and sun) to plants and from plants back to soils. A large portion of each vascular plant (with few exceptions, such as epiphytes) is below ground, in soil, and many animals live in soils. Considering all the interactions among components of geoecosystems, a landscape investigation that ignores any of the biotic or abiotic components of geoecosytems and their interactions is incomplete.

1.2 Ultramafic Rocks, Minerals, and Chemistry

Strictly speaking, serpentine is a mineral, or group of minerals, with the chemical formula $Mg_3Si_2O_5(OH)_4$. Rocks are aggregates of minerals, generally more than one. Rocks that are composed of serpentine or related magnesium silicate minerals are commonly called serpentine rocks, even though a geologist would more accurately call them ultramafic rocks. Peridotite and serpentinite (fig. 1-2) are the main serpentine, or ultramafic rocks, in western North America. The state rock of California was named serpentine before geologists decided that they would add "-ite" to distinguish the rock, serpentinite, from the mineral serpentine (Wagner 1991).

Soils may inherit minerals from rocks as the rocks decompose, or weather, and new minerals may be formed in soil environments. The former (inherited minerals) are sometimes called primary minerals and the latter called secondary minerals. Minerals common in ultramafic rocks and the soils derived from them are described in part I and in appendix A. Some minerals in ultramafic rocks are asbestos minerals, and since the hazards of asbestos fibers from ultramafic rocks and soils have been investigated extensively, we discuss environmental asbestos from a geological perspective.

Figure 1-2 Peridotite (left) and serpentinite (right) cobbles. The peridotite has a warty (or hobnail) appearance, because olivine has been weathered away leaving the less-weatherable pyroxene standing in distinct relief. Tectonic shearing has polished the surface of the serpentinite.

The origin of serpentine, as described in part I, lies at the heart of understanding the serpentine syndrome. The earth has three major layers, with the heaviest materials at the center. These layers are a core at the center of the earth, a crust at the surface, and a mantle below the crust. Ultramafic rocks are derived more directly from the mantle than are other rocks in the crust. They are found at the earth's surface primarily because of the geologic processes of subduction of ocean plates, and subsequent orogeny or mountain uplift. Also, there are some layered ultramafic bodies, such as the Stillwater complex in the Rocky Mountains, that have formed on continents. Volcanic ultramafic rocks, which were more common when the earth was young and hot are now sparse.

Given their origin, it is not surprising that ultramafic rocks have chemical compositions more similar to the mantle than to the crust. The mantle is composed predominantly of ferromagnesian silicates, so the main chemical elements in ultramafic rocks are magnesium, iron, silicon, and oxygen. Aluminum, calcium, sodium, potassium, and phosphorous are more concentrated in rocks of the crust. Plants and other organisms that have developed since a crust formed on the earth and oxygen became more readily available about 2 Ga (billion years ago), are generally adapted to higher concentrations of calcium, potassium, and phosphorous than are available in serpentine substrates. Those plants that have evolved tolerances to lower concentrations of these elements and higher concentrations of magnesium are the ones that prevail on ultramafic rocks and serpentine soils.

1.3 Serpentine Soils and the Serpentine Syndrome

Serpentine soils form by the weathering of ultramafic rocks, or they form in colluvium, alluvium, or glacial deposits derived from ultramafic rocks. In the initial stage of soil development, plant occupancy is sparse on very shallow or fragmental soils. At this stage, serpentine landscapes are "barrens" (some severely eroded serpentine soil landscapes are also barrens). As weathering proceeds, serpentine soils become deeper or finer textured, and they support more productive plant communities. Increased soil depth is a major benefit to serpentine plant communities, mainly because deeper and finer textured soils can retain more water for plants to extract as they need it. Nevertheless, the "serpentine syndrome" of soil infertility and impaired plant growth (Jenny 1980) persists even on deeper and finer textured soils, because of unfavorable nutrient balances in them. Only after millions of years of leaching and nutrient recycling by plants does the serpentine soil nutrient balance become favorable for the growth of a broader spectrum of plants and more productive plant communities. In some very old serpentine soils, nickel and cobalt are concentrated to economically extractable levels. These very old soils are essentially ore deposits, although they may be more valuable for the biotic communities that they support.

Given the broad range of serpentine soils—from shallow to deep, from coarse-loamy to clayey, from gray to red, from very dry to wet, and from barren to highly productive—there is no typical serpentine soil that can represent more than a small slice of the large spectrum of serpentine soils. Serpentine soils occur in 9 of the 12 soil orders, which is the highest level of soil classification in soil taxonomy (Soil Survey Staff 1999), and serpentine soils in all of these nine soil orders occur in western North America. Plant communities on these serpentine soils range from humid forest to chaparral (shrubs), grassland, and desert.

1.4 Plant Communities and Vegetation

Plant communities are unique combinations of plants that occur in particular locations under particular environmental conditions. They reflect the interaction of biogeographic history with prevailing environmental influences such as soils, temperature, elevation, solar radiation, slope gradient and aspect, and precipitation (Dunster and Dunster 1996). Here we use the terms "plant community" and "vegetation" synonymously, although the former is often used more in reference to the floristic properties (species identities and abundances), and the latter to the structural properties (for example, percentage cover by trees, shrubs, and herbs) of plant assemblages. Vegetation may be defined as all the plant species of a region and the natural patterns they form across the landscape (Sawyer and Keeler-Wolf 1995). Modern vegetation classification takes both floristics and structure into account. Its goal is to define groupings of species that have a degree of structural and compositional integrity, and that recur predictably under similar habitat conditions throughout a defined geographic area, though with inevitable variations. By focusing on the effects of serpentine on vegetation as well as on individual species, it is possible to obtain a synthetic view of the serpentine effect: for example, to compare groups of species on serpentine to groups of species on adjacent nonserpentine substrates under similar climatic conditions.

The boundaries between serpentine and nonserpentine are often immediately evident in the covering of vegetation. In general, a gradual shift from one suite of plant species to another is the rule in vegetation, with "fuzzy" boundaries the norm. Gradual shifts are brought about by concomitant gradual shifts in moisture, temperature, or soil type across the landscape. However, shifts in vegetation growing on serpentine and nonserpentine substrates generally create the less usual situation—a distinctive "break" in the gradient. These abrupt breaks in vegetation are perhaps most easily noticed by structural differences (e.g., open often stunted trees or shrubs with large openings of bare unstable fractured serpentine, contrasting with dense, relatively tall trees and shrubs off of serpentine). Secondarily, a closer look will reveal, in many situations, a contrasting species composition. In some cases the species differences may be very pronounced with the dominant plant species of adjacent sites being completely different.

The vegetation of serpentine is most distinctive in regions like California where climate and geographic isolation conspire to affect both a physiognomic and a floristic uniqueness compared to adjacent nonserpentine areas. In California, several vegetation types are largely endemic to serpentine with the principal cover of plants made up of species such as leather oak (*Quercus durata*) or coast whiteleaf manzanita (*Arctostaphylos viscida* ssp. *pulchella*) largely restricted to serpentine. Farther north and south in western North America substrate-related shifts in species composition may be more subtle, involving far fewer endemic species, shifting instead to suites of species that may be widespread on harsh sites (physiologically challenging climate or soils). In Baja California species diversity is limited by scant precipitation and erratic patterns of moisture availability. To the north, most of the mountains and valleys in coastal southern Alaska, British Columbia, and northern Washington were glaciated during the Holocene, leaving only several thousand years for the plant communities to accumulate a diversity of species. Extreme cold in interior and northern Alaska and the Yukon Territory have limited species diversification there. Nevertheless, there are some distinct plant community differences from serpentine to nonserpentine on south-facing slopes (locality 10-13, Livengood, in chapter 22). Serpentine barrens occur as far north as interior Alaska (Juday 1992).

Vegetation structural differences tend to be most pronounced in the drier, warmer settings on south- or southwest-facing slopes and less pronounced in cooler, moister settings on north- or northeast-facing slopes. Distribution patterns are described in more detail in part III. By focusing on the effects of serpentine on vegetation as well as on individual species, it is possible to obtain a synthetic view of the serpentine syndrome. Using this approach, we hope we can better understand the relative influence the serpentine syndrome compared to other environmental factors.

1.5 Plant Taxonomy

Because western North America spans three countries and several states or provinces, there is variability in the taxonomic treatment of its flora. Reinterpretation of taxonomy and discovery of new taxa continue to modify the floristic treatments of this area. Classic references such as Hultén (1968) for Alaska, Peck (1961) for Oregon, Hitchcock and Cronquist (1973) for Oregon and Washington, and Abrams (1923–1951) and Ferris (1960) for the Pacific States have been replaced by more recent taxonomic updates such as Kartesz (1994) and Kartesz and Meacham (1999) for all of North America, and Hickman (1993) for California. For British Columbia, Douglas et al. (1998–2002) is the current reference, and for the Yukon Territory, it is Cady (1996). For Baja California, Wiggins (1980) still remains the most consistent reference. Newer treatments are used if available, but we also tried to maintain taxonomic consistency for those species that range widely in the region. Because California is the heart of serpentine endemism in western North America we have chosen to use the most current taxonomy maintained by the Jepson interchange (http://ucjeps.berkeley.edu). This website maintains the most current taxonomic relationships of all California vascular plants, and numerous refinements have been made since the publication of the Jepson Manual (Hickman 1993).

References

Abrams, L.R. 1923–1951. Illustrated Flora of the Pacific States, Washington, Oregon, and California, vols. 1–3. Stanford University Press, Stanford, CA.

Cady, W.J. 1996. Flora of the Yukon Territory. National Research Council of Canada, Ottawa.

Douglas, G.W., D. Meidinger, and J. Pojar. 1998–2002. Illustrated Flora of British Columbia, vols. 1–8. British Columbia Ministry of Sustainable Resource Management and Ministry of Forestry, Victoria.

Dunster, J. and K. Dunster. 1996. Dictionary of Natural Resource Management. University of British Columbia Press, Vancouver.

Ferris, R.S. 1960. Illustrated Flora of the Pacific States, Washington, Oregon, and California, vol. 4. Compositae. Stanford University Press, Stanford, CA.

Hickman, J.C. (ed.). 1993. The Jepson Manual: Higher Plants of California. University of California Press, Berkeley.

Hitchcock, C.L. and A. Cronquist. 1973. Flora of the Pacific Northwest. University of Washington Press, Seattle.

Huggett, R.J. 1995. Geoecology. Routledge, London.

Hultén, E. 1968. Flora of Alaska and Neighboring Territories: A Manual of the Vascular Plants. Stanford University Press, Stanford, CA.

Jenny, H. 1980. The Soil Resource. Springer-Verlag, Berlin.

Juday, G.P. 1992. Alaska Research Natural Areas. 3: Serpentine Slide. USDA Forest Service General Technical Report PNW-GTR-271.

Kartesz, J.T. 1994. A Synonymized Checklist of the Vascular Flora of the United States, Canada, and Greenland. Timber Press, Portland, OR.

Kartesz, J.T. and C.A. Meacham. 1999. Synthesis of the North American Flora. CD-ROM, Version 1.0. Biota of North America Program. University of North Carolina, Chapel Hill.

Kruckeberg, A.R. 1984. California Serpentines: Flora, Vegetation, Geology, Soils, and Management Problems. University of California Press, Berkeley.

Kruckeberg, A.R. 2004. Geology and Plant Life. University of Washington Press, Seattle.

Peck, M.E. 1961. A Manual of the Higher Plants of Oregon. Binfords & Mort, Portland, OR.

Sawyer, J. and T. Keeler-Wolf. 1995. A Manual of California Vegetation. California Native Plant Society, Sacramento.

Soil Survey Staff. 1999. Soil Taxonomy—A Basic System for Making and Intrepreting Soil Surveys. USDA, Agriculture Handbook No. 436. U.S. Government Printing Office, Washington, D.C.

Troll, C. 1971. Landscape ecology (geoecology) and biogeocenology—a terminological study. Geoforum 8: 43–46.

Wagner, D.L. 1991. The state rock of California: serpentine or serpentinite? California Geology 44: 164.

Wiggins, I.L. 1980. Flora of Baja California. Stanford University Press, Stanford, CA.

Part I Geology and Hydrology

Geology is the science of the solid earth, or the lithosphere. It is the basis of serpentine geoecology. Serpentine soils develop from ultramafic rocks and plants get their nutrients, other than carbon, from soils. It is therefore appropriate that we begin part I with geology. The first chapter in part I (chapter 2) explains where ultramafic rocks come from and how some of them have reached the land surface. Chapter 3 describes the mineralogy and chemistry of ultramafic rocks and explains why these rocks and the soils developed from them are unique substrates for plants. Because water is an essential ingredient for life, chapter 4 explores the disposition and nature of water in serpentine landscapes.

2   Nature of Ultramafics

The earth is divided into three layers: the crust, the mantle, and the core (fig. 2-1). There are two principle regions within the crust: continents and ocean basins. The rocks that make up these layers differ from one another in chemical composition and density. The mantle is composed of dense ultramafic rocks, rich in magnesium-iron silicate minerals such as olivine and pyroxene. Ultramafic rock is the main source of serpentine soil in the continental crust. Most of the lighter crustal rocks are made up of silicate minerals that are enriched in the lighter elements sodium, calcium, and potassium, which have large cations, rather than magnesium (also a light element) and iron, which have smaller cations (table 2-1, appendix A). Over geological time living organisms have evolved on continents or in oceans with elemental concentrations dependent more on the crust than on the mantle.

In the oceanic realm, new oceanic crust forms at spreading centers between active plates where hot, decompressed mantle rock rising toward the surface partially melts to form basaltic magma (fig. 2-2). The spreading centers develop at mid-ocean ridges, behind volcanic arcs (back-arc basins), in front of volcanic arcs (forearc basins), or as continents rift apart, as with the Red Sea. Cracks formed between the spreading plates are intruded by basaltic magma that forms thin vertical sheets (sheeted dikes). New cracks and dikes are continually forming as the plates spread apart. Some of the magma rising into the cracks reaches the ocean floor and, as the hot lava is quenched by ocean water, it solidifies to form distinctive rounded, pillowlike structures. As magma above the partially melted mantle cools, some of the first crystals to form settle to the bottoms of liquid magma chambers, producing layered gabbros—a process called differentiation. The layered sequence of pillow lava, diabase dikes, and gabbro built upon the ultramafic mantle is typical of new ocean crust (fig. 2-2). New crust formed at spreading centers slowly

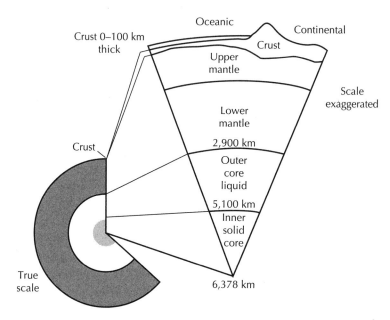

Figure 2-1 A cross-section of the earth showing its layers. An expanded view of the crust illustrates differences from oceans to continents.

migrates away from the spreading center and cools into a rigid oceanic crust that ranges in thickness from 6 to 12 km (fig. 2-2). The rigid new crust rests on mantle (ultramafic rock) that previously had been partially melted at depth. Eventually nearly all the oceanic crust that is formed at these spreading centers sinks beneath continental margins (a process called subduction), where it is returned back into the mantle (Coleman 1977). During subduction, small isolated slabs of the ultramafic rock are sometimes detached and embedded within trench sediments of continental margins. In other situations, complete sections of ocean crust are thrust onto the accreting continental margins as seen in the Circum–Pacific area. This is called obduction. Detached fragments of ocean crust attached to continents and imbedded within sediments along the continental margin of Circum-Pacific region are called ophiolites (fig. 2-3).

Table 2-1 Average chemical compositions of mantle, ocean crust, and continental crust.

	Mass (%)							CaO/MgO	
	SiO_2	Al_2O_3	FeO	MgO	CaO	Na_2O	K_2O	g/g	mol/mol
Mantle	46.0	4.1	7.5	37.8	3.2	0.33	0.026	0.08	0.06
Ocean crust	50.4	15.3	10.4	7.6	11.3	2.68	0.088	1.49	1.07
Continental	59.7	15.7	6.5	4.3	6.0	3.1	1.8	1.40	1.01

Data from Hofmann (1988) and Condie (1997).

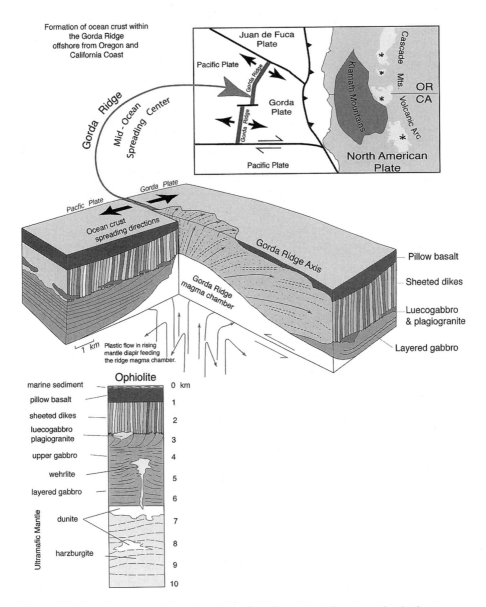

Figure 2-2 New Oceanic crust forming at Gorda Ridge. A typical section of ophiolite is very similar to the ocean crust now forming in Gorda Ridge. During subduction and/or obduction into or over the continental edge of North America, fragments of the rigid ocean crust have been accreted onto the continental margin of California. The rigid crust collides against the continental margin and either dives under (subducted) or is thrust over the continental margin (obducted) and is called ophiolite.

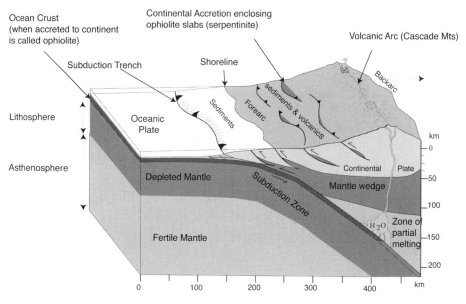

Figure 2-3 Pacific-type active margin for western North America illustrating subduction and accretion of ocean crust. Sediments in the subduction trench slowly are accreted to the continental margin. Continued subduction and accretion of the trench sediments underplate the leading edge of the continent. Infrequent slabs of ocean crust and mantle are wedged into these sediments and lifted to the surface as subduction continues. Most of the western North America serpentine is derived from the ultramafic mantle rocks. In some situations small isolated slabs of ultramafic rock are detached and embedded within the trench sediments of the margin. In other situations complete sections of ocean crust are obducted onto the accreting continental margins.

The ophiolites within orogenic zones (mountain belts) of western North America are rootless slabs imbricated within the sediments and volcanics of the continental margin (fig. 2-3). In the orogenic zones, the mantle ultramafic rocks, which are called peridotite (lherzolite, harzburgite, and dunite are common varieties of peridotite), are subjected to deformation that causes water to invade into fractured rock and slowly alter it to serpentinite, which is composed of serpentine minerals. The physical character of the ultramafic rock changes from a brittle, strong, dense peridotite to a weak, light, plastic serpentinite. This weak, plastic serpentine, or serpentinite, is easily squeezed into folds or faults and often rises as a diapir in a fashion similar to a salt dome in sediments, because the serpentinite is a buoyant rock mass within denser sediments. High-standing serpentinite masses that occupy fold crests are often ravished by rapid erosion and landslides, resulting in extensive serpentine-bearing deposits in adjacent sedimentary basins. Not all ultramafic slabs (such as dunite and harzburgite) incorporated into orogenic belts are completely serpentinized; that is, not completely altered to serpentinite. Infrequent large masses (75–150 km^2) surrounded with only a narrow serpentine border are found in the orogenic belts of western North America. Serpentine can also form in the oceanic realm where oceanic crust is segmented and uplifted along transform faults. The transform faults represent major structural discontinuities penetrating to great depths that can introduce seawater into the mantle ultramafic rocks.

The plate tectonic theory provided new ways of understanding the serpentines of western North America. The original idea that serpentine and its related ultramafic rock were igneous intrusions produced by molten magmas forming within the earth's crust or deeper in the mantle has been completely abandoned. Instead, the plate tectonic theory postulates that ultramafic mantle rocks resulted from partial melting of fertile mantle rock under oceanic spreading centers, producing vast amounts of basalt-diabase ocean crust from the melt. Ultramafic rock that is the residue of the partially melted mantle rock forms the basement upon which the oceanic crust is deposited. Constant movement of newly formed crust, or plate, toward the Circum–Pacific margin is caused by pull from the sinking, or subduction, of the plate under the continental margin (fig. 2-3). Obduction and underplating of small fragments of this oceanic crust along the western North America continental margin and incorporation into orogenic belts provides various tectonic scenarios for emplacement of ultramafic rock slabs into accretional zones along the continental margin.

In addition to accretion of oceanic crust along continental margins, there are restricted areas of serpentine related to basaltic magmas that have intruded the continental crust. During crustal thinning, basalt melts formed in the mantle are trapped in the continental crust and form closed liquid magma chambers. Slow cooling of the basaltic magma allows heavy magnesium–iron silicates to crystallize early and settle to the bottom, forming layers of ultramafic rock. These layers are very similar to those in sedimentary rocks and usually mark the lower part of the magma chamber. The ultramafic layers are often interlayered with calcium-rich silicates, such as gabbro and anorthosite. After the magma chamber crystallizes, it may become cracked by later tectonic action. Invading water within the cracks will react with ultramafic layers to form serpentine. The amount of ultramafic rocks and serpentine formed in layered intrusions is much smaller than in serpentine areas associated with ophiolites. The Stillwater complex in the Beartooth Mountains of Montana is an example of a layered intrusion (see part IV).

Minor amounts of ultramafic lava are present in rocks of Archean age >2.4 Ga and are called komatiites. These rocks were formed by partial melting of ultramafic mantle rocks at shallow depths when the earth's heat flow was much higher than it is today. These volcanic ultramafic rocks contain 18%–30% MgO, compared to a mean concentration of 38% MgO in the mantle (table 2-1).

3 Mineralogy and Petrology of Serpentine

3.1 Definitions and Nomenclature

"Serpentine" is used both as the name of a rock and the name of a mineral. Mineralogists use "serpentine" as a group name for serpentine minerals. Petrologists refer to rocks composed mostly of serpentine minerals and minor amounts of talc, chlorite, magnetite, and brucite as serpentinites. The addition of "-ite" to mineral names is common practice in petrologic nomenclature. For instance, quartzite is a name for a rock made up mostly of quartz. Serpentinites are rocks that form as a result of metamorphism or metasomatism of primary magnesium–iron silicate minerals. This entails the replacement of the primary silicate minerals by magnesium silicate serpentine minerals and the concentration of excess iron in magnetite.

"Mafic" is a euphonious term derived from magnesium and ferric that is used for dark colored rocks rich in ferromagnesian silicate minerals. "Ultramafic" is used when the magnesium–ferrous silicate minerals compose >90% of the total rock. Olivine, clinopyroxene, and orthopyroxene are the minerals in primary ultramafic rocks, with minor amounts of plagioclase, amphibole, and chromite (fig. 3-1). Ultrabasic has been used by some geologists in referring to ultramafic rocks. The most common ultramafic rocks are harzburgite, containing <75% olivine and 25% orthopyroxene; dunite, with 100% olivine; and lherzolite, which has 75% olivine, 15% orthopyroxene, and >10% clinopyroxene, with or without plagioclase. Very small amounts of chromite are present in all of the mantle ultramafic rocks (Coleman 1971).

The alteration of primary ultramafic rocks to serpentine mineral assemblages is incremental due to episodic invasion of water into the ultramafic rock. It is difficult to distinguish and map the gradations from primary ultramafic rock to serpentinite. Because of

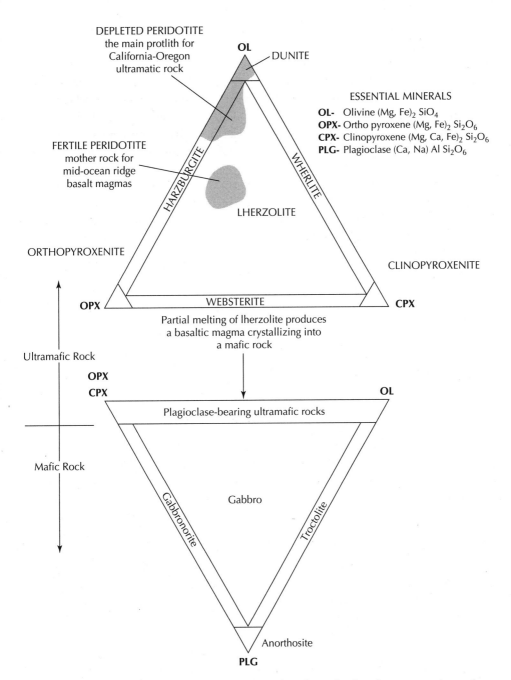

Figure 3-1 Modal classification of ultramafic and mafic rocks that form serpentine soils. Modes are determined by estimating the volume percentage of the essential minerals and normalizing to 100%. The volume percentages for OPX, OL, CPX, and PLG within each triangle can be estimated by measuring the relative distances along a line to the apex representing the mineral (100%) from the base of the triangle that is opposite that apex (0%).

this difficulty in distinction, we prefer to use the term ultramafic or serpentinized peridotite for all gradations to serpentinite.

Pedologists and botanists commonly group serpentinites with primary ultramafic rocks and refer to these substrates as serpentine because all of them have similar chemical compositions. As will become apparent later, there is great variability in the mineralogical compositions of these rocks and the soils derived from them. We attempt to clarify our usage of terms for ultramafic rocks in this book because soils developed on slightly serpentinized ultramafics are different from those derived from serpentinite.

Most of the ultramafic rocks in western North America are associated with assemblages of mafic and ultramafic rocks that are referred to as ophiolite (Coleman 1977). The term "ophiolite" was first introduced by Brongniart (1813) to describe an assemblage of rocks consisting of peridotite–serpentinite, gabbro, diabase–basalt, and chert. This association is now interpreted to be fragments of oceanic crust accreted into the continental margins (Coleman 1977). Characteristically the ophiolites of western North America are tectonically dismembered into a melange. Because of this mixing, ultramafics commonly crop out in very irregular patches, forming islands of serpentine soils.

3.2 What Do Serpentine Rocks Look Like?

The variety of metamorphic gradients and minerals from primary ultramafic rock to serpentine results in a wide variety of colors and textures. Mantle ultramafic rocks rich in olivine and pyroxene exhibit a deformation fabric that developed during slow plastic circulation within the mantle. Ultramafic rock containing <10%–20% serpentine minerals is light brown to reddish brown on weathered surfaces. When olivine is the predominant primary mineral of the ultramafic rock, the weathered surface is smooth and sometimes referred to as "buckskin" ultramafic rock. Where the more resistant pyroxenes are abundant and form a bumpy surface, they are called "hobnail" ultramafic rocks.

The textures of serpentinized rocks display amazing variety. Light to dark shades of green predominate, but serpentine rocks may be white, black, or red with all gradations in between. Where shearing occurs during tectonic movement, individual "fish scale" flakes develop with a brilliant smooth polish, whereas blocky serpentinized ultramafic has dull, dark gray-green coloration and rough, hackled fractured surfaces. In massive unsheared ultramafic rock, when serpentinization is nearly complete, the unweathered rock on fresh broken surfaces is almost black and cut by numerous cross cutting veins. These veins are often filled with cross-fiber asbestos serpentine (chrysotile). Inspection by petrographic thin-sections of the massive blocky serpentinized ultramafics under the microscope reveals a mesh texture showing multiple periods of serpentinization.

3.3 Serpentinization of Ultramafic Rocks

The alteration of primary ultramafic rock, or peridotite, to serpentinite, is called serpentinization. In the absence of water during emplacement of solid masses of primary ultramafic rock on continental margins, the mantle minerals (olivine and pyroxene; fig. 3.2) are preserved with only small amounts of serpentine minerals forming. Wherever water meets ultramafic rock of any origin at temperatures <500°C, the olivine and pyroxenes of the ultramafic rock will alter to serpentine minerals. Serpentinization is an alteration process, or metamorphism, in which olivine and pyroxene are replaced by serpentine. Serpentine minerals (chrysotile, lizardite, and antigorite), brucite, and magnetite are the most important products of serpentinization, with talc forming only in the presence of excess silica (fig. 3-2; appendix A; Coleman 1971). To completely serpentinize an ultramafic rock requires a dramatic influx of water that produces a huge expansion (up to

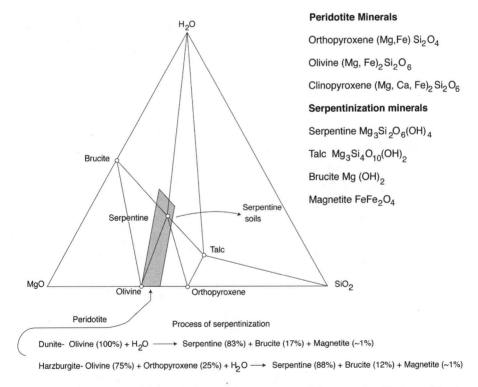

Figure 3-2 A triangular diagram showing the composition of serpentine (dark area) in the system $MgO-SiO_2-H_2O$. The common peridotite and serpentine minerals are also plotted on the diagram and their ideal compositions are listed on the right of the diagram. Simple reactions for metamorphism of a dunite and harzburgite to serpentinite are shown. Of course, other more complicated reactions are known to take place if the hydrothermal water is not pure H_2O.

33%, as well as a decrease in density from ~3.3 to ~2.5), as the newly formed serpentine minerals require more space than the primary minerals they replace (Hostetler et al. 1966). It is rare for complete serpentinization (change from primary ultramafic rock to serpentinite) to be accomplished in one metamorphic event because the huge expansion requires so much extra space for the newly formed serpentinite.

Serpentinization is generally accomplished with no change in the relative amounts of silica, aluminum, magnesium, chromium, manganese, iron, cobalt, and nickel (fig. 3-2). The only component removed during serpentinization is calcium. If carbon dioxide is present, it will combine with the calcium and precipitate as carbonate ($CaCO_3$). The common presence of minute amounts of native nickel-iron and scarcity of sulfides in serpentinites indicates limiting amounts of sulfur and oxygen and the pervasive reducing effect of hydrogen during serpentinization (Sleep et al. 2004). The iron (FeO) content of ultramafic rock is about 7%, and it is diluted in serpentinites to about 5%. Excess iron developed during serpentinization is either incorporated into brucite or forms magnetite. Nickel and cobalt are present in high amounts in most serpentinites and do not easily enter into the serpentine mineral structure. They are most likely sequestered in the native nickel-iron compounds or the newly formed magnetite. Small grains of chromite in the ultramafic remain mostly unaltered during serpentinization.

Mantle ultramafic rock accreted to the continental margin undergoes gradual changes to serpentinite because of episodic tectonic movements that allow the entry of water and

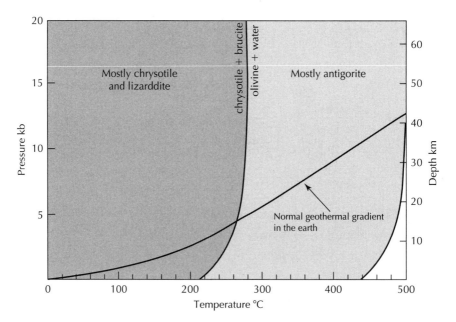

Figure 3-3 Temperature–pressure diagram illustrating the wide range under which serpentine minerals are stable within water-rich subducting zones along accretional continental margins. Note that on this diagram serpentine is not pressure sensitive and can survive the extreme pressures and low geothermal gradients found in subduction zones.

further expansion. The shiny, smooth surfaces of the completely tectonized serpentinized mantle ultramafic are evidence of this movement and expansion. These altered rocks from the mantle via ocean crust are called Alpine-type serpentinites.

Some serpentinites have origins quite different from those of the Alpine type. Small amounts of serpentinite are formed by hydrothermal alteration of dolomite, a magnesium-rich limestone. Cumulate layers of olivine–pyroxene produced in large basaltic magma chambers can be serpentinized, also.

In a general way, the relative proportions of products from Alpine-type serpentinization are fixed by the original olivine–pyroxene ratios in the primary ultramafic rock. The ratios of iron–magnesium–silica generally remain constant during progressive serpentinization, indicating that expansion accompanies the change from primary ultramafic rock to serpentinite (fig. 3-3). The expansion combined with a decrease in density accelerates the upward rise of serpentinite bodies through orogenic zones within the continental crust and explains their presence in some of the highest elevations in the Pacific Borderland mountain ranges.

3.4 Serpentine Mineralogy

Serpentine is not just one mineral, but a group of three minerals: chrysotile, lizardite, and antigorite (appendix A). It is commonly difficult to distinguish any one serpentine mineral from any other, because the crystals are fine grained and intergrown one with another. The individual minerals are light to dark green, commonly varied in hue and may be white. Serpentinite rocks commonly appear waxy or greasy and feel slippery. They can be scratched easily with a steel knife (the hardness of serpentine is 2.5 on Moh's scale). Chrysotile has a fibrous habit, often filling veins as cross-fibers that have a silky appearance. Chrysotile fibers in serpentinite are the main source of asbestos. Lizardite and antigorite have platy, rather than fibrous, crystal habits.

Serpentinite generally contains lesser amounts of minerals other than serpentine. Other minerals that are generally present in serpentinites are brucite and magnetite, and commonly, talc, carbonates, and amphibole. Magnetite is usually present in minute grains that impart a dark color to most serpentines. It produces a strong magnetic signature in serpentinized bodies. Brucite is present when the ultramafic rock is mostly olivine (dunite) and there is not enough silica available to form much serpentine or any talc. In the presence of carbon dioxide, much of the excess magnesium will form carbonate.

3.5 Serpentine Mineral Stability

Mantle ultramafic rocks are unstable in continental crust because they have formed at very high temperature and pressures in the absence of water. You might say that they are in a very hostile environment in the earth's crust, where lower temperatures and water destabilize the olivine and pyroxenes. The most important reaction for low temperature serpentinization in the crust is:

$$\text{olivine} + \text{water} = \text{chrysotile} + \text{brucite} + \text{magnetite} + \text{hydrogen}$$

At temperatures <300°C, both chrysotile and lizardite are stable (fig. 3-3). It is important to note that the above reaction is not pressure sensitive; therefore, serpentine can form deep within subduction zones. The formation of chrysotile at ambient temperatures has been reported (Barnes et al. 1967, Barnes and O'Neil 1978). Field observations have shown that during crustal metamorphism of serpentinites, with increasing temperatures, antigorite replaces both chrysotile and lizardite (Evans and Trommsdorff 1970). Often lizardite and antigorite coexist metastably. Above 500°C, antigorite converts to talc + olivine + water by dehydration. From these data, it is obvious that serpentine minerals are stable at the earth's surface and deep within the earth's crust in areas of low heat flow, such as subduction zones. The appearance of antigorite instead of chrysotile in serpentinites is a strong indicator of higher grade (i.e., higher temperature and pressure) crustal metamorphism (Evans and Trommsdorff 1970).

3.6 Serpentinization near the Land Surface

Water is an essential ingredient for the formation of serpentine. Where is the water from and how does it get into the ultramafic rocks to cause serpentinization? Ultramafic rock within the oceanic realm is exposed to seawater by faulting and uplift and is serpentinized beneath the ocean. In trenches along continental margins and in subduction zones, the dehydration of ocean sediments is another source of water. In the continental accretional areas, water invades ultramafic bodies during tectonic movement or by percolating meteoric water. Well-exposed serpentinized ultramafic rocks always display incremental, or multistage, serpentinization. Commonly right-angle fracture sets develop in the brittle zone of the crust, allowing invasion of water. The large volume increase during serpentinization effectively seals off the initial fractures making it difficult for additional water to invade the ultramafic rock, unless there is subsequent tectonism (fracturing and faulting). Cross-fiber serpentine (asbestos) may fill the fractures in the blocky serpentinite. Progressive increments of tectonic movement are necessary to open new channel ways for water. Continued shearing and serpentinization along planes of weakness will produce fish-scale aggregates characterized by lens shaped fragments with shiny surfaces. The absence of open cavities in veins cutting serpentinized ultramafic rock demonstrates the difficulty water has in penetrating these rocks because expansion and sealing accompanies each episode of serpentinization and effectively seals the rock.

In 1967, Ivan Barnes of the U.S. Geological Survey discovered that spring waters issuing from partially serpentinized ultramafic rock in the California Coast Ranges have unique compositions (Barnes et al. 1967, Barnes and O'Neil 1969). Careful chemical analyses of these waters revealed two distinct types: (1) magnesium bicarbonate water

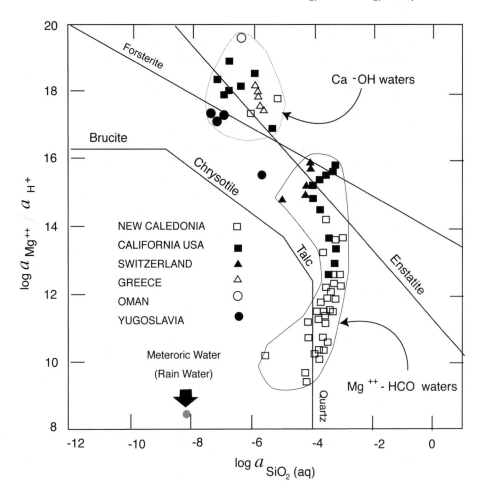

Figure 3-4 Water samples from 75 springs issuing from partially serpentinized peridotites. Aqueous silica and magnesian activities and mineral stability states are shown with respect to the common magnesium minerals found in mantle peridotites. Solid lines represent saturation at 25°C and 1 bar pressure for the designated minerals. The $Ca(OH)_2$ waters are issuing from partially serpentinized ultramafics and form travertine aprons. The $Mg(HCO_3)_2$ waters are found in sheared serpentinites with scarce primary iron–magnesium silicates remaining. Data from Coleman and Jove (1993).

and (2) calcium hydroxide water. Magnesium bicarbonate water is the most abundant and provides a perennial water supply in many of the California Coast Ranges serpentinites. The $Mg(HCO_3)_2$ water is present in completely serpentinized ultramafic rock as well as partially serpentinized rock. In these waters, magnesium (Mg^{2+}) is the predominant cation and bicarbonate ($HCO_3)^-$ is the most abundant anion. This water is present in ground water, springs, and in streams that drain many serpentinite areas. Calcium hydroxide water is much less abundant and present only within incompletely serpentinized ultramafic rock as low discharge seeps and springs. The water composition is consistent and characterized by pH values >11. Calcium (Ca^{2+}) is the predominant cation accompanied by small amounts of Mg^{2+}, Fe^{2+}, Mn^{2+}, Cr^{3+}, and silica (SiO_2), although silicon is apparently in solution as an oxyanion, SiO_4^{4-}, rather than as silica. The main anion is hydroxyl $(OH)^{-1}$ with no detectable carbonate. These waters result from low grade metamorphic reactions with primary ultramafic rock minerals to form serpentine (fig. 3-4; Barnes et al. 1967; Barnes and O'Neil 1969).

Understanding the significance of these unique waters found in many serpentine areas of the world brought new insights to the chemistry of serpentinite formation. The calcium hydroxide water is metamorphic water produced during the crystallization of serpentine minerals at ambient conditions and up to 500°C. Barnes has shown that these calcium hydroxide waters are supersaturated and ready to react almost instantaneously with any carbon dioxide, silica, or other available anions (Barnes et al. 1967, Barnes and O'Neil 1969, Neal and Stanger 1984) to form carbonates or calcium silicate minerals.

The presence of travertine ($CaCO_3$) aprons surrounding the calcium hydroxide springs and seeps demonstrates that the calcium hydroxide waters react with atmospheric carbon dioxide precipitating calcite ($CaCO_3$). The formation of secondary calcium silicate enclaves within serpentinites and along their borders can be directly tied to the movement of the highly reactive calcium hydroxide waters.

4   Water in Serpentine Geoecosystems

Water is continuously cycled from the atmosphere through geoecosystems to water bodies and, by evaporation and evapotranspiration, back to the atmosphere. Water is commonly transported long distances in the atmosphere. Eventually, it forms clouds that drop rain, snow, or dew on plants or the ground. The contributions of fog and dew to geoecosystems are generally minor, but they can be important factors along some coastlines.

Water from most of the precipitation that falls to the ground infiltrates soils. Some is intercepted by plants and evaporates before it can reach soils, and some runs overland to streams without entering soils. Soils are important stores of water for plants. Excess water in soils and permeable substrata drains gradually. This gradual draining of infiltrated water diminishes flooding from storms and supplies water to streams between rainfall events, helping maintain more constant stream levels.

4.1 Watersheds and Hydrology

The study of meteoric water, or water that is cycled through the atmosphere, is called "hydrology." Watersheds are basic units of hydrological investigations. A watershed is a drainage basin—an area from which water drains to a common point. All water falling on a watershed (and not lost by evapotranspiration) leaves through a single, joint location that can be monitored with a stream gauge. There are exceptions, however, in which water drains from watersheds through permeable substrata, rather than at the lowest point in the ground surface topography. These "leaky" watersheds are common in basalt, poorly consolidated sandstone, and limestone terrains. We can examine some of the data from watersheds that are not known to be leaky to learn about the runoff characteristics of serpentine streams and their chemistry.

Watersheds range in size from less than a hectare to large portions of continents (e.g., the Amazon River drains 6,475,000 km^2, about 35% of the South American continent). The smaller watersheds are drained by headwater streams with no tributaries, and the larger ones are drained by streams with many tributaries. Some of the most useful information can be gained from small watersheds because they have more uniform lithology, topography, soils, climate, and vegetation than larger ones. Runoff from a few watersheds that are small enough to have predominantly serpentine lithologies have been monitored. Clear Creek, a tributary of the San Benito River in the southern part of the California Coast Ranges, is the only example of a serpentine stream monitored in western North America. It has an elevation range from 792 m above sea level at the stream gauging station to 1579 m on San Benito Mountain.

Monthly proportions of runoff through Clear Creek are shown in figure 4-1A, along with those from four other streams in different climatic regimes (see also table 4-1). Streams in the California Coast Ranges have maximum discharges in February, or in March from Clear Creek, during or after the peak of the rainy season. Large snow packs in the higher elevations of the Sierra Nevada delay runoff to produce maximum water yields in late spring or early summer, as at Pohono Bridge on the Merced River where it passes though Yosemite Valley. Because stream flows are relatively low during late summer throughout California, relative discharges from these five streams are shown on a logarithmic scale (fig. 4-1B) to emphasize the differences from August through October (months 8–10). Redwood Creek is typical of the northern Coast Ranges, inland from the fog belt. Streams rise relatively early there, with relatively high discharges in November, indicating that winter storms reach the northern before the southern Coast Ranges. Big Sur River represents a stream that is largely within the coastal fog belt. Arroyo Mocho is typical of the southern Coast Ranges, inland from the fog belt. It practically ceases flowing in late summer, and most of its tributaries are certainly ephemeral streams.

Clear Creek is in a climatic regime similar to that of Arroyo Mocho, but Clear Creek has a relatively high flow at the end of summer (fig. 4-1B). That is, its flow is high compared with Arroyo Mocho and Redwood Creek because the discharge declines to 1.5% of the annual discharge during October compared to 0.3% in September for Redwood Creek and 0.1% in October for Arroyo Mocho. The more gradual decline in stream flow through Clear Creek during spring and summer is attributed to fractures in tectonically sheared serpentinite that retain infiltrated water and release it gradually. Delayed stream discharge is epitomized by Hat Creek in volcanic terrain on the eastern side of the Cascade Mountains that has nearly constant flow throughout the year (Rantz 1972). Peridotite is not as highly fractured as tectonically sheared serpentinite, and it is unlikely to cause the more delayed flow that is characteristic of Clear Creek.

The longer flow of water from sheared serpentinite than from watersheds with more massive rock was confirmed on a scale of days, rather than months, by Onda et al. (2001). They found that storm flow from Japanese watersheds with sheared serpentinite

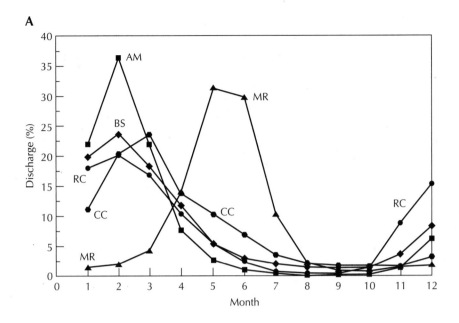
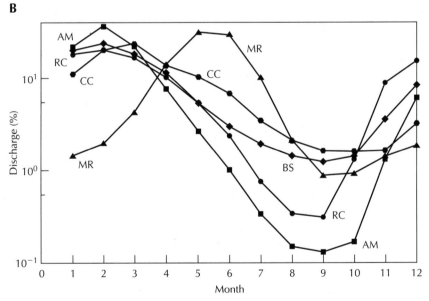

Figure 4-1 Runoff characteristics from Clear Creek (CC) and four other streams that are representative of California streams with different climatic regimes: AM, Arroyo Mocho, BS, Big Sur River, MR, Merced River, and RC, Redwood Creek. Mean monthly stream discharge is shown as a percentage of annual discharge, in order to compare the flow regimes of streams with watersheds of different sizes. The first month is January, rather than October, which is the first month of the water year in the California Region. (A) Linear scale for stream discharge. (B) Logarithmic scale for stream discharge.

Table 4-1 Some characteristics of representative California watersheds.

Watershed with gauging station	Station no.	Water years of record	Station location			Watershed area (km²)	Annual runoff (mm)	Dominant vegetation types
			Latitude (°N)	Longitude (°W)	Altitude (m)			
Arroyo Mocho	11-1760	1913–1930 1963–2000	37.626	121.704	227	98.9	49	Woodland-grass, shrub, forest
Big Sur River	11-1430	1950–2000	36.246	121.772	73	120.4	760	Forest, shrub
Clear Creek	11-1547	1994–2000	36.365	120.755	792	36.5	134	Bare soil, shrub, forest
Merced River	11-2665	1917–2000	37.717	119.665	1177	831.4	560	Bare rock, forest, meadow
Redwood Creek	11-4815	1953–1993 1997–2000	40.906	123.814	259	175.3	1210	Forest, grassland

Data from Webster et al. (2001).

was delayed, compared with flow from granite watersheds. The water balance at individual sites within watersheds is discussed in chapter 8 and in appendix E.

4.2 Watershed Chemistry

Water is pure when it enters the atmosphere by evaporation from land, although some salty water may enter the atmosphere as droplets blown from waves on oceans or lakes. Water generally condenses on nuclei of colloidal particles or salts before it falls to the ground. If it falls on plants, they influence composition of the water before it drips from leaves or drains down stems to the soil. The chemistry of water is thus altered before it reaches terra firma.

The main sources of watershed water chemistry data are springs, wells, and stream monitoring sites. Spring water chemistry is more completely dependent than is stream water on reactions in the regolith, or in the bedrock below the regolith. Stream water chemistry is greatly affected by atmospheric inputs, as well as by reactions with local regolith or bedrock. Its chemistry generally results from the integration of water chemistry over larger areas.

Serpentine spring waters are of two distinct types (Barnes et al. 1967; Barnes and O'Neil 1969, 1978). Meteoric water reacting with olivine and pyroxenes during their alteration to serpentine yields a calcium hydroxide [$Ca(OH)_2$]-dominated water. In the absence of CO_2, these waters attain alkaline reactions up to pH 11.8. They emerge in springs with low flow rates that remain active year around. The more common serpentine spring waters are dominated by magnesium bicarbonate, $Mg(HCO_3)_2$. They come from serpentine rocks in which serpentine is not being produced by the alteration of olivine and pyroxenes.

Because the weatherable minerals in serpentine rocks are essentially magnesium silicates, or hydrous magnesium silicates, with lesser amounts of iron, stream water in serpentine watersheds is dominated by Mg^{2+} and silica, or H_4SiO_4, and bicarbonate, HCO_3^- (table 4-2). Only the cations and silica are listed in table 4-2 because the anions are not derived from serpentine rocks. Bicarbonate, the dominant cation, is derived from atmospheric CO_2 and from CO_2 released by microbial respiration and oxidation of organic matter produced by plants. Water from the Yaou Basin contains appreciable Ca^{2+}, because the talc-serpentine bedrock in it is about one-tenth calcite. Water in Clear Creek is essentially saturated with Mg^{2+} and has a reaction of pH 8.9, while water in the wetter watersheds is more dilute and less alkaline.

The alkaline calcium hydroxide spring waters in serpentine terrain that have high concentrations of Ca^{2+}, Na^+, OH^-, and Cl^- and little Mg^{2+} and silica (fig. 4-2; cations only) do not contain carbonate and biocarbonate ions until they are exposed to the atmosphere. Then, upon exposure to air containing CO_2, a white to buff-colored scum appears on the spring water surfaces, and a travertine rim gradually builds up around some of the springs. Barnes and O'Neil (1969) suggested that serpentinization of olivine and pyroxene in peridotite is the source of the calcium. Even though there is very little calcium in olivine, little

Table 4-2 Chemical denudation, or mass loss, from serpentine watersheds.

Watershed[a]	Area (km²)	Watershed Precipitation (mm)	Runoff (mm)	Loss, mass[b] (kg/ha-year)						Water pH
				Ca	Mg	Na	K	SiO_2	Total	
Soldiers Delight	0.57	1140	153	—	59.1	—	—	24.5	83.6	8
Yaou basin		2410	458	5.2	16.8	0.5	1.8	57.7	82.0	6
Clear Creek, CA	36.5	—	134	4	189	1	1	7	202	8.9
New Caledonia										
Dumbéa Est	56.1	2618	1923	16	167	57	3	269	512	7.8
Dumbéa Nord	28.1	2405	1770	6	170	50	3	258	487	7.8
Couvelée	40.5	1662	1009	6	143	40	2	148	339	7.9

[a] Soldiers Delight is in Maryland, on serpentinite of the Piedmont of the Appalachian Mountain system (Cleaves et al. 1974); the Yaou basin watershed is in French Guiana, on talc-serpentine schist of the Amazon Basin (Freyssinet and Farah 2000); Clear Creek is in San Benito County, on serpentinized and tectonically fractured peridotite of the southern California Coast Ranges, and the New Caledonia watersheds are on peridotite on an island in the western Pacific Ocean.
[b] These are net losses from Soldiers Delight and Yaou basin, where additions from precipitation were subtracted from runoff losses, and gross losses from the other.

in orthopyroxenes, and clinopyroxenes are not major constituents of harzburgite (the most common kind of peridotite in western North America), the conversion of these minerals to serpentine and brucite (minerals lacking calcium) consumes the magnesium and releases enough calcium to make it the dominant cation, along with sodium, in the groundwater. Barnes and O'Neil did not suggest a source for the Na^+ and Cl^- in the spring water, but it is possible that it is from sea water that has been residing in the rock for many millions of years, or it might be derived from the atmosphere via rainfall.

Streams in serpentine terrain of the California Coast Ranges and short distances downstream generally have cemented gravels in their beds, which are sometimes called mortar beds because they resemble concrete (Barnes and O'Neil 1971). The cement is calcium carbonate, with undetermined amounts of silica, probably with greater proportions of silica at greater distances from springs. Cemented stream bed gravels in serpentine terrain of the Klamath Mountains are noncalcareous, therefore the cement is presumed to be silica (Alexander 1995). Prime examples of serpentine stream bed cementation in the Klamath Mountains are on High Camp Creek near the head of the Trinity River on the Trinity ultramafic body and on Hayfork Creek just downstream from Stringbean Creek in the Rattlesnake Creek terrane.

Neal and Stanger (1985) found springs in serpentine terrain of the Semail ophiolite in Oman having high concentrations of Ca^{2+}, Na^+, OH^-, and Cl^- and little Mg^{2+} and silica (fig. 4-2), much like those found by Barnes and O'Neil in California. The mean reactions of the spring waters were pH 11.2–11.6, while those of Mg^{2+}, Na^+, HCO_3^-, Cl^- surface waters were pH 8.4–9.1 in the Semail ophiolite. Neal and Stanger have a thorough discussion of the processes involved in the evolution of the spring waters.

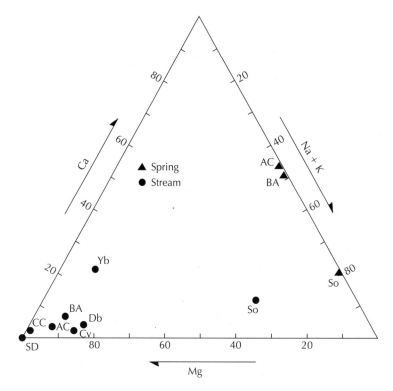

Figure 4-2 Relative amounts of calcium, magnesium, and sodium (actually sodium+potassium) in serpentine springs and streams. Sodium, rather potassium, is a major element in the serpentine springs. Sites: AC, Adobe Creek in western Stanislaus County, near Red Mountain; BA, Big Austin Creek in Sonoma County; CC, Clear Creek in San Benito County; Cv, Couvelée in New Caledonia; Db, Dumbéa Est and Nord (mean) in New Caledonia; SD, Soldiers Delight in Maryland; So, Semail ophiolite in Oman; and Yb, Yaou basin in French Guiana.

Subsurface waters in New Caledonia are dominated by Mg^{2+}, silica, and HCO_3^-, rather than by Ca^{2+} and OH^- ions, and the reaction ranges from pH 6.8 to 8.2 (Trescases 1975). Most of the samples reported by Trescases had appreciable Na^+ and Cl^-, but only one of eight had appreciable Ca^{2+}, and it was dominated by Mg^{2+} and HCO_3^- ions also.

Part I References

Alexander, E.B. 1995. Silica cementation in serpentine soils in the humid Klamath Mountains, California. Soil Survey Horizons 36(4): 154–159.

Barnes, I., and J.R. O'Neil. 1969. The relationship between fluids in some fresh alpine-type ultramafics and possible modern serpentinization, western United States. Geological Society of America Bulletin 80: 1947–1960.

Barnes, I., and J.R. O'Neil. 1971. Calcium-magnesium carbonate solid solutions from Holocene conglomerate and travertines in the Coast Ranges of California. Geochimica et Cosmochimica Acta 35: 699–718.

Barnes, I., V.C. LaMarche, and G. Himmelberg. 1967. Geochemical evidence of present day serpentinization. Science 156: 830–832.

Barnes, I., and J.R. O'Neil. 1978. Present day serpentinization in New Caledonia, Oman, and Yugoslavia. Geochimica et Cosmochimica Acta 42: 144–145.

Brongniart, A. 1813. Essai de classificacion minéralogique des roches melanges. Journal des Mines 34: 190–199.

Cleaves, E.T., D.W. Fisher, and O.P. Bricker. 1974. Chemical weathering of serpentine in the eastern Piedmont of Maryland. Geological Society of America Bulletin 85: 437–444.

Coleman, R.G. 1971. Petrologic and geophysical nature of serpentinites. Geological Society of America Bulletin 82: 897–918.

Coleman, R.G. 1977. Ophiolites. Springer-Verlag, New York.

Coleman, R.G., and C. Jove. 1993. Geological origin of serpentinites. Pages 1–17 in J. Proctor, A.J.M. Baker, and R.D. Reeves (eds.), The Vegetation of Ultramafic (Serpentine) Soils. Intercept, Andover NH.

Condie, K.C. 1997. Plate Tectonics and Crustal Evolution. Butterworth-Heinemann, Oxford.

Evans, B.W., and V. Trommsdorff. 1970. Regional metamorphism of ultramafic rocks in the central Alps. Pargenesis in the system $CaO-MgO-SiO_2-H_2O$. Swiss Bulletin of Mineralogy and Petrology (Schweiz. Min. Petr. Mitt.) 50: 481–492.

Freyssinet, P., and A.S. Farah. 2000. Geochemical mass balance and weathering rates of ultramafic schists in Amazonia. Chemical Geology 170: 133–151.

Hofmann, A.W. 1988. Chemical differentiation of the earth: The relationship between mantle, continental crust, and oceanic crust. Earth and Planetary Science Letters 90: 297–314.

Hostetler, P.B., R.G. Coleman, F.A. Mumpton, and B.W. Evans. 1966. Brucite in alpine serpentinites. American Mineralogist 51: 75–98.

Neal, C., and G. Stanger. 1984. Calcium and magnesium hydroxide precipitation from alkaline groundwaters in Oman and their significance to the process of serpentinization. Mineralogical Magazine 48: 237–241.

Neal, C., and G. Stanger. 1985. Past and present day serpentinization of ultramafic rocks; an example from the Semail ophiolite nappe of northern Oman. Pages 249–275 in J.I. Drever (ed.), The Chemistry of Weathering. Reidel Publishing, Dordrecht.

Onda, Y., Y. Komatsu, M. Tsujimura, and J. Fujihara. 2001. The role of subsurface runoff through bedrock on storm flow generation. Hydrological Processes 15: 1693–1706.

Rantz, S.E. 1972. Runoff Characteristics of California Streams. U.S. Geological Survey, Water-Supply Paper 2009-A.

Sleep, N.H., A. Meibom, T. Fridriksson, R.G. Coleman, and D.K. Bird. 2004. H_2-rich fluids from serpentinization: geochemical and biotic implications. Proceedings of the National Academy of Science USA 101: 1218–1223.

Trescases, J.J. 1975. L'évolution géochemique supergène des roches ultrabasiques en zone tropicale—formation des gesements nickéléferous de Nouvelle Calédonie. Mémoires O.R.S.T.O.M. No. 78 [English abstract].

Webster, M.D., S.W. Anderson, M.F. Friebel, L.A. Freeman, and J.R. Smithson. 2001. Water Resources Data—California, Water Year 2000, vol. 2, Pacific Slope Basins from Arroyo Grande to Oregon State Line, except Central Valley. Water Data Report CA-00-2, U.S. Geological Survey.

Part II Soils and Life in Them

Plant community distributions are commonly related to soil distributions. Soils are major factors in determining where different kinds of plants are propagated and how well they grow. Soils store water and are the major source of plants nutrients, other than carbon dioxide.

Soils are also of general interest for their hydrologic properties. They filter and store water. Microbes feed on organic pollutants in soil water, and many compounds harmful to people are adsorbed on soil particles, purifying water as it passes through soils. Water stored in soils reduces runoff from storms, minimizing flooding, and water stored in soils is essential for plants to grow robustly and to survive through long periods of dry weather.

Plant species distributions and growth, which are dependent on soil hydrology, among other things, are distinguishing aspects of serpentine soils. The investigation of plant growth in relation to soil properties is called "edaphology." To understand edaphology it is necessary to have some basic knowledge of "pedology," which includes soil morphology, development, and classification. Some principles of pedology and edaphology are introduced and discussed in this part about soils, with emphasis on serpentine soils.

Many of the same kinds of animals and microorganisms inhabit both serpentine soils and nonserpentine soils, but some in serpentine soils are unique. Although there are few published articles on animals and microorganisms in serpentine soils, there are enough to warrant a chapter on these organisms.

5 Nature of Soils and Soil Development

We walk on soils frequently, but we seldom observe them. Soils are massive, even though they are porous. Soil 1 m (40 inches) deep over an area of 1 hectare (2.5 acres) might weigh 10,000–15,000 metric tons. It is teeming with life. There are trillions, or quadrillions, of living organisms (mostly microorganisms), representing thousands of species, in each square meter of soil (Metting 1993). In fact, species diversity, or number of species, may be greater below ground than above ground. We seldom see these organisms because we seldom look below ground or dig into it. The many worms and insects one finds digging in a garden are a small fraction of the species in soils because the greatest diversity of soil-dwelling species exists among microscopic insects, mites, roundworms (or nematodes), and fungi. Even though individual organisms in soils are mostly very small or microscopic, the total mass of living organisms in a hectare of soil, excluding plant roots, may be 1–5 or 10 metric tons. More than one-half of that biomass is bacteria and fungi. Living microorganism biomass generally accounts for about 1%–5% of the organic carbon and about 2%–6% of the nitrogen in soils (Lavelle and Spain 2001).

5.1 The Character of Soils

The upper limit of soil is the ground surface of the earth. The lower limit is bedrock for engineers, or the depth of root penetration for edaphologists. Unconsolidated material that engineers call soil can be called "regolith" (Merrill 1897, Jackson 1997) to distinguish it from the soil of pedologists and edaphologists. Regolith may consist of disintegrated

38 Soils and Life in Them

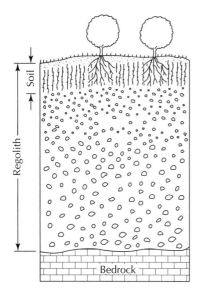

Figure 5-1 Relationships among bedrock, regolith, and soil.

bedrock, gravel, sand, clay, or other materials that have not been consolidated to form rock (fig. 5-1). Pedologists investigate the upper part of regolith, where changes are effected by exchanges of gases between soil and aboveground atmosphere and by biological activity. This soil of pedologists may coincide with that of edaphologists or include more regolith. In fact, the lower limit of soil that pedologists investigate is arbitrary, unless this limit is a contact with bedrock that is practically impenetrable with pick and shovel. A pedologist may investigate unweathered regolith beneath soils, which commonly represents "parent material" from which soils have formed, for better understanding of soil development.

Soils extend laterally over most of each continent, although regolith beneath ice sheets is not generally considered to be soil by pedologists and edaphologists. Lateral limits between different kinds of soils, or soil types, are commonly indistinct. They generally coincide with landform boundaries and changes in slope gradient, shape, aspect, or configuration (fig. 5-2). In flat terrain, soils may differ almost imperceptibly over long distances—for example, in response to gradual changes in the depth of a water table. The lateral limits between soil types may be arbitrary where differences between the soils depend on artificial limits that are based on the depth to mottles.

Soils develop through time, by weathering and vertical redistribution of constituents, to form layers that are generally parallel to the ground surface. These layers are called soil horizons. The layer of plant detritus that commonly accumulates on the ground surface is called an O horizon. Below ground, horizons are designated, from top to bottom, as A, B, and C horizons. Horizons bleached light gray to white that are called E horizons occur between A and B horizons in some soils; and between O and B horizons in others

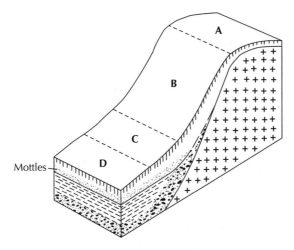

Figure 5-2 A hypothetical landscape with four soils showing relatively distinct boundaries from a shallow soil on the summit (A) to a moderately deep soil on a hill side (B) and to deep soils (C and D) on a footslope. An indistinct boundary is shown from soil C to soil D based on the depth to mottles, which decreases gradually toward the center of a valley where groundwater is nearer the ground surface.

that lack A horizons. The extent of soils with E horizons is minuscule in the California Region, but they are extensive in the boreal forests of northwestern North America. Hard bedrock is designated R horizon. All O and B horizons, and some A and C horizons have subhorizon designations to indicate different kinds of these horizons (fig. 5-3). The O, or organic horizon of plant detritus, has a subhorizon designation i in figure 5-3, indicating that the plant detritus has accumulated recently, without enough time elapsing for it decompose appreciably. The A horizon is dark colored because of humus accumulation. The B horizon in figure 5-3 has a subhorizon, designation t, indicating the accumulation of clay washed (or leached) down from the A horizon. The C horizon is the parent material of the soil.

Serpentine soils are distinguished from other soils by their parent materials, which in western North America are peridotite and serpentinite. These ultramafic rocks are nearly all in mountainous terrain. The serpentine soil parent materials are generally disintegrated bedrock that has commonly been moved downslope en masse by gravitational forces. Material that moves down slopes en masse or as falling particles, rather than suspended in water, is called "colluvium." Serpentine colluvial deposits are common in western North America. Material that moves downslope, or downstream, suspended in water is called "alluvium." Alluvial deposits in which the cobbles, gravel, and sand are dominated by peridotite and serpentinite are not extensive.

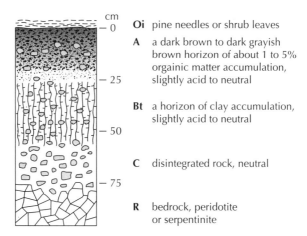

Figure 5-3 A typical serpentine soil profile showing the horizons, which form parallel to the ground surface.

5.2 Soil Development

Soil development begins when rocks or fresh (relatively unweathered) geologic materials are exposed by erosion, retreat of glaciers, uplift of land, or lowering of lake, stream, or ocean levels. Alteration of these geologic materials, called soil "parent materials," at the surface of the earth is soil development. Soil development proceeds mostly in the root zones of vascular plants.

Soil development involves many processes. Simonson (1959) grouped these processes into four categories: additions, losses, transfers, and transformations. Additions are mainly plant detritus (including dead roots), dust, and impurities in rainfall. Losses are by erosion, leaching, and plant harvest or burning. Plant nutrient losses from biomass harvesting or burning are essentially losses from soils because plants obtain the bulk of their nutrients, other than carbon dioxide, from soils. It is mainly the elements with gaseous oxides, such as carbon, nitrogen, and sulfur, that are lost in burning. These elements, however, are major constituents of proteins and essential to living organisms. Transfers are movements of soil constituents within soils. The constituents are transported in water or gasses or by animals. Transformations are changes in composition of soil constituents, for example, decay of organic matter or weathering of biotite to vermiculite en situ.

Here we discuss processes of soil development under three categories: (1) accumulation and incorporation of organic matter into soils, (2) weathering, and (3) soil horizon differentiation. In each of these categories, there are additions, losses, transfers, and transformations of materials.

Processes of soil development depend on the parent material, topography (or relief), organisms that are available to occupy a soil, and climate. These materials and conditions that control the processes of soil development have been called "factors of soil formation"

(Jenny, 1941). Beginning with regolith (or parent material), soils evolve (or change) through time. Some changes, such as accumulation of soil organic matter, are relatively rapid with a balance between gains and losses obtainable within a few hundred years, or less. Accumulation of clay minerals and their concentration in a subsoil is a much slower process requiring thousands of years. Clay minerals may continue to accumulate for tens or hundreds of thousands of years before losses begin to balance gains.

5.2.1 Incorporation and Accumulation of Organic Matter in Soils

"Organic matter" refers to complex carbon compounds that are produced by living organisms. Practically all organic matter that accumulates in a terrestrial ecosystem is derived primarily from photosynthesis in organisms containing chlorophyll. The first of these organisms to occupy a freshly exposed surface may be microorganisms and non-vascular plants, such as cyanobacteria, algae, lichens, mosses, and liverworts. But organic matter does not accumulate rapidly until vascular plants occupy a site. Vascular plants, especially seed-bearing plants, have massive root systems in soils. Much of their biomass is below ground—from one-quarter to one-third of the biomass of trees to about one-half of many perennial grasses. That is, bunch grasses may have as much biomass below ground as above ground. When roots die, microorganisms convert them to humus. Also, vascular plants yield much greater quantities of above ground biomass than do nonvascular plants. Leaves and other vascular plant detritus that fall on the ground are fed upon by insects, mites, worms, and microorganisms. Ants, worms, and larger animals mix plant detritus and humus from decaying ground floor detritus into soils.

The accumulation of plant detritus on ground surfaces and humus in soils depends on many factors. Basically, accumulation is controlled by a balance between additions and losses of organic matter. Additions depend mostly on plant productivity and incorporation of humus into soils by animals, as mentioned above. Losses are incurred by complete decomposition of organic matter to carbon dioxide and water. This carbon dioxide either dissolves in water or escapes to the atmosphere. Most nitrogen and sulfur and much of the phosphorus in soils are in organic matter. When soil organic matter decays, these elements become available for reuse by plants. They are recycled from the detritus of plants, through microorganisms that decompose the detritus, back to plants that take NH_4^+, NO_3^-, HSO_4^-, and $H_2PO_4^-$ in through their roots. Nitrogen and other elements released from decaying organic matter that are not reused quickly by microorganisms or plants may escape to the atmosphere or be leached from the soil to a groundwater table. Losses of N from most undisturbed ecosystems via groundwater are slight (Jenny 1980), but losses can be increased substantially when disturbance of the vegetative cover reduces nitrogen uptake by plants.

Many factors influence the activities of soil organisms and their effects on decay of organic matter. Three of the most important factors are soil temperature, precipitation and infiltration of water into soils, and soil drainage. Both plant productivity and decay of organic matter increase with temperature, but increases in decay are greater than increases in productivity. Therefore, soil organic matter (SOM) contents generally decrease as temperature increases. Both plant productivity and decay of organic matter increase as precipitation and soil moisture increases. In this case, increases in productivity are greater than increases in decay. Therefore, soil organic matter contents generally increase with greater precipitation—at least until precipitation is excessive and further increases do not incur increases in SOM contents. Excessive soil drainage, as in sand dunes, very shallow soils, and extremely stony stream terraces or alluvial fans, reduces productivity and excessively drained soils consequently have less organic matter than associated soils that hold more water. This drainage effect may be nullified in very humid climates where precipitation adds water to soils throughout the year and plants are less dependent upon water storage capacities of soils. Impeded drainage that causes soils to remain saturated with water long enough into the summer to become depleted of oxygen retards organic matter decay because anaerobic organisms are less effective decomposers of soil organic matter than are aerobic organisms. Therefore, organic matter contents of poorly drained soils are generally greater than those of associated well-drained soils. In extreme cases of very poor drainage and continual influx of water from springs or seeps, organic soils develop. Organic soils are those with more than 20% organic matter if the inorganic matrix is sand or more that 30% organic matter if the inorganic matrix is fine clay.

Table 5-1 shows soil organic matter contents of some California serpentine soils. A few nonserpentine soils are included in the table to show that serpentine soils have organic carbon (SOC) contents comparable to those in nonserpentine soils with similar climate (fig. 5-4), or at least not appreciably less. The organic C contents of these soils are related to mean annual precipitation (fig. 5-4, $r^2 = .500$). Adding altitude as a second independent variable, along with precipitation, the coefficient of determination is practically the same, $R^2 = .504$).

Because most of the nitrogen in soils is in organic matter, the C/N ratio is an indicator of the soil nitrogen status. This ratio is commonly >60 in forest and shrub detritus and about 30–60 in herbaceous detritus. It is generally about 15–45 in the surface of forest soils and 10–12 or 15 in the surface of grassland soils (table 5-1). It is generally lower in subsoils than in surface soils (fig. 5-5). Because most organic detritus is added to soils at, or near, the surface, the alteration of SOM, or its degree of transformation, increases with depth in soils. Therefore, the mean age of SOM in subsoils is greater than that in surface soils. The decrease in proportions of radiogenic carbon (^{14}C) in subsoils commonly indicates that some of the carbon has been in soils for thousands of years. Carbon/nitrogen ratios commonly decrease with depth in soils to <10 in subsoils. Ammonium-nitrogen is a substantial portion of the total nitrogen in some subsoils, contributing to lower carbon/nitrogen ratios.

The carbon/nitrogen ratios in serpentine soils are comparable to those in nonserpentine soils with similar vegetative cover. Thus, if the organic carbon contents of serpentine soils are not appreciably lower than those of nonserpentine soils, then neither are the nitrogen contents. Effendi et al. (2000) found that carbon and nitrogen contents in serpentine Inceptisols and Oxisols in tropical Indonesia were greater than in nonserpentine Ultisols in the same area. This may be an exception rather than a rule, but it emphasizes the risk of hypothesizing that serpentine soils contain less organic carbon and nitrogen than nonserpentine soils.

Table 5.1 Organic matter (or carbon, SOC) in soils of the California Region.

Pedon designation	Soil depth class	Soil subgroup	Bedrock	Altitude (m)	Mean precipitation (cm/year)	Vegetative cover	Pedon SOC (kg/m²) 50 cm	Pedon SOC (kg/m²) 100 cm	Surface soil exchange C/N	Surface soil exchange Ca:Mg	Subsoil exchange Ca:Mg	Reference[b]
Fresno County, CA												
58-CA-10-003	DP	Haploxeralf	Schist	335	50	Shrub/grass	5.0	5.5	13	nd	nd	1
58-CA-10-007	SH	Haploxeralf	Serpentinite	335	50	Grass	2.7	—	12	nd	nd	1
Humboldt County, CA												
60-CA-12-008X	MD	Haploxeralf	Serpentinite	685	190	Open forest	6.4	9.9	29	nd	nd	2
76-CA-12-001X	DP	Haploxeralf	Peridotite	640	180	Semi-dense forest	6.9	9.7	23	0.49	0.04	2
60-CA-12-004X	DP	Haploxerept	Greenstone	1160	145	Dense forest	6.0	7.2	34	nd	nd	2
60-CA-12-011X	VD	Haploxerept	Peridotite	610	230	Semi-dense forest	7.1	10.4	31	nd	nd	2
62-CA-12-026X	DP	Haploxerept	Peridotite	840	200	Semi-dense forest	8.4	10.1	37	0.38	0.08	2
Lake County, CA												
54-CA-17-010X	MD	Argixeroll	Serpentinite	1050	90	Shrub	7.4	—	21	nd	0.01	2
54-CA-17-003X	DP	Argixeroll	Andesite	915	125	Semi-dense forest	6.2	8.9	27	7.3	6.7	2
S78-CA-033-021	SH	Argixeroll	Serpentinite	360	100	Shrub/grass	3.4	—	nd	0.18	0.01	5
S78-CA-033-029	SH	Haploxeroll	Serpentinite	430	70	Grass	2.8	—	12	0.16	0.07	5

(*continued*)

Table 5.1 (continued)

Pedon designation	Soil depth class	Soil subgroup	Bedrock	Altitude (m)	Mean precipitation (cm/year)	Vegetative cover	Pedon SOC (kg/m^2) 50 cm	Pedon SOC (kg/m^2) 100 cm	Surface soil exchange C/N	Surface soil exchange Ca:Mg	Subsoil exchange Ca:Mg	Reference[b]
Nevada County, CA												
68CA-29-29	VD	Haploxeralf	Granodiorite	395	90	Open forest	5.9	7.7	13	3.2	1.5	4
68CA-29-36	MD	Hapoxeralf	Serpentinite	760	120	Open forest	3.2	—	22	0.44	0.03	4
Shasta County, CA												
65-CA-45-162X	VD	Haploxeralf	Peridotite	685	180	Dense forest	9.3	12.0	37	nd	nd	2
65-CA-45-163X	VD	Haploxeralf	Peridotite	855	165	Dense forest	10.4	14.1	21	1.0	0.27	2
Sonoma County, CA												
64-CA-49-022X	SH	Haploxeroll	Serpentinite	135	125	Grass	8.2	—	13	0.34	0.16	2
Tehama County, CA												
58-CA-52-015	MD	Haploxeralf	Serpentinite	1495	125	Open forest	7.0	—	33	0.41	nd	2
58-CA-52-018	SH	Argixeroll	Serpentinite	840	65	Shrub	5.6	—	26	nd	nd	2
Tuolumne County, CA												
73-CA-55-012X	SH	Rhodoxeralf	Serpentinite	355	65	Shrub/grass	3.9	—	11	0.46	0.16	1
77-CA-55-043X	SH	Rhodoxeralf	Serpentinite	360	60	Shrub/grass	4.0	—	13	0.98	0.73	1
73-CA-55-013X	SH	Haploxerept	Serpentinite	390	65	Grass	4.3	—	15	0.74	0.36	1
Siskiyou County, CA												
S74CA-093-014	MD	Haploxeralf	Serpentinite	1365	70	Open forest	6.0	7.5	42	0.84	0.11	4

Sample	Soil depth	Subgroup	Parent material			Vegetation						Ref
S74CA-093-015	VD	Haploxeralf	Serpentinite	1380	75	Open forest	7.2	8.1	22	0.64	0.05	4
S76CA-093-003	DP	Haplohumult	Alluvium	455	140	Dense forest	8.4	10.4	nd	9.0	1.1	4
S76CA-093-006	SH	Haploxerept	Peridotite	1005	150	Open forest	5.6	—	nd	0.35	0.16	4
S78CA-093-001	VD	Haplohumult	Peridotite	790	180	Dense forest	7.8	9.6	nd	10.8	1.0	4
S79CA-093-001	DP	Palexeralf	Peridotite	1585	125	Semi-dense forest	3.8	5.5	36	1.1	0.10	4
S79CA-093-002	MD	Palexeralf	Till, ultramafic	1735	125	Open forest	4.8	—	24	0.21	0.08	4
S79CA-093-003	MD	Argixeroll	Serpentinite	1235	90	Open forest	6.0	—	20	0.21	0.07	4
Yuba County, CA												
67-CA-58-125X	SH	Haploxeralf	Serpentinite	945	175	Open forest	10.0	—	26	0.48	0.37	3
68-CA-58-127X	MD	Haploxeralf	Serpentinite	945	180	Semi-dense forest	12.6	16.6	26	0.72	0.19	3
72-CA-58-131X	MD	Haploxerult	Schist	805	170	Dense forest	9.3	10.1	33	2.9	3.6	3
Curry County, OR												
S89OR-015-006	VD	Eutrudept	Peridotite	610	250	Semi-dense forest	7.9	11.7	—	0.45	0.07	5
S89OR-015-018	DP	Eutrudept	Peridotite	1247	360	Semi-dense forest	11.6	—	—	0.48	0.14	5
S92OR-015-007	DP	Eutrudept	Peridotite	823	280	Semi-dense forest	10.6	13.0	57	1.31	0.06	5

[a] Soil depth classes: SH, shallow, 25 to 50 cm; MD, moderately deep, 50 to 100 cm; DP, deep, 100 to 150 cm; and VD, very deep.
[b] References: 1. Allardice et al. 1983; 2. Begg et al. 1984; 3. Begg et al. 1985; 4. Soil Survey 1973; 5. Natural Resources Conservation Service (NRCS), unpublished data.

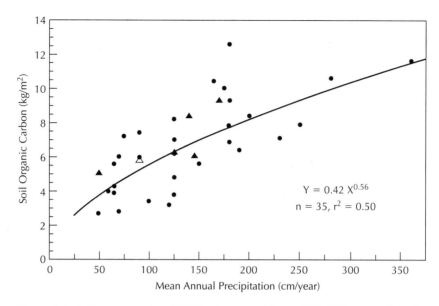

Figure 5-4 Soil organic carbon (SOC) in the upper 50 cm of 35 California and southwestern Oregon soils in relation to mean annual precipitation (MAP). Nonserpentine soils (there are only six) are represented by triangles.

Some chemical elements other than carbon and nitrogen accumulate in soils as organic matter is added to them. Plant detritus is a major source of carbon, nitrogen, and sulfur in soils as indicated by their concentrations in plants. Elements that are higher above the base of the triangle in figure 5-6 are more concentrated in plants than in rocks and soils. Notice that the most abundant elements in plants, including phosphorus, commonly form oxyanions (for example, HPO_4^{2-} or $H_2PO_4^-$) in soils. Also, notice that bromine, iodine, selenium, and arsenic, the other common elements that are more abundant in soils and plants than in rocks, are nonmetals. The proportions of carbon and nitrogen in rocks are negligible; the main source of these elements is the atmosphere. Elements along the base of the triangle have only slight, or negligible, roles in plant nutrition.

5.2.2 Weathering

Serpentine soil development begins with the weathering of peridotite and serpentinite. These rocks, upon disintegration, are the parent materials of serpentine soils. Weathering begins when rocks are exposed to the atmosphere and meteoric water. Alteration of rocks by connate (see Glossary) water deep within the earth is considered to be metamorphism, rather than weathering. Weathering penetrates from the ground surface to depths less than a meter in very dry uplands and to tens of meters in wet tropical climates.

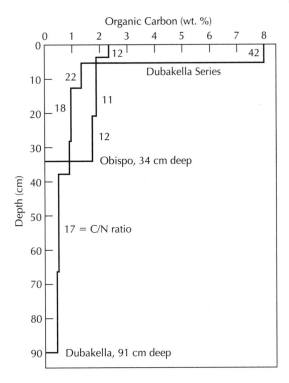

Figure 5-5 Comparisons of soil organic carbon and C/N ratios in profiles of forest and grassland soils. Dubakella (NRCS pedon S74CA-093-014 in Siskiyou County) is a Mollic Haploxeralf with an open forest plant community, and Obispo (pedon CR2 sampled in Santa Clara County for a 1999 Western Society of Soil Science field trip and analyzed in the laboratory of R. Southard, University of California-Davis) is a Lithic Haploxeroll with an herbaceous plant community.

Physical weathering occurs in all climates, but it is generally subordinate to chemical weathering. In many cases, it is not possible to differentiate clearly between physical and chemical processes. For example, hydration and crystallization of salts are chemical process that may contribute to the physical disintegration of rocks. Other causes of physical weathering are heating and cooling, wetting and drying, freezing and thawing, and abrasion by windblown sand. Chemical weathering commonly aids physical weathering by altering minerals to yield weaker components or by dissolving cementing agents that bind rocks. Weathering by thermal expansion and contraction is confined to the ground (or above ground) surface because diurnal and seasonal temperature changes are too slow below ground to cause any disintegration of rocks. Also, weathering by wetting and drying and by freezing and thawing are strictly surface and near-surface phenomena.

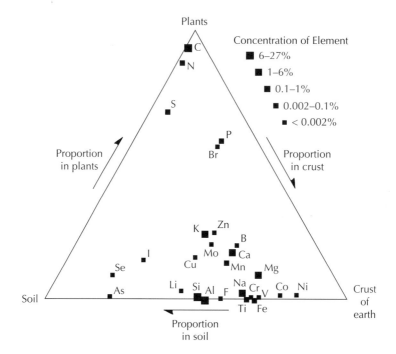

Figure 5-6 Distribution of chemical elements among equal masses of rock (continental crust, except sedimentary rocks), soils, and terrestrial plants. Sources of data are Romankevich (1990) for continental crust and plants and Shacklette and Boerngen (1984) for soils. Size of squares is proportional to the mean elemental concentration. For example, C averaged among the means for crust, soil, and plants is 16% (large square) and mean for plants is 95% of the sum of means for crust, soil, and plants (which locates the square near the plants apex of the triangle).

Once the ground is covered by soil, weathering of soil parent materials is almost exclusively chemical, including hydration with its physical aspects. Although minerals in a few kinds of rocks dissolve in neutral water, most weathering is promoted by acids. Some acids are formed by hydrolysis when sulphur and nitrogen oxides from the atmosphere react with water, but atmospheric carbon dioxide and plants are responsible for most acids in soil solutions. Both plant roots and decomposing plants are sources of organic acids and carbon dioxide. Carbon dioxide dissolves in water and forms a weak acid:

$$CO_2 + H_2O \rightleftarrows H_2CO_3,$$

which dissociates to form $H^+ + HCO_3^-$. This reaction is the main source of protons that are primarily responsible for weathering in most soils. Oxidation of ammonium from organic matter and oxidation of sulfides from marine sedimentary and some metamorphic rocks are other sources of acids in some soils and their parent materials.

There are several possible paths of mineral alteration and weathering losses in soils: (1) minerals, amorphous matrix, or cement can be dissolved and the constituents leached out of the soil; (2) dissolved constituents can be leached to a lower level in the soil and precipitated as mineral or amorphous material; (3) a mineral can be altered in place to another mineral, without dissolution; or (4) a mineral can be dissolved and another mineral formed in the same place from the products of dissolution. Weathering produces different soil materials in different weathering and leaching environments.

Olivine, pyroxene, and serpentine are the common minerals, or groups of minerals, in ultramafic rocks. They are unstable in well-drained soils that are being leached by water passing through them. This includes most serpentine soils in western North America, and at least the surface in the drier serpentine soils that do not receive enough water to leach their subsoils.

Olivine [$(Mg,Fe)_2SiO_4$] is composed of independent silicon tetrahedra with Mg^{2+} and Fe^{2+} ions between the tetrahedra (appendix A). It is readily weathered by acid solutions that exchange protons for the Mg^{2+} and Fe^{2+} ions and carry them away, or at least carry the Mg^{2+} ions away. In well-drained soils and other oxidizing environments, the Fe^{2+} is oxidized to Fe^{3+} and immobilized in insoluble compounds. Loss of mafic cations from olivine weakens its structure and promotes the loss of silicon. Silicon enters soil solution as acid silicate (H_4SiO_4), or it forms magnesium or iron smectites (clay minerals) if there are sufficient Mg^{2+} and Fe^{3+} ions to combine with the silicon and oxygen. Rapid leaching of serpentine soils may deplete readily available Mg^{2+} and Fe^{3+} ions and allow silicon that would be used to produce smectite minerals to be precipitated as silica. This silica is initially opal, but it may crystallize to produce a microcrystalline quartz called chalcedony (Thiry and Millot 1987).

Pyroxenes (of which there are many kinds) are composed of single chains of silicon tetrahedra, with mafic cations between parallel but independent chains. They are more stable than olivine. Orthopyroxenes, $(Mg,Fe)SiO_3$, are in the orthorhombic crystal system. They weather to form talc and smectites (Yatsu 1989). Pyroxenes containing Ca^{2+} ions (which are larger than Mg^{2+} and Fe^{2+} ions) between chains crystallize in the monoclinic crystal system. Therefore, they are called clinopyroxenes, $(Ca,Mg,Fe)SiO_3$. Clinopyroxenes are much less common than orthopyroxenes in ultramafic rocks formed on the sea floor and accreted to western North America. They are more common in Alaskan-type concentric bodies.

Serpentine minerals, $Mg_3Si_2O_5(OH)_4$, are composed of sheets of silicon tetrahedra and magnesium octahedra sandwiched together in alternating layers (appendix A). Although serpentine is more stable than olivine and pyroxenes in soil environments, it is weatherable. Serpentine appears to be stable in alkaline environments, but not in acid soils (Schreier 1989, Schreier et al. 1987). A variety of serpentine called chrysotile may even be generated in some current environments near the ground surface. Craw et al. (1987) found chrysotile in Quaternary debris flows from serpentine melange containing lizardite—apparently the chrysotile was produced under alkaline and reducing conditions following deposition of the debris flows. Wildman and Jackson (1968) found, experimentally, that increasing the amount of carbon dioxide in a weathering solution greatly increases the disintegration of serpentine from California soil parent materials (remember that the hydrolysis reaction of carbon dioxide and water produces protons, which are largely responsible for the weathering of silicate minerals).

Chromite ($FeCr_2O_4$), which is an important accessory mineral in peridotite, and magnetite ($FeFe_2O_4$), which is produced during the serpentinization of peridotite, are stable compared to olivine and pyroxenes. Even though they contain ferrous iron, they are not very susceptible to weathering in oxidizing environments, and they are even more stable in reducing environments.

Magnesite, $MgCO_3$, is present in some serpentine soil parent materials, but it is very susceptible to weathering in acid soils:

$$MgCO_3 + H_2O + CO_2 = Mg^{2+} + 2HCO_3^-.$$

Upon drying of some soils with excess magnesium, this reaction is reversed and magnesium precipitates as magnesite in nodules or in cracks in weathered bedrock. Magnesite is common in serpentine soils of Baja California. Much magnesite may be inherited from disintegrated bedrock, rather than formed within soils.

Concentrations of chemical elements in soils change as weathering is accompanied by biological activity and leaching. Figure 5-7 shows how elemental concentrations in

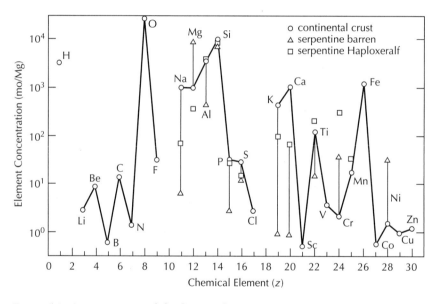

Figure 5-7 Concentrations of the first 30 chemical elements (Z=1–30), except inert gas elements (Z=2, 10, and 20) in continental crust (Romankevich, 1990), and some elements in a serpentine barren on Mt. Tamalpais in Marin County, California, and in the upper 50 cm of a Haploxeralf on serpentinite in Josephine County, Oregon (Robinson et al. 1935). Vertical lines link elemental concentrations in continental crust (circles) and in serpentine detritus of barrens (triangles). Relative to crust, aluminum, sodium, potassium, calcium, phosphorus, and titanium are depleted in the serpentine detritus, whereas magnesium, chromium, cobalt, and nickel are concentrated in the serpentine detritus, although no barrens data were available for cobalt and nickel. Weathering and leaching in the Haploxeralf (squares) removed much of the magnesium and concentrated titanium and chromium.

raw serpentine detritus, or rudimentary soil (serpentine barren), compared to concentrations in average igneous and metamorphic rocks (continental crust), and which elemental concentrations increase or decrease as weathering and leaching proceed to produce a Haploxeralf. Most notably, magnesium, chromium, and nickel concentrations are relatively high in the rudimentary serpentine soil of serpentine barrens and sodium, aluminum, phosphorus, potassium, and calcium are low. Weathering releases magnesium rapidly, and it is leached from the older serpentine soil (Haploxeralf) to greatly diminish the magnesium concentration in it. Concentrations of other elements utilized extensively by plants, such as phosphorus, potassium, and calcium, increase substantially from rudimentary to older soil. Elements that are not readily leached, such as aluminum, chromium, and iron, are concentrated as the total mass of soil is reduced by removal of magnesium and other readily leached elements. The increase in sodium in the older soil, relative to serpentine barrens, may be largely from atmospheric dust and precipitation.

5.2.3 Soil Horizon Formation

The first soil horizons to form when a ground surface is exposed are O horizons of plant detritus above ground and A horizons of humus accumulation below ground. O horizons have subscripts, which are "i" if the plant detritus, or litter, is fresh or only slightly altered by weathering; "e" if it is partially decomposed; and "a" if it has been reduced almost completely to fine organic matter, or humus (table 5.2). Most soils have Oi horizons, unless they are very dry and have sparse vegetative cover, but Oa horizons occur only in soils that are wet or cold enough to retard decay of organic detritus substantially. In the California Region conditions favorable for development of Oa horizons prevail only in high mountains or in fens, bogs, or marshes. Even Oe horizons are common in well-drained soils only at high elevations in mountains of this area. Although soil organic horizons are most commonly designated Oi, Oe, or Oa horizons, there are other systems of horizon designation (see below).

"A" horizons are generally the darkest colored horizons below ground. They commonly range from black with much organic matter to dark grayish brown or dark brown with less organic matter. The A horizons are commonly thicker under grass than

---

An Alternative Classification of Soil Organic Horizons

Some investigators have designated fresh organic detritus an "L-layer," partially decomposed detritus an "F-layer," and well decomposed detritus an "H-layer." These are similar to Oi, Oe, and Oa horizons, but not exactly the same. F-layers, for example, may include much organic detritus, or leaf litter, that has been discolored by weathering but not converted to humus. Consequently, F-layers may include parts of Oi horizons, as well as all of Oe horizons, and be much more common than Oe horizons in serpentine soils.

Table 5-2 Soil horizon suffix symbols

Suffix symbol	Master horizons[a]	Explanation
a	O	Highly decomposed organic material
b	A,E,B	Buried genetic horizon
c	A,E,B,C	Concentrations or nodules
d	B,C	Physical root restriction (e.g. dense till and plow pans)
e	O	Moderately decomposed organic material
f	A,B,C	Frozen soil
g	A,E,B,C	Strong gleying
h	B	Illuvial accumulation of organic matter
i	O	Fresh or slightly decomposed organic material
k	B,C	Accumulation of carbonates, commonly $CaCO_3$
m	B,C	Cementation or induration
n	B,C	Accumulation of sodium
o	B	Residual accumulation of sequioxides
p	O,A	Mixing by tillage or other disturbance
q	B,C	Accumulation of silica
r	C	Bedrock weakly consolidated or weathered soft
s	B	Illuvial accumulation of sesquioxides
ss	A,B,C	Presence of slickensides
t	B	Illuvial accumulation of silicate clay, t for *Ton* (clay in German)
v	B,C	Plinthite-soft iron concentrations that harden upon repeated wetting and drying
w	B	Development of color (higher chroma) or structure
x	B,C	Fragipan properties-firm (moist) with brittle failure when dry
y	B,C	Accumulation of gypsum
z	A,B,C	Accumulation of salts more soluble than gypsum

[a]Master horizons to which the suffix is applicable.

in forests, but there are many exceptions. Grass roots are generally concentrated in the upper 10–15 inches (25–38 cm) of soil and there is relatively rapid root turnover (cycle of root growth and decay) adding much organic matter to A horizons. Compared to grass, a larger proportion of tree and shrub detritus accumulates above ground in O horizons, and in thin A horizons, and may not be mixed deeply into subsoil horizons (fig. 5-5).

B horizons form in many different ways by many different processes. Some B horizons form simply by the concentration or local (within horizon) redistribution and oxidation of iron from weathered parent material. These conditions may be observed as reddening of soil or formation of reddish or yellowish mottles where iron is concentrated and oxidized. Reddish or yellowish mottles may be in a grayish matrix where Fe has been reduced and depleted when it moves to the reddish or yellowish spots. If there is sufficient clay in a soil to cause shrinking and swelling upon wetting and drying, prismatic or blocky structure will form. A B horizon that has either of these features (colors indicating concentration or redistribution of iron or development of soil structure) is designated a Bw horizon. A Bw horizon that has both features, and is not limited to the upper 25 cm of soil, is called a "cambic" horizon. It generally takes thousands of years for cambic horizons to form in serpentine soils of western North America.

Fine, colloidal clay forms as coarser soil particles weather to produce clay minerals. Some of these particles are washed downward and deposited in subsoils. Where this process is evident from coatings of clay in pores or on the faces of blocky or prismatic structural aggregates, the horizon is designated a Bt horizon. Bt horizons in which the accumulation of clay results in a minimum 3% increase, and more than 20% over the amount in the A horizon, are called "argillic" horizons. This criterion assumes uniform parent material; differences in clay concentration among alluvial strata (for example, stream deposits) are not grounds for argillic horizon designation. No horizons are designated Bw in soils with Bt horizons, unless the Bw and Bt horizons are separated by a surface that has been buried by alluvium, colluvium, or volcanic ash. It generally takes tens of thousands of years for argillic horizons to form in serpentine soils of western North America.

Figure 5-8 Cement Bluff in northeastern Trinity County, California. It is made of silica-cemented glacial till more than 100 m thick.

Prolonged weathering and leaching of basic cations and silicon produces subsoil horizons lacking weatherable minerals and with very low cation-exchange capacities (CEC). These are designated Bo horizons. If thick enough, they are "oxic" horizons. These horizons may take millions of years to form. They do occur in serpentine soils on surfaces <5 Ma (million years) old in the Greater Antilles (Beinroth 1982) and in the Sierra Nevada (Alexander 1988a).

Olivine, orthorhombic pyroxenes, and serpentine lack calcium, although some peridotite contains monoclinic pyroxenes and accessory actinolite or tremolite that have calcium. Thus, calcium is scarce in serpentine soils. Nevertheless, lime ($CaCO_3$) does accumulate in some serpentine soils where precipitation is too low to leach excess calcium through them. Subsoil horizons with calcium carbonate accumulation are designated Bk horizons.

Silica accumulates in some serpentine soils, both in humid areas where the source of silicon is from the weathering of olivine and other minerals in serpentine soils and in dry areas where the source of silicon may be from weathering of eolian silt or dust. The silica accumulates as opal, which eventually may be converted to microcrystalline quartz (Thiry and Millot 1987). Silica appears to accumulate in humid areas only where drainage is impeded by a groundwater table or by a dense substrate such as glacial till. Subsoil horizons with silica accumulation are designated Bq. If the silica is in hard nodules, the horizon is designated Bqc. If more than 90% of a silica-cemented horizon is hard, it is designated Bqm, or Cqm if the cementation is in soil parent material. Sequences of buried Bqm and Cqm horizons up to at least 100 m thick are found in the Klamath Mountains (Alexander 1995; fig. 5-8). Silica-cemented alluvium is found in many stream channels in ultramafic terrain in the Klamath Mountains.

Clayey soils that wet and dry annually develop large vertical cracks during summer. Shrink and swell in these soils forms smooth inclined planes over which blocks of soil slide as the soil is deformed. These planes are called "slickensides" and horizons in which they develop are designated Ass, Bss, or Css horizons. Most western North American serpentine soils with slickensides are on alluvial fans, in alluvial or lacustrine basins, or on floodplains or stream terraces. Some, however, are in serpentine melange of the Franciscan complex. The cracking clay soils are called Vertisols (Soil Survey Staff 1999).

Soils in which colloidal humus or iron and aluminum are leached from surface horizons and deposited in subsoils have horizons designated Bh, Bhs, and Bs. These horizons form only where intensive leaching has produce a strongly acid, bleached horizon (E horizon) from which iron and aluminum can be transported to the subsoil. These conditions exist in serpentine soils only where nonultramafic materials have been deposited over the ultramafic materials.

While many processes of soil development increase distinctions between horizons in a soil, shrink–swell, animal activity, and wind-thrown trees mix soils and tend to obscure differences between horizons in a soil. These are important processes in some serpentine soils.

6   Serpentine Soil Distributions and Environmental Influences

Serpentine soils occur in all but one of the twelve orders (Alexander 2004b), which is the highest level in Soil Taxonomy (Soil Survey Staff 1999), the primary system of soil classification utilized in this book (appendix C). They occur in practically every environment from cold arctic to hot tropical and from arid to perhumid (always wet). Thus the variety of serpentine soils is very great even though they occupy only a small fraction of the earth.

6.1 Kinds of Serpentine Soils and Their Distributions

Serpentine soils have been found in all states and provinces that are adjacent to the Pacific Ocean from Baja California to Alaska. They are most concentrated in the California Region, where they have been mapped in 34 counties in California and in 5 counties in southwestern Oregon. Serpentine lateritic (or "nickel laterite") soils, which have not been mapped separately from other soils, are economically significant in California and southwest Oregon, even though they are not widely distributed in western North America. A representative serpentine soil is shown in figure 6-1.

6.1.1 Soil Orders

Serpentine soils, or soils in magnesic (serpentine) families, are represented in 11 of the 12 soil orders. Spodosols and Histosols in magnesic families occur only where there is a thin cover of nonserpentine materials over the serpentine materials, and there are no serpentine Andisols.

Figure 6-1 A typical serpentine soil, a moderately deep Mollic Haploxeralf in the Klamath Mountains. Graduations on the tape indicate 10-cm increments.

Andisols

Andisols contain amorphous and poorly ordered aluminum-silicate minerals, which are responsible for andic soil properties of these soils. Serpentine soil parent materials do not contain enough aluminum for the development of andic soil properties that are definitive of Andisols.

Alfisols

Alfisols are soils with argillic (or natric) horizons having more than 35% exchangeable bases (Ca^{2+}, Mg^{2+}, Na^+, and K^+) on the cation exchange complex. Al^{3+} and H^+ are the common nonbasic (acidic) cations on the exchange complex. The Mg^{2+} that serpentine soil parent materials release upon weathering keeps the basic cation status of soils high, unless they are leached intensively. Some of the soil horizon sequences are A-Bt, A-Btn, and A-Bt-Btk in Alfisols.

Soils of Dubakella Series and other moderately deep Mollic Haploxeralfs with a mesic soil temperature regime are the most extensively mapped serpentine Alfisols in California and southwestern Oregon. Figure 6-1 is representative of the Mollic Haploxeralfs. The common horizon sequence is A-Bt-R or A-Bt-C-R. The common plant community is open conifer forest with scattered shrubs and grass in the understory.

Serpentine Alfisols at higher elevations, with thin A horizons, are Typic Haploxeralfs. They are represented by the Toadlake Series with A-Bt-R, A-Bt-C, or A-Bt-C-R soil horizon sequences. The common plant community is open to semi-dense conifer forest with shrubs and grass in the understory.

Serpentine soils of the Cornutt and Walnett Series (table C-7) are well-drained soils in wetter areas where leaching is more intense and the basic cation status is lower. Thus they are Ultic, rather than Typic or Mollic, Haploxeralfs. The common horizon sequences are A-Bt-C in colluvium and A-Bt-C-R on bedrock. The common plant communities are open to semi-dense conifer and broadleaf (madrone or tanoak) forest with shrubs and grass in the understory.

Serpentine Palexeralfs are common, but not extensive, in western North America. There are two distinct kinds of Palexeralfs that are not distinguished by different class names in U.S. Soil Taxonomy. One kind of Palexeralf is recognized by a distinct or clear boundary at the top of the argillic horizon, with a large increase in clay from the A to the Bt horizon. This kind of Palexeralf is represented by the Tangle Series and by serpentine Mollic Palexeralfs with A-Bt-Btss-Cr-R horizons (Alexander 2003). The Btss horizon is an argillic horizon with smooth, oblique (neither horizontal nor vertical) surfaces called "slickensides" along which soil is displaced when it swells and shrinks upon wetting and drying. These Mollic Palexeralfs have open conifer-forest plant communities with shrubs in the understory. Another kind of Palexeralf is recognized by a thick and deep argillic horizon in which the clay content is relatively high below 150 cm depth. This kind of serpentine Palexeralf is represented by Ultic Palexeralfs that occur on relatively old land surfaces (Alexander et al. 1989, 1990), but it appears to be represented more commonly by less intensively leached Haplic Palexeralfs in ultramafic colluvium of unknown age that might be no more than ~10,000 years old (Alexander 2003). The soil horizon sequence is A-Bt-C. These Haplic and Ultic Palexeralfs have semi-dense to dense conifer-forest plant communities with shrubs in the understories.

Serpentine Alfisols are represented by the Kingmont series in the eastern part of the Northern Cascade Range, adjacent to the Basin and Range Province. It is a Typic Palexeralf with an E-Bt horizon sequence. A serpentine Vitrandic Haploxeralf (Kenotrail series) with an A-Bt horizon sequence has been mapped in the Washington part of the Okanagan highland, which is at the southern end of the Intermomtane belt that is between the Rocky Mountains and the Coast Ranges in British Columbia.

Serpentine Alfisols are represented by the Chrome series in the Piedmont of the Appalachian Mountains from southwestern Pennsylvania though Maryland and Virginia. They are Typic Hapludalfs with A-Bt-C horizon sequences. Soils of the Aldino series occur over serpentinite in the same area (Rabenhorst and Foss 1981), but they have fragipans that may be in loess that has been deposited over the serpentine materials, or in a mixture of these materials.

Aridisols

Aridisols are dry most of the time when they are not very cold, and they have diagnostic soil horizons other than ochric epipedons only. Serpentine Argixerols in western North America generally have ochric epipedons and argillic horizons. Dry soils that lack diagnostic horizons other than ochric epipedons are Entisols, or Gelisols if they are frozen (or below freezing temperature) and do not thaw more than 1 or 2 m deep each year.

No serpentine Aridisols have been found in the California Region, and none has been described in Nevada. Serpentinite has been mapped in northern Nye County, in the Hot Springs Range of Humboldt County, and in the Candelaria mining district of Mineral County, near the Esmeralda County border. These areas in Nevada are very small and there is scarcely any ultramafic rock that outcrops at the ground surface.

Haplargids were observed and described on the Calmalli ophiolite at the south end of the Serrania Borja in Baja California and on the Sierra de San Andres ophiolite of the Vizcaína Peninsula in Baja California Sur. The serpentine plant communities are sparse shrubs and cacti, with scattered copalquín (*Pachycormus discolor*) trees in the Sierra of San Andreas and torote (*Bursera* spp.) on soils of the Calmalli ophiolite. Brittlebush (*Encelia farinosa*) is a predominant shrub and palo Adán (*Fouquieria diqueti*) and several species of cactus are commonly present on these serpentine soils.

Entisols

Entisols are soils that lack the diagnostic horizons present in other orders. They generally have A-C, A-C-R, or A-R soil horizon sequences. Ochric epipedons are the only diagnostic horizons. Other diagnostic horizons develop rapidly in California serpentine soils of the California Region, leaving few of them in the initial (Entisol) stage of soil formation (fig. 6-2). In grasslands, mollic epipedons commonly develop in serpentinite materials within a hundred (or a few hundred), years, making the soils Mollisols. In forests, weathering is

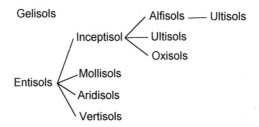

Figure 6-2 Likely paths of soil development in serpentine soils of western North America.

so rapid in peridotite materials, and so much iron is released in weathering, that cambic horizons develop in a few thousand years, making the soils Inceptisols. Recognized serpentine soil series in the Entisol order are generally very shallow (depth <25 cm) or, if deeper, in cold or arid climates where weathering is relatively slow.

Lithic Xerorthents, shallow Entisols, are represented by the Hiltabidel Series of serpentine soils, which were mapped and established in western Mendocino County, California. They have also been mapped in Trinity County, California (Alexander 2003) and probably occur in many other counties in the California Region. A typical Hiltabidel soil on serpentinized peridotite in the Klamath Mountains has moist colors of 5YR 3/4, making the chroma too high for a mollic epipedon, and it is too shallow to have a cambic horizon (lower boundaries of cambic horizons must be below 25 cm depth, except in very cold soils). The horizon sequence is A-R and the common plant community is shrubs, grass, and sparse conifer trees.

Cryorthents—cold Entisols, are represented by the Deadfall Series of serpentine soils, which were mapped in the Klamath and Shasta-Trinity National Forests. The soil horizon sequnce is A-C-R and the common plant community is shrubs, grass, and sparse conifer trees.

Torriorthents—dry Entisols that are either very shallow overy bedrock or very deep in recent ultramafic colluvium occur in the Sierra de San Andres ophiolite of the Vizcaína Peninsula in Baja California Sur.

Roberts (1980) described Cryorthents (Round Hill series) and Cryaquents (Blomidon series) in Newfoundland, which is at the northeastern end of the Appalachian Mountains. The vegetative cover is sparse and the O-horizons are discontinuous on these soils.

Gelisols

Gelisols are frozen soils. They have permafrost (permanently frozen ground) within 1 m of the ground surface or they have ice or evidence of cryoturbation in the upper meter and permafrost within 2 m of the ground surface. Ultramafic rocks are present in the Brooks Range of Alaska where permafrost prevails. Therefore, there must be serpentine Gelisols in the Brooks Range, but none have been described (D. Swanson, personal communication, 2001).

Histosols

The parent materials of all Histosols are organic detritus. Therefore, they are actually not serpentine soils, unless they contain a layer of serpentine materials more than 30 cm thick within 130–160 cm depth. These serpentine soils are in Terric subgroups of Histosols.

Many fens in the Klamath Mountains are surrounded by serpentinized peridotite, which has profound influences on their chemistry, even where the soils are not strictly serpentine soils (soils in magnesic families). The pitcher plant fens of Klamath Mountains are well known to botanists, especially Howell Fen on Eight Dollar Mountain in Josephine County, Oregon. The soils in Howell Fen are Saprists, because the organic detritus is highly decomposed; they have mesic soil temperature regimes. A serpentine fen in the Rattlesnake Creek terrain of Trinity County, California, lacking pitcher plants has soils similar to those in Howell Fen. In this case, the described Histosol, a Terric Haplosaprist, may be called a serpentine soil, because inorganic materials were encountered above 100 cm depth and were continuous to at least 130 cm depth (Alexander 2003). The Terric Haplosaprist is neutral to slightly acid in the organic portion and neutral in the inorganic portion.

Inceptisols

Inceptisols are young soils that will become Alfisols, Ultisols, or Oxisols if they remain undisturbed by severe erosion or mixing for thousands, or millions, of years. Serpentine Inceptisols of western North America have cambic horizons that may have taken thousands of years to form. Common soil horizon sequences are A-Bw-C and A-Bw-R, but there are many other possibilities.

Serpentine Inceptisols with xeric soil moisture regimes—Lithic, Typic, and Dystric Haploxerepts—are common in colluvium on very steep slopes. Weitchpec Series, the most extensively mapped of these Haploxerepts (table 6-1), is moderately deep, but mapping of serpentine soils in the Rattlesnake Creek terrain of the southern Klamath Mountains indicates that deep or very deep Inceptisols are more common on very steep slopes (Alexander 2003). Much of the Weitchpec soil mapped in the Klamath mountains was classified at the family level, which includes deep and moderately deep soils. The soil horizon sequence is generally A-Bw-C or A-Bw-C-R and the common plant community is open conifer forest with understory shrubs and grass.

Soils of the Pearsoll Series are the most extensive serpentine Lithic Haploxerepts. They commonly have A-Bw-R soil horizon sequences and the common plant community is open conifer forest with understory shrubs and grass. In Curry County, Oregon, the forest cover may include tanoak and madrone trees.

Serpentine Inceptisols that occur on older, or more stable, land surfaces are mostly in wetter parts of the Klamath Mountains where they are leached more intensively. Several serpentine soils series in the Lithic and Dystric Eutrochrept subgroups have been mapped in Coos and Curry counties, Oregon. The most extensive of these is the Serpentano Series (see table C-7). These soils have A-Bw-Cr soil horizon sequences and the common plant community is conifer and broadleaf (tanoak or madrone) forest with understory shrubs.

Mollisols

Mollisols are distinguished by having mollic epipedons. They may or may not have cambic, argillic, or natric horizons. Many combinations of soil horizons are possible, with A-Bw, A-Bt, A-Bt-Btk, and A-Btn being common sequences.

Most serpentine Mollisols mapped in western North America are dry during summers and have xeric soil moisture regimes. Therefore, they are mostly Xerolls. Very cold Mollisols occuring at higher evelations or higher latitudes are Cryolls.

Xerolls lacking argillic horizons are Haploxerolls and those with argillic horizons are Argixerolls. Warm Xerolls, those with thermic soil temperature regimes, generally have either shrub or grassland plant communities. Soils of the Henneke Series, the most extensively mapped serpentine soils in California Region, are Lithic Argixerolls with shrub plant communities. Soils of the Montara Series, another series of extensively mapped Xerolls with thermic soil temperature regimes, are Lithic Haploxerolls with grass plant communities. Deeper and cooler Xerolls commonly have thicker mollic epipedons. They are represented by the Kang Series of Pachic Argixerolls with mesic soil temperature regimes, generally with a A-Bt-R soil horizon sequence and plant communities of open forest with shrubs and grass in the understory. Also, serpentine Pachic Haploxerolls occur in warm valleys of the California Coast Ranges with grass plant communities. Examples of very deep Haploxerolls with thermic soil temperature regimes are in the Bearvalley and Leesville Series.

Cryolls lacking argillic horizons (Haplocryolls) have been described in ultramafic parent materials in the Klamath Mountains (Foster and Lang 1982) and in Alaska (chapter 22). Those with argillic horizons (Argicryolls) occur in the Blue Mountains.

Serpentine Mollisols with udic soil moisture regimes and isomesic soil temperature regimes, Udolls, have been mapped along the Pacific coast in southwestern Oregon. They are Lithic Hapludolls represented by the Sebastian Series and Typic Hapludolls represented by the Rustybutte Series. Soils in both series generally have forest plant communities with grass in their understories.

Small areas of serpentine Mollisols with aquic conditions have been mapped in colluvial and alluvial materials of landslides, drainageways, and topographic depressions in southwestern Oregon and California. They are represented by the Copsey Series of Vertic Haplaquolls, which generally have A-C horizon sequences and white oak grassland plant communities.

Oxisols

The representative pedon of the Forbes Series in Placer County, California, appears to be an Oxisol (Alexander, 1988a), but no Oxisols with xeric soil moisture regimes have been recognized in U.S. Soil Taxonomy (Soil Survey Staff 1999). There is no argillic nor kandic horizon in the typical pedon and a subsurface horizon has low cation-exchange capacity (CEC) and sparse weatherable minerals, characteristics of an oxic horizon. The oxic horizon in the Forbes pedon has about 50%–60% clay, estimated by multiplying the 1.5 MPa (megapascals of pressure) water retention by 2.5 or 3, and has a neutral NH_4-acetate CEC of only 4.7 me/100g (47 mmol/kg) of fine earth. Weatherable minerals in the oxic horizon are about 2% in the fine sand (0.075–0.25 mm) fraction, consisting of pyroxenes, amphibole (tremolite or actinolite), and serpentine (Alexander 1988a). The pH in neutral salt ($CaCl_2$) solution is higher than in distilled water, indicating that the oxic horizon has a net positive charge. This is another indicator of low clay activity, because smectite and other layer silicate clay minerals have negative charges. The net positive charge is presumably a result of high iron oxyhydroxide concentrations. These oxyhydroxides are amphoteric and have positive charges in acid soils.

The only Oxisols in western North America, other than those with peridotite parent materials, are those in kaolinitic clays of the Ione formation in central California (Singer and Nkedi-Kizza 1980). Sediments of the Ione formation, including kaolinite, were eroded from highly weathered landscapes during the Eocene (Bates 1945). Oxisols of the Ione formation developed in parent material that was already highly weathered and had low-activity clay minerals. Soils of the Forbes Series are presumed to have been formed at locations where peridotite disintegrated to produce the parent materials. There is no evidence that the materials have been transported to the sites from elsewhere. This part of the Sierra Nevada was covered by Pliocene lahars (volcanic mudflows). Soils of the Forbes Series may be either middle Tertiary soils that were buried and subsequently exhumed by erosion, or they may have developed from serpentinized peridotite during the Quaternary after the lahars were eroded to expose peridotite.

Serpentine Oxisols are represented by the moderately deep Rosario (Lithic Hapludox) and the very deep Nipe (Anionic Acrudox) series in the Greater Antilles of the Carribean region. These soils have A-Bo horizon sequences and are very red, with 7.5R and 10R hues in Bo horizons. The Anionic subgroup of the Acrudox great group indicates that the soils have net negative charges in the Bo horizons. Rosario soils are in a ferruginous family, which requires 18%–40% iron oxide in the fine earth (fraction <2 mm), and Nipe soils are in a sesquic family, which requires 18%–40% iron oxide and 18%–40% gibbsite in the fine earth. The Nipe series was established by Bennett and Allison (1928) and has been characterized more recently by Beinroth (1982), Fox (1982), and Jones et al. (1982). In Puerto Rico, serpentine Oxisols are found on Pliocene surfaces, which means that they are no more than 5 Ma old. No Oxisols this young are found on other parent materials in Puerto Rico (Beinroth 1982). Serpentine materials weather sufficiently to become Oxisols more rapidly than other materials, because the minerals in ultramafic rocks are either highly weatherable (olivine, pyroxenes, and serpentine) or highly resistant to weathering (chromite and magnetite), there are no moderately weatherable minerals such as feldspars.

Spodosols

Spodosols do not develop in serpentine parent materials unless the serpentine materials have been covered by more siliceous glacial drift, colluvium, alluvium, or eolian deposits (Alexander et al. 1994). Leaching of iron and aluminum from surface soils to be immobilized in subsurface layers to form spodic horizons requires very acid surface soils. High rates of weathering and release of magnesium keeps serpentine soil pH relatively high until practically all of the peridotite and serpentinite are gone. The end products may be Inceptisols with very high iron oxyhydroxide concentrations (Alexander et al. 1994), or Oxisols (Alexander 1988a), but not Spodosols.

Several soil series of serpentine Spodosols have been established in the Northern Cascade Range for soils with colluvial or alluvial deposits of volcanic ash, glacial till, or loess over the serpentine materials (for example, the Edfro and Jorgensen series). These soils are mostly Haplocryods and Humicryods, and some Orthods). They have O-E-Bhs-Bs and O-E-Bs horizon sequences, with the E horizons in nonserpentine (or mixed, mostly nonserpentine) materials.

Ultisols

Ultisols are soils with argillic (or kandic) horizons having <35% exchangeable bases (Ca^{2+}, Mg^{2+}, Na^+, and K^+) on the cation exchange complex. Al^{3+} and H^+ are the common nonbasic (acidic) cations on the exchange complex. These soils with low base saturation are leached intensively. In serpentine materials they are highly weathered, also; otherwise Mg^{2+} released from serpentine soil parent materials upon weathering keeps the basic cation status of soils relatively high. The Bt horizons in intensively leached serpentine soils may be kandic horizons, which are argillic horizons with very low CEC/clay ratios, but with higher base saturation (BS >35%) in C horizons. Thus, California Region serpentine soils with kandic horizons are commonly Alfisols, rather than Ultisols. Ultisols are sparse in the California Region, occurring in areas with high precipitation or on old land surfaces or benches below old land surfaces. They generally have A-Bt-C horizon sequences.

Most of the Ultisols in the Klamath Mountains, which is where serpentine Ultisols are most likely to be present, are Humults. The only serpentine Ultisol in western North America that has been confirmed by adequate laboratory data is a soil in the Littlered series. It is a fine, ferritic, mesic Xeric Kanhaplohumult.

Vertisols

Vertisols are cracking-clay soils. They have subsurface horizons with slickensides and, upon drying, have wide cracks (width >1 cm) that open to the ground surface. They do not occur in wet climates where the soils have udic soil moisture regimes, because shrink–swell cycles caused by wetting and drying are necessary to mix the soil and develop slickensides, which are smooth surfaces along which blocks of soil move when a soil wets or dries.

Most serpentine Vertisols are in dry valleys, particularly in the California Coast Ranges. The plant communities are grass, with annual forbs. Soils in the Maxwell Series are representative of Typic Haploxererts in ultramafic alluvium. Soil horizon sequences are commonly A-Ass-C or A-Bss-C. Soils of the Venado Series are representative of Aridic Endoaquerts in alluvium.

Although serpentine Vertisols are generally in basins, they do occur in Franciscan melange on hills in the California Coast Ranges. These upland Vertisols are represented by the Climara series of serpentine Aridic Haploxererts with grass plant communities.

6.1.2 Laterization and "Laterite"

Serpentine soil development resulting from weathering and leaching culminates with deep reddish brown soils. This ultimate stage has been widely investigated because products of the development process, called "laterization," include economically viable nickel ore, and in some cases cobalt and other elements (Golightly 1981). Some investigators have called the soils "laterite," but this term has been so widely applied, obscuring its original meaning given by Buchanan in 1807 (see Scrivenor 1930) for a reddish brown layer that hardens on drying and was used as building blocks in India, that Hotz (1964) has called these soils in California and Oregon "lateritic soils," or "ferruginous lateritic soils." He added "ferruginous" for the high iron content that distinguishes lateritic soils derived from serpentinized peridotite from those with other parent materials containing much more aluminum. Ferruginous lateritic soils are found in small areas on old, Tertiary land surfaces in the northwestern part of the Klamath Mountains, and there are remnants of these soils on the lower western slope of the Sierra Nevada (Rice 1957). Also, they have been found in the California Coast Ranges as far south as Red Mountain in northern Mendocino County (Chesterman and Bright 1979).

Ferruginous lateritic soil profiles are described with three zones, or five including fresh and weathered bedrock (Rice 1957, Trescases 1997). From the base, these zones are (1) fresh bedrock, (2) weathered bedrock, (3) yellowish brown saprolite with silica boxwork, (4) brownish yellow soil, and (5) reddish brown soil with iron pellets. All of these zones are present in ferruginous lateritic soils on Nickel Mountain in Douglas County, Oregon, but the saprolite with silica boxwork is missing from most other ferruginous lateritic soils in the Klamath Mountains (Rice 1957, Foose 1992). Some silica boxwork persists in remnants of ferruginous lateritic soils in the Sierra Nevada (Rice 1957).

Trescases (1997) described the development of ferrugenous lateritic profiles in some detail. A brief description of that development, illustrated in figure 6-3, follows. Upon exposure of serpentinized peridotite containing olivine and pyroxene to the atmosphere, olivine weathers rapidly. Most of the olivine disappears from weathered bedrock before it disintegrates to produce soil, or soil parent material. Silicon, magnesium, and iron from weathered olivine form smectite and iron oxyhydroxides. Much of the magnesium and silicon from olivine is leached away, but some silicon is precipitated as amorphous silica (opal) or cryptocrystalline quartz (chalcedony) along fractures in the bedrock. In some cases, magnesite is deposited in cracks in the bedrock, below silica deposits, or in veins. Long after bedrock has lost its integrity, leaving stones rounded by weathering as remnants, the silica veins persist to form the silica boxwork (fig. 6-3) found on Nickel Mountain and in the Sierra Nevada. The primary minerals (minerals inherited from bedrock) in this yellowish brown saprolite with boxwork are serpentine, pyroxenes diminished by weathering, and minor chlorite and chromite. Magnetite is present, along with chromite, in more highly serpentinized peridotite. The secondary minerals, those produced in weathering and soil development, are goethite, smectite, and minor talc. Nickeliferous phyllosilicates deposited along the veins of silica are called *garnierite*. Manganese oxyhydroxides containing nickel and cobalt accumulate in mineralogically nondescript black deposits, called ab-

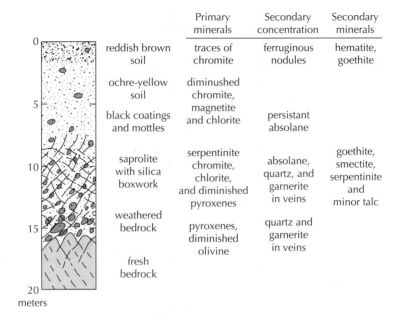

Figure 6-3 A conceptual ferruginous lateritic soil profile generalized from data of Rice (1957) and Trescases (1997). The mineralogy is from Trescases (1997), with slight modifications from data of Hotz (1964).

solane, along silica veins, or replacing them, in the upper part of the yellowish brown saprolite zone. Eventually, the silica boxwork collapses to form an ochre yellow soil similar to the C horizon of an Oxisol. Primary minerals in the ochre yellow soil are diminished chromite and minor amounts of pyroxenite and chlorite. Goethite accumulates as serpentine and smectites are lost. Absolane persists through the ochre yellow zone. At the top of the profile, above the ochre yellow zone, is reddish brown soil. Goethite, hematite, and iron pellets predominate in this zone. The pellets, or nodules, are mainly soil material cemented by iron oxyhydroxides, lacking concretionary structure. They are generally magnetic, because maghemite accumulates along with goethite and hematite in the nodules. These nodules protect primary minerals from weathering, allowing some chromite, pyroxene, and serpentine to persist in this highly weathered reddish brown soil zone.

There are wide deviations from the idealized model presented in the last paragraph. Foose (1992) analyzed samples from 11 ferruginous lateritic soil profiles in the Klamath Mountains of northwestern California and southwestern Oregon and detected smectite only in soils on slopes where water drained laterally from serpentinized peridotite upslope. He did not identify the kind of smectite, but assumed that it contains iron. However, if cations were transported downslope to occupy octahedral positions in smectites, they may contain more magnesium than iron, because magnesium is much more soluble in water, unless the iron is reduced to its ferrous form. The iron concentration range in the 11 ferruginous lateritic soils was 33.2%–77.6% (weight) Fe_2O_3.

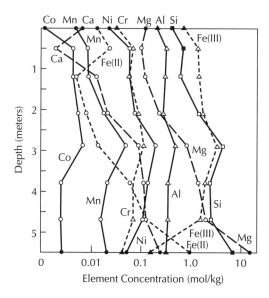

Figure 6-4 Molar contents of elements in samples from slightly serpentinized peridotite and ferrugious lateritic soil of Eight Dollar Mountain, Oregon (Hotz 1964), relative to titanium. Data for each element in all horizon samples were multiplied by the ratio of Ti in the sample for each horizon to Ti in unweathered bedrock, assuming that Ti losses during soil development have been negligible, compared to losses of other elements.

Data from elemental analyses of eight samples taken from fresh bedrock to the surface of the Eight Dollar Mountain ferruginous lateritic soil (Hotz 1964) help to understand the fate of chemical elements as serpentinized peridotite weathering and leaching progress. Results for each element were converted to moles per kilogram and the amount of each element in each horizon was multiplied by the ratio of titanium in fresh bedrock to titanium in the corresponding horizon (fig. 6-4). A straight vertical line would represent titanium in the figure and correspond to the assumption that vertical translocations and losses of titanium are negligible. Deviation of an element in any horizon to the left of its position in fresh bedrock at the bottom of the graph is a net loss and any deviation to the right is a net gain of the element from the horizon. Hotz (1964) assumed that aluminum was conserved, rather than titanium, in order to estimate elemental gains and losses. The line for aluminum is nearly vertical in figure 6-4, but it deviates to the left at the top, indicating that aluminum is lost from the surface soil. The offset to the right in sample four plotted at 2.9 m depth may be caused by a slightly inaccurate datum for titanium in that horizon. Silicon shows a first stage of dramatic loss from fresh to weathered bedrock as silicon is released from olivine and another stage of loss above the yellowish brown saprolite where the amounts of all silicate minerals diminish greatly and iron oxyhydroxides dominate the soil. Magnesium is lost continuously from 10.7 mol/kg in fresh bedrock to about 0.1 mol/kg in the ochre yellow soil, then it increases in the surface, possibly as a result of recycling by plants. Most iron in fresh bedrock, particularly that in olivine, is in the ferrous (Fe^{2+}) form. It is mostly oxidized to the ferric (Fe^{3+}) form in weathered bedrock, where it is concentrated to become codominant with silicon in the yellowish brown saprolite. Iron is further concentrated to become the dominant element in the ochre yellow soil. Calcium concentration is low in fresh bedrock and declines much further to the ochre yellow soil, above which the trend is reversed by plants that extract calcium from the soil and return it to the surface soil when their leaves fall to the ground and decay. Manganese is leached through the profile only when, or after, black absolane mottles disappear from the ochre yellow soil. Chromium is lost from the surface soil, where chromite is more weatherable and some Cr^{3+} may be oxidized to the more mobile Cr^{6+} form. It accumulates slightly in the remainder of the profile. Nickel accumulates in garnierite through the yellow brown saprolite. It is leached from the surface and accumulates along with Fe in goethite in the subsoil. Cobalt accumulates in absolane, but unlike nickel, it remains high throughout the ochre yellow soil even after absolane is gone.

Deviation from the idealized ferruginous lateritic soil profile may be related to degree of serpentinization of peridotite, climate, and topography (Golightly 1981). Lack of silica boxwork and lower smectite contents in ferruginous lateritic soils nearer the ocean in northwestern California and southwestern Oregon may be related to wetter climate and more intensive leaching than in the Sierra Nevada and on Nickel Mountain in Douglas County, Oregon.

Areas of ferruginous lateritic soils are too small to have been mapped in soil surveys in California and Oregon, and data have not been obtained to classify them according to U.S. Soil Taxonomy. The soils may be in any one or all of the Inceptisol, Alfisol, Ultisol, and Oxisol orders.

6.2 Environmental Influences on Serpentine Soils

Soils are in open systems allowing input and output of materials. We are seldom able to track the evolution of these systems in detail, so we generally relate the properties of the soils to the current conditions such as topography and climate.

6.2.1 Parent Material

Peridotite and Serpentinite

Mineralogically there are two main classes of rock that produce serpentine soils in western North America: peridotite containing olivine and pyroxenes and serpentinite containing serpentine minerals. Chemically they are similar. Most plants do not seem to distinguish between soils in peridotite and soils in serpentine materials, with one exception. Soils in pyroxenite materials, or at least in clinopyroxenite materials which contain

68 Soils and Life in Them

more calcium, appear to support the growth of some plants that do not grow on the usual serpentine soils with similar climate and topography (Alexander 2003).

Rock structure influences slope stability. Serpentinite is often sheared and fractured, reducing its resistance to mass failure. Serpentine melange is particularly susceptible to rupture and sliding, even on only moderately steep slopes. Landsliding is common in steep serpentinite terrain. A detail mapping of serpentine terrain in the southern exposure of the Rattlesnake Creek terrane (Alexander 2003) revealed that 1% of about 12,000 ha mapped is in peridotite landslides >1 ha and 11% is in serpentinite landslides. Because the serpentinite is less resistant to mass failure, there are fewer very steep slopes than in peridotite terrain (fig. 6-5).

Weathering rinds on serpentinite are commonly gray and those on peridotite are generally much redder. This is a consequence of iron being release by weathering of peridotite more rapidly than from serpentinite (chapter 5). Also, well-drained soils in peridotite materials may be redder than those of comparable climate and age in serpentinite (Alexander 2000, 2004a). Because well-drained soils in peridotite materials generally contain more secondary iron oxides (rather than magnetite, which is a primary iron oxide), the particles in them may be expected to be more well aggregated and the soils less erodible than those in serpentinite materials.

In the Rattlesnake Creek terrane in the southern Klamath Mountains, soils in peridotite materials generally have higher clay concentrations than soils in serpentinite mate-

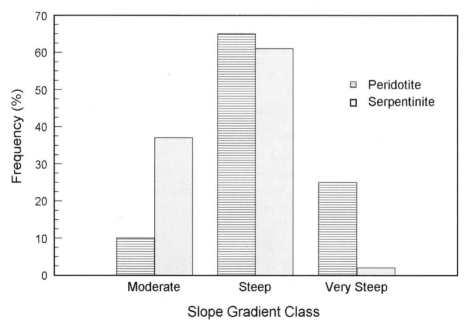

Figure 6-5 The slope distribution of serpentine soils in the southern exposure of the Rattlesnake Creek terrane that are not in large landslides. The slope classes are moderately steep (15%–30%), steep (30%–60%), and very steep (>60%).

rials (Alexander 2003). This observation is applicable only to this area with moderate precipitation and dry summers; it is possible, but unsubstantiated, that this soil parent material–clay concentration relationship may be reversed in areas with higher precipitation. The rationale for this hypothesis is based on the fact that peridotite weathers more rapidly than serpentinite in any climate of the Klamath Mountains, but the dominance of leaching over formation of smectite clay increases with wetter climate. In wetter climates most of the silicon and magnesium may be leached from the soils before these elements can be used to produce much smectite or other phyllosilicate (layered) clay minerals.

Gabbro

Gabbro and norite contain olivine, pyroxene, and calcium-feldspars. The pyroxenes are predominantly clinopyroxene in gabbro and orthopyroxene in norite. Gabbro and norite, with around 48%–52% or 54% silica, are near the low silica end of a sequence of plutonic rocks with a gradual trend in chemical composition from granite with about 70%–75% silica to peridotite with about 38%–45% silica (Le Maitre 1976, Alexander 1993). Aluminum and calcium increase gradually as silicon decreases from 75% to about 46% or 48% silica and then decrease drastically at about the same silica concentration that magnesium increases drastically (Le Maitre 1976). Soils on olivine gabbro, norite, or pyroxenite with 46%–48% silica may support plant communities more similar to those on serpentinized peridotite than to those on more silicic rocks. Whereas peridotite (and serpentinite) has much lower calcium concentration, gabbro with silica concentration >48% has higher calcium concentration than more silicic plutonic rocks. Also, magnesium concentrations are higher in gabbro than in more silicic plutonic rocks. Molar Ca:Mg ratios range widely about one in gabbro. Most first transition elements, from scandium to zinc, have relatively high concentrations in gabbro, but manganese and iron are slightly higher and chromium, cobalt, and nickel concentrations are much higher in peridotite. Some other first transition elements (scandium, titanium, and vanadium) have higher concentrations in gabbro than in either more silicic plutonic rocks or in peridotite.

Whittaker (1960) investigated plant distributions on quartz diorite, olivine gabbro, and serpentinized peridotite in the Klamath Mountains. He found that physiognomic aspects of the plant communities on gabbro were much more similar to those on quartz diorite, but that species compositions in plant communities on gabbro were intermediate, with no more affinity to those on quartz diorite than to those on serpentinized peridotite.

The Pine Hill intrusive complex, where Hunter and Horenstein (1992) recorded unique plant communities on soils with gabbro parent materials, is a layered body more similar to Alaskan-type ultramafic complexes than to ophiolites (Springer 1980). Lenses of pyroxenite are spread through the gabbro and may lead to lower Ca:Mg ratios in soils on the intrusive complex than in most soils with gabbro parent materials. Many plants previously considered to be serpentine endemics were found on soils of the intrusive complex, along with plants that are not present on serpentine soils. Thus species compositions of plant communities on soils of the intrusive complex are intermediate, with many similarities to those on either serpentine soils or soils with more silicic parent materials.

6.2.2 Climatic Influences

Solar radiation, size and configuration of land masses, and circulation of the atmosphere are major factors controlling climate. Even though climatic systems are very complex, most of the major climatic influences on soil development are related to temperature and precipitation. Wind can have substantial effects on evapotranspiration and other ecosystem processes, but its effects are generally subordinate to temperature and precipitation.

In western North America soil temperature regimes (STRs) range from hot, with very hot summers (hyperthermic STR) below about 300 m in southern California and in Baja California to very cold (cryic STR) on high mountains and high latitudes. Soils in the Brooks Range are permanently frozen, at least in subsoils, and permafrost is discontinuous to sporadic in central Alaska. Temperatures along the Pacific coast differ little from winter to summer; therefore some soils have isothermic STRs along the southern California coast and isomesic STRs along the northern California coast into Orgeon. Serpentine soils on the next coastal exposure of ultramafic rock, northward from Oregon, have cryic STRs. Thus serpentine soils in western North America are represented by frozen soils and those in hyperthermic, thermic, mesic, frigid, cryic, isothermic, and isomesic STRs. It is uncertain whether any of the serpentine soils on Cedros Island, or from there southward along the Pacific coast of Baja California, have isohyperthermic STRs.

Serpentine soils with aridic, xeric, udic, and aquic soil moisture regimes (SMRs) occur in western North America. No family class is recognized for soils with perudic SMRs, as those that occur in the wettest areas along the coast between Vancouver and Anchorage.

Precipitation has direct effects on weathering and leaching and indirect effects on soil processes through its influence on vegetation and living organisms within soils. Temperature modifies these effects through its influence on evapotranspiration. It has other effects on the vegetation and soil organisms. Perhaps the main effects of temperature in serpentine soils that are not frozen are on rates of soil processes rather than on the kinds of processes in the soils. Although differential rates of weathering and clay mineral formation can determine processes that form Andisols, there are no serpentine soils in this order.

The driest serpentine soils in western North America, those with aridic SMRs, are Entisols and Aridisols. The serpentine Aridisols have ochric epipiedons and argillic horizons and no other diagnostic horizons.

Many kinds of well-drained serpentine soils with xeric or udic SMRs are represented in the Klamath Mountains in which STRs range from thermic to cryic and include isomesic. The mean annual precipitation ranges from about 200 cm at low elevations along the Pacific coast to >300 cm in mountains near the coast and diminishes inland to about 100 cm, or less, around Scott Valley in Siskiyou County, California. It is up to about 200 cm south of Mt. Shasta, even at relatively low elevations, to about 100 cm farther south, near the Pit River in Shasta County, California. The distributions of well-drained cool soils (mesic or isomesic STRs) in relation to precipitation and time, or degree of soil development, are shown in figure 6-6. Very young soils are Entisols in any climate. Mafic or ultramafic parent materials shift soil classes to the left in figure 6-6, because more precipitation or time is required to leach basic cations from these materials that release them

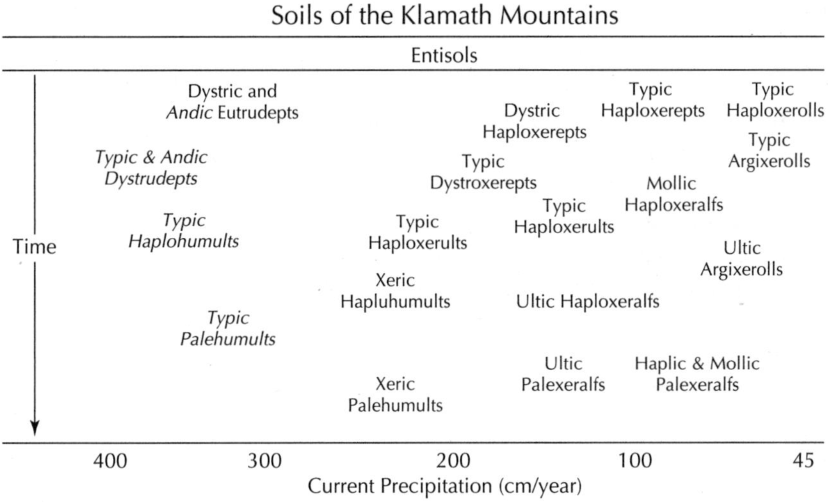

Soils of the Klamath Mountains

			Entisols			
		Dystric and *Andic* Eutrudepts			Typic Haploxerepts	Typic Haploxerolls
				Dystric Haploxerepts		
	Typic & Andic Dystrudepts		Typic Dystroxerepts			Typic Argixerolls
Time		Typic Haplohumults			Mollic Haploxeralfs	
			Typic Haploxerults	Typic Haploxerults		Ultic Argixerolls
			Xeric Hapluhumults	Ultic Haploxeralfs		
		Typic Palehumults				
				Ultic Palexeralfs	Haplic & Mollic Palexeralfs	
			Xeric Palehumults			
	400	300	200		100	45

Current Precipitation (cm/year)

Figure 6-6 Soils of the Klamath Mountains, excluding those that are shallow, poorly drained, or very cold. Subgroups with names in italic letters are not represented by serpentine soils.

rapidly and abundantly as the materials weather. Nearly all Eutrudepts have ultramafic parent materials, and there are some Hapludolls. The prevalence of Alfisols diminishes with higher elevation and colder temperatures to be replaced entirely by Entisols, Cryepts, and Cryolls at the highest elevations in the Klamath Mountains.

Mollisols with shrub and open woodland or open forest plant communities predominate in the driest areas of the Klamath Mountains. Serpentine Mollisols are represented by the Grell series of Lithic Haploxerolls, the Henneke Series of Lithic Argrixerolls, the Shadeleaf Series of Typic Argixerolls, and the Kang Series of Pachic Argixerolls. Ultic Argixerolls, which have been sampled in the Klamath Mountains (Alexander et al. 1990, site 001), occur where there is more precipitation or soils have been leached for a longer period of time. These Xerolls are in areas with about 50–100 or 150 cm of precipitation annually. Wetter serpentine Mollisols, Udolls with udic SMRs, have been mapped in Curry County, Oregon, but they are not extensive.

Inceptisols are widely distributed in all parts of the Klamath Mountains that are not dominated by Mollisols. From drier areas with about 100 cm to wetter areas with about 400 cm of precipitation annually, they are represented by the Weitchpec Series of Typic Haploxerepts, the Gravecreek Series of Dystric Haploxerepts, and the Serpentano Series of Dystric Eutrudepts. Although Dystrudepts have been mapped in the wetter parts of the Klamath Mountains or adjacent Coast Ranges, none have serpentine parent materials. On stable landforms, Alfisols develop from Typic Haploxerepts and either Alfisols or Ultisols develop from Dystric Haploxerepts. No serpentine Alfisols nor any serpentine Ultisols have been mapped in areas with serpentine Dystric Eutrudepts. Evidently, the high free-iron content and lack of drying in these Eutrudepts is not conducive to movement of clay particles. Soil water retention at 1.5 MPa (~15 atmospheres) of suction is 0.6–1 times the clay contents of surface (upper 60 cm) horizons in Eutrudepts (NRCS data for soils in Cedarcamp, Flycatcher, Redflat, Serpentano, and Snowcamp Series) compared to about 0.4 in most soils, indicating that the clay particles are not readily dispersed. Dispersion is necessary for the clays to be translocated (eluviated) and deposited (illuviated) in argillic horizons.

Alfisols occur along with dry Mollisols and also in slightly wetter parts of the Klamath Mountains with up to about 275 cm of precipitation annually. They develop on stable landforms where grass roots or mixing of other organic matter into soils do not produce the thick dark-colored surface horizons, or mollic epipedons, of Mollisols. Given enough time, serpentine Typic and Mollic Haploxeralfs, which are represented by the Toadlake and the Dubakella Series, will likely become Ultic Haploxeralfs or Mollic Palexeralfs, which are represented by the Ishi Pishi and the Tangle Series. Eventually, serpentine Ultic Haploxeralfs might become Ultic Palexeralfs in drier areas, and Palexeralfs have been sampled in the Klamath Mountains (Alexander et al. 1990, site 041); and in wetter areas they might become Xeric Kanhaplohumults, which are represented by the Littlered Series. Ultisols are not well represented by serpentine soils in the Klamath Mountains. In drier areas there is insufficient precipitation to leach magnesium from subsoils and in wetter areas high free-iron contents and continually moist climate inhibit the translocation of soil clay minerals. Conditions in the more moist serpentine soils are not conducive to the accumulation of smectites, which are the clay minerals that are most commonly translocated to form argillic horizons. In areas of intermediate precipitation, 150–275 cm/year, there are Humults of the Littlered series, which are Kanhaplohumults based on recent laboratory data. Palehumults are not represented by serpentine soil series, but they are represented by soils with other kinds of parent materials in the Klamath Mountains.

In the eastern Klamath Mountains the CEC of inorganic fine earth (particles <2 mm) from serpentine subsoils was inversely related to mean annual precipitation from 75 to 175 cm/year (Alexander 1999). Also, a weak but significant trend toward redder soils in wetter climates was found in the same investigation. These results were interpreted as indicating that drier climates in the Klamath Mountains favor the formation of smectites with relatively high CEC and wetter climates favor the formation of iron oxides (and oxyhydroxides) with reddish colors. Most iron oxide in serpentine soils is probably goethite, but only small amounts of hematite produce soils with 5YR or 2.5YR hues. Yellowish hues of 7.5YR in soils of the Serpentano Series, which have udic soil moisture regimes and are among the wettest well drained serpentine soils in the Klamath Mountains, may be because the cooler and moister climate where these soils occur in the northwestern Klamath Mountains inhibit the formation of hematite.

6.2.3 Topographic Influences

Landforms and topography have both local and regional influences. Locally they control the drainage of water and air. High mountains control the flow of air masses and have related temperature and precipitation effects. Regional influences and soil temperature and moisture regimes of serpentine soils have been discussed already. Local topographic influences include soil erosion, drainage of water, and slope aspect.

Serpentine soils on mountains and hills commonly have vegetative cover that is sparser than on other kinds of soils on these landforms. Sparse cover allows more rain drops to strike the ground and dislodge soil particles, and there may by more surface runoff by overland flow to carry dislodged soil particles downslope. Thus, more sparse cover on serpentine soils may lead to more erosion by overland flow of water than on many other kinds of soils. If this is what actually happens, then a larger proportion of serpentine soils should be shallow, compared with other kinds of soils. This is a common perception, but it has never been tested by a definitive study; it may not be true.

Mass failure, or landsliding, generally predominates on steep and very steep slopes. Material is transported downslope, creating more shallow soils near the top and deeper soils below. Soils on very steep slopes (~60%–75% gradient) are generally complexes of shallow and very deep soils. Commonly, shallow soils predominate on summit and shoulder slopes, both shallow and very deep soils are common on very steep sideslopes, and very deep soils predominate on very steep footslopes below very steep sideslopes.

The effects of topography and slope stability on local soil distribution are illustrated by topographic profiles across both very steep and steep mountains and hills on peridotite in the Rattlesnake Creek terrane of the southern Klamath Mountains (fig. 6-7). On very steep

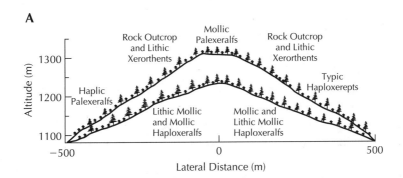

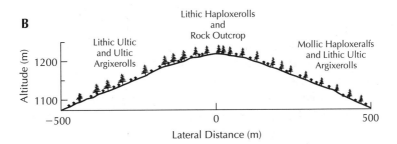

Figure 6-7 Profile sketches of very steep and steep slopes on peridotite (A) and steep slopes on serpentinite (B) in the Rattlesnake Creek terrane of the Klamath Mountains. Soil distributions were ascertained by detailed mapping of serpentine soils in the terrane (Alexander 2003).

peridotite slopes (gradient >60%), rock outcrop and Lithic Xerorthents predominate on shoulder slopes and very deep soils predominate in colluvium below them. These very deep soils are commonly Typic Haploxerepts on north-facing slopes and Haplic Palexeralfs on south-facing. Thus, the very steep topography includes a climatic effect, because soils on north-facing slopes are wetter and more susceptible to mass failure and soils on south-facing slopes are warmer and more favorable for the formation of smectite clay minerals and clay translocation and accumulation in argillic horizons. In exceptional cases, there are flat summit slopes with Mollic Palexeralfs. On steep slopes (30%–60% gradients) where erosion by overland flow of water predominates over mass movement, the soils are mostly Lithic Haploxeralfs and moderately deep Mollic Haploxeralfs, with the shallow Haploxeralfs predominating on south-facing slopes and the moderately deep Haploxeralfs predominating on north-facing slopes. Steep serpentinite slopes in the Rattlesnake Creek terrane have similar soil distributions, except that the shallow soils and some of the moderately deep ones are Argixerolls, rather than Haploxeralfs. Also, very steep serpentinite slopes lack Palexeralfs, or Palexerolls.

Drainage influences include both loss or concentration of water and the transfer of chemicals in water. Water drains from slopes and some accumulates in alluvial fan and floodplain soils in valleys. Most serpentine soils on slopes hold about as much water as other kinds of soils of comparable depth. Plants on serpentine soils are not exposed to any more drought than are plants on other kinds of soils of comparable depth, even though the serpentine plants may have more xeric appearances.

Silicon and magnesium, major elements in serpentine soils, accumulate where water accumulates in serpentine valleys, and they promote the formation of smectite clay minerals. Iron smectites, or nontronites, are formed in sloping serpentine soils where the Mg^{2+} is carried away in drainage water (Wildman et al. 1968), and Mg^{2+} smectites, or saponites, accumulate in alluvial fan and floodplain soils to which water drains from the sloping serpentine soils (Senkayi 1977). Smectite clays shrink and swell upon wetting and drying to produce the cracking-clay soils that are common in serpentine valleys of the California Coast Ranges.

A good example of drainage influences was reported by Dirven et al. (1974) for a catena, or slope sequence, of serpentine soils in the Matanzas Province of Cuba where the mean annual precipitation is 138 cm (54 inches). Water drained from a Chromic Phaeozem (Lithic Hapludoll) and a Chromic Cambisol (Dystric Eutrudept) in higher parts of the landscape and accumulated in a Mollic Gleysol (Aquoll) in a basin in the lower part. The main clay minerals were smectite and goethite, except in the surface of drained soil, where there was more kaolinite than smectite. Hematite was present in the drained soils, and magnesite ($MgCO_3$) and siderite ($FeCO_3$) were present in the soil in the basin. Formation of siderite requires very high CO_2 pressures, which may occur in very poorly drained soils. Small amounts of chlorite and illite were present in all of the soils, and vermiculite was present in subsoils of the drained soils. Quartz was present in all of the soils, presumably inherited from the parent material, which was a serpentinite with secondary quartz along fractures. Total Al_2O_3 was concentrated from only 6 g/kg in the serpentinite to 73–100 g/kg in the surface horizons of the drained soils. This concentration of aluminum was accomplished by loss of about one-third of the silica and nearly all of the magnesia, which together constituted 73% of the serpentinite. It is unlikely that appreciable aluminum was lost by leaching and some must have been added in dust from the atmosphere. Aluminum is a major constituent of kaolinite. With more intensive leaching, more aluminum accumulates in well-drained serpentine soils and conditions are favorable for the formation of gibbsite.

Ségalen et al. (1980) sampled two well-drained serpentine Ferralsols (Oxisols) on a much older land surface west of Habana and found gibbsite. For the surface of the soil developed from less serpentinized peridotite, they estimated about 16% kaolinite, no smectite, 13% gibbsite, 36% goethite, 24% hematite. The SiO_2/Al_2O_3 ratio was about 0.5, attesting to extensive leaching of silica over a long period, and confirmation of conditions favorable for the formation of gibbsite.

In a very intense weathering and leaching environment on the western Pacific island of New Caledonia, most of the silica is leached from well-drained soils on peridotite to leave soils consisting mostly of iron oxides—goethite, hematite, and maghemite (Schwertmann and Latham 1986). More chromium accumulated in soils containing more hematite, and nickel in soils containing predominantly goethite and no hematite. In one toposequence, soils below 1050 m altitude contained hematite, but goethite was the only crystalline iron oxide above 1050 m, implying the cooler temperature above that altitude are inimical to the formation of hematite. Gibbsite occured only in soils on those rocks with higher aluminum concentrations.

6.3 Rates of Soil Development

A rate is an amount of change in position or composition over time. Soil formation entails development of soil from bedrock and changes in regolith, or soil. A general approach to soil transformations is from a soil horizon perspective. It is much more general than recording specific material gains or losses (for example, potassium or organic carbon), for which there is scarcely any data for serpentine soils. It is convenient to discuss rates of soil development from bedrock separately from rates of soil horizon development. Because rates of soil horizon formation are not known with any precision, estimates are given in orders of magnitude (powers of ten).

6.3.1 Development of Soil from Bedrock

As bedrock is weakened by chemical weathering along fractures, the rock disintegrates and is converted to soil where chemical weathering continues. This process is most active where there is plenty of water to leach the products of chemical weathering from soils. The amount (mass) of soil (S) formed, excluding organic matter, is equal to the amount of bedrock disintegrated (W) minus the amount of material leached from the soil (D). Of these three variables, only the amount of dissolved materials in watershed runoff (D) is ever measured in watershed investigations (If the atmospheric input of materials from precipitation is measured, it can be subtracted from D). Nevertheless, the equation $S = W -$

D, or $W = S + D$, can be solved by transforming the S variable to an S/W ratio that can be ascertained analytically or estimated (Alexander 1985), giving $W = D/(1 - S/W)$. S/W ratios range from ~0.1 in limestone watersheds, where most the rock is lost in runoff, to $S/W > 0.9$ in highly silicic rocks where quartz and feldspars are released from rocks by weathering of biotite in plutonic rocks or a cementing medium in sedimentary rocks. S/W ratios in ultramafic watersheds are expected in be in the range from about 0.2 in tropical areas with deeply weathered soils to about 0.8 in temperate areas with shallow soils.

Cleaves et al. (1974) collected and analyzed runoff (~14 cm/year) from a serpentine watershed in Maryland to estimate the rate of serpentine weathering and soil formation. They computed the amount of serpentine lost by chemical weathering (59.3 kg/ha each year). Then, they subtracted the mass of basic cations and silica in runoff (D) from this to obtain what they called secondary minerals (32.4 kg/ha each year). The secondary minerals are mostly smectites and cryptocrystalline quartz. Iron minerals are not accounted for in this analysis. A more substantial contribution to soil formation is the accumulation of primary minerals released from the serpentinite bedrock by weathering. These primary minerals are serpentine (presumably antigorite) and magnetite. Soil formation from bedrock is the sum of these primary minerals and secondary minerals, including iron oxyhydroxides, added to the soil. Cleaves et al. (1974) did not provide bedrock and soil analyses from which an S/W could be obtained. Assuming an S/W ratio of 0.8 (Alexander 1988b), the rate of Maryland serpentine soil formation (S) is 236 kg/ha each year; or assuming $S/W = 0.7$, $S = 138$ kg/ha each year. Obviously, this method of estimating rates of soil formation form bedrock lacks precision when there are no bedrock and soils analyses from which to calculate an S/W ratio, but it is the best we have. Assuming that the rate of soil formation is 150 kg/ha annually and the soil has a bulk density of 1.2 Mg/m^3, then 0.0125 mm of soil will accumulate each year, minus erosion losses. At this rate it would take 800 years for 1 cm of soil to accumulate and 40,000 years for a moderately deep soil (depth >50 cm) to form. These soils, assuming that they are the Chrome series or similar soils, have no more than 3% organic matter in an A horizon that is about 18 cm thick. This organic matter will add <1 cm to the depth of the soil at 40,000 years of age.

Rates of weathering and soil formation from bedrock are highly dependent on rates on precipitation and watershed runoff. Runoff from ultramafic watersheds in western North America ranges from <10 cm to >200 cm; it is 214 cm/year from the Smith River that drains from the Klamath mountains where the precipitation is about 300 cm/year. Assuming S/W ratios of 0.8 in arid serpentine soils to 0.6 in wet ones in western North America, the annual rates of soil formation are ~0.05–0.2 Mg/ha (Alexander 1988a). This corresponds to about 0.005–0.020 mm/year, or 5–20 mm of soil in 1000 years. At these rates, it would take least 50,000 years for a meter of soil to form from ultramafic bedrock, assuming no erosion losses by overland flow of water.

6.3.2 Formation of Diagnostic Soil Horizons

Soils are generally Entisols, lacking diagnostic soil horizons, initially. Serpentine soils in California weather rapidly to produce the clay and yield the free iron oxides that produce the structure and colors that are indicative of cambic horizons. Within a few thousands of years cambic horizons commonly form in California serpentine soils that are deeper than 25 cm, converting Entisols to Inceptisols. Cambic horizons develop so quickly (on a soil age scale) that in the Rattlesnake Creek terrane of the southwestern Klamath Mountains, where thousands of hectares of serpentine soils have been mapped (Alexander 2003), practically all Entisols are very shallow. Very shallow soils (soils <25 cm deep) cannot have cambic horizons, because cambic horizons must extend below 25 cm (except in very cold soils).

Mollisols require only a mollic epidepon. A mollic epipedon can develop within a few hundred years, possibly less with optimum conditions.

Vertisols develop rapidly in any soils that are dominated by swelling clay and are alternatively wet and dry every year. Vertisols are known to form in hundreds of years and may form in tens of years. It does not take many shrink–swell cycles to produce the structure and slickensides that are indicative of Vertisols.

Inceptisols in serpentine materials have cambic horizons, because they lack other features that would place them in this order. It generally takes thousands of years for Inceptisols to form, but they may form in less time if they are hydric soils and the cambic horizon qualifier is mottling. Cambic horizons are very loosely defined in U.S. Soil Taxonomy. If a defining criterion is redder hues, it may take thousands of years to form cambic horizons in serpentine soils. It takes longer for reddish hues from iron oxides to develop in soils with serpentinite parent materials than in those with peridotite parent materials, because much of the iron in serpentinite is generally in minerals of the spinel group (mostly chromite and magnetite) that are resistant to weathering in soils (Alexander 2004a).

Alfisols have argillic horizons. Although some serpentine Aridisols may have cambic horizons, those in western North America generally have argillic horizons, which differentiates them from arid Entisols. It generally takes tens of thousands of years for argillic horizons to form.

Ultisols have argillic horizons that are more intensively leached than those in Alfisols. Ultisols may develop directly from Inceptisols in tens of thousands of years in climates with abundant precipitation, or take hundreds of thousands of years to develop in drier climates where they develop from Alfisols. Because serpentine materials supply abundant basic cations (mainly magnesium) to weathering soils, they must be highly weathered before those cations can be sufficiently depleted to form Ultisols. Therefore, it may take hundreds of thousands of years for serpentine Ultisols to develop.

Oxisols have oxic horizons that are practically devoid of weatherable minerals and active (high CEC) clay minerals. Millions of years of weathering and leaching are required to form oxic horizons, depending on soil temperature, precipitation, and soil parent material. Because the major minerals in serpentine materials are highly weatherable, oxic horizons may develop in serpentine materials within 5 Ma (Bienroth 1982). Because it takes such a long time for the development of oxic horizons in warm tropical climates, and even longer in cooler climates, they are generally not found in nontropical climates. In California, however, they are found on kaolinitic sediments that were deposited during the Eocene. Also, they are found in serpentine materials in the Sierra Nevada (domain 2) that must be on very stable landforms in order for the oxic horizons to have formed and survived for millions of years.

Acknowledgment

Major sources of information for soil series and families were soil survey reports of the U.S. Department of Agriculture (Natural Resources Conservation Service [NRCS, formerly Soil Conservation Service] and Forest Service) and soil series descriptions available from the NRCS website.

7   Animals, Fungi, and Microorganisms

Although plants are the major living components of terrestrial geoecosystems, other organisms are very important. Some animals move large amounts of soil, and many microoganisms promote the weathering of rocks and minerals in soils. Perhaps the greatest effects of animals, fungi, and microorganisms on geoecosystems are indirect through their effects on plants and plant communities. Mycorrhizal fungi are beneficial to plants in nutrient-limiting substrates where the fungi can scavenge phosphorous and nitrogen for plants. Many animals, from large ungulates (moose, elk, deer, etc.) to microscopic nematodes, graze on the leaves and roots of plants. Microorganisms cause many diseases in plants. A complete inventory of plant interactions with other organisms is virtually limitless. This chapter concentrates on organisms that live in serpentine soils, that live on ultramafic rocks, or that are dependent on plants that grow on serpentine soils.

There have been few field investigations of living organisms, other than plants, on serpentine soils. Many of the investigations on animals, fungi, and microorganisms in serpentine soils of the western North America have been conducted on Jasper Ridge in San Mateo County, and some have been on Coyote Ridge in Santa Clara County and on the McLaughlin Reserve in Napa and Lake counties, California. Some investigations of animals and other organisms for which there are no published accounts relating to serpentine soils in western North America (e.g., termites) are cited from other areas.

The associations of organisms with serpentine soils, whether utilization or avoidance, largely depend on the chemistry of the soil parent materials. Therefore, this chapter begins with a review of the effects of serpentine chemistry on living organisms.

7.1 Chemical Elements in Organisms: Utilization and Toxicity

Organisms are about 50% or more water. Moss plants that are less than 50% water when desiccated can absorb much more water than their dry weights to increase their weights several fold within hours.

About half of the biomass of living organisms that is not water is carbon. Other than water, carbon dominates the chemistry of all organisms. It forms large polymers that are far beyond the capabilities of other elements. Carbon dioxide in the atmosphere is the primary source of carbon in soils. It is fixed by plants and by some bacteria, mainly through photosynthesis. These carbon-fixing organisms are ingested by other organisms that utilize carbon and many of the other elements acquired along with carbon.

The chemical elements most utilized by living organisms are in the first 30 of the 92 naturally occurring elements (table 7.1). Few elements with atomic numbers greater than 30 have important functions in any kinds of living organisms. Molybdenum in plants, bromine in marine algae, and iodine in animals are notable exceptions. The most abundant elements in the ocean and in living organisms are among the first 20, and the next 10 are transition elements (elements with unfilled d-orbitals) and copper and zinc. Most of the first 20 elements are either soluble as cations or form oxyanions (anions formed by the combination of a cation and oxygen; for example, NO_3^- and SO_4^{2-}), with the notable exceptions of beryllium and aluminum, which form insoluble oxides and hydroxides (fig. B-1 in appendix B) and fluorine and chlorine, which are commonly anions. Helium, neon, and argon are inert gases. Chlorine is abundant as an anion in sea water and in organisms, and sulfur is common in a -2 oxidation state, as well as other states up to $+6$. Concentrations of phosphorus in living organisms are very high, relative to the low concentrations of this element in the lithosphere and ocean. Westheimer (1987) attributes the importance of phosphorus in organisms to the crucial role of phosphates (PO_4^{3-}) in mediating chemical reactions (e.g., in adenosine triphosphate; ATP) and in bonding organic compounds (e.g., in DNA).

Most of the cations among the first 30 elements that are either readily solvated or form soluble oxyanions are essential elements in a broad array of organisms. A notable exception is lithium, but its concentration is very low in all compartments of the earth (table 7-1). Some of the transition elements have multiple oxidation states in soils and plants and may be soluble in some oxidation states and not in others; for example, vanadium and chromium form soluble oxyanions in higher oxidation states and insoluble oxides or hydroxides in lower ones, and manganese and iron are soluble in $+2$ oxidation states and insoluble oxides or hydroxides in higher ones.

Many of the first 20 chemical elements that are essential or beneficial in small amounts are toxic to organisms in higher concentrations. Some of the first transition elements, from vandium to nickel, that are more concentrated in serpentine geoecosystems than in others, can be toxic in small concentrations, especially nickel. Many of the more toxic elements, such as cadmium and mercury, are not concentrated in ultramafic rocks. The toxicities of common elements relative to some basic properties, such as ionic potential and electronegativity, are presented in appendix B. Degrees of toxicity and sensitivities

Table 7-1 Concentrations of chemical elements in living organisms, the lithosphere, and the sea—means or medians.

Atomic no./ element	Bacteria/fungi (mg/kg dry matter)	Bryophytes (mg/kg dry matter)	Land plants (mg/kg dry matter)	Land animals (mg/kg dry matter)	Upper mantle (mg/kg)	Upper crust (mg/kg)	Ocean (µg/L)
1 H	74000	—	67000	74000	—	—	—
2 He	—	—	—	—	—	—	<1
3 Li	<1	—	2	<1	2	20	180
4 Be	—	1	<1	<1	<1	3	<1
5 B	6	18	25	1	<1	15	4500
6 —	540000	—	463000	510000	150	300	28000
7 N	96000	12000	19000	98000	10	20	420
8 O	230000	—	396000	268000	—	—	—
9 F	—	—	4	>150	16	500	1300
10 Ne	—	—	—	—	—	—	<1
11 Na	4600	800	1200	—	2600	28900	10800000
12 Mg	7000	1000	3200	1000	222000	13000	1300000
13 Al	210	3000	200	—	22000	80000	1
14 Si	180	3500	3000	>120	210000	308000	2500
15 P	30000	2000	2000	>17000	80	650	65
16 S	5000	1500	4800	5000	180	—	898000
17 Cl	2300	1200	2000	2800	1	100	18800000
18 Ar	—	—	—	—	—	—	<1
19 K	115000	6700	11000	7400	250	28000	390000
20 Ca	5100	50000	15000	>20	26000	30000	450000

(continued)

Table 7-1 (continued)

Atomic no./ element	Bacteria/fungi (mg/kg dry matter)	Bryophytes (mg/kg dry matter)	Land plants (mg/kg dry matter)	Land animals (mg/kg dry matter)	Upper mantle (mg/kg)	Upper crust (mg/kg)	Ocean (μg/L)
21 Sc	—	1	<1	<1	17	11	<1
22 Ti	—	300	32	<1	1100	3000	<1
23 V	4	11	2	<1	82	35	2
24 Cr	4	9	2	<1	3200	100	<1
25 Mn	260	240	<1	1000	600	<1	
26 Fe	170	1000	200	160	64000	35000	<1
27 Co	8	3	1	<1	100	10	<1
28 Ni	—	—	2	1	2000	20	1
29 Cu	150	20	10	2	28	25	<1
30 Zn	83	150	50	160	10	190	<1

Elemental concentrations in the earth's mantle, crust, and sea water are from Li (2000), concentrations in land plants and animals are from Romankovitch (1990), and concentrations in all other living organisms are from Bowen (1979). Concentrations for calcium, phosphorus, fluorine, and silicon in some animals are minima—they can be much higher in the skeletal parts of those animals.

to different elements differ among different kinds of organisms, but the general order of toxicity in water plants, where uptake is not influenced by root penetration and nutrient mobility as in soils, is Hg > Cu > Cd > Ni, Zn > Pb (Küpper and Kroneck 2005).

To comprehend the effects of chemical elements in ultramafic materials on organisms and adaptations of the organisms, it is instructive to review the history of the earth and its effects on the evolution of organisms that are adapted to serpentine geoecosystems. The disposition of the chemical elements through the physical and biochemical evolution of earth is the subject of many articles and some books (including Mason 1991, Wächtershäuser 1992, Williams and Fraústo da Silva 1996). Some general principles and pertinent details relative to ultramafic materials are given here.

The chemical elements in earth were generated by nuclear reactions before our planet was formed by the agglomeration of materials within our solar system. Thus, the mean composition of earth resembles that of our solar system, except that large portions of the volatile elements (hydrogen, helium, and other inert gasses) have escaped from our planet, including carbon, nitrogen, oxygen, and sulfur losses as oxides or hydrogenated compounds. Within earth, elements have been segregated from a heavy core (specific gravity, SG = 11) through a mantle of intermediate SG to a light crust (SG = 2.8, or 2.7 in upper crust). Even though the earth was assembled about 4.5 billion years ago (Ga), crust is still being produced at its interface with the mantle. At ocean spreading centers, mantle material is being fractionated by partial melting and recrystallization to produce gabbro and basalt with relatively high calcium, sodium, potassium, and aluminum concentrations, and peridotite with very low Ca and Al concentrations. Weathering of basalt on the sea floor releases silicon and calcium, which are major elements in the skeletal parts of marine organisms such as radiolarians (silica) and foraminifera (calcium carbonate). When these organisms die and their skeletons settle on the ocean floor, they become oozes that upon solidification are mostly chert and some limestone. The mafic rocks and sea floor sediments that are rafted on large plates to continental margins are either subducted and returned to the mantle or become parts of a continent. Many of them are further differentiated by tectonic and volcanic activity. The leftovers from these activities are more silicic than the mafic rocks of the sea floor and have a mean SG about 2.7, or less, compared to 3.0 for gabbro and 2.8 for basalt. Living organisms, or at least all eukaryotes, evolved in this more silicic environment of the upper crust in which all of the 20 lightest elements, other than magnesium, are more abundant than in the mantle (table 7-1). Because ultramafic rocks have compositions similar to that of the upper mantle, with very high concentrations of magnesium and relatively low concentrations of calcium, potassium, and phosphorus, most organisms find serpentine geoecosystems inimical to life. Also, some of the first transition elements from vanadium ($Z = 23$) to nickel ($Z = 28$) that are more abundant in ultramafic rocks than in the upper crust (table 7-1) can be toxic to many living organisms.

Life on earth began in the ocean, which was formed soon after the earth congealed and was anoxic (lacking free oxygen) for the first half of the earth's existence. During that time, sulfur was a dominant anion in the sea. Green and purple sulfur bacteria were the first organisms to practice photosynthesis—the use of solar energy to reduce the carbon in CO_2 and produce hydrocarbons. In the process they oxidized H_2S to elemental sulfur, without producing oxygen. Some oxygen was liberated from water by photolysis— the decomposition of water by radiant energy. Sulfur was oxidized in the upper part of the ocean and formed sulfate, but free oxygen was scarce in the ocean and atmosphere when life first evolved on earth. Iron and manganese were much more readily available than in the oxygenated environments of today. Sulfides were a source of nickel that prevailed in many proteins and enzymes, such as urease in Archean organisms (Nesbit and Sleep 2003). Cobalt was utilized by prokaryotes and is an element in cobalamine, or vitamin B_{12}, that is essential for nitrogen-fixing microorganisms. Cobalt is not an essential element for plants (Epstein and Bloom 2005), which evolved after free oxygen became more abundant. Plants, however, are dependent on microogansims for fixing nitrogen, and cobalt is particularly important for animals that are dependent on microorganisms in their rumen. Vanadium, and even tungsten, may have had greater roles in anoxic Archean environments (Williams and Fraústo da Silva 1996). Many of these roles have been taken over by molybdenum. Copper and zinc became more important in organisms after the seas became more oxygenated and these elements were more readily available. A good example of the changing roles of trace elements is in the superoxide dismutases (SODs). Three types of SODs may be distinguished: Mn-SOD, Fe-SOD, and Cu/Zn-SOD (Scandalios 1995). Those with manganese proteins are found only in prokaryotes and mitochondria, those with iron proteins are mostly in prokaryotes, and those with Cu/Zn-SODs are found only in eukaryotes outside of mitochondria. A Ni-SOD was discovered in two species of *Streptomyces* (Ragsdale 1998), which are prokaryotes. The eukaryotes evolved only after the atmosphere was oxygenated, and at the same time copper and zinc became more readily available in marine environments. Copper and zinc have very important roles in the metabolism of eukaryotes. Nevertheless, these elements, especially Cu(I), are very toxic in high concentrations.

7.2 Animals

Some animals are completely dependent on serpentine soils, and many others spend parts of their lives on serpentine soils. Serpentine soils and plant nutrition will have some effects on large herbivores, such as deer, large predators, such as coyote, and birds that feed on serpentine plants and animals at least part of the time; but the greatest effects are likely to be on animals that spend their entire lives on serpentine soils. Nevertheless, some insects that are mobile enough to leave serpentine soils are still dependent on them. Animals on serpentine soils have been studied for many reasons, but the coverage of phyla related to serpentine is very incomplete.

Animal diversity is high in serpentine soils, but there are few data on biomass. Leakey and Proctor (1987) and Thomas and Proctor (1997) sampled serpentine soils at 10 elevations ranging from 280 to 1540 m altitude in forests of southeastern Asia. Numbers of arthropod individuals ranged from 382 to $2000/m^2$ in soils and 236 to $3300/m^2$ in above-ground detritus (litter). Arthropod biomass ranged from 1.1 to $8.4\,g/m^2$ in soils and 1.6 to 7.4 g/m² in above ground detritus. Mean earthworm biomass was 8.6 g/m² in soils and 0.7 g/m² in litter. Ants were the most numerous arthropods in most soils, and termites were most numerous in some soils. Invertebrate biomass was dominated by earthworms in most soils, by beetles in a few soils, and by crabs in one soil. Litter had many of the same invertebrates plus more pulmonate gastropods than in soils. Also, litter had many more isopods, more spiders, daddy long-legs, and roaches, and in a few cases many springtails and stick insects.

Besides carbon, hydrogen, oxygen, and nitrogen, all animals require calcium, phosophorus, potassium, magnesium, sodium, chlorine, and sulfur for ionic balance and in amino acids, nucleic acids, and structural compounds. Also, iron, iodine, copper, manganese, zinc, cobalt, molybdenum, selenium, chromium, nickel, vanadium, silicon, and arsenic are required by higher vertebrates (Hopkin 1989).

Metal accumulations in serpentine plants can cause excessive accumulations in some insects that are collected above ground over serpentine soils. Davison et al. (1999) sampled Coleoptera, Diptera, Hemiptera, and Hymenoptera at four sites in Scotland and found that the Hemiptera from a serpentine site accumulated much more magnesium, manganese, and chromium than from the nonserpentine sites. Some ascidians (tunicates, or sea squirts) are hyperaccumulators of vanadium (Crans et al. 1998).

7.2.1 Pocket Gophers (Mammalia, Rodentia, Geomyidae)

Some pocket gophers (*Tomomys bottae*) spend their entire lives on serpentine soils in central California grassland (Proctor and Whitten 1971). They feed on below-ground parts of herbaceous plants and bring subsoil to the surface on mounds. The extent of gopher disturbance and mounds is much greater on serpentine soils than on adjacent nonserpentine soils of the Franciscan formation (Proctor and Whitten 1971). Plant productivity is lower on the mounds than between mounds (Koide et al. 1987), and species compositions are different (Hobbs and Mooney 1985, Koide et al. 1987). It is not surprising that plant growth is less in subsoil piled on the mounds than in topsoil between mounds. Fertilization increased plant biomass and altered species distributions greatly, and gophers preferred fertilized plots (Hobbs et al. 1988). Most of the serpentine soil area on Jasper Ridge is disturbed at least once every 3–5 years (Hobbs and Mooney 1995). Variations in rainfall add to variability in plant distribution caused by gopher disturbance, with a nonnative annual grass increasing greatly in extremely wet years, at the expense of forbs, and declining drastically in dry years (Hobbs and Mooney 1991).

7.2.2 Harvestmen (Chelicerata, Arachnida, Opiliones)

Seven species of harvestmen, or daddy long-legs (family Phalangogidae), are endemic to serpentine soils in the San Francisco Bay area. They include the Marin and Edgewood blind harvestman (*Calicina diminua* and *C. minor*) and the Edgewood, Hom's, Jung's, Fairmont, and Tiburon microblind harvestman (*Microcina edgewoodensis, M. homi, M. jungi, M. lumi,* and *M. tiburona*). All of these species have highly restricted ranges; most are known from fewer than 10 locations, and the entire *Microcina* genus is restricted to San Francisco Bay area serpentines. All of them are roughly 1 mm long as adults. Blind and microblind harvestmen generally favor microhabitats that are moist, warm, and totally dark. The adults live under large rocks and prey on springtails (primitive soil-dwelling insects). Eggs are laid directly in the soil, and adults are thought to estivate underground during the dry summer months. Blind and microblind harvestmen in the San Francisco Bay area are believed to be threatened by habitat loss, the degradation of serpentine grasslands by exotic grass invasions, and predation by the invasive Argentine ant (*Linepithema humile*) (Elam et al. 1998).

7.2.3 Ants (Mandibulata, Hexapoda, Hymenoptra)

Fisher (1997) found 20 species of ants in serpentine chaparral soils and 22 species in adjacent nonserpentine chaparral soils in Napa and Lake counties, California. Five of the species were found only in the serpentine soils. One of these is believed to be either *Formica xerophila*, a Great Basin species showing a range extension on serpentine, or an undescribed sister species that is confined (endemic) to serpentine soils in the California Coast Ranges. Seed-harvesting ants (*Messor andrei*) may alter grassland plant species compositions (Hobbs 1985). Hobbs found them distributed throughout serpentine grassland on Jasper Ridge, with an average of 1 mound per 100 m^2 (an area 10 m square, or a circle 11.3 m in diameter) and feeding paths 10–12 m long. Brown and Human (1997) found that seed harvesting by *Messor andrei* (fig. 7-1) had no effects on plant composition of a serpentine grassland, but nest mounds had much greater amounts of organic matter and higher temperatures than the surrounding soils. They found more grasses and fewer forbs on ant mounds. Boulton et al. (2003) and Boulton and Amberman (2004) found that *Messor andrei* activity in serpentine grasslands at the McLaughlin Reserve, Napa and Lake counties, California, affected the soil microfauna. Food material brought by the ants into their underground nests led to higher soil nutrient levels and elevated abundances of bacteria, fungi, nematodes, and other soil-dwelling animals. Boulton et al. (2004) found slight differences in ant species diversity and relative abundance between serpentine and nonserpentine grasslands and showed that these were direct responses to soil rather than to the plant community differences.

Figure 7-1 A seed-harvesting ant (*Messor andrei*) grasping an insect pupa on Jasper Ridge in San Mateo County, CA. Photo by A. Wild.

7.2.4 Termites (Mandibulata, Hexapoda, Isoptera)

There are no published accounts of termites in serpentine soils of western North America, but Hopkin (1989) reported an interesting investigation in ultramafic deposits of Zimbabwe where the soil contains up to 10% Ni and 0.6% Cr (Wild 1975). Worker termites, which feed directly on plant material, contained 5000 μg/g Ni and 1500 μg/g Cr. Soldiers contained only 100 μg/g Ni and 300 μg/g Cr, and queens contained even less of these elements. Progeny of the queens lacked detectable nickel and chromium until they began to feed.

7.2.5 Butterflies (Mandibulata, Hexapoda, Lepidoptra)

Butterflies can fly from one plant community to another, but their larvae are less mobile. Larvae of a few species feed only on plants restricted to serpentine soils, and several others feed on plants that are more abundant on serpentine soils than on other soils.

There are at least nine species and subspecies of butterfly associated with serpentine in the northern California Coast Ranges (Harrison and Shapiro 1988). Five of these occur in the Sierra Nevada as well (Gervais and Shapiro 1999). These associations arise because larvae of these species feed on host plants that are either entirely restricted to, or much more abundant on, serpentine. Two of the species most completely restricted to serpentine soils in California are Muir's hairstreak, *Mitoura muiri*, and the lacustra skipper, *Erynnis brizo lacustra* (Gervais and Shapiro 1999). Larvae of Muir's hairstreak feed on Sargent and McNab's cypress north of San Francisco and on McNab's cypress in the Sierra Nevada, but south of San Francisco they feed on California juniper growing on nonserpentine soils (fig. 7-2). The serpentine race is strongly differentiated ecologically, in that it shows genetically based differences in adult host preference and larval adaptation, despite the fact that the genetic data available thus far suggests it is closely related to neighboring nonserpentine races. The different phenology that comes from being associated with serpentine has almost certainly been important in the divergence of the serpentine *Mitoura* (Forister 2004). Larvae of lacustra skipper feed exclusively on leather oak, which is almost completely restricted to serpentine soils, in the northern part of their range, but some live beyond the southern limit of leather oak, where they feed on other plants and are not restricted to serpentine soils.

Figure 7-2 A Muir's hairstreak (*Mitoura muri*) caterpillar that mimics the branch of Sargent cypress. Photo by E. Ross.

Figure 7-3 An adult Bay checkerspot (*Euphydryas editha bayensis*) butterfly. Photo by E. Ross.

One of the most extensively studied butterflies is the Bay checkerspot, *Euphyrdryas editha bayensis* (Ehrlich et al. 1975; Elam et al. 1998; fig. 7-3). Its primary host plant in the San Francisco Bay area is dwarf plantain, *Plantago erecta*, and larvae may feed on purple owl-clover, *Castilleja densiflora*, also. These butterflies do not readily fly more than a few hundred meters. They appear to be dependent on dwarf plantain growing on serpentine soils, even though dwarf plantain also grows on other soils. The greatest threats to Bay checkerspot butterflies, an endangered species, are the direct and indirect effects of urbanization that include the loss of most of their habitat and the overgrowth of the remainder by exotic grasses. Exotic grasses such as *Lolium multiflorum* (ryegrass) have become more prevalent on Bay Area serpentine grasslands, probably because of the cessation of cattle grazing and the deposition of nitrogen oxides from air pollution (Elam et al. 1998, Weiss 1999). The butterfly has gone locally extinct at Jasper Ridge and Edgewood Park, where its populations were studied for over 40 years by Stanford biologists, and it has declined catastrophically at Coyote Ridge, south of San Jose, where its largest population is still located.

Opler's longhorn moth (*Adela oplerella*, Adelidae) is a lesser-known inhabitant of the same Bay area serpentine grasslands that support *Euphydryas editha bayensis* (Elam et al. 1998). It is a small brown moth with very long antennae and weak flying ability. The larval host plant of the moth is *Platystemon californicus*, Papaveraceae (cream cups), which like *Plantago erecta* is a small native grassland forb that grows both on and off of serpentine. However, the moth, like the butterfly, is almost completely confined to serpentine soils, perhaps because of the higher density of plants and/or the more favorable microclimate these soils provide. In addition to serpentine grasslands on the San Francisco Peninsula, the moth is also known from one location in Sonoma County, two locations (including the Ring Mountain Preserve) in Marin County, and one location in Santa Cruz County (its only known nonserpentine occurrence). The moth populations appear to be in decline for substantially the same reasons as the butterfly populations (Elam et al. 1998).

Barro et al. (2004) reported 303 species in 21 families of Lepidoptera from serpentine localities in Cuba. Seventeen of the species are endemic on serpentine, and they expect more endemics to be found.

7.2.6 Bugs (Mandibulata, Hexapoda, Hemiptera)

A species of plant-bug (Miridae family), *Melanotrichus boydi*, is believed to be completely dependent on the milkwort jewel flower plant, *Streptanthus polygaloides* (Schwartz and Wall 2001), which is endemic on serpentine soils in the Sierra Nevada and is a nickel hyperaccumulator. Adult plant-bugs feed on milkwort and lay their eggs on it. Larvae from the eggs then live and develop on the plants. The plant bugs accumulate nickel to concentrations of about 750 µg/g. A second kind of plant bug (*Coquillettia insignis*) prefers to feed on milkwort jewel flower plants, rather than on plants that are not nickel hyperaccumulators, and were found to have concentrations of 270–910 µg Ni/g of dry weight (Boyd et al. 2004).

Wall and Boyd (2002) collected and analyzed insects that fed on both milkwort jewel flower plants that grow only on serpentine and on toyon (*Heteromeles arbutifolia*) plants growing on nonserpentine soils in the same area. They found that the individuals of two species of bees (Hymenoptera, Apidae) that fed only on the milkwort jewel flower plants accumulated significantly more nickel than those that fed on toyon shrubs, but much less nickel than two bugs (*M. boydi* and *Lygus hesperus*, Heteroptera, Miridae) that fed on milkwort jewel flower plants.

After finding that nickel hyperaccumulation protects mountain pennycress (*Thlaspi montanum montanum*) from grazing by an insect (Boyd and Martens 1994), Boyd and Moar (1999) tested the hypothesis that high nickel concentrations in bugs fed on nickel hyperaccumulating plants protect them from infection by pathogens and consequential mortality. They fed adults and nymphs of *M. boydi* on terminal portions of milkwort jewel flower inflorescence. Neither the adults nor the nymphs were protected from infection, and mortality was greater when they were exposed to spores of an imperfect fungus or to either of two parasitic nematodes. Boyd and Wall (2001) found, however, that *M. boydi* bugs fed on milkwort jewel flowers were toxic to spider crabs, the flowers giving them a possible protective effect against these predators.

7.2.7 Beetles (Mandibulata, Hexapoda, Coleoptera)

A South African leaf beetle (*Chrysolina pardalina*, Chrysomelidae family) feeds only on *Berkheya coddii* (Asteracae), which is a nickel hyperaccumulating species that grows only on serpentine soils. The leaf beetle has been observed by Mesjasz-Przybylowicz and Przybylowicz (2001) to complete several life cycles on *B. coddii*. Nickel concentrations in dried leaf beetle specimens were 2650 mg/kg in adults ($n = 30$), but only 310 mg/kg in larvae

Table 7-2 Phyla, classes, and orders of some animals associated with serpentine geoecosystems.

Phylum	Class	Order	Suborder or infraorder	Representatives in serpentine soils
Craniata	Mammalia	Rodentia		Pocket gophers
Mollusca	Gastropoda	Pulmonata		Slugs and snails
Annelida	Oligochaeta	Haplotaxida	(Lumbricina)	Earthworms
Crustacea	Malacostraca	Isopoda		Isopods (pillbugs etc.)
		Amphipoda		Amphipods (sand flies etc.)
		Decapoda	(Brachyura)	Crabs
Mandibulata	Hexapoda	Collembola		Springtails
		Hymenoptera	(Apocrita)	Ants (Formicidae)
		Isoptera		Termites
		Coleoptera		Beetles
		Blattaria		Roaches
		Phasmida		Stick insects
		Lepidoptera		Butterflies and moths
	Myriapoda	Diplopoda		Millipedes
		Chilopoda		Centipedes
Chelicerata	Arachnida	Aranae		Spiders
		Orpiliones		Longlegs (harvest men)
Nematoda				Roundworms

Classification hierarchy based on Margulis and Schwartz (1998) and Pechenik (1996).

($n = 35$). They discovered that *C. pardalina* eliminated excess nickel from the hemolymph through Malphighian tubules to the gut (Przybylowicz et al. 2003).

7.2.8 Earthworms (Annelida, Oligochaeta)

Mariño et al. (1995) isolated 10 species of earthworms in 5 genera (*Allobophora, Dendrobaena, Eisenia, Lumbricus,* and *Octolasion*) from serpentine soils in Galacia, Spain, although the 2 species of *Octolasion* were isolated only from soils with exchangeable Ca/Mg ratios >1.0 mol/mol. They found that zinc was more concentrated in these 10 species than in the serpentine soils and copper and nickel were less concentrated in the earthworms than in the soils. Evidently these species were able to concentrate nickel in their feces and prevent its accumulation in their bodies.

Serpentine soils, with their low calcium and high magnesium and nickel concentrations, are not favorable environments for all earthworms. Schreier and Timmenga (1986) introduced earthworms (*Lumbricus rubellus*) from the Fraser River delta in British Columbia to ultramafic flood deposits along the Sumas River, a tributary of the Fraser River. During 3 weeks to 1 month of preliminary trials, the earthworms in ultramafic deposits accumulated up to 4 times as much magnesium (4 mg/gm) and 17 times as much nickel (40 μg/g) as in nonserpentine soil. No earthworms survived a 297-day field trial in the ultramafic deposits.

Hubers et al. (2003) found that the abundance of earthworms was low in serpentine Oxosols of the Nipe series in Puerto Rico. *Pontoscolex corethrurus* was the most abundant species, and *Onychochaeta borincana* and *Neotrigaster rufa* were the only other species present.

7.2.9 Nematodes, or Roundworms (Nematoda)

Numbers of nematodes in grassland communities of the Coast Ranges were generally greater in serpentine soils than in adjacent nonserpentine soils of the Franciscan formation (Hungate et al. 2000). Fungus-feeding nematodes were most numerous, followed by bacteria-feeding and root-feeding nematodes. Predatory nematodes were sparse.

7.3 Fungi, Mycorrhiza, and Lichens

Fungi are sedentary eukaryotes, generally with chitinous cell walls, that form chitinous spores and lack means of locomotion that are characteristic of many protoctists (Margulis and Schwartz 1998). They range from microscopic unicellular yeast to filamentous bodies with extensive hyphal networks and large spore-producing structures such as puffballs and mushrooms. Fungi are heterotrophs, none fix carbon by photosynthesis, and perhaps none fix it by other chemical reactions either. Fungi live on or in plants, plant detritus and animal wastes, soils, and rocks, mostly in aerobic environments. Many fungi have symbiotic relationships with bacteria, protoctists, and plants, and some are parasitic in plants, animals, and other organisms.

Phyla within the fungi kingdom are Zygomycota, Basidiomycota, and Ascomycota. Hyphae of Zygomycota are aseptate and those of Basidiomycota and Ascomycota are septate, having partitions called "septa." Basidiomycota bear spores on parent cells, and Ascomycota bear spores within parent cells. Many morphological characteristics and metabolic functions are shared by members of the Basidiomycota and the Ascomycota, making these categories of limited usefulness for ecological groupings of fungi.

Many fungi have low or insignificant requirements for calcium (Nicholas 1963). Many tolerate concentrations of metallic cations that would be highly toxic to vascular plants (Brown and Hall 1989). Cations can be adsorbed rapidly at exchange sites on cell surfaces, but intracellular uptake requiring metabolic energy is greater. Although nickel is toxic in high concentrations, nonmetallic cations (including magnesium) reduce its toxicity.

An inventory of fungi in soils with Douglas-fir and subalpine fir forests in the Wenatchee Mountains of the Cascade Mountains revealed dozens of species in both serpentine and nonserpentine soils with different suites of plant species, but only 18% of the fungal species were shared by both kinds of soils (Maas and Stuntz 1969). The majority of those found were Basidiomycota. Nearly one-quarter of the species in serpentine soils were mycorrhizal symbionts. Hungate et al. (2000) found as much (more at certain times) fungal biomass and activity in a serpentine soil as in a sandstone soil of the Franciscan formation in the California Coast Ranges.

7.3.1 Mycorrhiza

Many fungi form symbiotic associations with plant roots. These associations are called mycorrhiza. The mycorrhizal fungi extract nutrients and water from soils and transfer them to the plants. Seedlings of some plants will not survive unless they are infected with mycorrhizal fungi. Mycorrhiza are particularly helpful in occupying larger volumes of soils than roots do and in scavenging for phosphorus and nitrogen. Plants that can get plenty of phosphorus without mycorrhiza are less likely to be colonized by mycorrhizal fungi.

Mycorrhiza are grouped into several categories based on their modes of root infection and morphology. The arbuscular (or vesicular-arbuscular) mycorrhiza form arbuscules (shrublike forms) within cells of infected plants and, in many cases, vesicles. These fungi are Zygomycota in the order Glomales. They are dependent on their symbionts; they will not grow apart from their plant hosts. They Arbuscular mycorrhiza infect many different kinds of both nonvascular and vascular plants. Fungi in all other types of mycorrhiza are either Ascomycota or Basidiomycota, or both. The ectomycorrhiza form sheaths around fine roots, and hyphae grow from them inward between roots cells to form complex intercellular systems. They infect trees and shrubs. The arbutoid mycorrhiza with a sheath and the ericoid mycorrhiza lacking a sheath extend hyphae into cells of roots. They infect plants of the Ericaceae (heath family). Orchids (Orchidaceae) and monotropoids of the Ericaceae, which lack chlorophyll, and have unique symbiotic fungal relationships.

Different kinds of mycorrhiza dominate in different kinds of soils (Read 1991). Ericoid mycorrhiza are associated with plants growing in very acid soils containing humus with very high C/N ratios. Ectomycorrhiza are commonly associated with trees growing in strongly to moderately acid soils containing humus with high to moderate C/N ratios. Arbuscular mycorrhiza are generally associated with herbaceous plants growing in moderately acid to neutral soils containing humus with moderate to low C/N ratios. These associations are related to the abilities of the different kinds of mycorrhiza to produce enzymes that degrade organic polymers, allowing them to decompose organic matter in plant detritus and humus to scavenge phosphates and nitrogen compounds for the plants infected by the mycorrhizal fungi (Read and Perez-Moreno 2003). Because most serpentine soils are slightly acid to neutral, arbuscular mycorrhiza can be expected to dominate in them, along with arbutoid mycorrhiza in the California region where manzanita bushes are common on serpentine soils.

Hopkins (1987) found that 26 of 27 species of grass and forbs in a serpentine grassland south of San Jose, California, were infected with arbuscular mycorrhizal fungi. Roots of 23 of the species representing 78% of the herbaceous cover were highly colonized by these fungi. Rillig et al. (1999a) found the several herb and grass species they investigated on Jasper Ridge to be well colonized by arbuscular mycorrhiza in both serpentine and sandstone (graywacke) soils. Chiariello et al. (1982) found that the same hyphae from the same mycorrhiza infected the roots of two or more plants in a serpentine soil on Jasper Ridge, forming bridges between plants. Radioactive phosphorus added to a dwarf plantain (*Plantago erecta*) plant was transferred by mycorrhiza to forb and grass plants of several species within a radius of several centimeters.

Turnau and Mesjasz-Przybylowicz (2003) reported arbuscular mycorrhiza growing on several species of nickel hyperaccumulating plants in the Asteraceae. *Berkheya coddii* plants with extensive arbuscule development had greater growth and accumulated more nickel than those lacking arbuscule development. In contrast, Gonçalves et al. (2001) found no increase in plant nickel related to arbuscular mycorrhizal colonization of *Festuca brigantina* growing in a serpentine soil containing high levels of Ni concentration, but found that P concentration in the plant roots was related to colonization.

Moser et al. (2005) found abundant ectomycorrhizas on white oak (*Quercus garryana*) tree roots in both serpentine and nonserpentine soils of southwestern Oregon. Of 74 fungal species they characterized, 46 were found in serpentine soils, 32 of which were found only in the serpentine soils. Panaccione et al. (2001) investigated an ectomycorrhizal fungus growing on Virginia pine (*Pinus virginiana*) seedlings in Maryland. They found that those individuals isolated from serpentine soils were genetically more similar to each other than to those from nonserpentine soils.

Fungi of arbuscular mycorrhiza in the Glomales order of Zygomycota produce a glycoprotein called "glomalin" that is very effective in enhancing soil aggregate stability (Wright and Upadhyaya 1996). The occurrence of glomalin in serpentine soils remains to be investigated.

7.3.2 Lichens

Lichens are essentially fungi with coexisting symbiotic algae that are incorporated into lichens. The algal partner is usually either a green algae (Chlorophyta) or a blue-green algae (Cyanobacteria). Cyanobacteria are eubacteria, and Chlorophyta are protoctists (Protoctista kingdom).

The majority of lichens are epiphytes, growing on plants, but many grow on rocks (saxicolous lichens) and on soils (terricolous lichens). Many lichen fungi have rhizines, which are bundled strands of hyphae, that can penetrate into soils and cracks in rocks. These rhizines have limited conducting capacities, but they do allow for some exploitation of water and nutrients from substrates on which the fungi grow. Nevertheless, lichens get most of their mineral (inorganic) nutrients from the atmosphere—from dust particles and rainfall or other precipitation. Some fungi may not need calcium in significant quantities (Bowen 1979), and calcium requirements of green and blue-green algae were found to be very low compared to their requirements for magnesium (Gerloff and Fishbeck 1969), even though calcium concentrations are generally higher than magnesium in lichens (Nieboer et al. 1978). Therefore, ultramafic rocks and soils with low calcium concentrations may be suitable chemical substrates for at least some lichens. An overgrowth of plants limits the growth of saxicolous and terricolous lichens, because the main source of carbohydrates is from photosynthesis by the bacterial or algal component. Lichens commonly accumulate large quantities of metallic elements that would be toxic to plants (Hale 1983).

Even though most of their mineral nutrients are from the atmosphere, saxicolous lichens commonly grow on specific kinds of rocks. For example, lichens on "basic" rocks are commonly dominated by species in the Collemataceae, Physicaceae, and Teloschistaceae and those on "acid" rocks are commonly dominated by species in the Parmeliaceae and Umbilicariaceae (Hale 1983). Both calcicole (favoring calcium-rich substrates) and calcifuge (avoiding calcium-rich substrates) species occur on ultramafic rocks at many sites (Purvis and Halls 1996). Gilbert (2000) found large populations of many species, including some found only on serpentine, on serpentinite of the Lizard Peninsula in Cornwall, but found lichen impoverishment on harzburgite and dunite of Unst in the Shetlands and on the Island of Rhum. Many species reach their northern limits of distribution on ultramafic rocks (Purvis and Halls 1996). Ryan (1988) found 61 species of lichens on ultramafic rocks in Washington Park on Fidalgo Island, Washington, but none of them were restricted to ultramafic rocks. The only serpentine indicator species in Washington Park was *Aspidotis densa*, a fern.

Baltzo (1970, 1989) inventoried the lichens on Mount Diablo in Contra Costa County, California. She found species diversity to be greater on clastic sedimentary rocks than on serpentine. *Caloplaca decipiens* and at least two other species in the same genus were found on ultramafic rocks, but on other kinds of rocks, also. Sigal (1989) found a total of 76 species of lichens on five serpentine sites in the California Coast Ranges. Fourteen percent of them were cyanolichens. Eleven percent of them were endolithic lichens. Six species were particularly characteristic of the serpentine sites in the Coast Ranges: *Aspicilia caesiocinerea*, *Lecidella carpathica*, *Rhizocarpon bolanderi*, *Caloplaca squamosa*, *Candelariella vitellina*, and *Acarospora fuscata*. Two of these species (*A. caesiocinerea* and *C. vitellina*) were among the most conspicuous species on ultramafic rocks in Bosnia (Ritter-Studnička and Klement 1968).

Gough and Jackson (1988) analyzed two kinds of lichens growing on Douglas-fir trees in serpentine soils in the Klamath Mountains of Del Norte County, California. Some of the results are in table 8.5. The magnesium and nickel concentrations in *Hypogymnia enteromorpha* were about two times as high as those commonly reported for epiphytic species (Gough and Jackson 1988). Dust and water flowing or dripping through the Douglas-fir canopy are the likely sources of these chemical elements.

Lichens produce many acids that are more or less effective in the dissolution of rocks. Etch pits and furrows in rock surfaces are found beneath the thalli of many lichens (Wilson 1995). It is commonly suggested that oxalic acid from fungi in lichens is a major agent of weathering. Crystals of calcium-oxalate are generally found at the rock–lichen interface—usually the monohydrate, whewellite, and seldom the dihydrate, weddellite (Adamo et al. 1993). A magnesium-oxalate, glushinskite ($MgC_2O_4 \cdot 2H_2O$), has also been identified among very fine silt-size crystals at the lichen–ultramafic rock interface (Wilson et al. 1981), although Adamo et al. (1993) found whewellite to be the main precipitate of oxalate beneath *Caloplaca* sp. and *Ochrolechia parella* at a lichen–serpentinite interface. They also found amorphous silica at the interface. It is the insolubility of calcium-oxalate that causes it to precipitate at the lichen-rock interface, even on ultramafic rocks containing very little calcium, while magnesium is washed away.

7.4 Protozoans (Protoctista)

Protoctists are eukaryotic microorganisms other than fungi (Margulis and Schwartz 1998). They are readily dispersed in air and water and are persistent as inert cysts or spores. Therefore the individuals of most species are easily distributed around the globe. The ratio of global to local species diversity is very small compared to that of animals and vascular plants (Finlay 2002). The total number of species is likely in the range between 10^4 and 10^5, much less than the number of insect species.

Protozoans are generally unicellular protoctists. They may be grouped informally (not taxonomically) by modes of locomotion. The common groups are amoebae, ciliates, and flagellates. Amoebae and flagellates are much more abundant (thousands per gram) in soils of the Franciscan formation on Jasper Ridge, San Mateo County, than are ciliates. Numbers of amoebae and flagellates were comparable in serpentine and nonserpentine soils. Sampling in January, Hungate et al. (2000) found more ciliates in serpentine soils; however, sampling in April, Rillig et al. (1999b) found more ciliates in nonserpentine soils. It is not known whether the differences are related to time of sampling or another variable.

Terlizzi and Karlander (1979) found algal communities in serpentine soils of Soldiers Delight in Maryland similar to those on other kinds of soils. They were mostly green algae (Cyanophyta), some Cyanobacteria (bacteria rather than protoctists), and a few chrysomonada (Chrysophyta) and diatoms (Bacillariophyta). These are the same kinds of algae that are most abundant in nonserpentine soils.

7.5 Bacteria (Prokarya)

Bacteria are prokaryotic microorganisms, which means that they lack a nucleus enclosed by a nuclear membrane. Prokarya are split into two domains, kingdoms, or subkingdoms, depending on preference. The subkingdoms (Margulis and Schwartz 1998) are eubacteria and archaebacteria, separated by differences in their RNA molecules. A peptoglycan layer that is characteristic of cell walls in eubacteria is absent from archaebacteria. Eubacteria are much more abundant than archaebacteria, except in extreme temperature and chemical environments. Our discussion of bacteria in terrestrial geoecosystems is based on data from eubacteria.

There is little information about bacteria in serpentine soils. Bacteria require little or no calcium. They evolved in a maritime environment containing greater concentrations of magnesium than calcium. Magnesium is required in phosphorylation systems, whereas no definite function has been discovered for calcium (Nicholas 1963). Therefore, serpentine soils lacking calcium may be suitable for many different species of bacteria. Hungate et al. (2000) found that bacterial biomass was similar in grassland soils on both serpentine and sandstone (graywacke) of the Franciscan Formation in the California Coast Ranges.

Molybdenum is another chemical element, besides calcium, that bacteria can get along without. Many nitrogen-fixing bacteria can use vanadium and iron, or iron only, as substitutes for molybdenum and iron in nitrogenase, which catalyzes the reduction of nitrogen to ammonia. *Azotobacter chroococcium*, an aerobic bacterium in soils, can utilize these alternative nitrogenases when molybdenum is scarce (Madigan et al. 2000). When molybdenum is as available, the molybdenum-nitrogenase will be used, rather than the others.

Organic molecules in the cell walls of eubacteria dissociate to yield negatively charged sites which exchange cations. Cations bound to sites on the cell walls attract anions and may form complexes that precipitate on the surfaces of bacteria. These precipitates may serve as nuclei for more precipitation, building large concentrations of nonessential inorganic cations around the bacteria. Dead or inactive bacteria may build greater depositions of metallic cations than active bacteria which pump protons to the cell walls where they compete with other cations for the negatively charged sites (Beveridge 1995).

Although there may be a high diversity of bacteria in serpentine soils and mine sites, as found by Mengoni et al. (2001) in Italy and Abin et al. (2004) in Cuba, most of the investigations of bacteria in serpentine soils have been related to nickel tolerance and plant interactions. Abin et al. (2004) found that heterotrophic gram negative bacteria were most resistant to cobalt and nickel. Serpentine soils in New Caledonia that are very rich in nickel-resistant bacteria of many kinds under nickel hyperaccumulating trees and shrubs have fewer resistant and less resistant bacteria where there are no nickel hyperaccumulating trees and shrubs (Schlegel et al. 1991). Amir and Pineau (2004) found that readily extractable cobalt and nickel were significantly greater in serpentine soils of new Caledonia containing native microorganisms than in sterilized soils, especially when degradable organic matter was added to the soils. Cobalt and nickel accompany the dissolution of manganese and iron oxides in anaerobic reactions mediated by bacteria in serpentine soil of New Caledonia (Quantin et al. 2001). Schlegel et al. (1991) proposed a nickel-cycling system, with the plants accumulating it and the bacteria releasing it from plant detritus and making it available for re-uptake by the plants. Stoppel and Schlegel (1995) characterized nickel-resistant bacteria in several genera isolated from serpentine soils in New Caledonia.

Each of three species of bacteria in two genera from a serpentine soil in the Illinois Valley in Josephine County, Oregon, increased the uptake and concentration of nickel in *Alyssum murale* without affecting plant biomass (Abou-Shanab et al. 2003). The greatest nickel accumulation was in soil with an acid-producing bacterium that is not a siderophore producer. A siderophore in one of the three treatments mobilized iron in the rhizosphere, which may have inhibited the uptake of nickel by the plants. Burd et al. (2000) found that seeds inoculated with a siderophore overproducing stain of a bacterium mitigated the inhibitory effects of high concentrations of nickel, lead, and zinc on plants. Quantin et al. (2002) found that iron, manganese, nickel, and cobalt were mobilized by bacterial reduction of iron and manganese oxides in Ferralsols of New Caledonia containing labile organic matter, while only iron, manganese, and cobalt were mobilized in the Oregon soil containing a siderophore; no nickel was mobilized (Abou-Shanab et al. 2003).

Symbiotic bacteria associated with the roots of plants are discussed in chapter 8, along with nitrogen fixation.

8   Serpentine Soils as Media for Plant Growth

Plants and animals require water, energy sources, and nutrients to make tissues and perform vital functions. The primary source of energy is the sun. Green plants use solar energy to manufacture organic compounds that are later oxidized to produce energy for both plants and animals. Many microorganisms produce energy by inorganic chemical reactions, but that source of energy is minor compared to the very large amounts of solar energy used by green plants. The major source of water and nutrients (other than CO_2) for green plants is soil (fig. 8-1).

Barren rocks, including ultramafic rock outcrop and talus, are colonized by lichens, which are symbiotic alliances of fungi and either cyanobacteria or green algae. These and other small organisms promote weathering and contribute to soil formation. Once soils are deep enough to support vascular plants (plants with roots), plants are the primary users of soils and producers of ecosystem biomass. Vascular plants send roots into soils and exploit both a high soil particle surface area and soil solutions, neither of which are available to lichens growing on rock surfaces. The surface area of particles in a soil 10 cm deep is about a thousand times greater than a planar bedrock surface if the soil is coarse sand, or about a billion times greater if the soil is clayey. With these dramatic increases in surface area accompanying soil formation, and lack of water retained on rock surfaces, it is easy to understand that ecosystem productivity is relatively low on rock surfaces and increases greatly with soil depth in very shallow soils. Annual plants approach maximum productivity in moderately deep soils and trees in deep or very deep soils.

Ecosystems with serpentine soils are generally less productive than ecosystems with other kinds of soils, and they have unique plant species distributions. Therefore,

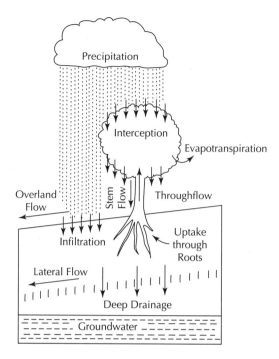

Figure 8-1 Flow of water from precipitation through a geoecosystem.

serpentine soils attract attention from botanists who are interested in the profound effects that serpentine soils have on plant distributions and growth. These effects include those that affect the supply of water (section 8.1) and those that affect the supply of nutrients (section 8.2) to plants. These in turn affect plant growth and productivity (section 8.3). The effects of serpentine soil water supply have scarcely been addressed previously, but serpentine soil effects on nutrient supply have been addressed by many edaphologists and ecologists. Some of the attempts to revegetate serpentine soils are summarized in section 8.4. Bryophytes are not neglected completely (section 8.5), although there is little information about their relationships to ultramafic rocks and soils.

8.1 Water in Serpentine Soils

Water is essential for the growth and survival of living organisms. Few plants can survive solely on water taken directly from the atmosphere. Most of the water that vascular plants use comes from soil, through their roots. Soil water supply is a limiting factor in most terrestrial ecosystems. Both plant growth and species distributions are affected by the availability of soil-water. Are serpentine soils more droughty than others? We need to review some aspects of soil-water before addressing this question.

Soil water supply depends on the amount of precipitation, the proportion of that precipitation that enters the soil, the amount of water that can be stored in a soil, water transport to plant roots, and the energy required to take water from the soil into roots. Major soil properties and characteristics that affect the supply of water to roots are the rate of infiltration of water into the ground, movement of water within and between soils, storage capacity for water within a soil, movement of water to plant roots, and solute content of water in the root zone.

The water status of a soil is defined by the amount of water and by the energy with which it is held in a soil. The energy is referred to as a "potential," because the "kinetic" energy of water in soils is negligible. Sources of the potential energy are gravity, adsorption of water on soil surfaces, capillary forces, and solutes in water. Soil-water potentials are negative, except in saturated soils where the pressure is equal or greater than atmospheric pressure. Pressure is force/area, therefore the Standard International (SI) Units are newtons per square-meter (N/m^2), or pascals (Pa). A pressure of 1 bar (a 10-m column of water), which is approximately 1 atmosphere, is equivalent to 10^5 Pa, a very large number; therefore, it is more convenient to use kilopascal (kPa) or megapascal (MPa) units. Energy terms are sometimes substituted for pressure terms, but only the latter will be used here (box 8-1).

Water flows from zones of higher potential to zones of lower potential. Water potentials range from zero at the surface of saturated soil, or greater where water stands over the ground surface, to -1.5 MPa or less (perhaps -3 to -7.5 MPa) in dry soil and much lower in transpiring plants.

8.1.1 Infiltration

Most water enters soils from rain or melting snow. Some precipitation is intercepted and retained by plants, but most of it falls to the ground, drips through plants, or flows down stems to the ground (fig. 8-1). Most water that reaches the ground enters, or "infiltrates," soils. Water that infiltrates soils is either retained in them or flows through to other soils in lower topographic positions or to a groundwater table.

Infiltration is promoted by plant detritus, or litter, that absorbs the impact of rain drops and retards overland flow of water. Where the ground is not protected by plant litter or by rock fragments, raindrop impact and splash from saturated ground ruptures soil aggregates and carries soil particles downslope. Loss of aggregation in wet soils is called "puddling." In puddled soils, larger pores are filled with finer particles, plugging pores

---

Correspondence of Water Pressure and Energy
    Energy is force times distance, or pressure times volume. For a unit volume of water (or for a unit mass of water, because water is practically incompressible), energy is equivalent to pressure; for example, 1 joule per kg (J/kg) is equal to 1 kilopascal (kPa).

that might otherwise carry water through soils. This reduces infiltration rates and allows more water to flow over the ground. Water that is lost by overland flow will not be available to plants where it might have infiltrated soils. And water that flows overland carries suspended particles with it, causing soil loss, or erosion.

Serpentine soils are commonly stony (containing rock fragments >2 mm), but not exceptionally stony for soils of western North America. Rain drops falling on stones do not cause puddling, and serpentine soils have high iron contents. Iron released by weathering of peridotite forms iron oxyhydroxides that promote soil aggregation, reducing the incidence of puddling. Thus, even though the vegetative cover is commonly thinner on serpentine soils, they are commonly stony and appear to have relatively high infiltration rates (at least those derived from peridotite rather than from serpentinite) and are not more susceptible to particulate erosion than other soils. There is considerable variability among serpentine soils, however, and some are more erodible than others.

8.1.2 Soil Water Storage

Water that infiltrates soils is held in the smaller pores with capillary forces that are greater than the force of gravity. These pores have diameters <0.03 mm, approximately. When these smaller pores are filled, water in larger pores drains downward though soils having free drainage that is unimpeded by drainage restrictions. In a day or two after a rain, when the larger pores have drained, the water that remains in the smaller pores is referred to as the "field capacity" of a soil. At this time the soil-water is held with a potential of approximately −0.1 bars. Because drainage is not as definite as the concept of field capacity implies, different agencies choose potentials from −0.05 to −0.3 bars to represent field capacity: −0.1 bars seems to be a reasonable compromise.

Plants readily draw water from soils that are at field capacity, and they continue to take up water, although with increasing difficulty as the potential decreases, until the potential reaches −15 bars, or less. At this time water is held in very fine pores with nominal diameters <0.005 mm and as hygroscopic water. Even though plants may extract some water from soils with potentials <15 bars, the amount is very much smaller than the amount available between −0.1 and −15 bar potentials. Therefore, the amount of water held in a soil between −0.1 and −15 bar potentials is approximately the amount available to plants. It is called the "available-water capacity" (AWC) of a soil.

Available-water capacities of soils are closely related to soil particle-size distributions, or texture (table 8-1). Soil organic matter increases the AWC, especially in sandy soils with low clay contents. Soil parent material effects on AWC are generally small, except that soils with volcanic ash commonly hold more available water. Therefore, serpentine soils are not expected to hold less available water than other soils lacking volcanic ash. They will hold less available water only if they are either shallower or stonier than other soils in an area. No data have been collected to verify that serpentine soils are shallower or stonier than other soils.

Available-water capacities were computed from pedon descriptions for soils mapped on those parts of the Shasta-Trinity and the Klamath National Forests that are in the

Klamath Mountains (Foster and Lang 1982, Lanspa 1993). Mean AWCs of soils with mesic (cool) temperature regimes on the Shasta-Trinity National Forest are 6.6 cm in map units with granitic soils (28,000 ha), 7.4 cm in those with serpentine soils (72,000 ha), and 7.8 cm in all other map units with soils that have parent materials from other pre-Cenozoic rocks or their colluvium (468,000 ha). Means for these categories on the Klamath National Forest are 13.5 cm in granitic soils (60,000 ha), 6.4 cm in serpentine soils (37,000 ha), and 7.8 cm in all other map units with soils that have parent materials from other pre-Cenozoic rocks or their colluvium (275,000 ha). Along with the soils, 9000 ha of ultramafic and 87,000 ha of nonultramafic rock outcrop were mapped on the Klamath National Forest, but ultramafic rock outcrop was not differentiated from other kinds on the Shasta-Trinity National Forest. The frequency distributions, or areas of soils, in different AWC categories, from low in shallow stony soils to high in very deep weathered soils with maximum AWCs, are similar for both serpentine soils and other soils in the Klamath Mountains portions of the two Forests (fig. 8-2). This evidence, even though it is based on soil mapping, which is a somewhat subjective procedure, supports the contention that the availability of water from serpentine soils is about the same as from other kinds of soils on similar landforms. Cenozoic volcanic soils in the Southern Cascade Mountains are exceptions because they generally have higher AWCs. Also, soils in cool, wet climates commonly have soils with higher AWCs. Burt et al. (2001) measured AWCs of clods from 13 horizons in three southwestern Oregon soils having udic (moist) soil moisture regimes. Nongravelly parts of the clods with mostly loam and clay loam textures had mean AWCs of 0.26 cm/cm, which is appreciably more than the 0.19 for clay loam and 0.21 cm/cm for loam recognized as means for most California soils (table 8-1). Burt et al. (2001) sampled clods from both serpentine and nonserpentine soils and the mean AWCs were practically the same (no significant difference).

8.1.3 Soil Drainage

Most plant roots are in the upper 1–1.5 m of soil. A soil that has free drainage to that depth is considered to be well drained. Soils with impeded drainage may remain saturated long enough for oxygen to become depleted by the respiration of plant roots and soil microorganisms. Drainage may be impeded by a high groundwater level or by high clay content and lack of soil structure. Pedologists recognize several arbitrary degrees of soil drainage, from very poor to poor, somewhat poor, moderately well, well, somewhat excessive, and excessive (Soil Survey Staff 1993).

Most serpentine soils in western North America are in mountainous terrain where relatively rapid external drainage, or lateral drainage to soils in lower landscape position, assures that most of them are well drained. Some serpentine soils on colluvial footslopes and in alluvial valleys, however, are poorly drained and may be in Aquoll or Aquert suborders. Very poorly drained Histosols in Terric subgroups with serpentinitic (magnesic) families are present, but sparse, in the Klamath Mountains.

Only specialized plants that are adapted to flooded or saturated soil conditions are able to take up all of the nutrients required for growth and survival in soils with scarce oxygen.

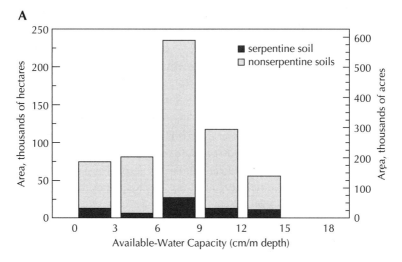

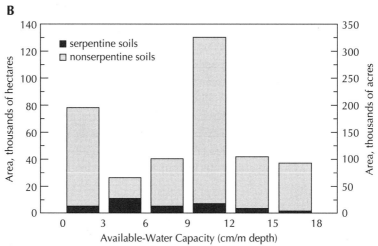

Figure 8-2 Available-water capacities (AWCs) in the upper meter (100 cm) of soils that have mesic temperature regimes and parent materials from pre-Cenozoic rocks or their colluvium. Soil data based on soil surveys of two National Forests in California. (A) Shasta-Trinity National Forest. (B) Klamath National Forest. The lower parts of bars indicate serpentine soils and entire bars represent all soils.

Therefore, plant species distributions are generally quite different in very poorly drained soils compared with well-drained soils.

8.1.4 Soil–Water Balance

The amount of water in a soil and the amount available to plants in an ecosystem is dependent on the amount and distribution of precipitation, the amount stored in root

Table 8-1 Available-water capacities of mineral soils by field grade, or soil textural class, based on water retention between 0.1 and 15 bars of suction.

Field grade	AWC[a] volume (%)
Ashy[b]	>21
Clay	
Coarse	16
Fine (clay > 60%)	12
Clay loam	19
Loam	21
Loamy sand	
OM < 1%	10
OM > 1%	12
Sand	
Coarse	
OM < 1%	4
OM > 1%	6
Fine	
OM < 1%	8
OM > 1%	10
Sandy clay	14
Sandy clay loam	15
Sandy loam	15
Coarse sandy loam	12
Fine sandy loam	18
Silt loam	
Coarse	24
Fine (clay < 18%)	22
Silty clay	15
Silty clay loam	18

[a] Means from more than 400 soil horizons (Alexander 1982).
The available-water capacity (AWC) in percent (%) divided by 100 gives the AWC in cm/cm or in./in. so that the available water retained in a column of soil is readily computed from the field grade or textures in a soil description.
[b] The AWCs of soils dominated by volcanic ash and cinders are highly variable and unpredictable. AWCs are generally >21% and can be more than double this lower limit.

zones, and the amount used by plants. It is commonly assumed that all precipitation enters soils, which is a reasonable approximation where only small proportions are intercepted and retained on plants or lost by runoff overland. Water is lost from soils by drainage, evaporation, and through transpiring plants. The aggregate of evaporation and transpiration is called "evapotranspiration."

The water budget in freely drained soils is a balance between precipitation, water intercepted and retained by plants above ground, drainage completely through soils, and evapotranspiration. To compute a soil–water balance, surplus water, where precipitation is greater than the storage capacity of the soil, is assumed to drain from the soil; and evapotranspiration is estimated from atmospheric conditions and the estimated amount of plant-available water in the soil.

Evapotranspiration is the most difficult component of the soil–water balance to evaluate. Many models have been developed to estimate evapotranspiration. Some require data that are difficult to obtain. One of the most frequently used models was developed by Thornthwaite (1948), refined by Thornthwaite and Mather (1955), and illustrated by Mather (1974). It requires only mean monthly precipitation and temperature data, which are relatively easy to obtain, to estimate potential evapotranspiration. Also, estimation of soil AWC and an assumed rate of soil-water depletion are necessary to compute actual evapotranspiration. This has been done for many stations in California (Major 1977). Newhall and Berdanier (1996) described a program for computing soil-water losses and predicting soil moisture regimes. The soil–water balance for Auburn, California, is evaluated by the Thornthwaite method in appendix E.

Walter (1974) used temperature and precipitation diagrams in which 2 mm on the precipitation scale is equivalent to 1°C on the temperature scale. These diagrams are crude simulations of water balances, where water is presumed to be lost by evapotranspiration in months when the temperature line is above the line precipitation. They are reasonably accurate in cool, wet months but grossly underrepresent potential water losses in warm, dry months (appendix E). Nevertheless, they are widely used because they are simple and easy to reproduce. We use these diagrams in part IV to represent climates in the serpentine domains of western North America.

8.1.5 Drought Tolerance in Plants

Although some serpentine soils hold little water that is available to plants, others hold plenty of water. The distribution of soils with low to high available-water capacities (AWCs) is not appreciably different in serpentine landscapes than in nonserpentine landscapes (section 8.1.2), although there are exceptions, such as the New Idria area (locality 3-12 in chapter 15).

A detailed soil survey of 12,000 ha of serpentine soils in low-to-middle elevation mountains in the Rattlesnake Creek terrain of the Klamath Mountains (Alexander 2003) provides a more accurate assessment of serpentine soils than the more general soil surveys in other serpentine areas. Of the 12,000 ha, 7% is rock outcrop (more peridotite than serpentinite); 1% is coarse colluvium, or rubble land (more peridotite than serpentinite); 9% is very shallow soils (AWC < 3 cm, more serpentinite than peridotite); 38% is shallow soils (AWC = 2–6 cm); and 45% is deeper soils (AWC > 5 cm). This distribution of soils may represent a slight shift toward more shallow serpentine soils than other kinds of soils in the area, but it is not an unusual distribution for steep nonserpentine mountainous terrain in the Klamath Mountains.

The conclusion from soil mapping is that the greater abundance in serpentine landscapes than in nonserpentine landscapes of plants with structural features common to drought-tolerant species cannot be attributed to serpentine soils having less available water than other soils, unless the differential availability of water is caused by differential responses of plants on serpentine and nonserpentine soils. Metal ions concentrated in plants have been implicated in the alteration of plasma membranes, increased stomatal resistance, reduced root growth, and other effects that inhibit the uptake of water from soils (Poshenrieder and Barceló 2004). Some plants have higher transition metal concentrations, such as nickel, in serpentine soils than in other soils. Is it possible that in serpentine soils the drought intolerance effects of nickel, or other transition elements, are counterbalanced by anatomical features such as sclerophyllous or hairy leaves that promote drought tolerance? Plants in many serpentine soils may have drought resistant appearances that are not attributable to low water content in the soils.

8.2 Serpentine Soil Fertility

The availability of nutrient elements that plants must obtain from soils for their growth and survival is called *soil fertility*. It is a dynamic feature because plants and other organisms interact with soils to alter the availability of nutrient elements. Fungi and lichens exude organic acids, particularly oxalic acid (Graustein et al. 1977, Wilson et al. 1981, Adamo et al. 1993, Wilson 1995, Sterflinger 2000), that act on minerals to cause their disintegration and release of nutrient elements. Plants alter root environments, or *rhizospheres*, increasing the availability of some nutrient elements (Marschner and Römheld 1996). Deeply rooted plants extract nutrients from subsoils and return them to surface soils, where they are more readily accessible, when leaves fall to the ground and decay. Many plants harbor symbiotic bacteria that fix nitrogen, converting atmospheric N_2 to organic forms. Some algae and free-living bacteria also fix nitrogen. Soil bacteria convert organic forms of nitrogen to ammonia, or ammonium ions, and nitrate, which are available to plants. These phenomena and many other soil processes that affect fertility are discussed in more detail in this chapter.

There are several good reviews of the effects of serpentine soils on plants (Rune 1953, Walker 1954, Proctor and Woodell 1975, Brooks 1987). Accumulated data from the serpentine soils of western North America confirm the concepts Walker (1954) stated more than 50 years ago in an article that is still a pertinent review of the ecological effects of serpentine soils. We can do little more than expand and elucidate those concepts, based on current data.

Plants require at least 16 chemical elements (Marschner 1995) and about as many more may be required in very small amounts or be beneficial (Bargagli 1998). Water and the atmosphere are the major sources of the most abundant elements in plants (carbon, hydrogen, oxygen, ntirogen, and sulfur). These elements plus phosphorus are major constituents of plant tissues (table 8-2). Also, calcium, magnesium, potassium, iron, manganese, copper, zinc, boron, molybdenum, and chlorine are required by plants.

Table 8-2 Chemical elements essential for plant growth.

Element	Major sources	Major functions in plants
C	Atmosphere	Major constituent of plant tissues
H	Atmosphere, water	Major constituent of plant tissues
O	Atmosphere, water	Major constituent of plant tissues
N	Atmosphere via soil microorganisms	Component of all amino acids and proteins
S	Sulfides, or mainly atmosphere via soil	Component of some essential proteins
P	Apatite (mineral)	Component of nucleic acids, including DNA and RNA, and phosphates (ATP and $NADP^+$) that are major links in energy transfers
K	Feldspars, micas	An abundant cation in cytoplasm, it regulates osmotic potential, controls pH, and is involved in many cellular processes
Ca	Feldspars, amphiboles, clinopyroxenes	Important component of cell walls, essential for cell growth, an important signal conductor
Mg	Olivine, pyroxenes, amphiboles	Component of chlorophyll and an activator of many enzyxes, it is essential for protein synthesis
Fe	Olivine, pyroxenes, amphibolles	Component in many enzymes, involved in oxidation-reduction systems, important for synthesis of proteins, and essential for chloroplast development
Mn	Substituted for Fe^{2+} in ferromagnesian silicate minerals	An enzyme activator, a component of some proteins, and involvement in oxidation–reduction systems
Cu	Sulfide minerals and organic C-rich sedimentary rocks	Enzymatically bound Cu is involved in oxidation–reduction reactions (Cu^{2+} can be reduced to Cu^+, which is unstable in ecosystem environments)
Zn	Sulfide minerals	Required for activities of some enzymes
B	Tourmaline	Lignification, root elongation
Cl	Salts, ubiquitous	Electrical charge balance and regulation of osmotic pressure
Mo	—	Essential component of enzymes required for biological N-fixation and for nitrate reduction

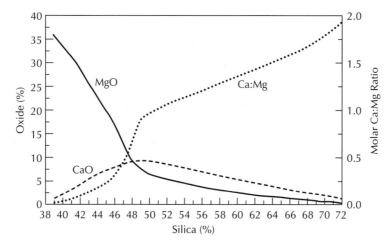

Figure 8-3 Calcium and magnesium in plutonic rocks, from peridotite with $SiO_2 < 48\%$ to granite with $SiO_2 > 70\%$ weight (data from Le Maitre 1976).

Four more of the first transition elements, other than iron and manganese, specifically vanadium chromium, nickel, and cobalt, are either required in very low concentrations or are beneficial to plants in low concentrations but can be toxic to them in higher concentrations (Bargagli 1998). Also, selenium can be toxic to plants. Algae require, in addition to previously mentioned elements required by plants, arsenic, bromine, cadmium, cobalt, iodine, sodium, nickel, silicon, strontium, and vanadium (Bargagli 1998). Animals require, in addition to most of the elements required by plants, sodium, chromium, fluorine, iodine, and selenium (Bogden 2000), and Bowen (1979) includes tin as an essential element.

The major elements in serpentine soil parent materials are hydrogen, oxygen, silicon, and magnesium. Magnesium is much more concentrated in ultramafic rocks (rocks with $SiO_2 < 48\%$) than is is calcium (fig. 8-3). Also, many of the first transition elements, especially iron, manganese, chromium, nickel, and cobalt, are more concentrated in ultramafic materials than in others (table 8-3). Streit et al. (1993) sampled soils and plants along an 8-m transect from serpentine soils to those in nonserpentine substrate of the Franciscan complex. They found gradual increases in acid extractable soil magnesium, chromium, cobalt, and nickel from nonserpentine soils to the substrate transition at the midpoint of the transect and relatively constant amounts in the serpentine soils. Gasser and Dahlgren (1994) placed the extractability of free cations, those not in crystalline minerals, from well-drained serpentine soils in three categories: calcium and magnesium are predominantly exchangeable; manganese and nickel are both exchangeable and nonexchangeable; and aluminum, chromium, and iron are nonexchangeable. Presumably, cobalt is predominantly in the nonexchangeable category and thus less readily available to plants than the more exchangeable cations.

Table 8-3 Nutrient element concentrations in rocks.

	Plutonic rocks[a] (mg/kg)			Sedimentary rocks[a] (mg/kg)			Soils[b] (USA) (mg/kg)
Element/atomic no.	Ultramafic	Mafic	Silicic	Sandstone	Shale	Carbonate	
Li 3	0.1	17	40	15	66	5	24
Be 4	0.1	1	3	0.1	3	0.1	1
B 5	3	5	10	35	100	20	33
C 6	(200)	(100)	(300)	—	—	—	25000
N 7	6	20	20	—	—	—	—
F 9	20	300	800	—	—	—	430
Na 11	4200	18000	25800	3300	9600	400	12000
Mg 12	204000	46000	1600	7000	15000	47000	9000
Al 13	20000	83000	73000	—	—	—	72000
Si 14	205000	230000	347000	368000	73000	24000	310000
P 15	220	1100	600	170	700	400	430
S 16	300	300	300	240	2400	1200	1600
Cl 17	85	60	200	10	180	150	—
K 19	40	8300	42000	10700	26600	2700	15000
Ca 20	25000	76000	5100	39100	22100	302000	24000
Sc 21	15	30	7	1	13	1	9
Ti 22	300	13800	1200	1500	4600	400	29000
V 23	40	250	44	20	130	20	80
Cr 24	1600	170	4	35	90	11	54
Mn 25	1620	1500	390	100	850	1100	550

Fe	26	94300	86500	14200	9800	47200	3800	26000
Co	27	150	48	1	0.3	19	0.1	9
Ni	28	2000	130	5	2	68	20	19
Cu	29	10	87	10	1	45	4	25
Zn	30	50	105	39	16	95	20	230
As	33	1	2	2	1	13	1	7
Se	34	0.05	0.05	0.05	0.05	0.6	0.08	0.4
Br	35	1	4	1	1	4	6	1
Rb	37	0.2	30	170	60	140	3	67
Sr	38	1	465	100	20	300	610	240
Mo	42	0.3	1.5	1.3	0.2	2.6	0.4	1.0
I	53	0.5	0.5	0.5	1.7	2.2	1.2	1.2
W	74	0.3	0.6	1.5	1	1.8	0.5	1.5

[a]Turekian and Wedepohl, 1961; Al, F, and W from Reimann and de Caritat (1998); numbers in parenthesis from Vinogradov (1959).
[b]Shacklette and Boerngen (1984).

8.2.1 Calcium and Magnesium

The main problem plants have in serpentine soils is obtaining sufficient calcium in soils with low calcium and high magnesium concentrations. This is such a dominant problem that other serpentine soil fertility problems are generally difficult to substantiate. Most plants have favorable responses to nitrogen, phosphorus, and potassium fertilization only when they have enough calcium, and the availability of calcium is very low in most serpentine soils.

Because magnesium is one of the main chemical elements in ultramafic rocks and calcium concentrations are relatively low, Ca/Mg ratios are very low (fig. 8-3). Molar Ca/Mg ratios are generally <0.1 in ultramafic rocks. These ratios increase as soils develop from Inceptisols to Ultic Palexeralfs and Haplohumults (Alexander et al. 1989, Alexander 1988a). The Ca/Mg ratios are higher in surface soils than in subsoils because much magnesium is lost by leaching and calcium is concentrated by plants that draw calcium from subsoils and return it to the surface when leaves fall to the ground and decay. Molar Ca/Mg ratios in leaves of plants growing in serpentine soils are generally greater than in the soils, but many nutrient elements are removed from senescent leaves before they fall, diminishing the effects of plant nutrient cycling in raising Ca/Mg ratios in surface soils. Amounts of calcium and magnesium on the cation-exchange complex (negatively charged sites on clay minerals and SOM) are generally considered to be good indicators of availability to plants.

The main cause of low serpentine soil fertility may be calcium deficiencies (Vlamis 1949), magnesium toxicities (Proctor 1970), or imbalances of calcium and magnesium indicated by very low Ca/Mg ratios. Both low calcium and high magnesium, and consequently very low Ca/Mg ratios, are prevalent in serpentine soils.

Vlamis (1949) and Walker et al. (1955) found that low calcium limited annual plant growth only with calcium saturation <20% (exchangeable Ca/CEC ratio < 0.2) for nonnative species, but Walker et al. (1955) found that a native sunflower (*Helianthus bolanderi*) grew well at calcium saturation <10% (<0.1). Also, McMillan (1956) found that growth of Sargent cypress, which is adapted to serpentine soils, was unaffected by exchangeable Ca/Mg ratios to values well below 20%. Apparently, results obtained from investigations of nonserpentine species do not apply to all serpentine species. McMillan (1956) also investigated within-species differences in Ca/Mg ratio tolerances. Kruckeberg (1954) found that the minimum calcium saturation for healthy nonserpentine strains of facultative (*bodenvag*) serpentine species was 25%, but that serpentine strains of the same species grew well in soils with only one-half as much exchangeable calcium.

Madhok and Walker (1969) investigated the effects of magnesium concentrations in solutions and found that high concentrations that were detrimental to a nonnative sunflower were beneficial to a native serpentine sunflower. Proctor (1970) found that a serpentine species of bentgrass was much more tolerant of very high magnesium concentrations than a nonserpentine species and suggested that stress symptoms of the nonserpentine species planted in serpentine soil were caused by magnesium toxicity. He found that the magnesium toxicity could be ameliorated by adding calcium. White (1971) attributed the exclusion of some native species from serpentine soils in Oregon to high magnesium saturation, and, to lesser extent, high nickel and chromium concentrations in serpentine soils. His evidence is circumstantial, and from what we know now, his implication of the effects of high nickel is questionable and that of high chromium is doubtful.

Conifer foliage was sampled on the lower south sides of trees on 22 plots (Alexander et al. 1989) on serpentine soils in the eastern Klamath Mountains. On these plots, incense-cedar foliage contains much more calcium and magnesium and higher molar Ca/Mg ratios than yellow pine and Douglas-fir foliage (table 8-4, fig. 8-5). There are large species-related differences in calcium and magnesium accumulation irrespective of soil similarities. Molar Ca/Mg ratios are all <1 in Douglas-fir (one exception) and yellow pine, but not in incense-cedar foliage. Monterey pine trees were found to have mean foliar Ca/Mg ratios of 0.4 on serpentine soils in New Zealand (Lee et al. 1991). Data in table 8-4 show that ponderosa and Jeffrey pine foliage have comparable calcium and magnesium concentrations, but the ponderosa pine are in more thoroughly leached serpentine soils with higher Ca/Mg ratios. Malley (1973) found that shoots of Jeffrey pine seedlings had significantly lower magnesium and higher calcium concentrations than ponderosa pine seedlings when both were grown in the same serpentine soils. He found that both ponderosa and Jeffrey pine seedlings take more magnesium and less calcium from serpentine soil than from nonserpentine soil.

Comparisons of Tree Growth in Relation to Exchangeable Calcium, Exchangeable Magnesium, and Ca/Mg Ratio

The facts that very high magnesium concentrations are detrimental to plant health and that detrimental magnesium effects are mitigated by increased calcium concentration suggest that Ca/Mg ratios may be the best indicators of plant responses in serpentine soils with high magnesium and low calcium saturation (calcium or magnesium saturation is exchangeable calcium or magnesium divided by CEC). The timber site index (TSI) of old trees (Dunning 1942) in serpentine soils of the eastern Klamath Mountains (Alexander et al. 1989) is positively correlated with both calcium saturation and exchangeable Ca/Mg ratios and negatively correlated with magnesium saturation in surface soils (0–30 cm depth). The coefficient of determination for TSI as a function of these variables are $r^2 = .41$ for calcium saturation, $r^2 = .38$ for magnesium saturation, and $r^2 = .63$ for the exchangeable Ca/Mg ratio (fig. 8-4). These coefficients of determination increase to $r^2 = .58, .52$, and $.68$ when available-water capacity (AWC) is added as a second independent variable. Coefficients of determination are even higher when TSI is replaced by a timber yield index (TYI) consisting of a product of TSI and basal area (Alexander et al. 1989). Conifer tree growth is limited at 40% (0.4) calcium saturation and severely limited below 30% (0.3) calcium saturation in soils that have low exchangeable Ca/Mg ratios, or high magnesium concentrations.

There have been many foliar and shoot analyses of native shrubs and herbaceous plants (table 8-5), but less experimentation than with native trees. Shrubs that are practically restricted to serpentine soils, such as *Ceanothus pumilus*, have foliar Ca/Mg ratios >1, although many shrubs that are facultative (*bodenvag*) serpentine species, such as *Adenostoma fasciculatum* and *Chrysothamnus nauseosus*, have molar Ca/Mg ratios <1 when growing in serpentine soils (table 8-5). Most forbs have molar Ca/Mg ratios <1 in serpentine soils—exceptions being *Calochortus* spp. and *Phacelia corymbosa* (table 8-5)— and Ca/Mg ratios >1 in nonserpentine soils.

O'Dell et al. (2006) compared root and shoot Ca/Mg ratios in serpentine and nonserpentine species of manzanita (*Arctostaphylos viscida* and *A. manzanita*), ceanothus (*Ceanothus jepsonii* and *C. cuneatus*), and oak (*Quercus durata* and *Q. berberidifolia*) in both unfertilized and fertilized serpentine soil in 3.8-L pots. All three serpentine species, but not the nonserpentine species, had leaf Ca/Mg ratios >1 in the pots with serpentine soil. All species had leaf Ca/Mg ratios ≤1 in fertilized pots. Root Ca/Mg ratios were <1 in all species, and these ratios were practically the same in fertilized and unfertilized plots. Greater calcium concentrations in the leaves of serpentine species in serpentine soil were attributed to the preferential transport of calcium to leaves, relative to magnesium. Fertilization increased leaf nickel and manganese concentrations in both serpentine and nonserpentine species. Shoot to root concentration of manganese were −1 compared to concentrations significantly ≤1 for nickel and other transition elements (chromium, iron, cobalt, and molybdenum). Evidently the transport of manganese from roots to shoots is not restricted in these plants, as it is for other transition elements that are concentrated in serpentine soils and for molybdenum.

Both tissue Ca/Mg ratios and productivity of cultivated plants are increased by adding calcium in the form of gypsum to the serpentine soils in which the plants are grown

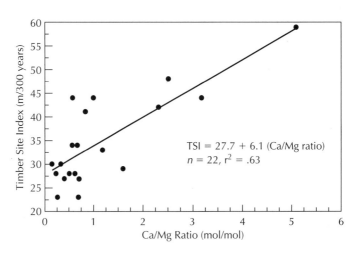

Figure 8-4 Timber site index (Dunning 1942) in relation to surface soil (0–30 cm) exchangeable Ca/Mg ratios in 22 plots on serpentine soils in the eastern part of the Klamath Mountains.

Table 8-4 Mean element concentrations in current-year yellow pine and Douglas-fir foliage from conifers on 22 serpentine soils of the Trinity ophiolite (Alexander and Zinke, unpublished) and from a representative sample of these conifers on all kinds of soils (Zinke and Stangenberger 1979).

Conifer species	Soil PM	Sample size (n)	Sample weight[b] (g)	N (mg/g)	P (mg/g)	Ca (mmol[+]/g)	Mg (mmol[+]/g)	K (mmol[+]/g)	Fe (μg/g)	Mn (μg/g)	Zn (μg/g)
Yellow pine											
Pinus ponderosa	All kinds	78	1.55	10.8	1.26	0.073	0.084	0.159	70	96	23
Pinus ponderosa	Ultramafic	4	1.79	10.0	1.46	0.070	0.122	0.141	43	126	22
Pinus jeffreyi	Ultramafic	14	1.86	10.0	1.38	0.058	0.118	0.172	35	72	26
Douglas-fir											
Pseudotsuga menziesii	All kinds	82	0.04	11.7	1.34	0.176	0.094	0.162	193	231	24
Pseudotsuga menziesii	Ultramafic	16	0.05	9.1	1.44	0.126	0.189	0.197	68	124	15

Element, median[a]

[a]The 50th percentile of sample populations of Zinke and Stangenberger (1979) are considered to be medians for samples from Douglas-fir and ponderosa pine populations on all kinds of soil parent material (PM).
[b]Weight of 10 leaf fascicles and sheaths.

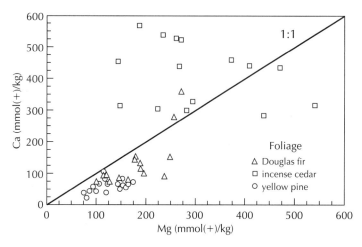

Figure 8-5 Concentrations of Ca and Mg in current year conifer foliage from 22 plots on serpentine soils in the eastern Klamath Mountains. Yellow pine (mostly Jeffrey pine) trees were sampled on 18 of the plots, incense cedar on 17, and Douglas fir on 16.

(Marten et al. 1953; R. Meyer 1999, unpublished data). Crushed limestone ($CaCO_3$) can be added to acid soils to increase soil pH, but most serpentine soils in California are neutral or only slightly acid, such that higher pH will not be beneficial to most plants. Marten et al. (1953) found that 4–8 tons of gypsum/acre (5–10 Mg/ha) increased calcium saturation and Ca/Mg ratios in infertile serpentine cracking-clay soils sufficiently to produce favorable yields of barley, oats, and vetch in Lake County, California.

8.2.2 Nitrogen and Molybdenum

Soil parent materials are not significant sources of nitrogen for plants—most rocks contain negligible amounts. Nitrogen enters soils directly from the atmosphere, and in rain water. Small amounts of nitrogen are oxidized to N_2O and NO by lightning, and automobiles and industrial pollution are sources of nitrogen oxides (NO_x). These oxides are further oxidized to nitrate or nitric acid, which is washed out of the atmosphere by precipitation or collects on the ground as dry deposition. Most atmospheric nitrogen that enters soils, however, enters as N_2 and is fixed as ammonia (NH_3) by bacteria, hydrolyzed to produce ammonium (NH_4^+), and oxidized via nitrite (NO_2^-) to nitrate (NO_3^-) by nitrifying bacteria. Plants acquire nitrogen from soil solutions by taking in either ammonium or nitrate ions through their roots.

Bacteria that fix nitrogen may be either free-living or live in symbiotic relations with other organisms. Cyanobacteria, photosynthetic bacteria that are associated with fungi in lichens in some habitats, can fix up to 50 kg N/ha annually in flooded soil habitats, where they may not have symbiotic fungal partners (Marschner 1995). Terlizzi and Karlander (1979) found that algae (including cyanobacteria) on well-to-poorly drained ser-

Table 8-5 Calcium, magnesium, and trace element concentrations in foliage of California and southwest Oregon wild (noncultivated) plants sampled on serpentine soils

Lifeform	Species[a]	N^b or n	Ca (mmol(+)/g)	Mg (mmol(+)/g)	Cr (µg/g)	Mn (µg/g)	Fe (µg/g)	Ni (µg/g)	Cu (µg/g)	Zn (µg/g)	Mo (µg/g)	Reference
Tree	Abies concolor	2	0.17	0.16	—	433	65	4	—	20	—	Alexander and Zinke, unpubl.[c]
	Calocedrus decurrens	17	0.47	0.21	—	43	91	—	—	10	—	Alexander and Zinke, unpubl.[c]
	Calocedrus decurrens	4	0.38	0.18	2	—	90	26	3	—	1.6	White, 1971
	Calocedrus decurrens	(3)[d]	0.55	0.11	2	—	124	12	4	—	0.6	White, 1971
	Cupressus sargentii	15	0.39	0.29	<1	—	—	10	—	—	—	Koenigs et al. 1982
	Cupressus sargentii	18	0.40	0.30	<1	21	53	8	3	10	—	Wallace et al. 1982
	Lithocarpus densiflorus	1	0.13	0.15	4	—	111	23	3	—	trace	White 1971
	Lithocarpus densiflorus	(2)	0.30	0.07	4	—	150	10	4	—	trace	White 1971
	Pinus contorta contorta	2	0.04	0.09	—	292	41	—	—	35	—	Alexander et al. 1989
	Pinus contorta contorta	(2)	0.05	0.08	—	200	52	—	—	37	—	Alexander et al. 1989
	Pinus jeffreyi	14	0.06	0.12	—	70	38	12	—	26	—	Alexander and Zinke, unpubl.[c]
	Pinus jeffreyi	2	0.07	0.17	2	—	173	8	2	—	0.2	White 1971
	Pinus ponderosa	4	0.07	0.14	—	182	42	12	—	22	—	Alexander and Zinke, unpubl.[c]
	Pinus ponderosa	10	0.05	0.16	—	—	—	—	—	—	—	Powers 1981
	Pinus ponderosa	(78)	0.08	0.09	—	117	88	—	—	29	—	Zinke and Stangenberger 1979

(continued)

Table 8-5 (continued)

Lifeform	Species[a]	N^b or n	Ca (mmol(+)/g)	Mg (mmol(+)/g)	Cr (μg/g)	Mn (μg/g)	Fe (μg/g)	Ni (μg/g)	Cu (μg/g)	Zn (μg/g)	Mo (μg/g)	Reference
	Pinus radiata (N.Z.)	50	0.09	0.22	—	117	35	15	4	28	—	Lee et al. 1991
	Pinus sabiniana	91	0.13	0.20	40	11	91	40	10	10	—	Gough et al. 1989
	Pseudotsuga menziesii	16	0.10	0.16	—	235	70	7	—	15	—	Alexander and Zinke, unpubl.[c]
	Pseudotsuga menziesii	(82)	0.18	0.09	—	258	219	—	—	30	—	Zinke and Stangenberger 1979
	Pseudotsuga menziesii	2	0.11	0.22	1	—	142	12	4	—	0.6	White 1971
	Pseudotsuga menziesii	(5)	0.21	0.15	8	—	341	20	3	—	1.7	White 1971
	Quercus chrysolepis	1	0.12	0.23	5	—	143	21	3	—	trace	White 1971
	Quercus chrysolepis	(5)	0.23	0.21	7	—	188	25	2	—	trace	White 1971
Shrub[e]	Adenostoma fasciculatum	22	0.15	0.34	1	16	110	13	8	33	—	Wallace et al. 1982
	Arctostaphylos benitoensis	21	0.48	0.26	6	—	—	2	—	—	—	Arianoutsou et al. 1993
	A. franciscana	6	0.40	0.19	<1	—	—	2	—	—	—	Arianoutsou et al. 1993
	A. hispidula	3	0.27	0.13	5	—	—	7	—	—	—	Arianoutsou et al. 1993
	A. manzanita bakeri	12	0.23	0.18	3	—	—	2	—	—	—	Arianoutsou et al. 1993
	A. nevadensis	2	0.24	0.25	12	—	167	14	3	—	0.8	White 1971
	A. nevadensis	(2)	0.33	0.23	6	—	74	14	4	—	0.4	White 1971
	A. obispoensis	15	0.49	0.19	2	—	—	3	—	—	—	Arianoutsou et al. 1993
	A. stanfordiana	18	0.36	0.17	1	—	—	1	—	—	—	Arianoutsou et al. 1993

	Species										Reference
	A. viscida	112	0.33	0.28	40	29	28	21	40	30	—	Gough et al. 1989
	A. viscida	34	0.27	0.21	1	20	57	5	5	32	—	Wallace et al. 1982
	Berberis aquifolium repens	1	0.09	0.15	1	—	44	30	3	—	0.4	White 1971
	Ceanothus cuneatus	4	0.26	0.10	5	—	136	21	4	—	trace	White 1971
	Ceanothus ferrisae	31	0.39	0.33	1	—	—	5	—	—	—	Arianoutsou et al. 1993
	Ceanothus jepsonii	35	0.31	0.28	1	—	—	8	—	—	—	Arianoutsou et al. 1993
	Ceanothus papillosus	1	0.48	0.24	—	—	—	16	—	—	—	Sánchez-Mata et al. 2004
	Ceanothus pumilus	18	0.68	0.29	8	—	—	25	—	—	—	Arianoutsou et al. 1993
	Ceanothus pumilus	7	0.27	0.20	15	—	312	47	3	—	1.4	White 1971
	Chrysothamnus nauseosus	1	0.22	0.34	38	—	1730	40	7	—	trace	White 1971
	Cupressus macnabiana	2	0.48	0.25	1	18	85	10	2	9	—	Wallace et al. 1982
	Eriodictyon californicum	5	0.36	0.30	2	—	139	11	4	—	0.2	White 1971
	Eriodictyon californicum	(6)	0.35	0.25	<1	—	69	8	5	—	trace	White 1971
	Eriogonum ternatum	2	0.27	0.64	51	—	1835	104	4	—	1.7	White 1971
	Quercus durata	36	0.16	0.24	1	45	185	11	10	18	—	Wallace et al. 1982
	Quercus durata	9	0.34	0.18	1	—	—	5	—	—	—	Arianoutsou et al. 1993
	Quercus vaccinifolia	2	0.19	0.14	5	—	116	16	4	—	3.1	White 1971
	Quercus vaccinifolia	(6)	0.27	0.13	<1	—	46	11	3	—	0.2	White 1971
Forb[f]	Acanthomintha lanceolata	1	0.71	0.47	—	—	—	25	—	—	—	Sánchez-Mata et al. 2004
	Achillea millefollium	1	0.28	0.78	61	—	2020	62	6	—	1.8	White 1971
	Arabis breweri	1	0.54	0.12	—	—	—	0	—	—	—	Sánchez-Mata et al. 2004

(continued)

Table 8-5 (continued)

Lifeform	Species[a]	N^b or n	Ca (mmol(+)/g)	Mg (mmol(+)/g)	Cr (μg/g)	Mn (μg/g)	Fe (μg/g)	Ni (μg/g)	Cu (μg/g)	Zn (μg/g)	Mo (μg/g)	Reference
	Calochortus albus	(5)	1.58	0.55	<1	29	157	296	317	77	—	Fiedler 1985
	Calochortus obispoensis	5	0.10	0.15	<1	37	481	391	207	44	—	Fiedler 1985
	Calochortus pulchellus	(5)	1.29	0.69	<1	52	386	241	291	68	—	Fiedler 1985
	Calochortus striatus	(5)	0.80	0.30	<1	37	211	325	267	79	—	Fiedler 1985
	Calochortus tiburonensis	5	0.68	0.09	<1	31	435	434	259	51	—	Fiedler 1985
	Cerastium arvense	1	0.11	0.34	36	—	1580	40	3	—	2.1	White 1971
	Chorizanthe membranacea	1	0.48	0.22	—	—	—	3	—	—	—	Sánchez-Mata et al. 2004
	Chorizanthe obovata	1	0.50	0.47	—	—	—	62	—	—	—	Sánchez-Mata et al. 2004
	Chorizanthe membranacea	1	0.48	0.22	—	—	—	3	—	—	—	Sánchez-Mata et al. 2004
	Clarkia concinna	1	0.83	0.19	—	—	—	3	—	—	—	Sánchez-Mata et al. 2004
	Claytonia gypsophiloides	1	0.56	0.54	—	—	—	6	—	—	—	Sánchez-Mata et al. 2004
	Collinsia heterophylla	1	0.59	0.20	—	—	—	2	—	—	—	Sánchez-Mata et al. 2004
	Collinsia sparsiflora	1	0.67	0.13	—	—	—	0	—	—	—	Sánchez-Mata et al. 2004
	Cordylanthus tenuis	1	0.59	0.30	—	—	—	11	—	—	—	Sánchez-Mata et al. 2004

Cuscuta californica parasitic on *Lessingia nemaclada*	4	0.11	0.18	6	18	158	11	5	27	—	Boyd et al. 1999
Cuscuta californica parasitic on *S. polygaloides*	4	0.11	0.17	2	14	111	800	3	40	—	Boyd et al. 1999
Eriophyllum lanatum	3	0.25	0.68	66	—	3960	95	8	—	3.1	White 1971
Heliathus bolanderi	1	0.65	1.00	—	—	—	—	—	—	—	Walker et al. 1955
Lasthenia californica, A	22	0.25	0.73	22	76	1951	84	8	67	—	Rajakaruna and Bohm 1999
Lasthenia californica, C	22	0.27	0.56	18	81	1808	87	8	63	—	Rajakaruna and Bohm 1999
Lessingia nemaclada	8	0.18	0.20	6	23	382	22	6	28	—	Boyd et al. 1999
Linanthus androsaceous	5	0.42	0.46	7	—	—	>49	—	—	—	Woodell et al. 1975
Linanthus androsaceous	(7)	0.71	0.32	1	—	—	1	—	—	—	Woodell et al. 1975
Lotus purshianus	1	0.65	0.80	9	—	706	53	8	—	—	White 1971
Lotus purshianus	(7)	0.42	0.28	8	—	717	24	9	—	—	White 1971
Malacothrix floccifera	1	0.44	0.45	—	—	—	6	—	—	—	Sánchez-Mata et al. 2004
Mimulus layneae	1	0.10	0.93	—	—	—	89	—	—	—	Sánchez-Mata et al. 2004
Minuartia douglasii	1	0.21	0.25	—	—	—	5	—	—	—	Sánchez-Mata et al. 2004
Minuartia obtusiloba (WA)	1	0.67	1.20	16	—	1220	104	—	—	—	Kruckeberg et al. 1993
Minuartia rubella (WA)	3	0.26	5.24	213	—	28900	(1360)	—	—	—	Kruckeberg et al. 1993
Minuartia rubella (WA)	(1)	0.94	0.12	3	—	600	3	—	—	—	Kruckeberg et al. 1993
Phacelia corymbosa	1	0.60	0.57	45	—	1330	34	3	—	0.4	White 1971

(continued)

Table 8-5 (continued)

Lifeform	Species[a]	N[b] or n	Ca (mmol(+)/g)	Mg (mmol(+)/g)	Cr (µg/g)	Mn (µg/g)	Fe (µg/g)	Ni (µg/g)	Cu (µg/g)	Zn (µg/g)	Mo (µg/g)	Reference
	Phlox diffusa	1	0.31	1.81	196	—	15800	303	12	—	3.8	White 1971
	Phlox diffusa	(1)	0.24	0.68	87	—	6120	146	10	—	3.1	White 1971
	Polygala californica	1	0.13	0.42	9	—	281	26	5	—	trace	White 1971
	Sedum laxum	2	0.49	0.84	70	—	1975	83	4	—	2.9	White 1971
	Streptanthus barbiger	6	(0.55)	(2.11)	(2)	(130)	(172)	(12)	(7)	(53)	—	Kruckeberg and Reeves 1995
	Streptanthus breweri	2	0.42	1.60	2	88	99	10	4	30	—	Kruckeberg and Reeves 1995
	Streptanthus breweri	1	0.44	0.29	—	—	—	21	—	—	—	Sánchez-Mata et al. 2004
	Streptanthus glandulosus	2	0.52	0.91	2	26	150	11	5	26	—	Kruckeberg and Reeves 1995
	Streptanthus glandulosus	1	1.06	0.25	—	—	—	0	—	—	—	Sánchez-Mata et al. 2004
	S. glandulosus pulchellus	1	0.52	1.21	—	—	—	—	—	—	—	Walker 1954
	Streptanthus hesperidis	2	0.30	1.81	2	61	82	20	4	94	—	Kruckeberg and Reeves 1995
	Streptanthus insignis	1	0.55	0.82	—	—	—	14	—	—	—	Sánchez-Mata et al. 2004
	Streptanthus morrisoni	1	0.46	1.18	2	101	98	5	2	29	—	Kruckeberg and Reeves 1995
	Streptanthus polygaloides	8	0.15	0.18	2	11	97	2875	3	44	—	Boyd et al. 1999
	Streptanthus polygaloides	8	(0.67)	(1.54)	(30)	(124)	(948)	(10515)	(7)	(56)	—	Kruckeberg and Reeves 1995

Group	Species	n									Reference
	Streptanthus polygaloides	1	0.06	—	—	—	2410	—	—	—	Sánchez-Mata et al. 2004
	Streptanthus tortuosus	1	0.60	8	146	130	12	4	48	—	Kruckeberg and Reeves 1995
	Streptanthus tortuosus	1	0.21	—	—	—	0	—	—	—	Sánchez-Mata et al. 2004
	Thlaspi montanum	3	0.16	3	23	324	3833	3	82	—	Boyd and Martens 1998a
	Thlaspi montanum	(2)	0.42	1	37	480	470	4	320	—	Boyd and Martens 1998a
	Thysanocarpus laciciantus	1	0.39	—	—	—	11	—	—	—	Sánchez-Mata et al. 2004
Grass	*Festuca californica*	3	0.08	25	—	228	27	2	—	0.7	White 1971
	Festuca californica	(2)	0.13	28	—	258	18	2	—	0.7	White 1971
Fern	*Cheilanthes siliquosa*	2	0.09	6	—	350	13	3	—	1.7	White 1971
	Polystichum munitum imbricans	2	0.13	43	—	122	13	4	—	2.8	White 1971
Lichens, epiphytes on Douglas-fir trees	*Hypogymnia enteromorpha*	26	0.19	5	89	850	11	4	25	—	Gough and Jackson 1988
	Usnea spp.	26	0.16	1	97	170	6	3	21	—	Gough and Jackson 1988

[a] Botanical authorities for these species are given in Hickman, 1993.
[b] Sample size, number (*n*) of individual plants (or sites for *L. californica*)–Monterey pine (*P. radiata*) sampled in New Zealand. Entire above-ground parts were sampled for some herbs. Sample populations with their numbers (*n*) in parenthesis represent nonserpentine soils.
[c] Samples collected by E. Alexander and C. Adamson and analyzed by J. Bertinshaw for P. Zinke. Some of these data are reported in Alexander and Zinke (1994).
[d] Koenigs et al. (1982) reported results practically identical to those of Wallace et al. (1982) for three of these shrubs.
[e] Samples of small annual plants may include stems and branches. *L. californica* A and C are different races of goldfields.

pentine soils in Maryland were similar to those on other soils. In the California Region, where most serpentine soils are well drained and their surface soils are dry through summer, amounts of nitrogen fixed by photosynthetic bacteria (cyanobacteria) may be negligible. Nonphotosynthetic free-living bacteria fix <1 kg N/ha each year. These bacteria are present in most soils, but Lipman (1926) found them to be scarce in serpentine soils of Mt. Diablo in Contra Costa County, California. Some bacteria living in root zones (rhizospheres) of grass plants (especially those with C_4 photosynthesis) fix up to 300 kg N/ha annually (Evans and Barber 1977). We lack data for nitrogen-fixation in serpentine grasslands. The greatest potential for nitrogen fixation in serpentine soils may be symbiotic bacteria that form root nodules on legumes (*Rhizobium*, *Bradyrhizobium*, and *Azorhizobium* spp., Vance 1996) and actinorhizal bacteria (*Frankia* spp.) that form root nodules on many species in several genera of woody plants, including *Ceanothus* spp. Legumes are sparse in most serpentine soils. Nevertheless, *Lotus wrangelianus* is common in some warmer soils, and *Lupinus* spp. and *Astragalus* spp. are common in some colder soils. Only a few species of ceanothus are widely distributed in serpentine soils—most notably buckbrush (*Ceanothus cuneatus*). Other genera (and some species) with common western North American serpentine plants that have symbiotic nitrogen-fixing actinorhizal bacteria are alder (*Alnus* spp.), dryas (*Dryas drummondii* and *D. octapetala*), bitterbrush (*Purshia tridentata*), buffaloberry (*Shepherdia canadensis*), and mountain mahogany (*Cercocarpus* spp.) (Benson et al. 2004). White (1967) found that actinorhizal nodules were sparse on *Ceanothus cuneatus* roots in serpentine soils in parts of California and Oregon. O'Dell et al. (2006) found that most of 11 ceanothus (*Ceanothus cuneatus* and *C. jepsonii*) plants grown in unfertilized serpentine soil had actinorhizal nodules, but the same species grown in fertilized serpentine soil lacked actinorhizal nodules. Nodulated *Ceanothus* spp. fix up to about 25–30 kg of N/ ha/yr in dense stands (Conard et al. 1985). In New Caledonia, several species of *Frankia* that are symbiotic on *Gymnostoma* spp. are associated exclusively with serpentine soils (Navarro et al. 1999).

Deposition of nitrogen directly from the atmosphere in dust and rainfall is about 2–3 kg/ha annually in wildlands of northern California (McColl et al. 1982). The largest amounts are near large urban areas such as Sacramento and downwind from them (Blanchard and Tonnessen 1993).

Molybdenum is mentioned along with nitrogen here because nitrogenase and nitrate reductase contain molybdenum (Marschner 1995). Microorganisms that fix nitrogen require nitrogenase, and generally molybdenum, because molybdenum is essential for this enzyme to function in the fixation of nitrogen in these organisms, with few exceptions. Nitrate reductase is most important in plants whose main source of nitrogen is nitrate. Many woody plants take ammonium ions in through their roots and are less dependent on nitrate reductase, and therefore less dependent on molybdenum (Römheld and Marschner 1991). Some conifers have marked preferences for ammonia ions rather than for nitrate (Kronzucker et al. 1997). Low levels of nitrate reductase in plants of the Ericaceae (heath family, which includes manzanita) may indicate a nitrogen metabolism based on ammonium, rather than on nitrate sources of nitrogen (Lee and Stewart 1978).

Molybdenum is present in minute concentrations in both rocks and plants. Other than minor molybdenite in some silicic plutonic rocks, molybdenum-bearing minerals are sparse. Soil molybdenum is derived from traces of molybdenum that are present in rock minerals, rather than from minerals in which molybdenum is a dominant element. Although molybdenum concentrations are low in serpentine soils (table 8-3; Walker 2001), only minute amounts of molybdenum are required by plants, especially in those taking ammonium ions in through their roots. Plant tissues contain much less molybdenum than any other element required for plant growth (Allaway 1976). Molybdenum is generally present in soils as an oxyanion of a weak acid. It dissociates to HMO_4^- in moderately to weakly acid soils and to molybdate (MO_4^{2-}) in slightly acid to neutral soils. Molybdenum is most readily available to plants in neutral to slightly acid soils. It is converted to oxycations in strongly acid soils and adsorbed on iron and manganese oxyhydroxides (Stiefel 2002). Molybdenum deficiencies generally occur in well-drained, acid soils (Allaway 1976), and most serpentine soils in western North America are only slightly acid or neutral. Walker (1948) found that crop plants grown in serpentine soils showed molybdenum deficiencies, suggesting that low molybdenum concentrations may exclude some plants from serpentine soils. Johnson et al. (1952) grew many crop plants in serpentine soil, finding that none of seven legume species nor any of five grass species showed molybdenum deficiencies, whereas all species in the goosefoot, composite, mustard, nightshade, buckwheat, and carrot families showed molybdenum deficiencies. Deficiencies of molybdenum have not been observed in people (Adriano 2001), although molybdenum enzymes are important in the sulfur metabolism of organisms from bacteria to humans.

Molybdenum is reduced to Mo(IV) and forms thiomolybdate ions in anaerobic environments containing sulfides (Stiefel 2002). These anaerobic conditions prevailed for most of the first half of the earth's history. Stiefel suggested that vanadium and all-iron nitrogenases might have been used before there was enough O_2 in the environment to make molybdenum sufficiently available for nitrogen reduction. Vanadium and all-iron nitrogenases are generally less efficient, but they are retained in some organism as a back-up for molybdenum nitrogenase (Stiefel 2002). It is unknown if any organisms inhabiting serpentine soils use vanadium or all-iron nitrogenases. Tungsten (a third transition element) is chemically very similar to molybdenum (a second transition element), except that it is not as readily reduced and immobilized in a sulfide environment; tungsten enzymes found in hyperthermic archaebacteria that inhabit suboceanic vents, or "black smokers," may have been the first to perform functions that are now common among molybdenum enzymes.

Plants in all kinds of soils benefit from additional nitrogen, although the benefits of additional nitrogen are recognized in some soils only after remediation of other elemental deficiencies. Turitzin (1982) planted annual fescue (*Vulpia microstachys*) and soft chess (*Bromus mollis*) in soil from Jasper Ridge in San Mateo County, California, in a greenhouse. He found that soft chess responded to nitrogen-fertilization only after adding phosphorous. With annual fescue, he found a significant shoot-weight response to addition of nitrogen alone, a highly significant response with nitrogen and phosphorus together, and a very highly significant response with nitrogen, phosphorus, and potassium. Carbon contents of serpentine soils are not particularly lower than in other California soils, and organic C/N ratios are not noticeably different either (chapter 5). Therefore, nitrogen contents of serpentine soils do not appear to be unequivocally lower than is characteristic of other soils in California. Nevertheless, nitrogen concentrations in needles of conifers growing in serpentine soils are much lower than are concentrations in needles of the same species on nonserpentine soils (table 8-4). Therefore, low nitrogen concentrations in plant tissues, as noted by White (1971), appear to be related to factors other than lack of nitrogen in serpentine soils. For example, lack of another element in soils or plants may affect the uptake or metabolism of nitrogen, or intraspecies varieties that are adapted to serpentine soils may inherently store less nitrogen in their foliage.

8.2.3 Phosphorous (and Sulfur)

Apatite, $Ca_5(PO_4)_3F$, in which F^- can be replaced by OH^-, Cl^-, or a CO_3^{2-} fraction, is the main source of phosphorus in soils derived from plutonic rocks (Deer et al. 1966). Some ultramafic rocks contain little or no apatite and phosphorus concentrations are considerably lower in peridotite than in more silicic plutonic rocks (table 8-3) and in most sedimentary rocks. Phosphorous is released slowly by weathering of apatite and other minerals containing traces of phosphorus. Once phosphorus is released by weathering, or accumulates from dust, only small amounts of it are lost by leaching from well-drained soils. It accumulates as calcium-phosphate in alkaline soils and by adsorption on iron and aluminum oxyhydroxides in acid soils. Bonifacio and Berberis (1999) found that the largest fraction of phosphorus in slightly to moderately weathered serpentine soils in northern Italy is that adsorbed or occluded by iron oxyhydroxides. They found that weathering accompanied by silica and magnesium leaching losses concentrated phosphorus from only 10 mg/kg in the serpentine rock (considerably less than Burch [1968] found in peridotite from Burro Mountain in the California Coast Ranges) up to about 500 mg/kg in soil horizons with the greatest free iron concentrations. These soil phosphorus concentrations are comparable to those in nonserpentine soils (Stevenson and Cole 1999). Much of the phosphorus in well-drained serpentine soils, however, is unavailable to plants. Burt et al. (2001) found that serpentine soils in southwestern Oregon containing high concentrations of free iron oxides are capable of retaining large amounts of phosphorus. Plants survive on serpentine soils by using soil phosphorus that is not in apatite or bound by iron oxides and recycling it. Organic phosphorus from plant remains and other organisms is decomposed throughout the growing season to supply plants with $H_2PO_4^-$ in neutral soils and HPO_4^{2-} in acid soils (Stevenson and Cole 1999).

Nucleic acids, phospholipids, phosphorylated polysaccharides, phytin, and phosphoropyridine nucleotides are among the most important phosphorus compounds in plants (Stevenson and Cole 1999). Adenosine triphosphate (ATP) has a major role in energy

Phosphorous in Atmospheric Dust, in Peridotite, and in Ecosystem Biomass

Although dust is a minor source of phosphorus, it may be important in natural ecosystems with peridotite containing little or no apatite. Phosphorous, unlike nitrogen and sulfur, does not form any compounds that are common atmospheric gases that can be washed from the atmosphere in precipitation. Williams and Melack (1997) found that dust falling in the Sequoia National Forest deposited phosphorus at an average rate of 2 mg/m^2 (0.02 kg/ha) annually.

Zinke et al. (1979) found phosphorus contents of 118 kg/ha in trees plus plant detritus (litter) of a 100-year-old lodgepole pine forest in the Tahoe Basin and 92 kg/ha in a redwood forest in Humboldt County, California—most of the phosphorus was in litter in the pine forest and in trees in the redwood forest. Woodmansee and Duncan found phosphorus contents about 16 kg/ha in herbaceous plants and litter of annual grassland on the San Joaquin Experimental Range (Jones and Woodmansee 1979, Woodmansee and Duncan 1980). There is likely more phosphorus in these ecosystems than in those on serpentine (10–50 kg P/ha should be close to the range for most California serpentine ecosystems, herbaceous to forested ecosystems). It would take 500–2,500 years for soils in the Sierra Nevada to accumulate 10–50 kg P/ha from dust deposition.

Burch (1968) reported 300 mg P_2O_5 (130 mg P)/kg in peridotite on Burro Mountain in the California Coast Ranges. With a peridotite density of 3.12 Mg/m^3, a slab of bedrock 1 mm thick would contain 4 kg P/ha. Peridotite weathering rates, based on a formula modified from that of White and Blum (1995) for chemical loss and on Alexander (1988b) for estimation of weathering from chemical loss and soil formation, are 400 years/mm in a warm dry (40 cm precipitation/year) climate representative of the San Joaquin Experimental Station and 300 years/mm in a cool moist (150 cm precipitation/year) climate higher in the mountains. Release of phosphorus from weathering of peridotite to supply the amounts in ecosystem plants and litter requires 1600 years in a warm dry climate and 9000 in a cool mountain forest climate.

Phosphorous is deposited in dust in the Sierra Nevada at least three times faster than it is expected to be released by weathering of peridotite. Deposition of phosphorus may be more rapid in other parts of California, but even where it is slow, it supplies more ecosystem phosphorus than weathering of peridotite and serpentinite bedrock.

---

storage and transfer within plants, mainly because of the energy release upon hydrolysis of the pyrophosphate bond in ATP (Marschner 1995). Phosphorus deficiencies reduce plant growth and hinder maturation of seeds.

Phosphorous added, along with calcium, to two California serpentine soils increased both the yield of bush bean (*Phaseolus vulgaris* L.) and foliar phosphorus concentrations (Wallace et al. 1977). This may or may not be relevant for native plants. Koide and Mooney (1987) found that phosphorus added to a serpentine grassland topsoil in Santa

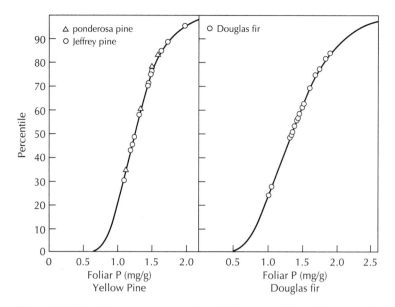

Figure 8-6 Phosphorous concentrations in first year foliage from trees in serpentine soils of the Trinity ophiolite plotted on frequency curves for trees growing in all kinds of soils. The frequency distributions are from Zinke and Stangenberger (1979). Medians (50th percentile) are 1.263 mg P/g for ponderosa pine and 1.342 mg P/g for Douglas-fir foliage.

Clara County, California, increased production of annual grasses and some, but not all, native forbs. Subsoil from the same grassland was so infertile that phosphorus fertilization did not increase herb production significantly.

Phosphorous concentrations in foliage collected by Alexander et al. (1989) from trees in serpentine soils are relatively high in both yellow pine (ponderosa and Jeffrey pine) and Douglas-fir foliage (table 8-4). Few of these trees have foliar concentrations much below the median (50th percentile) for foliage from trees of the same species growing in non-serpentine soils (fig. 8-6). This implies that phosphorus is not limiting tree growth on these serpentine soils. It might become limiting if gypsum were applied to increase exchangeable Ca/Mg ratios, but as long as the productivity is low from other causes, phosphorus does not appear to be limiting tree growth.

Jones et al. (1977) found that subterranean clover (a non-native plant) grown in topsoils from Henneke soils responded favorably to phosphorus fertilization, but yields were increased significantly only after adding sulfur along with phosphorus. Most sulfur in soils is derived from the atmosphere, rather than from soil parent materials, except in fine-grained marine sediments, which may have higher sulfur concentrations. Therefore, sulfur is no more likely to be deficient in serpentine soils than in most other soils. Annual additions of sulfur from rainfall and dry deposition average 1.1 kg/ha in wildlands of northern California (McColl et al. 1982). The largest additions are along the northern coast of California (Blanchard and Tonnesson 1993).

8.2.4 Potassium

Silicic plutonic rocks generally contain much potassium feldspar and micas and consequently have relatively high potassium concentrations (table 8-3). Peridotite lacks these minerals and has very low potassium concentrations. Olivine and pyroxenes, the main minerals in peridotite, are practically devoid of potassium, but some hornblende and other amphiboles have potassium concentrations up to 20 g/kg (Deer et al. 1966). Dust and precipitation add another 0.1–0.6 kg K/ha annually to northern California ecosystems (McColl et al. 1982).

As potassium is released by weathering, it either accumulates on the cation exchange complexes of clay minerals and organic matter, is fixed between layers in silicate clay minerals, is taken up by plants, or is leached from the soil. Only small fractions of the potassium in soils are exchangeable and available to plants. Plants conserve potassium by recycling it. Most soils contain more potassium in the surface, where there is more organic matter, than in subsoils. The amounts of exchangeable K^+ range from 1 to 11 mm mol K^+/kg in the surface to K^+ <1–4 mm mol/kg in subsoils of serpentine Alfisols and Mollisols on the Trinity ophiolite (Alexander et al. 1989). This is comparable to amounts in nonserpentine Alfisols, Mollisols, and Ultisols in northern California (Soil Survey 1973). Potassium concentrations are relatively higher in foliage collected from yellow pine (ponderosa and Jeffrey pine) and Douglas-fir trees growing in serpentine soils from these kinds of trees growing in nonserpentine soils (Alexander et al. 1989). In this respect, potassium is similar to phosphorus and dissimilar from nitrogen (table 8-4). Although Proctor and Woodell (1975) reported contrary results, with lower concentrations of phosphorus and potassium in conifer foliage on serpentine soils, there is little evidence that potassium limits plant growth in California. This lack of evidence may be partly due to the fact that few agricultural crops are grown in serpentine soils. According to Alfred Cass, potassium-deficiency symptoms are common in wine grapes on serpentine soils in the California Coast Ranges (personal communication, 2002).

8.2.5 First Transition Elements

The first transition chemical elements are in the fourth row, or period, of the periodic table. They are more concentrated in mafic or ultramafic rocks than in silicic plutonic rocks (table 8-3). Plants growing in serpentine soils do not appear to be deficient in any of these elements, but some of these elements are toxic to plants in serpentine soils. The elements most commonly mentioned as being toxic to plants growing in serpentine soils are chromium, nickel, and cobalt. Manganese and iron are required by plants and vanadium is either essential or at least beneficial to algae and may be beneficial to plants (Bargagli 1998).

Vanadium

Although it is a minor element, vanadium is more abundant than either copper or zinc in the crust of the earth. It is most prevalent in mafic igneous rocks and shale (table 8-3). Trivalent and tetravalent vanadium substitute for Al^{3+}, Fe^{3+}, Cr^{3+}, and Ti^{4+} in common minerals of igneous rocks, especially in titaniferous spinel minerals and ilmenite (Hopkins et al. 1977). Pentavalent vanadium may substitute for phosphorus in apatite.

Vanadium released by weathering of primary minerals is oxidized to the pentavalent state in neutral to alkaline, well-drained soils. It may be reduced to the tetravalent state in poorly drained soils, especially if they are acid. Several species of bacteria reduce V(V), or vanadate (Ehrlich 1996). The trivalent ion (V^{3+}) prevails only in extremely acid, reduced soils and forms insoluble hydrated oxides. Pentavalent vanadium in dilute solutions occurs as VO_2^+ in very acid environments, $VO_2(OH)_2^-$ in acid to neutral, and VO_3OH^{2-} in alkaline ones (Baes and Mesmer 1976), with (VO_4^{3-}) occuring only in extremely alkaline solutions. Although monomeric $H_2VO_4^-$ and HVO_4^{2-}, generally presented as $VO_2(OH)_2^-$ and VO_3OH^{2-}, are prevalent in dilute solutions, as are common in soils, oligomeric forms of vanadate such as $V_4O_{12}^{4-}$ and $V_{10}O_{28}^{6-}$ prevail in more concentrated solutions (Baes and Mesmer 1976, Crans et al. 1998). Vanadate is readily reduced to vanadyl, VO^{2+}, in acid solutions and hydrolyzed to $VOOH^+$. Vanadyl is an oxycation that may be strongly bound to clays. Tetravalent vanadium is precipitated as $VO(OH)_2$ in neutral solutions. Both V(V) and V(IV) forms are strongly chelated by organic ligands. In some cases, V(V) may be reduced to V(IV) in the process of chelation (Morrell et al. 1986). Vanadium concentrations in soils generally reflect concentrations in their parent materials because vanadium is not readily transported in soils or leached from them (Kabata-Pendias 2001). Nevertheless, vanadium may be leached from the surface horizons of Spodosols, which may be related to the affinity of vanadium for organic ligands that are translocated from O or A horizons through E horizons to B horizons in the process of podzolization. V(III) substitutes for aluminum in dioctahedral smectites and, in reducing environments, for ferric iron in goethite (Taylor and Eggleton 2001).

Vanadium is important to prokaryotic nitrogen-fixing organisms as an alternative to molybdenum in nitrogenase, and vanadium concentrations are high in root nodules on some legumes (Kabata-Pendias 2001). Vanadium is also an alternative for heme iron in peroxidase in plants, animals, and bacteria. Plants generally acquire either V(V) or V(IV) from soils (Morrell et al. 1986). Pentavalent vanadium can replace phosphorus in phosphates and interfere with metabolic processes. Toxic V(V) is reduced to less toxic V(III) and V(IV) forms in plants and animals. Few plants transport vanadium from their roots and few contain V > 1 ppm in their shoots or leaves. Some accumulators contain V > 10 ppm and a few contain V > 100 ppm (Cannon 1963, Hopkins et al. 1977). Apparently, vanadium is noticeably toxic to plants only in contaminated soils. Mosses and some fungi and animals accumulate vanadium. It is an essential element for some algae.

Chromium

In ultramafic rocks, most of the chromium-spinel is the chromium-end member, chromite (FeCr$_2$O$_4$), and Cr(III) commonly substitutes for Fe(III) in silicate minerals. The chromium concentration is about 1–5 g/kg in ultramafic rocks. Chromite and magnetite weather more slowly than other minerals in serpentine soils. Because of slow weathering, chromite persists as the main repository of chromium in serpentine soils (Suzuki et al. 1971). Chromium released from chromite is present in soils in trivalent, Cr(III), and hexavalent, Cr(VI), states. Oxidation of Cr^{3+} is very slow and is retarded by soil organic matter. Presence of Mn(IV) facilitates the oxidation of Cr^{3+} (Bartlett and James 1988). The oxidized, or Cr(VI), forms are (CrO$_4$)$^{2-}$ in neutral to alkaline soils and H(CrO$_4$)$^-$ in acid soils. Several species of bacteria reduce Cr(VI), mostly anaerobically (Ehrlich 1996). Reduction of Cr(VI) to Cr(III) is mediated by facultative anaerobes that are common in soils (Wang 2000)—a very effective process in soils containing organic matter. Ferrous iron may be more effective reducing Cr(VI) in well-drained, neutral soils and H$_2$S more effective in very acid soils (Fendorf et al. 2000). In an intensive investigation of chromium in serpentinite and its soils on Jasper Ridge, San Mateo County, California, Oze et al. (2004) detected no Cr(VI). They found most of the chromium in both serpentinite and soils to be in chromite, chromium-magnetite, and a chromite-silicate mixture.

Plants and animals require chromium in small amounts, but it is toxic in larger amounts. Hexavalent chromium can replace sulfur in sulfate, and free radicals are produced in the reduction of Cr(VI) to Cr(III). Although Cr(VI) is highly toxic to plants, it is generally subordinate to Cr(III) in soils (Bartlett and James 1988). Plants acquire chromium as CrO$_4^{2-}$ or HCrO$_4^-$ ions. It is then reduced to Cr(III) and concentrated in roots; little is transported to other parts of plants (Streit and Stumm, 1993). Chromium concentrations are most commonly <10 μg Cr/g in foliage of California serpentine plants (table 8.5), exceeding 70 μg Cr/g only in spreading phlox (*Phlox diffusa*). Brooks (1987) suggested that foliage concentrations >100 μg Cr/g are likely caused by foliage sample contamination by dust from chromium-rich soils. He found no chromium-hyperaccumulator plants.

Trivalent chromium forms hydrated oxides in soils similar to those of trivalent aluminum and iron. It is soluble only in extremely acid soils. Hexavalent chromium is much more soluble and available to plants. Because Cr(III) is oxidized very slowly in soils, and Cr(VI) does not persist in the presence of active soil organic matter, Cr(VI) does not appear to be much of a threat to plants. Drying of organic matter increases its activity as a reducing agent (Bartlett and James 1988). Severe ground fires that consume most of the aboveground biomass and leave ash on the ground that increases surface-soil pH upon infiltration of water through the ash create an environment for the oxidation of Cr^{3+}, but we are unaware of any reports of Cr(VI) toxicities following severe fires.

Whereas Cr(III) is essential for human nutrition, Cr(VI) is very toxic (Adriano 2001). Chromium toxicities and health hazards are practically always associated with polluted environments rather than with natural ones.

Manganese

Manganese is required by both plants and animals (Adriano 2001). It is more concentrated in ultramafic and mafic rocks than in more silicic plutonic rocks (table 8-3). Manganese is not a dominant element in any primary minerals in ultramafic rocks. It substitutes for Fe^{2+} in the dominant minerals in peridotite (Gilkes and McKenzie 1988). Manganese(II) released by weathering of olivine and pyroxenes accumulates in secondary iron-bearing minerals, such as goethite, and as Mn(III) and Mn(IV) oxides and oxyhydroxides that may occur along with iron oxides in nodules, films, or coatings, and disseminated through soils. Crystallization of manganese oxides is generally poor in soils, and it is difficult to identify specific minerals. Birnessite, vernadite, lithiophorite, and hollandite are the most common manganese oxide minerals in soils (McKenzie 1989), but there are many others. Only Mn(II) is mobile in soils, and it does not form insoluble oxyhydroxides in acid soils. Trivalent manganese is relatively unstable and is readily reduced to Mn(II) or oxidized to Mn(IV) or bound in oxide minerals.

In well-drained, neutral-to-slightly acid soils, which predominate among California serpentine substrates, only very small proportions of the manganese are in the mobile divalent form. In poorly drained soils, manganese is reduced to Mn(II) and is then available to plants. Acid and reducing microenvironments around plants roots provide conditions for the reduction of manganese in well-drained soils, such that deficiencies are unlikely even in neutral well-drained serpentine soils. Amounts of manganese available to plants are difficult to discern because of the transitory nature of manganese valence states in soils (Robson 1988). In the xylem, manganese is transported as a divalent cation and also chelated with citric and malic acids (Rengel 2000). Foliar analyses of California

---

Foliar Manganese and Conifer Tree Growth on the Trinity Ophiolite

There are no apparent reasons for manganese deficiencies in plants on serpentine soils, nor have there been any reports of such deficiencies. Nevertheless, the timber site index and a timber yield index (TYI), which is the product of timber site index and basal area (Alexander et al. 1989), are closely related to concentrations of manganese in conifer foliage (fig. 8-7) for trees on the Trinity ophiolite. The coefficients of determination (r^2) for TYI as a function of the logarithm of foliar manganese concentration are .690** for Douglas-fir ($n=16$), .348* for incense-cedar ($n=17$), and .425** for yellow pine ($n=18$), where one asterisk (*) indicates a significant relationship ($\alpha<0.05$) and two (**) indicate a highly significant relationship ($\alpha<0.01$). One might expect an interaction between iron and manganese, because foliar concentrations of neither are low, but correlations between foliar concentrations of these elements are insignificant ($\alpha>0.05$). There is no apparent explanation of the positive relationship of tree growth to foliar manganese concentration. Excess nickel and cobalt may interfere with manganese metabolism (Hewitt 1953), but the foliar analyses did not include these elements.

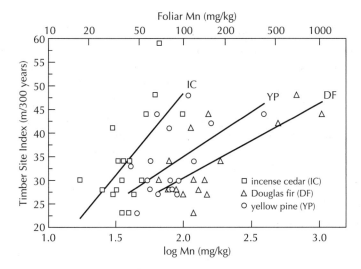

Figure 8-7 Timber site index (Dunning 1942) in relation to current year foliar manganese in trees on 22 serpentine soils in the eastern part of the Klamath Mountains. The coefficients of determination (r2) are .39, .35, and .52 for the logarithms of manganese in incense-cedar, yellow pine (mostly Jeffrey pine), and Douglas-fir foliage.

serpentine plants (table 8-5) indicate that all have manganese concentrations in the 20–500 µg/g range that Reeves and Baker (2000) suggest is normal. Nevertheless, conifer tree growth on serpentine soils in the eastern part of the Klamath Mountains is correlated positively with foliar manganese concentrations (box 8-4).

According to Adriano (2001), manganese is generally more likely to be toxic to plants than deficient, at least in acid soils. Moderate concentrations of manganese in plants reduce their susceptibility to fungal infections (Adriano 2001). Multiple oxidation states and high redox potential make manganese an important element for producing O_2 (e.g., in photosynthesis) and for mopping up free radicals.

Iron

Iron is required by both plants and animals (Adriano 2001). It is a major constituent of olivine, orthopyroxene, and chromite and is the main constituent of magnetite. It is plentiful in amphiboles and biotite, which are common rock-forming minerals. Thus, most kinds of rocks contain appreciable amounts of iron. Iron decreases in abundance from the core of earth to the crust and increases in oxidation state. Along with an increase in silica concentration from mafic to silicic plutonic rocks, iron decreases from about 12% (120 g/kg) in peridotite and 10% in gabbro to about 3% in granite (Le Maitre 1976). Plutonic rocks and mafic volcanic rocks have more ferrous than ferric iron, but the silicic volcanic rocks generally have more ferric than ferrous iron. Iron is readily oxidized or reduced in soils where its oxidation state depends on the soil environment.

The main iron-minerals in serpentine soils are ferrihydrite, goethite (FeOOH), and hematite (Fe_2O_3), which are well characterized by Cornell and Schwertmann (1996). Iron is a major constituent in some smectites that are common in serpentine soils (Wildman et al. 1968).

Other than in minerals, most of the iron in well-drained, neutral-to-slightly acid soils, which are common among serpentine substrates of western North America, is in the immobile trivalent state. Only very small proportions of the iron are in mobile divalent forms. In poorly drained soils, iron is reduced to mobile divalent forms and may be available to plants. Acid and reducing microenvironments around plants roots provide conditions for the reduction of iron in well-drained soils. In well-drained soils that are not acid, most of the mobile iron is in chelated form (Marschner 1995). The organic compounds that chelate iron in these soils are derived from humus, microorganisms, and root exudates. Iron is transported in chelated form within plants.

Although the average concentration of iron is only 0.2 g/kg in terrestrial plants, compared to 57 g/kg in the continental crust, it is essential for plants and most other organisms (Romankevich 1990). It is required by all microorganisms, except lactobacilli, which lack heme and use a ribonucleotide reductase containing cobalt (Hersman 2000). Much of the iron in plants and other organisms resides in the porphyrin ring of heme. As such, it is in the cytochromes of redox systems in chloroplasts and mitochondria as well as in catalases and peroxidases, which are so important in controlling the oxygen and oxygen radicals in organisms. Iron is also a constituent of iron–sulfur proteins in nitrogenase that catalyzes the reduction of N_2 in nitrogen fixation (Marschner 1995). Only prokaryotes contain nitrogenase, making bacteria essential for the fixation of nitrogen that is used in biological systems.

Foliar analyses of California serpentine plants (table 8-5) indicate that all have iron concentrations >35 µg/g. The most likely effect of serpentine soil iron on plants is in fixing phosphates and limiting the supply of phosphorus that is available to plants.

Mean concentrations of iron are about two to three times greater in marine than in terrestrial organisms (Romankevich 1990). This may be related to the reducing nature of early oceans and greater availability of divalent iron 3 or 4 Ga. Only about half way through the evolution of the earth did the oceans become oxygenated and oxidize much of the iron to its trivalent form and cause its precipitation. Many kinds of marine organisms may not have lost their higher iron requirements. Currently, the low availability of iron is a major factor limiting the productivity of marine ecosystems (Martin and Fitzwater 1988, Boyd 2004). When environments became more oxidizing, microoganisms developed the capacity to produce siderophores that chelate Fe(III) and surface proteins that transport the siderophore-Fe(III) complexes across cell membranes (Hersman 2000).

Cobalt

Cobalt is not a dominant element in any minerals in ultramafic rocks, yet it is much more concentrated in them than in more silicic rocks. Divalent cobalt substitutes for magnesium and iron in olivine and pyroxenes. Average cobalt concentrations are about 150 mg/kg (0.15 g/kg) in ultramafic compared to about 1 mg/kg in silicic plutonic rocks. Cobalt is present in both divalent, Co(II), and trivalent, Co(III), oxidation states in minerals and organic molecules. In soils, Co^{3+} is such a strong oxidizer that it will take an electron from water and release oxygen; therefore, the trivalent ion cannot exist in soil solution. The divalent ion (Co^{2+}) is mobile in acid soils, but it is strongly adsorbed by iron and manganese oxides such that its mobility is greatly reduced as pH increases; it is immobile in neutral soils, although it may be mobilized by chelation with organic compounds (Kabata-Pendias 2001). The lack of cobalt mobility is manifest in serpentine soils, which have high iron and manganese concentrations and are commonly slightly acid to neutral. McKenzie (1970) found that most of the Co^{2+} in soils is adsorbed on manganese oxides and that much of it displaces Mn^{2+} and becomes fixed or is not readily extractable. Apparently, the fixation of cobalt by its reaction with manganese minerals involves its oxidation to Co(III) and the replacement of Mn(III) (McKenzie 1970). McKenzie (1975) found that Co^{2+} adsorption was greatly diminished at lower pH, which would make it susceptible to leaching in acid soils. McLaren et al. (1986) found that humus adsorbs substantial amounts of Co^{2+}, but less than manganese oxides, and clay minerals adsorb much less cobalt. Moreover, Co^{2+} is readily displaced from humic and fulvic acids, in contrast to practically irreversible adsorption on manganese oxides. Bibak et al. (1995) confirmed that manganese oxides and organic matter retain more cobalt than iron oxides in an Oxisol and that cobalt is less readily extracted from the manganese oxides.

Cobalt is a constituent of vitamin B_{12}, or cobalamin, and cobalt is essential for nitrogen-fixing microorganisms. Three valence states of cobalt (I, II, and III) are involved in the enzymatic action of cobalamin. Plants acquire cobalt as a divalent cation. It is highly concentrated in roots relative to shoots (Marschner 1995). Average concentrations of cobalt in plants are about 1 µg/g on nonserpentine and 10 µg/g on serpentine soils (Baker and Brooks 1989). Species in at least 18 genera of flowering plants are hyperaccumulators with Co >1000 µg/g in above ground parts (Reeves and Baker 2000). Hyperaccumulators of cobalt are highly restricted geographically and none are native in western North America. All 26 known cobalt hyperaccumulators occur in the Shaban Copper Arc, Zaire (Baker et al. 2000). One plant (*Haumaniastrum robertii*) accumulates enough cobalt that it might be used for phytomining from soils rich in cobalt (Anderson et al. 1999). Chlorosis, or yellowing of leaves, caused by iron deficiency, is a typical symptom of cobalt toxicity (Adriano 2001). There are no known toxicities of cobalt in serpentine soils of western North America.

Nickel

Although the major commercial sources of nickel are sulfides, there is enough nickel in ultramafic ferromagnesian silicate rocks (peridotite and serpentinite) that it is concentrated to commercially exploitable quantities in highly weathered serpentine soils (chapter 6). Its concentration is generally about 0.5–3 g/kg in ultramafic rocks, where nickel is substituted for Mg^{2+} in olivine, pyroxenes, and serpentine. Nickel released from the weathering of these primary minerals can substitute for magnesium in smectites, and nickel released from the weathering of pyroxenes can substitute for magnesium in garnierite (Nahon et al. 1992). [Garnierite is a mixture of nickeliferous serpentine and nickeliferous talc (Trescases 1997)]. Mostly, however, nickel accumulates along with iron and manganese in oxides. Its affinity for manganese oxides is less than that of cobalt, but its affinity for organic ligands is greater. Nickel is most mobile in acid soils, in which it exists as a hydrated divalent cation (Uren 1992), but only small proportions of the Ni^{2+} are in solution or exchangeable by other cations. Nevertheless, it is more mobile in soil solutions than are trivalent ions of aluminum, iron, and chromium that form hydrated oxides. Lee et al. (2001) found nickel to be more mobile than aluminum and chromium on a landscape scale, and iron was relatively mobile where it could be reduced to divalent iron.

About three-quarters of the way through the twentieth century, nickel was found to be a component of urease in legumes and responsible for its activity in catalyzing the hydrolysis of urea to produce ammonia (Marschner 1995). Therefore, nickel is highly beneficial to legumes that use urea as a source of nitrogen. More recently, nickel was shown it to be essential in barley (*Hordeum vulgare* L.) plants. Nickel is required in such minute amounts that only after being deprived of nickel for three generations did barley plants lack viable seeds (Brown et al. 1987). Since its discovery in urease, nickel has been found in other enzymes—for example, in nickel CO dehydrogenase of anaerobic microorganisms (Hausinger 1994, Thauer 2001). Stewart and Schmidt (1998) suggest that a sole function for nickel in plants rather than its several functions in archaebacteria indicates that some of the catalytic functions of nickel may have been transferred to zinc. The role of nickel in the nutrition of organisms has apparently diminished with the addition of oxygen to the atmosphere about 2 Ga. Serpentine soils contain so much nickel that the threat to plants is toxicity, rather than deficiency.

Although nickel is toxic to many plants in many serpentine soils, there are great differences in susceptibility among different plant species, and even within species (Proctor and Baker 1994, Nagy and Proctor 1997). Calcium can ameliorate nickel toxicity, and magnesium can either ameliorate or exacerbate it under different conditions (Proctor and Baker 1994). Some plants tolerate high soluble-nickel concentrations in soils by limiting uptake, and others accumulate nickel and immobilize it. Nickel may be concentrated in the roots of plants that do not accumulate large quantities, but in nickel hyperaccumulators it is transported to leaves, possibly combined with histidine (Baker et al. 2000), where it is concentrated in epidermal and subepidermal tissues (Salt and Kramer 2000). Mesjasz-Pryzbylowicz et al. (2001) showed that nickel transported to apical leaves of *Berkheya coddii* was concentrated in the epidermis of the midrib (1250 µg/g), mesophyll (1700 µg/g), and the leaf margin (2200 µg/g). Heath et al. (1997) found by energy-dispersive X-ray microanalysis (EDAX) that nickel was concentrated in subsidiary cells around guard cells in pennycress (*Thlaspi montanum siskiyouense*), a nickel hyperaccumulator native to the Klamath Mountains, but Salt and Kramer (2000) suggested that this might be an artifact from specimen preparation in the EDAX procedure. Many plants accumulate nickel and other heavy metals in trichomes, or leaf hairs (Salt and Kramer

2000). Much of the nickel in hyperaccumulators is soluble in water extracts, in which it is bound to citric, malic, and malonic acids, and in extreme hyperaccumulators, some is even present as hydrated Ni^{2+} ions (Baker and Brooks 1989). Some nickel may be associated with pectate in cell walls, and within vacuoles it may be precipitated as an oxalate (Salt and Kramer 2000) or chelated with histidine (Kramer et al. 1996). Also, glutamine and proline are major nickel-binding ligands. The typical symptom of nickel toxicity is a chlorosis, or yellowing of leaves, caused by iron deficiency (Adriano 2001).

Most plants have foliar nickel concentrations in the 0.1–1 μg/g range; serpentine species that are not endemic to the substrate generally have higher concentrations, up to 100 μg/g; and nickel hyperaccumulators have concentrations in the 1000–10,000 μg/g range, or higher (Baker and Brooks 1989, Baker et al. 2000; table 8.5). Sap from *Sebertia acuminata*, a tropical tree in the Sapotaceae family, contained 11.2% nickel, or 25.7% when dried (Jaffré et al. 1976). Sagner et al. (1998) confirmed that most of the nickel in this tree is in the latex, where it is present mainly as a free or hydrated ion and as a citrate. Boyd and Jaffré (2001) found that very high nickel concentrations in plant litter under *S. acuminata* trees were reflected in comparatively high nickel concentrations in the surface soil (0–5 cm) layer, but not in subsurface soil (below 15 cm) layers.

Baker and Brooks (1989) tallied 145 hyperaccumulators of nickel in 22 plant families and 38 genera, but only a decade later Reeves and Baker (2000) reported more than 300 nickel-hyperaccumulating plants in 37 families. About 130 nickel-hyperaccumulating species were found in Cuba and neighboring islands (Reeves et al. 1999). Most of the nickel-hyperaccumulating species are in the Brassicaceae family in temperate regions, in the Flacourtaceae and the Violaceae families in tropical regions on the western margin of the Pacific Ocean, and in the Buxaceae and Euphorbiaceae families in the Caribbean area. In the Brassicaceae, large numbers of nickel-accumulating species are in the *Alyssum* genus, centered around Anatolia in Asia Minor, and in the *Thlaspi* genus, mainly in southern Europe. Reeves et al. (1983) analyzed herbarium specimens from 57 kinds of serpentine plants from California to Washington and found that only a milkwort jewel flower, *Streptanthus polygaloides*, in the Sierra Nevada and three varieties of pennycress, *Thlaspi montanum* (varieties *californicum*, *montanum*, and *siskiyouense*), in the Klamath Mountains were hyperacululators of nickel and that wedge-leaved violet, *Viola cuneata*, accumulated modest amounts of nickel (257 μg/g, mean of 37 samples). Kruckeberg et al. (1993) reported nickel hyperaccumulation in ruby sandwort (*Minuartia rubella*, Caryophyllaceae) of the northern Cascade Mountains (table 8–5), but the very high concentration of nickel was later found to have been a result of sample contamination (A. Kruckeberg, personal communication, 2003).

Kruckeberg and Reeves (1995) found in greenhouse studies with seven species of jewel flower (*Streptanthus*) sampled in California that only *S. polygaloides* is a nickel hyperaccumulator, although *S. barbiger* accumulated modest amounts of nickel from serpentine substrates. Reeves et al. (1981) found from herbarium specimens that milkwort jewel flower (*S. polygaloides*) accumulated about 2000–16,000 μg/g Ni in leaves and flowers, compared to 1000–6000 in stems and seed pods and about 2000 μg/g Ni in roots. Nicks and Chambers (1995) sampled milkwort jewel flower plants from three sites in the Red Hills of Tuolumne County, California, and found that the mean nickel concentration ranged from 5306 µg/g 1 week before flowering to 3372 µg/g at the onset of flowering and 2186 µg/g later. They explored possibilities of growing jewel flower to accumulate nickel for commercial harvesting and extraction, although a higher yielding nickel accumulator such as *Alyssum bertolonii* (southern Europe) or *Berkheya coddii* (southern Africa) might be more feasible for commercial production. Robinson et al. (1997) investigated the feasibility of *Berkheya coddii* for phytomining. Dodder (*Cuscuta californica*), a parasitic plant, accumulated 800 µg/g Ni when growing on a hyperaccumulating jewel flower in the Red Hills of Tuolumne County compared to only 11 µg/g Ni when growing on a nonhyperaccumulating lessingia (*Lessingia* sp., Asteraceae) in the same area, all on serpentine soils (Boyd et al. 1999). Although nickel-hyperaccumulator plants are generally restricted (endemic) to serpentine soils, Boyd and Martens (1998a) found that populations of pennycress (*T. montanum montanum*) sampled both on and off of serpentine soils accumulated comparable amounts of nickel (~3 mg/g) from nickel-amended soils in greenhouse studies—they suggested that this pennycress lacks different races or populations that respond differently to serpentine and nonserpentine soils.

Do plants derive any benefits from nickel hyperaccumulation? Boyd et al. (1994) found that milkwort jewel flower (*Streptanthus polygaloides*) with high concentrations of nickel was toxic to a bacterium (*Xanthomonas campestris*), an imperfect fungus (*Altenaria brassisicola*), and a powder mold (*Erisyphe polygoni*). Ghaderian et al. (2000) found that increased nickel concentrations in two nickel-accumulating species of *Alyssum* reduced seedling mortality from damping-off disease caused by the fungi *Pythium mamillatum* and *P. ultimum*. Larvae of cabbage butterfly (*Pieris rapae*) feeding on pennycress (*T. montanum montanum*) increased glucosinolate concentrations in the plants, but no toxic effects were identified (Boyd and Martens 1994). Boyd and Moar (1999) evaluated the effects of nickel concentration in leaves of milkwort jewel flower on larvae of butterfly (*Spodoptera exigua*) feeding on them and found that nickel concentrations >963 µg/g were lethal. Concentrations of an organic chemical toxic to cabbage butterfly larvae increased in milkwort jewel flower leaves in response to the larvae feeding on them, but nickel concentrations did not increase (Davis and Boyd 2000). It is still uncertain whether plants evolved with nickel-hyperaccumulating properties because of protection against pathogenic microorganisms and insects or whether this protection is an inadvertent effect of nickel hyperaccumulation. Aphids circumvented the nickel defense of milkwort jewel flower, presumably by feeding on phloem fluid that may have lower nickel concentrations than xylem sap (Boyd and Martens 1998b). Field trials showed that nickel hyperaccumulation provided some protection of milkwort jewel flower plants against grazing by insects, but that even nickel-rich leaves were largely consumed after 6 wk (Martens and Boyd 2002).

8.3 Plant Productivity

Relatively low biomass is readily apparent in most forest communities on serpentine soils because trees are generally farther from one another and smaller than in forest communities on other kinds of soils. Biomass differences from serpentine soils to others are less pronounced in shrub and grassland plant communities, even though there are distinct differences in plant species distributions related to these soil parent material differences. Measurements confirm these nonquantitative judgments of greater biomass and productivity on nonserpentine than on serpentine soils, with diminishing differences on drier sites. Productivity differences from serpentine to nonserpentine soils are minimized on dry soils, where water is the major factor limiting biomass production.

Productivity may be judged by the rate of biomass production or by the quantities of fruit or seed production. It is dependent on both soil-water and nutrient supplies. Also, productivity depends on the plant product used to gauge productivity; for example, some soils that are highly productive for timber may be practically nonproductive for annual crop plants. A common measure is the net primary productivity (NPP) of natural plant communities. NPP is the amount of biomass, or dry weight, added annually. It ranges from <1 Mg/ha in deserts to >30 Mg/ha in some evergreen tropical forests and in tropical wetlands. In intermediate stages of succession, NPP is commonly greater than it is in mature ecosystems, or those in steady state. In mature ecosystems, gains from NPP are balanced by losses from death and decay of plant tissues.

8.3.1 Grassland

McNaughton (1968) measured standing biomass and rates of production in dry grassland on serpentine and sandstone soils of Jasper Ridge in San Mateo County, California, in 1966. Standing biomass was greatest in May. It was about $0.5\,kg/m^2$ (5 Mg/ha) on sandstone soils, except only half that on southwest-facing slopes, and $0.2\,kg/m^2$ (2 Mg/ha) on serpentine soils. Annual productivity on sandstone soils was about $0.3\,kg/m^2$ (3 Mg/ha) on north-facing slopes and diminished to $0.2\,kg/m^2$ on southeast-facing slopes and $0.1\,kg/m^2$ on southwest-facing slopes. Annual productivity on serpentine soils was about $0.1\,kg/m^2$ (1 Mg/ha) on all slopes. On the warmest and driest slopes (southwest-facing), productivity was the same on both serpentine and nonserpentine soils.

8.3.2 Forest

Productivity, or rate of biomass production, is difficult to measure in forests. There have been few measurements of NPP in forests of western North America, and perhaps none on serpentine soils. Timber productivity, or the potential yield of timber, is somewhat closely related to a timber site index (TSI), which is based on the heights of representative trees at a certain age, or reference age. TSI is relatively easy to measure. Trees need not be measured at the reference age, because curves have been developed to show tree height–age relationships over a range of tree ages. Also, tables have been developed to estimate timber yield from TSI for many different kinds of trees. These tables assume that the basal area of trees is closely related to the TSI and timber-stand age. This assumption is not valid for serpentine soils, where forests generally have smaller basal areas than forests of the same TSI on other soils. The data of Alexander et al. (1989) indicate that basal area is poorly related to Dunning's site index in forests on serpentine soils in the eastern part of the Klamath Mountains. Basal areas were low, as expected, on most serpentine soils, but just as great on an old leached serpentine soil with a relatively high TSI (site 41, TSI = 59 m, or 194 ft, in 300 yrs) as on nonserpentine soils with a similar TSI.

Dunning's site index for old-growth trees is a convenient TSI, although it is seldom used in forest management. The index is convenient because it can be applied in old forests with any mixture of five different conifers that are common in the Sierra Nevada and much of the Klamath Mountains and California Coast Ranges (Dunning 1942). No yield tables have been developed for Dunning's site index for old forests because managed timber stands are much younger than 300 yr. To get a better index of productivity in forests on serpentine soils, for which timber yield tables are not applicable, Alexander et al. (1989) multiplied the Dunning site index (m/300 yr) by the basal area (m^2/ha) to obtain a timber yield index (TYI, m^3/ha yr). This is not an actual estimate of yield but an index that is expected to be more closely related to the productivity of old forests on serpentine soils than any TSI. It is applicable only in old forests in which nearly constant (or steady state) basal area is maintained as young trees replace deceased ones (those that succumb to wind, insects, diseases, etc.).

Alexander et al. (1989) found Dunning's old-growth TSI for trees on a sample of 22 serpentine soils in the eastern part of the Klamath Mountains to be closely related to exchangeable Ca/Mg ratios in the soils (fig. 8-4). Although no nonserpentine soils were included in the investigation, trees growing on nonserpentine soils generally have higher exchangeable Ca/Mg ratios and higher TSIs. Dunning's old-growth TSIs ranged from 23 m (75 feet in 300 yrs) on two serpentine soils with mean molar Ca/Mg ratios of 0.68 in surface and 0.22 in subsoils to 59 m (194 feet in 300 yrs) on a serpentine soil with molar Ca/Mg ratios of 5.9 in its surface and 3.1 in its subsoil. Enough magnesium has been leached from the most productive forest soil that its TSI is comparable to TSIs on nonserpentine soils with similar AWCs and climate. The relationship of the TYI to the exchangeable Ca/Mg ratio in the upper 30 cm of the sample of 22 soils is highly significant ($r^2 = .867$, $n = 22$), even though there were no controls on AWC or climate. AWCs ranged from 4 to 17 cm, altitude from 720 to 2140 m, and mean annual precipitation from 75 to 175 cm. Adding AWC as a second independent variable, after exchangeable Ca/Mg ratio, raised the coefficient of determination to 0.935.

Timber productivities on serpentine soils in the central part of the Klamath Mountains are closely related to AWC and also to exchangeable Ca/Mg ratios. A representative sample of serpentine soils from 600 m to >1200 m altitude with 120–180 cm of precipitation annually indicates how Dunning's TSI increases slightly (30–37 m/300 yr) with increasing soil depth (shallow to deep) and AWC; then it increases to 48 m/300 yr on old soils that have been leached of excess magnesium (Alexander 1988a, 2004b). Because only two of the soils were analyzed for exchangeable cations, pH was assumed to be an indicator of leaching in these soils. A shallow soil with mean pH values of 6.4 in the surface and 7.0 in its subsoil had exchangeable Ca/Mg ratios of 0.24 in the surface and 0.19 in its subsoil. A very deep Ultisol (Xeric Haplohumult) with mean pH values of 6.0 in the surface and 5.7 in its subsoil had exchangeable Ca/Mg ratios of 5.8 in the surface and

1.3 in its subsoil. The very deep Ultisol, which is more acid than the other soils, especially in the subsoil, is leached of magnesium sufficiently to raise the exchangeable Ca/Mg ratio substantially and make the soil favorable for timber production.

A chronosequence of serpentine soils on outwash terraces along Rough and Ready Creek where it discharges into the Illinois Valley in Josephine County, Oregon, indicates how soils and plant communities evolve and tree growth increases as the soils age (Alexander 2004b). Soils on two lower terraces are assumed to be younger than the last major glacial stage, which ended between 12 and 14 ka. They are extremely cobbly, slightly acid soils that support open shrub and herbaceous plant communities with only sparse tree cover. Soils on an intermediate terrace are assumed to date from the penultimate major glacial stage, which ended about 125,000 years ago. They are very cobbly, slightly acid soils with moderately acid surface horizons that support open forest with semi-dense shrub cover. Soils on the highest terrace are assumed to date from an older major glacial stage and may be on the order of 1 Ga old. They are cobbly, moderately acid soils, with strongly acid surface horizons, that support dense forest. Dunning's TSI increases significantly ($\alpha < 0.05$) from 22 m (72 feet in 300 years) on the younger terraces to 32 m (105 feet in 300 years) on the intermediate one as the AWC doubles. Also, the TSI increases significantly ($\alpha < 0.05$) from the intermediate to the highest terrace as soil pH declines. The pH decline is concomitant with steady increases in exchangeable Ca/Mg ratio from 0.7 to 1.6 in surface soils and from 0.3 to 1.2 in subsoils as excess magnesium is leached from the soil. Although the TSI increase as cobbles and pebbles weather to produce finer particles and increase the AWC within $\sim 10^4$ years is significant, the TSI increase to 55 m (180 feet in 300 years) as the older soils are leached of excess magnesium is greater.

Most soils in leaching environments decline in fertility and productivity as they age (Kronberg and Nesbitt 1981). Minerals weathered from soil parent materials are the main sources of phosphorus, potassium, calcium, and most other elements required by plants. As these elements are released by weathering, portions of each are used by plants and other organisms and portions are leached from soils. Over thousands or millions of years the required elements are gradually depleted and soils become infertile. Serpentine soils, in contrast, are infertile because they have large amounts of magnesium and low exchangeable Ca/Mg ratios. Therefore, leaching which removes excess magnesium increases the fertility of serpentine soils, at least for natural plant communities in the examples from the eastern, central, and northern parts of the Klamath Mountains. Although trees with their extensive root systems grow well on old, leached serpentine soils, annual crops may grow poorly on them. With extreme age (millions of years) timber productivity may decline on very old serpentine soils. Low productivity in very old serpentine soils might be because of extremely low available plant nutrient concentration or because of nickel toxicity.

8.4 Site Revegetation

Reestablishing plant cover on serpentine soils that have been disturbed and denuded is a challenge, because of the low fertility of the substrate. Five trials in the California Coast

Ranges and Klamath Mountains indicate some of the challenges and possibilities for success over a range of climates from warm and dry with about 40 cm/year rain to cool with 200 or 300 cm/yr precipitation (snow and rain).

Koide and Mooney (1987) planted seeds from native grassland plants in serpentine subsoil and topsoil material that was stockpiled for covering sanitary landfill in Santa Clara County, California. Revegetation was successful only in topsoil material. Most of the plants that grew on the experimental plots and dominated the plant cover were volunteers from seeds in the topsoil material, rather than planted ones. Fertilization with sodium acid phosphate increased plant cover and biomass production substantially.

Smith and Kay (1986) seeded native and nonnative herbs on serpentine subsoil and topsoil near the McLaughlin Gold Mine in Lake County, California, where the mean annual precipitation is about 75–80 cm. To establish plants and ground cover quickly to minimize soil erosion, they fertilized with nitrogen, phosphorus, potassium, sulfur, and calcium (gypsum). Blando brome, red brome, and lana vetch provided ground cover quickly. Fertilization was necessary, particularly on subsoil material. They suggested that fertilization should be continued after the first year to maintain the ground cover, and that native perennial grasses could eventually replace the non-native species.

O'Dell and Claassen (2006 and unpublished data) added compost and fertilizer to tectonically sheared serpentinite detritus from a barren highway cutslope in Colusa County, California, attempting to find a combination that would favor annual and perennial native grass species over non-native species. There was enough calcium in the compost that, after adding sufficient compost and fertilizer containing nitrogen, phosphorus, and potassium to increase grass productivity, adding more calcium was not beneficial to the plants. Unfortunately, no compost and fertilizer combination was found that was more beneficial for the native than for the non-native grasses. An alternative approach to establishing native plants on warm serpentine barren cutslopes was suggested that involved the preliminary establishment of native forbs (dicots) and shrubs, while suppressing nonnative monocots with herbicides, followed by the introduction of native grasses to supplement the shrub and forb cover. Yarrow (*Achillea millifolium*) was found to have serpentine and nonserpentine ecotypes that responded quite differently in the serpentine detritus; survivial of serpentine ecotype seedlings was good in unamended serpentine detritus whereas seedlings of the nonserpentine ecotype did not survive (O'Dell and Claassen 2006).

Hoover et al. (1999) planted native trees and shrubs to revegetate serpentine soils in Humboldt County, California, disturbed by mining. Seedlings were grown from seeds collected from local populations of Jeffrey pine (*Pinus jeffreyi*), Port Orford cedar (*Chamaecyparis lawsoniana*), coffeeberry (*Rhamnus californica*), and azalea (*Rhododendron occidentale*) and planted in both dry sites and seeps. They had controls, fertilizer treatments (nitrogen, phosphorus, and potassium), and topsoil additions. Plant survival was good (survival > 70%) for all species, with and without fertilizer and topsoil treatments, except for azalea. Survival of azalea was good only on fertilized plots.

Bertenshaw et al. (1983) investigated the feasibility of using non-native grasses and native trees and shrubs for short- and long-term revegetation of disturbed serpentine soils and stabilization of proposed mine tailings in the Gasquet Mountain area of Del Norte County, California. The soils are serpentine Inceptisoils, Alfisols, and Ultisols with about 200–300 cm/year precipitation. They found that shrub tanoak (*Lithocarpus densiflorus echinoides*) and coffeeberry are suitable for transplanting from adjacent sites and that knobcone pine (*Pinus attenuata*), lodgepole pine (*P. contorta*), Port Orford cedar, and coffeeberry can be regenerated successfully from nursery stock.

These five examples indicate that disturbed serpentine soils can be revegetated with herbaceous plants in dry areas and with native trees and shrubs in wetter areas. Grasses and forbs may require fertilization, particularly with phosphorus, for good growth. Hoover et al. (1999) and Koide and Mooney (1987) avoided fertilizers containing calcium because it has been reported to make soils more favorable for non-native species. Annual legumes (*Lotus wrangelianus* and *Astragalus gambelianus*) were a small, but important, component of plant cover on revegetated plots in Santa Clara County, California (Koide and Mooney 1987).

8.5 Nonvascular Plants (Bryophytes)

Although vascular plants are the major living components of terrestrial geoecosystems, bryophytes may have more biomass and fix more carbon than vascular plants in some Arctic and Antarctic geoecosystems. Bryophytes lack vascular conducting tissues, xylem and phloem, that carry water from roots to leaves and that distribute the metabolic products through vascular plants. There are three phyla of Bryophytes: Bryophyta (mosses), Hepatophyta (liverworts), and Anthocerophyta (hornworts). Bryophytes have no roots, but only rhizoids to bind them to the soils, rocks, or plants that they inhabit. Lack of roots requires bryophytes to rely largely on precipitation and atmospheric fallout for water and mineral elements they need to grow. They become desiccated during drought but survive in a dehydrated condition. When rewetted, which can take a few minutes or a few hours for different species (Larson 1981), bryophytes may quickly become active again, respiring within an hour and carrying on photosynthesis within a few hours. Thus bryophytes can live in some very dry habitats, although they are more active and abundant in wet habitats. They require practically the same chemical elements as vascular plants.

The cell walls of bryophytes have many negatively charged sites, giving them very high capacities for attracting metallic cations from precipitation and dust that fall from the atmosphere.

Bryophytes commonly colonize sites contaminated by metals and survive on some highly toxic substrates. Many species accumulate nickel, but accumulation of chromium is uncommon, even for bryophytes growing over serpentine rocks. Lee et al. (1977) reported that an epiphytic moss (*Acrobryopsis longissima*) common on some tropical trees of serpentine soils contained up to 0.75% (mean 0.47%) chromium in its ash (mean 384 µg/g based on dry weight). Shacklette (1965) found that even in nonserpentine areas, bryophytes accumulate more chromium and nickel than vascular plants, but less cobalt, calcium, magnesium, and phosphorus. The mean Ca/Mg ratio in the 38 moss samples that he analyzed was 2.8 mol/mol. Sigal (1975) found the molar Ca/Mg ratio in *Didymodon tophaceus* to be 0.01 on a serpentine substratum and 0.13 on a nonserpentine substratum in the northern California Coast Ranges. Lounamaa (1956) found much more cobalt, and very much more nickel, in mosses growing on ultramafic rocks than in those on other kinds of rocks.

Bryophytes grow on rocks, soils, plants, and detritus from plants (logs and leaves). Some species grow only on rocks, but not necessarily on specific kinds of rocks, although some species are restricted to rocks and soils containing calcium carbonate. Sigal (1975) failed to find any endemic species at five serpentine sites dispersed through the California Coast Ranges. No moss species, other than *Barbula vinealis* and *Grimmia tricophylla*, were found at more than two of the five sites. Sigal (1975) mentioned three common soil mosses that were not found on the serpentine sites. Intermediate alluvial terraces in a sequence of them on serpentine (ultramafic rock) outwash along Rough and Ready Creek in Josephine County, Oregon, have about one-quarter of the ground covered by moss. The ground moss is predominantly *Racomitrium lanuginosum* among sparse trees and common shrubs on the lower intermediate terrace and predominantly *Aulacomnium androgynum* among common trees and semi-dense shrubs on the higher intermediate terrace. These two mosses are not restricted to serpentine habitats. Only *Pseudoleskeella serpentinensis* is a serpentine endemic in the region, restricted to serpentine (Shevock 2003).

Five of the more common liverworts observed by Sigal (1975) at five dispersed serpentine sites in the California Coast Ranges were *Ricca dictyospora*, *R. crystallina*, *Fossombronia* sp, *Cephaloziella divaricata*, and *Cephaloziella* sp. Serpentine substrates are not known to have distinctive bryophyte flora (Lepp 2001). Perhaps *Pseudoleskeella serpentinensis* is a rare exception.

Mosses and liverworts are well represented on rocks and soils in Cuba (Motito et al. 2004). In Cuba, 61% of the bryophyte species occur in serpentine areas. They represent 195 genera and 582 species. None of them is endemic on serpentine substrates.

Part II References

Abin, L., O. Coto, Y. Gómez, S. Cortes, and J. Marrero. 2004. Characterization of indigenous microbiota from nickeliferous deposits of Moa, Cuba. Pages 205–207 *in* R.S. Boyd, A.J.M. Baker, and J. Proctor (eds.), Ultramafic Rocks: Their Soils, Vegetation, and Fauna. Science Reviews, St. Albans, Herts, UK.

Abou-Shanab, R.A., J.S. Angle, T.A. Delorme, R.L. Chaney, P van Berkum, H. Moawad, and H.A. Ghozlan. 2003. Rhizobacterial effects on nickel extraction from soil and uptake by *Alyssum murale*. New Phytologist 158: 219–224.

Adamo, P., A. Marchetiello, and P. Violante. 1993. The weathering of mafic rocks by lichens. Lichenologist 25: 285–297.

Adriano, D.C. 2001. Trace Elements in Terrestrial Environments. Springer-Verlag, Berlin.

Alexander, E.B. 1985. Rates of soil formation from bedrock or consolidated sediments. Physical Geography 6: 25–42.

Alexander, E.B. 1988a. Morphology, fertility, and classification of productive soils on serpentinized peridotite in California (U.S.A.). Geoderma 41: 337–351.

Alexander, E.B. 1988b. Rates of soil formation: implications for soil loss tolerance. Soil Science 145: 37–45.

Alexander, E.B. 1993. Gabbro and its soils. Fremontia 21(4): 8–10.

Alexander, E.B. 1995. Silica cementation in serpentine soils in the humid Klamath Mountains, California. Soil Survey Horizons 36: 154–159.

Alexander, E.B. 1999. Cation-exchange capacities of some serpentine soils in relation to precipitation. Soil Survey Horizons 40: 89–93.

Alexander, E.B. 2000. Reexamination of soil redness bedrock specific gravity relationships. Soil Survey Horizons 41: 24–26.

Alexander, E.B. 2003. Trinity Serpentine Soil Survey. Shasta-Trinity National Forest, Redding, California. Unpublished manuscript, revised in 2004. Available: http://www.fs.fed.us/r5/shastatrinity/publications/trinity-serpentine.shtml

Alexander, E.B. 2004a. Serpentine soil redness, differences among peridotite and serpentinite materials, Klamath Mountains, California. International Geology Review 46: 754–764.

Alexander, E.B. 2004b. Varieties of ultramafic soil formation, plant cover, and productivity. Pages 9–17 in Boyd, R.S., A.J.M. Baker, and J. Proctor (eds.). Ultramafic Rocks: Their Soils, Vegetation, and Fauna. Science Reviews, St. Albans, Herts, UK.

Alexander, E.B., C. Adamson, R.C. Graham, and P.J. Zinke. 1990. Mineralogy and classification of soils on serpentinized peridotite of the Trinity ophiolite, California. Soil Science 149: 138–143.

Alexander, E.B., C. Adamson, P.J. Zinke, and R.C. Graham. 1989. Soils and conifer productivity on serpentinized peridotite of the Trinity ophiolite, California. Soil Science 148: 412–423.

Alexander, E.B., C.L. Ping, and P. Krosse. 1994. Podzolization in ultramafic materials in southeast Alaska. Soil Science 157: 46–52.

Alexander, E.B., and P.J. Zinke. 1994. Serpentine soil fertility, foliar analyses, and conifer productivity in the Klamath Mountains, California. Fifteenth International (World) Congress of Soil Science Transactions 5b: 298–299.

Allardice, W.R., S.S. Munn, E.L. Begg, and J.I. Mallory. 1983. Laboratory Data and Descriptions for Some Typical Pedons of California Soils. Vol. I. Central and Southern Sierras. Land, Air, and Water Resources. University of California, Davis.

Allaway, W. 1976. Perspectives on molybdenum in soils and plants. Pages 319–339 in W.R. Chappell and K.K. Petersen (eds.), Molybdenum in the Environment. Dekker, New York.

Amir, H., and R. Pineau. 2004. Release of Co and Ni by microbial activity in New Caledonian ultramfic soils. Pages 209–214 in R.S. Boyd, A.J.M. Baker, and J. Proctor. Ultramafic Rocks: Their Soils, Vegetation, and Fauna. Science Reviews, St. Albans, Herts, UK.

Anderson, C.W.N., R.R. Brooks, A. Chiarucci, C.J. LaCoste, M. Leblanc, B.H. Robinson, R. Simcock, and R.B. Stewart. 1999. Phytomining for nickel, thallium, and gold. Journal of Geochemical Exploration 67: 407–415.

Arianoutsou, M., P.W. Rundel, and W.L. Berry. 1993. Serpentine endemics as biological indicators of soil elemental concentrations. Pages 179–189 in B. Markert (ed.), Plants as Biomonitors: Indicators for Heavy Metals in the Terrestrial Environment. VCH Publishers, Weinheim, Germany.

Baes, C.F., and R.E. Mesmer. 1976. The Hydrolysis of Cations. Wiley, New York.

Baker, A.J.M., and R.R. Brooks. 1989. Terrestrial higher plants which hyperaccumulate metallic elements: a review of their distribution, ecology, and phytochemistry. Biorecovery 1: 81–126.

Baker, A.J.M., S.P. McGrath, R.D. Reeves, and J.A.C. Smith. 2000. Metal hyperaccumulator plants: a review of the ecology and physiology of a biological resource for phytoremediation of metal-polluted soils. Pages 85–107 *in* N. Terry and G. Bañuelos (eds.), Phytoremediation of contaminated soil and water. Lewis Publishers, Boca Raton, FL.

Baltzo, D.E. 1970. A study of the lichens of Mount Diablo State Park. M.A. thesis, San Francisco State University.

Baltzo, D.E. 1989. Lichens of Mount Diablo State Park, Contra Costa County, California. Mycotaxon 34: 37–47.

Bargagli, R. 1998. Trace Elements in Terrestrial Plants. Springer, Berlin.

Barro, A., R. Nuñez, and K. Rodrígues. 2004. The Lepidoptera of plant formations on Cuban ultramafics: a preliminary analysis. Pages 221–226 *in* R.S. Boyd, A.J.M. Baker, and J. Proctor (eds.), Ultramafic Rocks: Their Soils, Vegetation, and Fauna. Science Reviews, St. Albans, Herts, UK.

Bartlett, R.J., and B.R. James. 1988. Mobility and bioavailability of chromium in soils. Pages 267–304 *in* J.O. Nriagu and E. Nieboer (eds.), Chromium in the Natural and Human Environments. Wiley, New York.

Bates, T.F. 1945. Origin of the Edwin clay, Ione, California. Geological Society of America Bulletin 56: 1–38.

Begg, E.L., W.R. Allardice, S.S. Munn, and J.I. Mallory. 1984. Laboratory Data and Descriptions for Some Typical Pedons of California Soils. Vol. II. North Coast. Land, Air, and Water Resources, University of California, Davis.

Begg, E.L., W.R. Allardice, S.S. Munn, and J.I. Mallory. 1985. Laboratory Data and Descriptions for Some Typical Pedons of California Soils. Vol. III. Southern Cascades and Northern Sierras. LAWR, University of California, Davis.

Beinroth, F.H. 1982. Some highly weathered soils of Puerto Rico, 1. Morphology, formation, and classification. Geoderma 27: 1–73.

Bennett, H.H., and R.V. Allison. 1928. The Soils of Cuba. Tropical Plant Research Foundation, Washington, D.C.

Benson, D.R., B.D. Vander Heuvel, and D. Potter. 2004. Actinorhizal symbiosis: diversity and biogeography. Pages 97–127 *in* M. Gillings and A. Holmes (eds.), Plant Microbiology. BIOS Scientific Publishers, London.

Bertenshaw, J., P. Zinke, and D. Rhodes. 1983. Summary of Gasquet Mountain Revegetation Program. Dames and Moore, Los Angeles, CA.

Beveridge, T.J. 1995. Hyperaccumulation of metals by prokaryotic microorganisms including the blue-green algae (Cyanobacteria). Pages 133–151 *in* R.R. Brooks (ed.), Plants that Hyperaccumulate Heavy Metals. CAB International, Wallingford, UK.

Bibak, A., J. Gerth, and O.K. Borggaard. 1995. Retention of cobalt by an Oxisol in relation to the content of iron and manganese oxides. Communications in Soil Science and Plant Analysis 26: 785–798.

Blanchard, C.L., and K.A. Tonnesson. 1993. Precipitation chemistry measurements from the California acid deposition monitoring program, 1985–1990. Atmospheric Environment 27A: 1755–1763.

Bogden, J.D. 2000. The essential elements and minerals. Pages 3–9 *in* J.D. Bogden and L.M. Klevay (eds.), Clinical Nutrition of the Essential Trace Elements and Minerals. Humana Press, Totowa, NJ.

Bonifacio, E., and E. Barberis. 1999. Phosphorous dynamics during pedogenesis on serpentinite. Soil Science 164: 960–968.

Boulton, A.M., B.A. Jaffre, and K.M. Snow. 2003. Effects of harvester ant (*Mesor andrei*) on richness and abundance of soil biota. Applied Ecology 23: 257–265.

Boulton, A.M., K.F. Davies, and P.S. Ward. 2004. Species richness, abundance, and composition of ground-dwelling ants in northern California grasslands: the role of plants, soil, and grazing. Environmental Entomology 34: 96–104.

Bowen, H.J.M. 1979. Environmental Chemistry of the Elements. Academic Press, London.

Boyd, P. 2004. Ironing out algal issues in the southern ocean. Science 304: 396–397.

Boyd, R.S., and Jaffré, T. 2001. Phytoenrichment of soil Ni content by *Sebertia acuminata* in New Caledonia and the concept of element allelopathy. South African Journal of Science 97: 535–538.

Boyd, R.S., and S.N. Martens. 1994. Nickel hyperaccumulated by *Thlaspi montanum* var. *montanum* is acutely toxic to an insect herbivore. Oikos 70: 21–25.

Boyd, R.S., and S.N. Martens. 1998a. Nickel hyperaccumulation by *Thlaspi montanum* var. *montanum* (Brassicaceae): a constitutive trait. American Journal of Botany 85: 259–265.

Boyd, R.S., and S.N. Martens. 1998b. The significance of metal hyperaccumulation for biotic interactions. Chemoecology 8: 1–7.

Boyd, R.S., S.N. Martens, and M.A. Davis. 1999. The nickel hyperaccumulator *Streptanthus polygaloides* (Brassicaceae) is attacked by the parasitic plant *Cuscuta californica* (Cuscutaceae). Madroño 46: 92–99.

Boyd, R.S., and W.J. Moar. 1999. The defensive function of Ni in plants: response of the polyphagus herbivore *Spodoptera exigua* (Lepidoptera: Noctidae) to hyperaccumulator and accumulator species of *Streptanthus* (Brassicaceae). Oecologia 118: 218–224.

Boyd, R.S., J.J. Shaw, and S.N. Martens. 1994. Nickel hyperaccumulation defends *Streptanthus polygaloides* (Brassicaceae) against pathogens. American Journal of Botany 8: 294–300.

Boyd, R.S., and M.A. Wall. 2001. Responses of generalist predators fed high-Ni *Melanotrichous boydi* (Heteroptera: Miridae): elemental defense against the third trophic level. American Midland Naturalist 146: 186–198.

Boyd, R.S., M.A. Wall, and M.A. Davis. 2004. The ant-mimetic bug, *Coquillettia insignis* (Heteroptera: Miridae), feeds on the Ni hyperaccumulator plant, *Streptanthus polygaloides* (Brassicaceae). Pages 227–231 *in* R.S. Boyd, A.J.M. Baker, and J. Proctor. Ultramafic Rocks: Their Soils, Vegetation, and Fauna. Science Reviews, St. Albans, Herts, UK.

Brooks, R.R. 1987. Serpentine and Its Vegetation. Dioscorides Press, Portland, OR.

Brown, M.T., and I.R Hall. 1989. Metal tolerance in fungi. Pages 95–104 *in* A.J. Shaw (ed.), Heavy Metal Tolerance in Plants: Evolutionary Aspects. CRC Press, Boca Raton, FL.

Brown, M.J.F., and K.G. Human. 1997. Effects of Harvester ants on plant species distribution and abundance in a serpentine grassland. Oegologia 112: 237–243.

Brown, P.H., R.M. Welch, and E.E. Cary. 1987. Nickel: a micronutrient essential to higher plants. Plant Physiology 85: 801–803.

Burch, S.H. 1968. Tectonic emplacement of the Burrow Mountain ultramafic body, Santa Lucia Range, California. Geological Society of America Bulletin 79: 527–544.

Burd, G.I., D.G. Dixon, and B.R. Glick. 2000. Plant growth-promoting bateria that decrease heavy metal toxicity in plants. Canadian Journal of Microbiology 46: 237–245.

Burt, R., M. Fillmore, M.A. Wilson, E.R. Gross, R.W. Langridge, and D.A. Lammers. 2001. Soil properties of selected pedons on ultramafic rocks in Klamath Mountains, Oregon. Communications in Soil Science and Plant Analysis 32: 2145–2175.

Cannon, H.L. 1963. The biochemistry of vanadium. Soil Science 96: 196–204.

Chesterman, C.W., and J.H. Bright. 1979. Nickel and cobalt in California. California Geology 32: 266–274.

Chiariello, N., J.C. Hickman, and H.A. Mooney. 1982. Endomycorrhizal role for intraspecific transfer of phosphorous in a community of annual plants. Science 217: 941–943.

Cleaves, E.T., D.W. Fisher, and O.P. Bricker. 1974. Chemical weathering of serpentinite in the eastern Piedmont of Maryland. Bulletin of the Geological Society of America 85: 437–444.

Conard, S.G., A.E. Jaramillo, K. Cromak, and S. Rose. 1985. The role of *Ceanothus* in western forest ecosystems. USDA Forest Service, Gen. Tech. Report PNW-182.

Cornell, R.M., and U. Schwertmann. 1996. The Iron Oxides. Wiley-VCH, Weinheim.

Crans, D.C., S.S. Amin, and A.D. Keramidas. 1998. Chemistry of revelance to vanadium in the environment. Pages 73–95 *in* J.O. Nriagu (ed.), Vanadium in the Environment. Part I: Chemistry and Biochemistry. Wiley, New York.

Craw, D., C.A. Landis, and P.I. Kelsey. 1987. Authigenic chrysotile formation in the matrix of Quaternary debris flows, northern Southland, New Zealand. Clays and Clay Minerals 35: 43–52.

Davis, M.A., and R.S. Boyd. 2000. Dynamics of Ni-based defense and organic defenses in the Ni hyperaccumulator, *Streptanthus polygaloides* (Brassicaceae). New Phytologist 146: 211–217.

Davison, G., C.L. Lambie, W.M. James, M.E. Skene, and K.R. Skene. 1999. Metal content in insects associated with ultramafic and non-ultramafic sites in the Scottish Highlands. Ecological Entomology 24: 396–401.

Deer, W.A., R.A. Howie, and J. Zussman. 1966. Introduction to the Rock-Forming Minerals. Wiley, New York.

Dirven, J.M.C., J. van Schuylenborgh, and N. van Breemen. 1974. Weathering of serpentinite in Matanzas Province, Cuba: mass transfer calculations and irreversible reaction pathways. Soil Science Society of America Journal 40: 901–907.

Dunning, D. 1942. A site classification for mixed selection conifer forests of the Sierra Nevada. USDA, Forest Service, Pacific Southwest Forest and Range Experiment Station, Forest Research Note 28.

Effendi, S., S. Miura, N. Tanaka, and S. Ohta. 2000. Serpentine soils on catena in the southern part of East Kalimantan, Indonesia. Pages 79–88 *in* E. Guhardj, M. Fatawi, M. Sutisna, and T. Mori (eds.), Rainforest Ecosystems of East Kalimantan. Springer, Tokyo.

Ehrlich, H.L. 1996. Geomicrobiology. Marcel Dekker, New York.

Ehrlich, P.R., R.R. White, M.C. Singer, S.W. McKechnie, and L.E. Gilbert. 1975. Checkerspot butterflies: a historical perspective. Science 188: 221–228.

Elam, D.R., B. Goettle, and D.H. Wright. 1998. Draft Recovery Plan for Serpentine Soil Species of the San Francisco Bay Area. U.S. Fish and Wildlife Service, Portland, OR.

Epstein, E., and A.J. Bloom. 2005. Mineral Nutrition of Plants. Sinauer, Sunderland, MA.

Evans, H.J., and L.E. Barber. 1977. Biological nitrogen fixation for food and fiber production. Science 197: 332–339.

Fendorf, S., B.W. Wielinga, and C.M. Hansel. 2000. Chromium transformations in natural environments: the role of biological and abiological processes in chromium (VI) reduction. International Geology Review 42: 691–701.

Fiedler, P.L. 1985. Heavy metal accumulation and the nature of edaphic endemism in the genus *Calochortus* (Liliaceae). American Journal of Botany 72: 1712–1718.

Finlay, B.J. 2002. Global diversity of free-living microbial eukaryote species. Science 296: 1061–1063.

Fisher, B.L. 1997. A comparison of ant assemblages (Hymenoptera, Formicidae) on serpentine and nonserpentine soils in northern California. Insectes Sociaux 44: 23–33.

Foose, M.P. 1992. Nickel—Mineralogy and chemical composition of some nickel-bearing laterites in southern Oregon and northern California. U.S. Geological Survey Bulletin 1877(E): 1–24.

Forister, M. 2004. Oviposition preference and larval performance within a diverging lineage of lycaenid butterflies. Ecological Entomology 29: 264–272.

Foster, C.M., and G.K. Lang. 1982. Soil Survey of Klamath National Forest Area, California. USDA Forest Service, Klamath National Forest, Yreka, CA.

Fox, R.L. 1982. Some highly weathered soils of Puerto Rico, 3. Chemical properties. Geoderma 27: 139–176.

Gasser, U.G., and R.A. Dahlgren. 1994. Solid-phase speciation and surface association of metals in serpentine soils. Soil Science 158: 409–420.

Gerloff, G.C., and K.A. Fishbeck. 1969. Quantitative cation requirements of several green and blue-green algae. Journal of Phycology 5: 109–114.

Gervais, B.R., and A.M. Shapiro. 1999. Distribution of edaphic-endemic butterflies in the Sierra Nevada of California. Global Ecology and Biogeography 8: 151–162.

Ghaderian, Y.S.M., A.J.E. Lyon, and A.J.M. Baker. 2000. Seedling mortality of metal hyperaccumulating plants resulting from damping-off by *Pythium* spp. New Phytologist 146: 219–224.

Gilbert, O. 2000. Lichens. HarperCollins, London.

Gilkes, R.J., and R.M. McKenzie. 1988. Geochemistry of manganese in soil. Pages 23–58 *in* R.D. Graham, R.J. Hannam, and N.C. Uren (eds.), Manganese in Soils and Plants. Kluwer Academic, Dordrecht.

Golightly, J.P. 1981. Nickeliferous laterite deposits. Economic Geology 75: 710–735.

Gonçalves, S.G., M.A. Martins-Loução, and H. Freitas. 2001. Arbuscular mycorrhizas of *Festuca brigantina*, an endemic serpentinophyte from Portugal. South African Journal of Science 97: 571–572.

Gough, L.P., and L.L. Jackson. 1988. Determining baseline element composition of lichens. Water, Air, and Soil Pollution 38: 169–180.

Gough, L.P., G.R. Meadows, L.L. Jackson, and S. Dudka. 1989. Biogeochemistry of a highly serpentinized, chromite-rich ultramafic area, Tehama County, California. U.S. Geological Survey Bulletin 1901.

Graustein, W.C., K. Cromack, and P. Sollins. 1977. Calcium oxalate: occurrence in soils and effect on nutrient and geochemical cycles. Science 798: 1252–1254.

Hale, M.E. 1983. The Biology of Lichens. Arnold, Melbourne, Victoria.

Harrison, S., and A.M. Shapiro. 1988. Butterflies of northern California serpentines. Fremontia 15(4): 17–20.

Hausinger, R.P. 1994. Nickel enzymes in microbes. Science of the Total Environment 148: 157–166.

Heath, S.M, D. Southworth, and J.A. D'Allura. 1997. Localization of nickel in epidermal subsidiary cells of leaves of *Thlaspi montanum* var. *siskiyouense* (Brassicaceae) using energy-dispersive X-ray microanalysis. International Journal Plant Sciences 158: 184–188.

Hersman, L.E. 2000. The role of sidophores in iron oxide dissolution. Pages 145–157 *in* D.R. Lovely (ed.), Environmental Microbe-Metal Interactions. ASM Press, Washington, DC.

Hewitt, E.J. 1953. Metal interrelationships in plant nutrition. 1. Effects of some metal toxicities on sugar beet, tomato, oat, potato, and narrowstem kale grown in sand culture. Journal of Experimental Botany 4: 59–64.

Hickman, J.C. (ed.), 1993. The Jepson Manual, Higher Plants of California. University of California Press, Berkeley.

Hobbs, R.J. 1985. Harvester ant foraging and plant species distribution in annual grassland. Oecologia 67: 519–523.

Hobbs, R.J., S.L. Gulmon, V.J. Hobbs, and H.A. Mooney. 1988. Effects of fretilizer addition and subsequent gopher disturbance on a serpentine annual grassland community. Oecologia 75: 291–295.

Hobbs, R.J., and H.A. Mooney. 1985. Community and population dynamic of serpentine grassland annuals in relation to gopher disturbance. Oecologia 67: 342–351.

Hobbs, R.J., and H.A. Mooney. 1991. Effects of rainfall and gopher disturbance on serpentine annual grassland dynamics. Ecology 72: 59–68.

Hobbs, R.J., and H.A. Mooney. 1995. Spatial and temporal variability in California annual grassland: results from a long-term study. Journal of Vegetation Science 6: 43–56.

Hoover, L.D., J.D. McRae, E.A. McGee, and C. Cook. 1999. Horse Mountain Botanical Area serpentine revegetation study. Natural Areas Journal 19: 361–367.

Hopkin, S.P. 1989. Ecophysiology of Metals in Terrestrial Invertebrates. Elsevier, Amsterdam.

Hopkins, N.A. 1987. Mycorrhiza in a California serpentine grassland community. Canadian Journal of Botany 65: 484–487.

Hopkins, L.L., H.L. Cannon, A.T. Miesch, R.M. Welch, and F.H. Nielsen. 1977. Vanadium. Pages 93–107 in W. Mertz (ed.), Geochemistry and the Environment, vol. 3. The Relation of Other Trace Elements to Health and Disease. National Academy of Sciences, Washington, DC.

Hotz, P.E. 1964. Nickeliferous laterites in southwestern Oregon and northwestern California. Economic Geology 59: 355–396.

Hubers, H., S. Borges, and M. Alfaro. 2003. The oligochaetofauna of the Nipe soils in the Maricao State Forest, Puerto Rico. Pedobiologia 47: 475–478.

Hungate, B.A., C.H. Jaeger, G. Gamara, F.S. Chapin, and C.F. Field. 2000. Soil microbiota in two annual grasslands: responses to elevated atmospheric CO_2. Oecologia 124: 589–598.

Hunter, J.C. and J.E. Horenstein. 1992. The vegetation of the Pine Hill area (California) and its relation to substratum. Pages 197–206 in A.J.M. Baker, J. Proctor, and R.D. Reeves (eds.), The Vegetation of Ultramafic (Serpentine) Soils. Intercept, Andover, Hampshire, UK.

Jackson, J.A. (ed.), 1997. Glossary of Geology. American Geological Institute, Falls Church, VA.

Jaffré, T., R.R. Brooks, J. Lee, and R.D. Reeves. 1976. *Sebertia acuminata*: a hyperaccumulator of nickel from New Caledonia. Science 193: 579–580.

Jenny, H. 1941. Factors of Soil Formation. McGraw-Hill, New York.

Jenny, H. 1980. The Soil Resource. Springer-Verlag, New York.

Johnson, C.M., G.A. Pearson, and P.R. Stout. 1952. Molybdenum nutrition of crop plants. II. Plant and soil factors concerned with molybdenum deficiencies in crop plants. Plant & Soil 4:178–196.

Jones, M.B., and R.G. Woodmansee. 1979. Biogeochemical cycling in annual grassland ecosystems. Botanical Review 45: 111–144.

Jones, M.B., W.A. Williams, and J.E. Ruckman. 1977. Fertilization of *Trifollium subterraneum* L. growing on serpentine soils. Journal of the Soil Science Society of America 41: 87–89.

Jones, R.C., W.H. Hudnall, and W.S. Sakai. 1982. Some highly weathered soils of Puerto Rico, 2. Mineralogy. Geoderma 27: 75–137.

Kabata-Pendias, A. 2001. Trace Elements in Soils and Plants. CRC Press, Boca Raton, FL.

Koenigs, R.L., W.A. Williams, M.B. Jones, and A. Wallace. 1982. Factors affecting vegetation on a serpentine soil. II. Chemical composition of foliage and soil. Higardia 50: 15–26.

Koide, R.T., L.F. Huenneke, and H.A. Mooney. 1987. Gopher mound soil reduces growth and affects ion uptake of two annual grassland species. Oecologia 72: 284–290.

Koide, R.T., and H.A. Mooney. 1987. Revegetation of serpentine substrates: response to phosphate application. Environmental Management 11: 563–567.

Kramer, U., J.D. Cotter-Howels, J.M. Charnock, A.J.M. Baker, and J.A.C. Smith. 1996. Free histidine as a metal chelator in plants that accumulate nickel. Nature 379: 635–638.

Kronberg, B.I, and H.W. Nesbitt. 1981. Quantification of weathering, soil geochemistry, and soil fertilty. Journal of Soil Science 32: 453–459.

Kronzucker, H.J., M.Y. Siddiqi, and A.D.M. Glass. 1997. Conifer root discrimination against soil nitrate and the ecology of forest succession. Nature 385: 59–61.

Kruckeberg, A.R. 1954. Plant species in relation to serpentine soils. Ecology 35: 267–274.

Kruckeberg, A.R., P.J. Peterson, and Y. Samiullah. 1993. Hyperaccumulation of nickel by *Arenaria rubella* (Caryophyllaceae) from Washington state. Madroño 40: 25–30.

Kruckeberg, A.R., and R.D. Reeves. 1995. Nickel accumulation by serpentine species of *Streptanthus* (Brassicaceae): field and greenhouse studies. Madroño 42: 458–469.

Küpper, H., and P.M.H. Kroneck. 2005. Heavy metal uptake by plants and cyanobacteria. Pages 97–144 *in* A. Sigel, H. Sigel, and R.K.O. Sigel (eds.), Biogeochemistry, Availability, and Transport of Metals in the Environment. Taylor and Francis, Boca Raton, FL.

Lanspa, K.E. 1993. Soil Survey of Shasta-Trinity Forest Area, California. USDA Forest Service, Shasta-Trinity National Forest, Redding, CA.

Larson, D.W. 1981. Differential wetting of some lichens and mosses: the role of morphology. The Bryologist 84: 1–15.

Lavelle, P., and A.V. Spain. 2001. Soil Ecology. Kluwer, Dordrecht.

Leakey, R.J.G., and J. Proctor. 1987. Invertebrates in the litter and soil at a range of altitudes on Gunung Silam, a small ultrabasic mountain in Sabah. Journal of Tropical Ecology 3: 119–129.

Lee, B.D., R.C. Graham, T.E. Laurent, C. Amrhein, and R.M. Creasy. 2001. Spatial distributions of soil chemical conditions in a serpentinitic wetland and surrounding landscape. Soil Science Society of America Journal 65: 1183–1196.

Lee, J., R.R. Brooks, and R.D. Reeves. 1977. Chromium-accumulating bryophyte from New Caledonia. The Bryologist 80: 203–206.

Lee, J.A., and G.R. Stewart. 1978. Ecological aspects of nitrogen assimilation. Advances in Ecological Research 6: 1–43.

Lee, W.G., R.P. Littlejohn, and P.G. Prema. 1991. Growth of *Pinus radiata* in relation to foliar element concentrations on ultramafic soil, New Zealand. New Zealand Journal of Botany 29: 163–167.

Le Maitre, R.W. 1976. The chemical variability of some common igneous rocks. Journal of Petrology 17: 589–637.

Lepp, N.W. 2001. Bryophytes and pteridophytes. Pages 159–170 *in* M.N.V. Prasad (ed.), Metals in the Environment. Marcel Dekker, Basel.

Li, Yuan–Hui. 2000. A Compendium of Geochemistry, from Solar Nebula to the Human Brain. Princeton University Press, Princeton, NJ.

Lipman, C.B. 1926. The bacterial flora of serpentine soils. Journal of Bacteriology 12: 315–318.

Lounamaa, J. 1956. Trace elements in plants growing wild on different rocks in Finland. Annales Botanici Societatis Zoologicae Botanicae Fennicae "Vanamo" 29(4): 1–196.

Maas, J.L., and D.E. Stuntz. 1969. Mycoecology on serpentine soil. Mycologia 61: 1106–1116.

Madhok, O.P., and R.B. Walker. 1969. Magnesium nutrition of two species of sunflower. Plant Physiology 44:1016–1022.

Madigan, M.T., J.M. Martinko, and J. Parker. 2000. Brock Biology of Microoganisms. Prentice Hall, Upper Saddle River, NJ.

Major, J. 1977. California climate in relation to vegetation. Pages 11–75 *in* M.G. Barbour and J. Major (eds.), Terrestrial Vegetation of California. Wiley, New York.

Malley, D.F. 1973. The differential capacity of Jeffrey pine and ponderosa pine to take up nitrogen, phosphorous, potassium, calcium, and magnesium from ultramafic soils. M.S. thesis, University of California, Berkeley.

Margulis, L., and K.V. Schwartz. 1998. Five Kingdoms. Freeman, New York.

Mariño, F., A. Ligero, and D.J. Dias Cosin. 1995. Metales pesados en varias especies de lombrices de tierra de suelos desarrollados sobre serpentinas. Nova Acta Cientifica Compostelana (Bioloxia) 5: 245–250.

Marschner, H. 1995. Mineral Nutrition in Higher Plants. Academic Press, London.

Marschner, H., and V. Römheld. 1996. Root-induced changes in the availability of micronutrients in the rhizosphere. Pages 557–579 in Y. Waisel, A. Eshel, and U. Kafkafi (eds.), Plant Roots: The Hidden Half. Marcel Dekker, New York.

Marten, W.E., J. Vlamis, and N.W. Tice. 1953. Field correction of calcium deficiency on a serpentine soil. Agronomy Journal 45: 204–208.

Martens, S.N., and R.S. Boyd. 2002. The defensive role of Ni hyperaccumulation by plants: a field experiment. American Journal of Botany 89: 998–1003.

Martin, J.H., and S.E. Fitzwater. 1988. Iron deficiency limits phytoplankton growth in the northeast Pacific subarctic. Nature 331: 341–343.

Mason, S.F. 1991. Chemical Evolution: Origin of the Elements, Molecules, and Living Systems. Clarendon Press, Oxford.

Mather, J.R. 1974. Climatology. McGraw-Hill, New York.

McColl, J.G., L.K. Monette, and D.S. Bush. 1982. Chemical characteristics of wet and dry atmospheric fallout in northern California. Journal of Environmental Quality 11: 585–590.

McKenzie, R.M. 1970. The reaction of cobalt and manganese dioxide minerals. Australian Journal of Soil Research 8: 97–106.

McKenzie, R.M. 1975. The mineralogy and chemistry of soil cobalt. Pages 83–93 in D.J.D Nicholas and A.R. Egan (eds.), Trace Elements in Soil-Plant-Animal Systems. Academic Press, New York.

McKenzie, R.M. 1989. Manganese oxides and hydroxides. Pages 439–465 in J.B. Dixon and S.B. Weed (eds.), Minerals in Soil Environments. Soil Science Society of America, Madison, WI.

McLaren, R.G., D.M. Lawson, and R.S. Swift. 1986. Sorption and desorption of cobalt by soil components. Journal of Soil Science 37: 413–426.

McMillan, C. 1956. The edaphic restriction of *Cupressus* and *Pinus* in the Coast Ranges of central California. Ecological Monographs 26: 177–212.

McNaughton, S.J. 1968. Structure and function in California grasslands. Ecology 49: 962–972.

Mengoni, A., R. Barzanti, C. Gonnelli, R. Gabrielli, and M. Bazzicalupo. 2001. Characterization of nickel-resistant bacteria isolated from serpentine soil. Environmental Microbiology 3: 691–698.

Merrill, G.P. 1897. A Treatise on Rocks, Rock Weathering, and Soils. Macmillan, New York.

Mesjasz-Przybylowicz, J., and W.J. Przybylowicz. 2001. Phytophagous insects associated with the Ni-hyperaccumulating plant *Berkheya coddii* (Asteraceae) in Mpumalanga, South Africa. South African Journal of Science 97: 596–598.

Mesjasz-Przybylowicz, J., W.J. Przybylowicz, and C.A. Pinada. 2001. Nuclear microprobe studies of elemental distribution in apipcal leaves of the Ni hyperaccumulator *Berkeya coddii*. South African Journal of Science 97: 591–593.

Metting, F.B. 1993. Structure and physiological ecology of soil microbial communities. Pages 3–25 in F.B. Metting (ed.), Soil Microbial Ecology. Marcel Dekker, Basel.

Morrell, B.G., N.W. Lepp, and D.A. Phipps. 1986 Vanadium uptake by higher plants: some recent developments. Environmental Geochemistry and Health 8: 14–18.

Moser, A.M., C.A. Petersen, J.A. D'Allura, and D. Southworth. 2005. Comparison of ectomycorrhizas of *Quercus garryana* (Fagaceae) on serpentine and nonserpentine soils in southwestern Oregon. American Journal of Botany 92: 224–2005.

Motito, A., K. Mustelier, M. Potrony, and A. Vicario. 2004. Caractización de la brioflora de las áreas ultramáficas cubanas. Pages 19–23 *in* Rocas Ultramáficas: Sus Suelos, Vegetación y Fauna. Science Reviews, St. Albans, Herts, UK.

Nagy, L., and J. Proctor. 1997. Soil Mg and Ni as casual factors of plant occurrence and distribution at the Meikle Kilrannoch ultramafic site in Scotland. New Phytologist 135: 561–566.

Nahon, D.B., B. Boulangé, and F. Colin. 1992. Metallogeny of weathering: an introduction. Pages 445–471 *in* I.P. Martini and W. Chesworth (eds.), Weathering, Soils, and Paleosols. Elsevier, Amsterdam.

Navarro, E., T. Jaffre, D. Gauthier, F. Gourbiere, C. Rinaldo, P. Simonet, and P. Normand. 1999. Distribution of *Gymnostoma* spp. microsymbiotic *Frankia* strains in New Caledonia is related to soil type and to host-plant species. Molecular Ecology 8: 1781–1788.

Nesbit, E.G., and N.H. Sleep. 2003. The physical setting of life. Pages 3–24 *in* L.J. Rothschild and A.M. Lister (eds.), Evolution of Planet Earth: The Impact of the Physical Environment. Academic Press, San Diego, CA.

Newhall, F., and C.R. Berdanier. 1996. Calculation of soil moisture regimes from the climatic record. USDA, Natural Resources Conservervation Service, Soil Survey Investigations Report 46.

Nicholas, D.J.D. 1963. Inorganic nutrition of microorganisms. Plant Physiology 3: 363–447.

Nicks, L.J., and M.F. Chambers. 1995. A pioneering study of the potential of phytomining for nickel. Pages 313–325 *in* R.R. Brooks (ed.), Plants that Hyperaccumulate Heavy Metals. CAB International, Wallingford, UK.

Nieboer, E., D.H.S. Richardson, and F.D. Tomassini. 1978. Mineral uptake and release by lichens: an overview. The Bryologist 81: 247–257.

O'Dell, R.E., and V.P. Claassen. 2006. Serpentine and nonserpentine *Achillea millifolium* accessions differ in serpentine substrate tolerance and response to organic and inorganic amendnents. Plant & Soil 279: 253–269.

O'Dell, R.E., J.J. James, and J.H. Richards. 2006. Cogeneric serpentine and nonserpentine shrubs differ more in leaf Ca:Mg than in tolerance of low N, low, P, or heavy metals. Plant and Soil 280: 49–64.

Oze, C., S. Fendorf, D.K. Bird, and R.G. Coleman. 2004. Chromium geochemistry in serpentinized ultramafic rocks and serpentine soils from the Franciscan complex of California. American Journal of Science 304: 67–101.

Panaccione, D.G., N.L. Sheets, S.P. Miller, and J.R. Cummings. 2001. Diversity of *Cenococcum geophilum* isolates from serpentine and non-serpentine soils. Mycologia 93: 645–652.

Pechenik, J.A. 1996. Biology of the Invertebrates. Brown, Dubuque, IA.

Poschenrieder, Ch., and J. Barceló. 2004. Water relations in heavy metal stressed plants. Pages 249–270 *in* M.N.V. Prasad (ed.), Heavy Metal Stress in Plants. Springer, Berlin.

Powers, R.F. 1981. Nutritional ecology of ponderosa pine and associated species. Ph.D. dissertation, University California, Berkeley.

Proctor, J. 1970. Magnesium as a toxic element. Nature 227: 742–743.

Proctor, J., and A.J.M. Baker. 1994. The importance of nickel for plant growth in ultramafic (serpentine) soils. Pages 417–432 *in* S.M. Ross (ed.), Toxic Metals in Plant-Soil Systems. Wiley, New York.

Proctor, J., and K. Whitten. 1971. A population of the valley pocket gopher (*Thomomys bottae*) on a serpentine soil. American Midland Naturalist 85: 517–521.

Proctor, J., and S.R.J. Woodell. 1975. The ecology of serpentine soils. Advances in Ecological Research 9: 255–365.

Przybylowicz, W.J., J. Mesjasz-Przybylowicz, P. Miguka, E. Glowacka, M. Nakonieczny, and M. Augustyniak. 2003. Functional analysis of metals distribution in organs of the beetle *Crysolina pardalina* exposed to excess nickel by Micro-PIXE. Nuclear Instruments and Methods in Physics Research B 210: 343–348.

Purvis, O.W., and C. Halls. 1996. A review of lichens in metal-enriched environments. Lichenologist 28: 571–601.

Quantin, C., T. Becquer, J.H. Rouiller, and J. Berthelin. 2001. Oxide weathering and trace metal release by bacteria in a New Caledonia Ferralsol. Biogeochemistry 53: 323–340.

Quantin, C., T. Becquer, J.H. Rouiller, and J. Berthelin. 2002. Redistribution of metals in a New Caledonia Ferralsol after microbial weathering. Journal of the Soil Science Society of America 66: 1797–1804.

Rabenhorst, M.C., and J.E. Foss. 1981. Soils and geologic mapping over mafic and ultramafic parent materials in Maryland. Journal of the Soil Science Society of America 45: 1156–1160.

Ragsdale, S.W. 1998. Nickel biochemistry. Current Opinion in Chemical Biology 2: 208–215.

Rajakaruna, N., and B.A. Bohm. 1999. The edaphic factor and patterns of variation in *Lasthenia californica* (Asteraceae). American Journal of Botany 86: 1576–1596.

Read, D.J. 1991. Mycorrhizas in ecosystems. Experientia 47: 376–391.

Read, D.J., and J. Perez-Moreno. 2003. Mycorrhizas and nutrient cycling in ecosystems—a journey towards relevance? New Phytologist 157: 475–492.

Reeves, R.D., and A.J.M. Baker. 2000. Metal-accumulating plants. Pages 193–229 *in* I. Raskin and B.D. Ensley (eds.), Phytoremediation of Toxic Metals. Wiley, New York.

Reeves, R.D., A.J.M.. Baker, A Borhidi, and R. Berazaín. 1999. Nickel hyperaccumulation in the serpentine flora of Cuba. Annals of Botany 83: 29–38.

Reeves, R.D., R.R. Brooks, and R.M. Macfarlane. 1981. Nickel uptake by Californian *Streptanthus* and *Caulanthus* with particular reference to the hyperaccumulator *S. polygaloides* Gray (Brassicaceae). American Journal of Botany 68: 708–712.

Reeves, R.D., R.M. Macfarlane, and R.R. Brooks. 1983. Accumulation of nickel and zinc in western North American genera containing serpentine-tolorant species. American Journal of Botany 70: 1297–1303.

Reimann, C., and P. de Caritat. 1998. Chemical Elements in the Environment. Springer, Berlin.

Rengel, Z. 2000. Manganese uptake and transport in plants. Pages 57–87 *in* A. Sigel and H. Sigel, (eds.), Manganese and Its Role in Biological Processes. Marcel Dekker, Basel.

Rice, S.J. 1957. Nickel. California Division of Mines and Geology Bulletin 176: 391–399.

Rillig, M.C., C.B. Field, and M..F. Allen. 1999a. Fungal root colonization responses in natural grasslands after long-term exposure to elevated atmospheric CO_2. Global Change Biology 5: 577–585.

Rillig, M.C., C.B. Field, and M.F. Allen. 1999b. Soil biota responses to long-term atmospheric CO_2 enrichment in two California annual grasslands. Oecologia 119: 572–577.

Ritter-Studnička, H., and O. Klement. 1968. Über flechtenarten und deren Gesellschaften auf Serpentin in Bosnien. Österreichische Botanische Zeitschrift 115: 93–99.

Roberts, B.A. 1980. Some chemical and physical properties of serpentine soils from western Newfoundland. Canadian Journal of Soil Science 60: 231–240.

Robinson, B.H., R.R. Brooks, A.W. Howes, J.H. Kirkman, and P.E.H. Gregg. 1997. The potential of a high-biomass nickel hyperaccumulator *Berkheya coddii* for phytoremediation and phytomining. J. Geochemical Exploration 60:115–126.

Robinson, W.O., G. Edington, and H.B. Byers. 1935. Chemical studies of infertile soils derived from rocks high in magnesium and generally high in chromium and nickel. USDA Technical Bulletin 471.

Robson, A.D. 1988. Manganese in soils and plants—an overview. Pages 329–333 in R.D. Graham, R.J. Hannam, and N.C. Uren (eds.), Manganese in Soils and Plants. Kluwer Academic, Dordrecht.

Romankevich, E.A. 1990. Biogeochemical problems of living matter of the present-day biosphere. Pages 39–51 in V. Ittekket, S. Kempe, W. Michaelis, and A. Spitzy (eds.), Facets of Modern Biogeochemistry. Springer-Verlag, Berlin.

Römheld, V., and H. Marschner. 1991. Function of micronutrients in plants. Pages 297–328 in Micronutrients in Agriculture. Book Series no. 4. Soil Science Society of America, Madison, WI.

Rune, O. 1953. Plant life on serpentines and related rocks in the north of Sweden. Acta Phytographica Suecica 31: 1–139.

Ryan, B.D. 1988. Marine and maritime lichens on serpentine rock on Fidalgo Island, Washington. The Bryologist 9: 186–190.

Sagner, S., R. Kneer, W. Gerhard, J-P. Cosson, B. Deus-Neumann, and M.H. Zenk. 1998. Hyperaccumulation, complexation, and distribution of nickel in *Sebertia acuminata*. Phytochemistry 47: 339–347.

Salt, D.E., and U. Kramer. 2000. Mechanisms of metal hyperaccumulation. Pages 231–246 in I. Raskin and B.D. Ensley (eds.), Phytoremediation of Toxic Metals. Wiley, New York.

Sánchez-Mata, D., N. Rodríguez, M. del Pilar, R. Amils, and V. de la Fuente. 2004. Studies on California's ultramafic flora: a selected microanalytical screening. Pages 259–261 in R.S. Boyd, A.J.M. Baker, and J. Proctor. Ultramafic Rocks: Their Soils, Vegetation, and Fauna. Science Reviews, St. Albans, Herts, UK.

Scandalios, J.G. 1995. Oxygen stress and superoxide dismutases. Plant Physiology 101: 7–17.

Schlegel, H.G., J.-P. Cosson, and A.J.M. Baker. 1991. Nickel-accumulating plants provide a niche for nickel resistant bacteria. Botamica Acta 104: 18–25.

Schreier, H. 1989. Asbestos in the Natural Environment. Elsevier, Amsterdam.

Schreier, H., J.A. Omueti, and L.M. Lavkullich. 1987. Weathering processes of asbestos-rich serpentinitic sediments. Journal of the Soil Science Society of America 51: 993–999.

Schreier, H., and H.J. Timmenga. 1986. Earthworm response to asbestos-rich serpentine sediments. Soil Biology & Biochemistry 18: 85–89.

Schwartz, M.D., and M.A. Wall. 2001. *Melanotrichus boydi*, a new species of plant bug (Heteroptera: Miridae: Orthotylini) restricted to the nickel hyperaccumulator *Streptanthus polygaloides* (Brassicaceae). Pan-Pacific Entomologist 77: 39–44.

Schwertmann, U., and Latham, M. 1986. Properties of iron oxides in some New Caledonian Oxisols. Geoderma 39: 105–123.

Scrivenor, J.B. 1930. Laterite. The Geological Magazine 67: 24–28.

Ségalen, P., D. Bosch, A. Cardenas, E Camacho, H Guénin, and D. Rambaud. 1980. Aspects minéralogiques et pédogénétiques de deux sols dérivés de péridotites. Cahier ORSTM, série Pédologique 18: 273–284.

Senkayi, A.L. 1977. Clay mineralogy of poorly drained soils derived from serpentinite rocks. Ph.D. dissertation, University of California, Davis.

Shacklette, H.T. 1965. Element content of bryophytes. U.S. Geological Survey Bulletin 1198–D.

Shacklette, H.T., and J.G. Boerngen. 1984. Element Concentrations in Soils and Other Surficial Materials of the Conterminous United States. Professional Paper 1270, U.S. Geological Survey.

Shevock, J.R. 2003. Moss geography and floristics in California. Fremontia 31(3): 12–20.
Sigal, L.L. 1975. Lichens and Mosses of California Serpentines. MA thesis, University of California, San Francisco.
Sigal, L.L. 1989. The lichens of serpentine rocks and soils in California. Mycotaxon 34: 221–238.
Simonson, R.W. 1959. Outline of a generalized theory of soil development. Soil Science Society America, Proceedings 23:152–156.
Simonson, R.W. 1968. Concept of soil. Advances in Agronomy 20: 1–45.
Singer, M.J., and P. Nkedi-Kizza. 1980. Properties and history of an exhumed Tertiary Oxisol in California. Journal of the Soil Science Society of America 44: 587–590.
Smith, R.F., and B.L. Kay. 1986. Revegetation of serpentine soils: difficult but not impossible. California Agriculture 40: 18–19.
Soil Survey. 1973. Soil Survey Laboratory Data and Descriptions of some Soils of California. Soil Conservation Service, Soil Survey Investigations Report No. 24, USDA.
Soil Survey Staff. 1993. Soil Survey Manual. USDA, Agriculture Handbook No. 18. U.S. Government Printing Office, Washington, DC.
Soil Survey Staff. 1999. Soil Taxonomy—a Basic System for Making and Intrepreting Soil Surveys. USDA, Agriculture Handbook No. 436. U.S. Government Printing Office, Washington, DC.
Springer, R.K. 1980. Geology of the Spring Hill intrusive complex, a layered gabbroic body in the western Sierra Nevada foothills, California. Geological Society of America Bulletin 91: 381–385.
Sterflinger, K. 2000. Fungi as geological agents. Geomicrobiology Journal 17: 97–124.
Stevenson, F.J., and M.A. Cole. 1999. Cycles of Soil: Carbon, Nitrogen, Phosphorous, and Micronutrients. Wiley, New York.
Stewart, G.R., and S. Schmidt. 1998. Evolution and ecology of plant mineral nutrition. Pages 91–114 in M.C. Press, J.D. Scholes, and M.G. Barker (eds.), Physiological Plant Ecology. Blackwell Science, Oxford.
Stiefel, E.I. 2002. The biogeochemistry of molybdenum and tungsten. Pages 1–29 in A. Sigel and H. Sigel (eds.), Metal Ions in Biological Systems. Marcel Dekker, Basel.
Stoppel, R.D., and H.G. Schlegel. 1995. Nickel-resistant bacteria from anthropogenically polluted and naturally nickel-percolated ecosystems. Applied and Envionmental Microbiology 61: 2276–2285.
Streit, B., and W. Stumm. 1993. Chemical properties of metals and the process of bioaccumulation in terrestrial plants. Pages 31–62 in B. Markert (ed.), Plants as Biomonitors. VCH, Weinheim, Germany.
Streit, B., R.J. Hobbs, and S. Streit. 1993. Plant distributions and soil chemistry at a serpentine/non-serpentine boundary in California. Pages.167–178 in B. Markert (ed.), Plants as Biomonitors: Indicators for Heavy Metals in the Terrestrial Environment. VCH Publishers, Weinheim, Germany.
Suzuki, S., N. Mizuno, and K. Kimura. 1971. Distribution of heavy metals in serpentine soil. Soil Science and Plant Nutrition 17: 195–198.
Taylor, G., and R.A. Eggleton. 2001. Regolith Geology and Geomorphology. Wiley, Chichester, UK.
Terlizzi, D.E., and E.P. Karlander. 1979. Soil algae from a Maryland serpentine formation. Soil Biology and Biochemistry 11: 205–207.
Thauer, R.K. 2001. Nickel to the fore. Science 293: 1264–1265.
Thiry, M., and G. Millot. 1987. Mineralogical forms of silica and their sequences of formation in silcretes. Journal of Sedimentary Petrology 57: 343–352.
Thomas, L., and J. Proctor. 1997. Invertebrates in the litter and soil on the ultramafic Mount Giting-Giting, Philippines. Journal of Tropical Ecology 13: 125–133.

Thornthwaite, C.W. 1948. An approach toward a rational classification of climate. Geographical Review 38: 55–94.

Thornthwaite, C.W., and J.R. Mather. 1955. The water balance. Drexel Institute of Technology, Laboratory of Climatology, Pubications in Climatology 8: 1–104.

Trescases, J.J. 1997. The lateritic nickel-ore deposits. Pages 125–138 in H. Paquet and N. Clauer (ed.), Soils and Sediments. Springer, Berlin.

Turekian, K.K., and K.H. Wedepohl. 1961. Distribution of the elements in some major units of the earth's crust. Geological Society of America Bulletin 72:175–192.

Turitzin, S.N. 1982. Nutrient limitations to plant growth in a California serpentine grassland. American Midland Naturalist 107: 95–99.

Turnau, K., and J. Mesjasz-Przybylowicz. 2003. Arbuscular mycorrhiza of *Berkeya coddii* and other Ni-hyperaccumulating members of Asteraceae from ultramafic soils in South Africa. Mycorrhiza 13: 185–190.

Uren, N.C. 1992. Forms, reactions, and availability of nickel in soils. Advances in Agronomy 48: 141–203.

Vance, C.P. 1996. Root—bacteria interactions: symbiotic nitrogen fixation. Pages 723–755 in Y. Weiss, A. Eshel, and U. Kafkafi (eds.), Plant Roots: The Hidden Half. Marcel Dekker, New York.

Vinogradov, A.P. 1959. The Geochemistry of Rare and Dispersed Chemical Elements in Soils. Consultants Bureau, New York.

Vlamis, J. 1949. Growth of lettuce and barley as influenced by degree of calcium saturation of soil. Soil Science 67: 453–466.

Wächtershäuser, G. 1992. Groundworks for an evolutionary biochemistry: the iron-sulphur world. Progress in Biophysics and Molecular Biology 58: 85–201.

Walker, R.B. 1948. Molybdenum deficiency in serpentine barren soils. Science 108: 473–475.

Walker, R.B. 1954. Factors affecting plant growth on serpentine soils. Ecology 35: 258–266.

Walker, R.B. 2001. Low molybdenum status of serpentine soils of western North America. South African Journal of Science 97: 565–568.

Walker, R.B., H.M. Walker, and P.R. Ashworth. 1955. Calcium-magnesium nutrition with special reference to serpentine soils. Plant Physiology 30: 214–221.

Wall, M.A., and R.S. Boyd. 2002. Nickel accumulation in serpentine arthropods from the Red Hills, California. Pan-Pacific Entomologist 78(3): 168–176.

Wallace, A., M.B. Jones, and G.V. Alexander. 1982. Mineral composition of native woody plants growing on serpentine soil in California. Soil Science 134: 42–44.

Wallace, A., E.M. Romney, and J.E. Kinnear. 1977. Metal interactions in bush bean plants grown in a glasshouse in amended sepentine soils from California. Communications in Soil Science and Plant Analysis 8: 727–732.

Walter, H. 1974. Vegetation of the Earth in Relation to Climate and the Ecophysiological Conditions. Springer-Verlag, New York.

Wang, Y.-T. 2000. Microbial reduction of chromate. Pages 225–235 in D.R. Lovley (ed.), Environmental Microbe-Metal Interactions. ASM Press, Washington, DC.

Weiss, S.B. 1999. Cars, cows, and checkerspot butterflies: nitrogen deposition and management of nutrient-poor grasslands for a threatened species. Conservation Biology 13: 1476–1486.

Westheimer, F.H. 1987. Why nature chose phosphates. Science 235: 1173–1178.

White, A.F., and A.E. Blum. 1995. Effects of climate on chemical weathering in watersheds. Geochimica et Cosmochimica Acta 59: 1729–1747.

White, C.D. 1967. Absence of nodule formation on *Ceanothus cuneatus* in serpentine soils. Nature 215: 875.

White, C.D. 1971. Vegetation—Soil Chemistry Correlations in Serpentine Ecosystems. Ph.D. dissertation, University of Oregon, Eugene.

Whittaker, R.H. 1960. Vegatation of the Siskiyou Mountains, Oregon and California. Ecological Monographs 30: 279–338.

Wild, H. 1975. Termites and the serpentine of the Great Dyke of Rhodesia. Rhodesian Scientific Association Transactions 57: 1–11.

Wildman, W.E., and M.L. Jackson. 1968. Serpentine rock dissolution as a function of carbon dioxide pressure in aqueous solution. American Mineralogist 53: 1252–1263.

Wildman, W.E., M.L. Jackson, and L.D. Whittig. 1968. Iron-rich montmorillonite formation in soils derived from serpentinite. Soil Science Society of America Proceedings 32: 787–794.

Williams, M.R., and J.M. Melack. 1997. Atmospheric deposition, mass balances, and processes regulating stream water solute concentrations in mixed-conifer catchments of the Sierra Nevada, California. Biogeochemistry 37: 111–144.

Williams, R.J.P., and J.J.R. Fraústo da Silva. 1996. The Natural Selection of the Chemical Elements. Oxford University Press, Oxford.

Wilson, M.J. 1995. Interactions between lichens and rocks; a review. Cryptogamic Botany 5: 299–305.

Wilson, M.J., D. Jones, and W.J. McHardy. 1981. The weathering of serpentine by *Lecanora atra*. Lichenologist 13: 167–176.

Woodell, S.R.J., H.A. Mooney, and H. Lewis. 1975. The adaptation to serpentine soils in California of the annual species *Linanthus androsaceus* (Polemoniaceae). Torrey Botanical Club Bulletin 102: 232–238.

Woodmansee, R.G., and D.A. Duncan. 1980. Nitrogen and phosphorous dynamics and budgets in annual grasslands. Ecology 61: 893–904.

Wright, S.F., and A. Upadhyaya. 1998. A survey of soils for aggregate stability and glomalin, a glycoprotein produced by hyphae of arbuscular mycorrhizal fungi. Plant and Soil 198: 97–107.

Yatsu, E. 1989. The Nature of Weathering. Sozosha, Tokyo.

Zinke, P.J., and A.G. Stangenberger. 1979. Ponderosa pine and Douglas-fir foliage analyses arrayed in probability distributions. Pages 221–225 *in* Proceedings, Forest Fertilization Conference. University of Washington, Institute of Forest Resources, Contribution No. 40.

Zinke, P.J., A. Stangenberger, and W. Colwell. 1979. The fertility of the forest. California Agriculture 33: 10–11.

Part III Plant Life on Serpentine

The world's major vegetation types are broadly shaped by the combination of temperature, moisture, and the seasonality of these factors (e.g., Holdridge 1967, Whittaker 1975, Walter 1994). For example, temperature and precipitation contributed 46%–60% to variation in plant composition in Oregon, followed by geology (11%–19%), disturbance (6%–12%), and topography (4%–8%) (Ohmann and Spies 1998). One way to view unusual geologic substrates such as serpentine is as factors that give rise to vegetation atypical for the particular climatic zone: vegetation that is often sparser, smaller in stature, composed of a different set of life forms, or that contains some unique species. Substrates that act similarly to serpentine include limestone, gabbro, gypsum, alkaline flats, and outcrops of granitic or volcanic rock (Huggett 1995, Anderson et al. 1999). Serpentine is perhaps the most extensively studied unusual plant substrate in the world in terms of how it influences plant life, from the level of physiological tolerance mechanisms to the level of broad patterns in flora and vegetation (for review, see Proctor and Woodell 1975, Kruckeberg 1984, Brooks 1987, Baker et al. 1992, Roberts and Proctor 1992). Serpentine is also the terrestrial substrate that has contributed most to plant diversity in western North America (Stebbins and Major 1965, Raven and Axelrod 1978, Kruckeberg 1984, Skinner and Pavlik 1994).

This part discusses the responses of plants to serpentine soils, first at the level of species (chapter 9) and then at the level of whole plant assemblages (chapter 10). In these chapters we attempt to place western North America in a global context by drawing on the worldwide literature. We then focus on western North America, beginning with an overview of broad latitudinal and climatic gradients in the serpentine flora and vegetation (chapter 11), and following this with a detailed description of plant associations on serpentine (chapter 12).

9 Responses of Individual Plant Species to Serpentine Soils

In this chapter we summarize current knowledge of the physiological, evolutionary, and distributional effects of serpentine on individual plant species (see also the excellent review of the evolutionary ecology of serpentine plants by Brady et al. [2005]) A useful terminology for discussing plant responses to serpentine is Kruckeberg's classification of species as serpentine avoiders, indifferents, endemics, and indicators (Kruckeberg 1954, 1984). Serpentine avoiders are taxa that are seldom or never found on serpentine, whereas indifferent (or *bodenvag*) taxa are found with roughly equivalent frequencies on and off of serpentine. Endemics are species or subspecific taxa that are entirely or almost entirely restricted to serpentine. Serpentine indicators are taxa that are either locally more common on serpentine than on other substrates or restricted to serpentine in only parts of their geographic or ecological ranges. As with any classification system, this one is only a starting point for understanding natural variation. Plant responses form a continuum from complete restriction to serpentine on one hand and complete avoidance of it on the other, and many taxa lie in between, with markedly higher or (more often) lower abundances on serpentine. It is especially common for species to be partly restricted to serpentine but also to occur on other unusual substrates (e.g., limestone, acid soils, or scree; Rune 1953), or for species to be restricted to serpentine at lower but not at higher elevations or latitudes (Rune 1953, Kruckeberg 1984, Brooks 1987). The degree of restriction to serpentine also commonly depends on the strength of other influential environmental variables such as climate, and on the history and diversity of the surrounding region. Moreover, "indifferent" taxa often show divergence into serpentine-tolerant and intolerant ecotypes, which may or may not represent the early stages of formation of new endemic species. Many fascinating ecological and evolutionary questions thus lie at the transitions between avoidance, indifference, indicator status, and endemism.

Serpentine endemics have been further classified in terms of their inferred evolutionary history as either neo- or paleoendemics. Neoendemics (called "true serpentinophytes" by Rune 1953) are species that are thought to have originated through a localized shift onto serpentine, as evidenced by their narrow present-day geographic distributions and the proximity of closely related taxa on other substrates. Paleoendemics (Rune's "serpentinicolous relics") are more widespread species that are believed to have existed off of serpentine in their evolutionary past and to have become restricted through the extinction of all or nearly all of their nonserpentine populations (Kruckeberg 1954, 1984; Raven and Axelrod 1978). However, as we shall see, this is also a simplification; none of the few endemic-containing taxa whose evolutionary histories have been analyzed with modern phylogenetic methods fits neatly into these categories. Despite decades of interest in the subject, a great deal remains to be learned about the evolutionary origins of serpentine endemism.

9.1 Reasons for Inability to Grow on Serpentine

Serpentine soils prevent the growth of some plants and reduce the growth of most others—even many serpentine endemics (Kruckeberg 1984, Brooks 1987). Chapter 8 describes in detail the chemical characteristics of serpentine soil and the relationship of chemical composition to plant physiology. Here we briefly review the material presented earlier as an introduction to discussing variation, adaptation, speciation, and geographic distributions in the serpentine flora. It is clear that different plant species on serpentine are affected by different soil factors, and that many species are capable of growing in some serpentine soils but that others are not. Thus, the emphasis in this chapter is on understanding the possible sources of variation among populations, species, and environments in the responses of plants to serpentine.

9.1.1 Calcium and Magnesium

A variety of methods have been used to identify the specific factors that affect particular species, the most common being experimental additions of nutrients to serpentine soils in pots or in the field, solution culture experiments to examine metal toxicity, and analyses of plant distributions along natural soil gradients. As noted in chapter 8, the preponderance of evidence points to low levels of calcium, high levels of magnesium, and/or low Ca/Mg ratios. High absolute levels of magnesium and low Ca/Mg ratios are a consistent feature of serpentine soil chemistry. High magnesium levels reduce the availability of calcium to plants. Conversely, the adverse effect of high levels of magnesium is greatest when calcium levels are low. Nutrient amendment studies have demonstrated strong effects of calcium addition on the growth of both adapted and nonadapted species on serpentine (reviewed in Brooks 1987). Differential performance under variable calcium and magnesium treatments often distinguishes serpentine-tolerant from intolerant species or ecotypes (e.g., Walker 1954, Madhok and Walker 1969, MacNair 1992, Brady et al.

2005). Natural variation in Ca/Mg ratios is often a strong predictor of changes in the vegetation, such as the prevalence and biomass of exotic versus native grasses in Californian serpentine grasslands (Armstrong and Huenneke 1992, McCarten 1992, Harrison 1999b). When serpentine is enriched in calcium by hydrothermal or other processes, it nearly always supports different vegetation from that found on typical calcium-deficient serpentine (Proctor and Nagy 1992).

9.1.2 Nutrients

Several experimental studies have shown that nutrient amendment alone can significantly enhance the ability of plants to grow on serpentine. Fertilizing a Californian serpentine grassland with nitrogen and phosphorus led to increased biomass and increased richness and dominance of alien grasses (Huenneke et al. 1990). Although this could be attributed to the serpentine soil being relatively mild (i.e., not extremely calcium-poor or metal-rich), another study found that nitrogen and phosphorus (especially phosphorus) fertilization of an extremely nickel- and magnesium-rich serpentine barren in Scotland also led to strong increases in plant growth (Nagy and Proctor 1997a). Some evidence from experiments with crop plants suggests that calcium and nutrient addition may have superadditive effects on plant growth, such that nutrient addition is only effective if the Ca/Mg ratio is not too low (for review, see Brooks 1987, Kruckeberg 1992).

9.1.3 Nickel

Levels of available nickel are high enough in some serpentine soils to cause symptoms of toxicity in nonadapted plants, such as reduced mitotic activity and root growth, impaired photosynthesis, and damage to plasma membranes (Gabbrielli and Pandolfini 1984, Gabbrielli et al. 1990, Nagy and Proctor 1997a). Levels of nickel in solution cultures that mimicked the levels in serpentine soils severely reduced the growth of plants from nonserpentine populations of *Festuca rubra* and *Cochlearia pyrenaica*, but not of *Cerastium fontanum*. In the same experiment, levels of magnesium comparable to those in serpentine soils reduced the growth of nonserpentine plants of all three species (Nagy and Proctor 1997a). Lime addition to reduce nickel availability has sometimes been used successfully to grow crops on serpentine (Brooks 1987). There is some correlative evidence for the influence of nickel content on serpentine vegetation in Europe (Brooks 1987) and in New Zealand (Robinson et al. 1997), but in other places in Europe (e.g., Chiarucci et al. 2001) the nickel level was almost never high enough to influence the distribution of plants. No negative effects of nickel have been found in plants in Californian chaparral (Koenigs et al. 1982).

There are complex interactions among levels of the nutrients calcium, magnesium, and nickel in their potential effects on plants. Magnesium may block the uptake of calcium (Brooks 1987), while calcium may reduce the effects of excess magnesium (Gabbrielli and Pandolfini 1984). Both calcium and magnesium may reduce the toxicity of nickel, but nickel may reduce the uptake of calcium and magnesium (Gabbrielli and Pandolfini 1984;

Gabbrielli et al. 1990; Yang et al. 1996a,b; Nagy and Proctor 1997a). Nickel and magnesium may reduce the availability of phosphorus in the soil by forming insoluble phosphates (Nagy and Proctor 1997b). The adverse effects of nickel and magnesium on nutrient uptake may be ameliorated by raising calcium levels or pH (Gabbrielli and Pandolfini 1984, Brooks 1987). Given the potential for interactive effects, the varied results from experiments and field analyses are not surprising. Instead, it is to be expected that the relative importance of levels of calcium, magnesium, metals, nutrients, and moisture will depend greatly on the particular setting, species, and plant response being studied.

9.1.4 Water

The drought-adapted appearance of serpentine vegetation has led many to assume that it is a drier substrate than other soils. No experimental studies have shown that water addition alone can reverse the infertility of serpentine. However, it is clear that the ability of many species to grow on serpentine depends strongly on soil depth, aspect, and shading, all of which control plant water balance (Whittaker 1960, McCarten 1992, Proctor and Nagy 1992, Tyndall and Hull 1999). Water availability is likely to be a strong contributing factor to poor plant growth on serpentine soils that are shallow and rocky and hence have a low water storage capacity. In such cases, water stress may exacerbate the primary effects of nutrient shortage and high cation concentrations. This is consistent with the evidence, discussed below, that drought tolerance is an important component of adaptation to serpentine soils.

Although it appears that multiple factors interact to inhibit the growth of plants on serpentine, perhaps the most important question from an evolutionary perspective is what special adaptations are found in plants capable of growing on serpentine, and what consequences these adaptations may have for speciation. This is discussed in the next section. From an ecological viewpoint, perhaps the more relevant issue is what natural gradients in soil composition, or in other environmental variables, are responsible for observed patterns in the vegetation and flora of serpentine. This is discussed in chapters 10 and 11.

9.2 Mechanisms of Serpentine Tolerance

9.2.1 Metal Exclusion

In some species, tolerance of excess magnesium and nickel is based on the ability to exclude them or to reduce their translocation, while in others tolerance is based on the ability to accumulate these elements in a nontoxic form. As discussed in chapter 8, these two strategies can be distinguished by comparing levels of cations in the growth medium and the plant: exclusion is evidenced by low foliar concentrations despite high levels in the environment, and accumulation is evidenced by adequate plant growth despite high foliar concentrations. Exclusion is a common form of metal tolerance, but its physiological basis has not been well explored (Gabbrielli et al. 1990). Four Californian chaparral species (*Adenostoma fasciculatum, Cupressus sargentii, Arctostaphylos viscida,* and *Quercus durata*) maintained low to moderate foliar magnesium while growing on serpentine soils with high to extreme magnesium levels (Koenigs et al. 1982, Wallace et al. 1982). Likewise, serpentine-adapted *Cerastium fontanum, Festuca rubra,* and *Cochlearia pyrenaica* maintained lower internal nickel concentrations than did their nonadapted conspecifics when grown in solutions with high levels of nickel (Nagy and Proctor 1997a). Tolerance may also involve limiting metal translocation from roots to shoots (Yang et al. 1996a,b).

In an elegant study, O'Dell et al. (2006) compared nutrient and metal uptake in six shrubs, of which three were serpentine endemics and three were congeneric to the endemics (*Arctostaphylos viscida* and *A. manzanita, Ceanothus jepsonii* and *C. cuneatus, Quercus durata* and *Q. berberidifolia*). When all six were grown on serpentine soils with and without fertilizer, the endemics consistently maintained higher Ca/Mg ratios in their shoots, although not in their roots. This was achieved by selective calcium transport or magnesium exclusion operating at the level of root-to-shoot translocation. In contrast, nutrient uptake did not differ between the endemics and their congeners.

Bradshaw (2005) discovered a gene that confers tolerance to the Ca/Mg ratios typical of serpentine soils in artificially generated mutants of *Arabidopsis thaliana*. He screened a library of *A. thaliana* mutants on a nutrient solution with a low Ca/Mg ratio that was similar to natural serpentine soils and that was lethal to wild-type individuals. The survivors were mutants with a nonfunctional allele of the *CAX-1* gene. These mutants also exhibited lower foliar magnesium and peak growth rates at higher levels of magnesium than in wild-type plants under the same range of calcium and magnesium concentrations. *CAX-1* in normal plants moves excess calcium from the cytoplasm into vacuoles. Bradshaw (2005) argued that the mutant nonfunctional *CAX-1* causes plants to reduce the activity of a general-purpose ion channel in the cell membrane, which causes the selective exclusion of magnesium in the solutions with excess magnesium relative to calcium.

9.2.2 Sequestration

Uptake and sequestration of excess cations is an alternative route to serpentine tolerance. Its physiological basis is poorly understood, although metal tolerance in general appears to be enabled by chelating agents or cell-membrane transport mechanisms (Tilstone and McNair 1997). The genus *Thlaspi* may have a generalized mechanism for sequestering multiple metals (Reeves and Baker 1984). However, nickel tolerance and copper tolerance in *Mimulus guttatus* appear to be under the control of different genes (Tilstone and MacNair 1997).

Two edaphic ecotypes of *Lasthenia californica*, distinguishable by biochemical and morphological characters, demonstrate a mixture of strategies for cation tolerance (Rajakaruna and Bohm 1999, Rajakaruna et al. 2003). The two races co-occur in central Californian serpentine grasslands, where race A is found in clay-rich soils with higher levels of magnesium, Na, and other ions, higher water capacity, and lower water availability late in the growing season. Race C is found in rocky serpentine soils that are lower in magnesium but higher in nickel. Both races appear to selectively exclude excess nickel, but race A tolerates high foliar levels of magnesium and Na, whereas race C maintains constant foliar levels of these elements. The sulfated flavonoids that distinguish race A may play some role in ion uptake and sequestration.

9.2.3 Selective Calcium or Nutrient Uptake

Adaptation to low levels of calcium can arise through selective uptake, in which plants maintain high foliar calcium levels despite low levels in the soil, or through the ability to grow with low internal calcium levels. Classic work by Walker et al. (1955) showed that selective uptake enabled serpentine-adapted plants to grow in extremely calcium-poor solutions. Comparisons of calcium concentrations in the plant and the soil revealed selective calcium uptake in *Cupressus sargentii* and tolerance of low foliar calcium in *Arctostaphylos viscida*; *Adenostoma fasciculatum* showed neither ability and was absent from the most calcium-poor serpentine sites (Koenigs et al. 1982). Tolerance may also relate to the ability to mobilize nutrients in a calcium-poor or metal-rich environment. The enzyme root phosphatase makes phosphorus in soil organic matter available to the plant. Root phosphatase activity in serpentine-tolerant grasses was highest at low calcium concentrations approximating that in serpentine soil, whereas in nontolerant species this enzyme was inhibited at low calcium concentrations (Willett and Batey 1977). Later work on *Festuca rubra* found that root phosphatase activity was differentially sensitive to magnesium and nickel concentrations in serpentine-adapted and unadapted populations (Johnston and Proctor 1981, 1984).

9.2.4 Drought Tolerance

Serpentine plants typically display a suite of morphological adaptations associated with dry environments, such as small, thick, hairy, pale, and/or evergreen foliage. Such "xeromorphic" features are often found in nutrient-poor soils, as well as in dry climates, and it remains unclear to what extent they are direct adaptations to increased water stress on serpentine, versus part of a physiological strategy for coping with excess metals and low nutrients. According to Borhidi (1996), Cuban serpentine plants are as xeromorphic on average as are nonserpentine plants from climates with 50 cm lower annual precipitation. However, Borhidi also argued that serpentine plants lack the inner anatomical features associated with water stress, and thus that their xeromorphism is probably related to nutrient deficiency rather than to water deficiency. There have been few attempts to test the functional significance of xeromorphic traits in plants on serpentine soil; for example, investigating whether altering the level of water, calcium, or magnesium in the soil affects the relative performance of smaller versus larger-leaved plants.

In a study of five species of annual monkeyflowers in the *Mimulus guttatus* group, Hughes et al. (2001) found that serpentine endemics were more drought-tolerant than generalists or avoiders and concluded that drought was the main driving force in the evolution of serpentine tolerance in these species. Several other studies linking serpentine tolerance to drought tolerance have been reviewed by Brady et al. (2005).

9.2.5 Costs of Tolerance

Serpentine-adapted plants, whether they are ecotypes or endemic species, may grow as well or better on nonserpentine than serpentine soils in the absence of competitors and pathogens. Thus, the avoidance of competition or disease is often the proximate factor excluding endemics from other soils (Kruckeberg 1967, Brooks 1987). To the extent this is the case, it is to be expected that "endemics" will also appear on other soils, especially other shallow, rocky substrates where competition is reduced. However, in other cases there may be costs or tradeoffs associated with serpentine tolerance. The serpentine endemics *Helianthus exilis* and *Mimulus nudatus* were not only more tolerant of low calcium and high magnesium than their nonendemic congeners *Helianthus annuus* and *Mimulus guttatus*, but were also incapable of normal growth at high calcium and/or low magnesium levels (Madhok and Walker 1969, MacNair and Gardner 1998). To the extent that selective uptake or exclusion abilities come at the cost of reduced functionality in more normal soils, the propensity of genetically distinct ecotypes or species to evolve will be enhanced.

9.3 Formation of Serpentine Ecotypes

Serpentine populations of "indifferent" species are sometimes visibly different from nonserpentine conspecifics, typically in the direction of xeromorphism. They may have smaller stature, smaller or hairier leaves, different flowering times, and fewer flowers or fruits. Transplant experiments generally show that these differences are at least partly genetically based. Whether such morphological differences are present, reciprocal transplants sometimes show that plants taken from serpentine soils survive and grow better on serpentine than do their nonserpentine conspecifics, although plants of both origins typically survive and grow better on nonserpentine than on serpentine (e.g., Kruckeberg 1951, 1954, 1967; Nagy and Proctor 1997a; MacNair and Gardner 1998; Harrison et al. 2001). Experimental evidence for the existence of serpentine-adapted ecotypes was found in 12 of 21 Californian species tested by Kruckeberg (1954), and in 13 of 18 of those tested by Kruckeberg (1967).

Serpentine-adapted ecotypes may be just as well adapted to serpentine soil as are reproductively isolated serpentine-endemic species, as MacNair and Gardner (1998) showed for the nonendemic *Mimulus guttatus* and the endemic *Mimulus nudatus*. Serpentine ecotypes can develop rapidly, as shown by their existence in *Avena fatua* and *Bromus hordeaceus*, two Mediterranean grasses that were introduced to California within the past 200 years (Harrison et al. 2001). Grown in a common soil, *Avena fatua* from serpentine populations produce more but smaller seeds and have higher root-to-shoot ratios than their nonserpentine conspecifics (fig. 9-1).

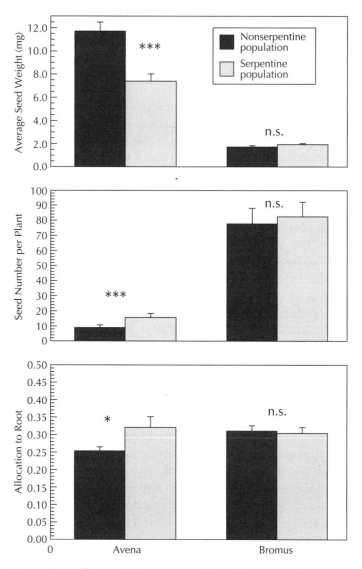

Figure 9-1 Differences between serpentine and nonserpentine populations of the exotic annual grasses *Avena fatua* and *Bromus hordeaceus* in their seed characteristics and root allocation. Avena from serpentine populations produce more but smaller seeds and allocate more to roots than do *Avena* from nonserpentine populations. For linear contrasts between populations within each species, ***significant difference at $P<.001$, *significant difference at $P<.05$, and n.s., not significant at $P>.05$. From Harrison et al. (2001).

Two studies found evidence for ecotype formation in plants growing on harsher versus less harsh serpentine soils. The annual forb *Lasthenia californica*, as described earlier, is divided into races that occur on wet clay-rich soils (race A) and shallow rocky soils (race C) within a central Californian serpentine grassland (Rajakaruna and Bohm 1999, Rajakaruna et al. 2003). Race A is also found on other soils that have high clay, magnesium, and sodium content, ranging from central California to northwestern Mexico; race C is found on soils with higher calcium and potassium, and ranges from southern Oregon to north-central California. Race A flowers 7–10 days earlier than race C, and they are not completely interfertile. The native annual grass *Vulpia microstachys* shows ecotypic differentiation between populations on rocky serpentine outcrops and those in grasslands on either serpentine or nonserpentine soil. When seeds from all three sources are grown in all three habitats in the field, plants produced by seeds from rocky serpentine habitat perform significantly better in rocky serpentine habitat than do plants grown from seed taken from either serpentine or nonserpentine grasslands. This suggests that the serpentine ecotype is actually adapted to rocky conditions (e.g., low soil-water capacity) rather than to strictly serpentine soil characteristics such a low Ca/Mg ratio (fig. 9-2; Jurjavcic et al. 2002).

Many questions about natural selection remain to be addressed in species with serpentine ecotypes. For example, what are the specific adaptations found in serpentine populations, and what are the costs of these traits in other environments? Over what spatial scales relative to gene flow can natural selection produce serpentine adaptation? Does adaptation to serpentine tend to arise many times within a species, which would be indicated by serpentine populations being most closely related to nearby nonserpentine conspecifics, or does it tend to arise once or a few times and then spread, indicated by widely separated serpentine populations being more closely related to each other than to adjacent nonserpentine populations?

Lack of evidence for ecotypes has been found in some natives (e.g., *Eriogonum compositum*, *Agropyron spicatum*, *Pinus contorta*) and exotics (e.g., *Bromus tectorum*) (Kruckeberg 1967). So-called general-purpose genotypes have been credited for the ability to grow under a wide variety of ecological conditions without differentiation into ecotypes, but the genetic and physiological basis of such tolerance remains unknown. Reeves and Baker (1984) reported that serpentine and nonserpentine populations of *Thlaspi goesingense* showed equal abilities to hyperaccumulate nickel and other metals (Zn, Co, and Mn) and suggested that a nonspecific metal-binding ligand was responsible for the widespread metal tolerance of *Thlaspi* spp. However, since nickel hyperaccumulation is rare in western North America, other explanations must be sought for the serpentine tolerance of truly indifferent taxa in this region.

9.4 Evolution of Serpentine Endemism

Although serpentine soils can provide a strong selective gradient that leads to the development of locally adapted populations, it is far less clear how serpentine-adapted populations become reproductively isolated species. As in all attempts to understand speciation,

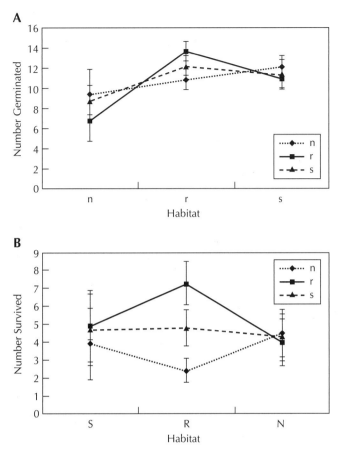

Figure 9-2 Ecotype formation in the native annual grass *Vulpia microstachys*. Habitats into which seeds were planted are nonserpentine grassland (n), serpentine grassland (s), and rocky serpentine slopes (r). Sources of the seeds were *Vulpia* populations in nonserpentine grassland (n), serpentine grassland (s), and rocky slopes (r). Ecotype formation is shown by the fact that both germination (A) and survival (B) in the rocky slope habitat were higher for seeds that came from rocky slopes. From Jurjavcic et al. (2000).

the problem lies in explaining the fixation of genes that confer reproductive isolation, because this isolation can be prevented by gene flow between populations. Geographic isolation (allopatry) provides the simplest route to the evolution of reproductive isolation. Alternatively, speciation with little geographic isolation (parapatric or sympatric speciation) can occur if strong selection on an adaptive trait produces reproductive isolation as a byproduct, but how frequently this happens is not well known.

Based on studies of the localized serpentine endemics *Mimulus nudatus* (fig. 9-3A) and *M. pardalis* and their widespread sympatric ancestor *Mimulus guttatus* (fig. 9-3B), it has

Figure 9-3 The serpentine endemic monkeyflower *Mimulus nudatus* (A) and its widespread ancestor *M. guttatus* (B). Photos copyrighted by S. Schoenig and G.W. Hartwell, used with permission.

been argued that adaptation to serpentine may lead to reproductive isolation without geographic isolation (MacNair 1992, MacNair and Gardner 1998, Gardner and MacNair 2000). *M. nudatus* tolerates higher nickel and magnesium and lower calcium than nonserpentine populations of *M. guttatus*. Both in the field and in a common environment, *M. nudatus* is smaller in stature, leaf area, and seed production, and it flowers earlier than *M. guttatus*. Flower size and shape in *M. nudatus* are more attractive to an early pollinator (sweat bees), while *M. guttatus* flowers are more attractive to later-flying bumblebees. In addition, there is strong post-mating isolation between *M. guttatus* and *M. nudatus*. The other endemic, *M. pardalis*, also flowers early in the season and is completely self-fertilizing.

As envisioned by McNair and Gardner (1998), the first step in the origin of *M. nudatus* and *M. pardalis* was magnesium and nickel tolerance, which has been shown to involve a small number of genes. Once this tolerance allowed colonization of serpentine, further selection favored multiple adaptations to drought and low nutrients, including early flowering time. In *M. nudatus*, changes in flowering time then selected for flowers that attracted different pollinators. Under the resulting reduced gene flow, further local adaptation in *M. nudatus* led to the fixation of genes for intersterility with *M. guttatus*, perhaps by linkage between genes for tolerance and intersterility. In *M. pardalis* the process was simpler; flowering at a season too early for reliable pollination led to selection for self-fertilization, ending gene flow with *M. guttatus*. MacNair and Gardner (1998) suggest that selection for metal tolerance may lead to speciation if the traits that are adaptive in the metal environment happen to cause reduced gene flow as a byproduct.

Some serpentine endemics do not fit such a model, however, because they are more closely related to other serpentine endemics than to nonserpentine taxa. In these species the allopatric speciation model may be more applicable, with isolated serpentine outcrops providing the separation necessary for reproductive isolation to develop. For example, the genus *Harmonia* (Asteraceae) consists of three species endemic to serpentine and one endemic to volcanic outcrops within the Klamath Mountains and Northern California Coast Ranges (Baldwin 1999). Molecular evidence indicates that the serpentine endemics form a clade, and that the species that now grow sympatrically are not sister taxa but probably represent secondary contact following allopatric divergence (B. Baldwin, personal communication). The same may be true of the genus *Hesperolinon* (Linaceae), which consists of eight serpentine endemics and five species that grow on and off serpentine (McCarten 1988, Fiedler 1992).

The genus *Streptanthus* consists of 40 species, of which 15 are partly or entirely restricted to serpentine (Kruckeberg 1984). Experimental, electrophoretic, and molecular analysis of the *Streptanthus glandulosus* complex suggests a strong role for allopatric divergence (Kruckeberg 1957, Mayer et al. 1994, Mayer and Soltis 1999). This complex includes the endemics (*S. glandulosus* ssp. *pulchellus*, *S. albidus*, and *S. niger*), and several taxa with serpentine ecotypes (*S. glandulosus* ssp. *glandulosus* and ssp. *secundus*). Allozyme frequencies in 56 populations of these taxa indicated a high degree of differentiation, and genetic distance increased with geographic distance (Mayer et al. 1994), just as did intersterility (Kruckeberg 1957). Genetic variation within populations appeared too high to be consistent with colonization by long-distance dispersal (Mayer et al. 1994).

Molecular analyses of 31 populations largely confirmed these results (Mayer and Soltis 1999). These authors conclude that the complex originated via the fragmentation and reduction of a wide-ranging species similar to *S. glandulosus* ssp. *glandulosus*, followed by increasing divergence among populations on isolated serpentine outcrops.

A molecular analysis of *Caulanthus amplexicaulus barbarae* and its relatives shows that serpentine endemism has arisen multiple times in the Streptanthoid complex, which includes *Streptanthus*, *Caulanthus*, and *Guillenia* (Pepper and Norwood 2001). These three genera are each nonmonophyletic but together form a single clade. Pepper and Norwood suggest that *Caulanthus amplexicaulus barbarae* diverged from its nearest relative, the nonserpentine *Caulanthus amplexicaulus amplexicaulus*, through the progressive restriction of the taxon during post-Pleistocene climate change. They also suggest that the prevalence of serpentine endemism in the Brassicaceae is related to some form of preadaptation to rocky, exposed environments.

9.5 Neoendemism, Paleoendemism, and Serpentine in California

The flora of the California Floristic Province is outstanding for its uniqueness, with nearly half its species (2125 of 4452) found nowhere else (Raven and Axelrod 1978). Of these Californian endemic (as opposed to serpentine endemic) species, many belong to certain genera that have their centers of diversity in the province. Most such groups are annual dicots (e.g., *Clarkia, Hemizonia, Lasthenia, Linanthus, Mimulus, Navarretia, Sidalcea*), although some are shrubs (e.g., *Arctostaphylos, Ceanothus*). Many of the species in these genera are poorly differentiated from one another and may hybridize, presumably because they have not had time to evolve complete reproductive isolation. These are considered classic "neoendemics," groups that are thought to have diversified in California since the present dry climate began to develop in the middle Pliocene (Stebbins 1942, Stebbins and Major 1965, Raven and Axelrod 1978).

Other California Floristic Province endemics are taxonomically isolated and have their nearest relatives in distant, wetter climates (e.g., *Fremontodendron, Sequoia, Umbellularia*). These are considered classic "paleoendemics," species whose present-day ranges are thought to be the result of contraction from more widespread distributions during the wet climate of the Tertiary (Stebbins 1942, Raven and Axelrod 1978). The greatest number of paleoendemics are found in the Klamath-Siskiyou Mountains, the Santa Cruz Mountains, and other regions of the province that have summer rainfall and moderate winters (Stebbins and Major 1965, Raven and Axelrod 1978).

Neoendemism and paleoendemism are different pathways by which species become restricted to serpentine. Concepts applied by Kruckeberg (1954, 1984) and Raven and Axelrod (1978). Serpentine neoendemics are species presumed to have arisen recently on serpentine from neighboring nonserpentine ancestors, whereas serpentine paleoendemics are taxa thought to have once been more widespread, and to have become restricted through the disappearance of their nonserpentine populations. Raven and Axelrod (1978) argued that serpentine paleoendemics are relicts from the Tertiary period,

Figure 9-4 The serpentine endemic shrubs *Cupressus macnabiana* and *Quercus durata*. Photo by S. Harrison.

although the reason for their disappearance from other substrates during Plio-Pleistocene climatic drying is not well explained.

The majority of Californian serpentine endemics show characteristics associated with neoendemism, such as small geographic ranges, many closely related taxa nearby, and often problematic taxonomic status (Kruckeberg 1984, Skinner et al. 1995). A minority, such as *Cupressus sargentii*, *C. macnabiana*, and *Quercus durata* (fig. 9-4) show the broad distributions, if not always the taxonomic isolation, associated with paleoendemism. However, as the *Harmonia*, *Streptanthus*, and *Caulanthus* cases illustrate, real taxa that have been well studied do not fit neatly into these two hypothetical modes of origin.

In any case, it is clear that serpentine endemism has been an important contributor to the high endemism of the California Floristic Province. Stebbins and Major (1965) conclude that the regions that are richest in Californian neoendemics combine diverse geology—especially serpentine—with strong moisture gradients, and Kruckeberg (1984) notes that 10% of endemics to the province are also endemics to serpentine. Of the 1742 plants that have been designated rare or endangered in California, 612 are associated with special substrates: 282 with serpentine, 103 with granite, 89 with carbonate rocks (limestone and dolomite), 77 with volcanics, and 61 with alkaline soils (Skinner and Pavlik 1994).

9.6 Shifts in Species Distributions on Serpentine

Serpentine and other unusual soils often affect the geographic distributions of individual species. Species that occur exclusively on serpentine in some parts of their ranges but on other substrates elsewhere have been termed *regional indicators* (Kruckeberg 1984) or *regional endemics* (Rune 1953). Examples from western North America include *Pinus jeffreyi*, *Calocedrus decurrens*, *Arctostaphylos nevadensis*, and *Quercus vaccinifolia*, which are widespread in the Sierra Nevada and/or Great Basin but are largely confined to serpentine in the westernmost parts of their distributions in the Klamath-Siskiyou Mountains and the California Coast Ranges (Whittaker 1960, Kruckeberg 1984). There are also numerous examples of species from the southern California deserts that reach their northward range limits on California Coast Ranges serpentines. These include *Juniperus californicus*, *Eriogonum wrightii*, *Fremontodendron californicum*, and *Arctostaphylos glauca*. In central Alaska, *Agropyron spicatum*, *Artemisia alaskana*, and *Carex eburnea* show northward range extensions on serpentine (Juday 1992).

Many similar examples are found in eastern North America, northern and central Europe, Japan, and Cuba (Rune 1953, Brooks 1987, Borhidi 1996). The most common pattern is for species to be confined to serpentine in the more climatically favorable parts of their ranges, but to also be found on other soils in the less favorable parts. For example, species often shift their elevational distributions downward on serpentine, sometimes by 1000 m or more (Brooks 1987, Borhidi 1996). Boreal and subarctic species often reach their southerly range limits on serpentine (Rune 1953, Brooks 1987). In the mesic

Klamath-Siskiyou region, range disjunctions on serpentine often involve plants whose main distributions are in colder or more arid climates (Whittaker 1960). Kruckeberg (1969, 1992) also notes the general tendency for xeric species to shift onto serpentine in more mesic settings.

One general explanation for this pattern involves an interaction among competition, climate, and substrate (Rune 1953, Raven and Axelrod 1978, Brooks 1987). Brooks (1987) summarized this explanation for disjunctions in Europe: because of competitive pressure, some plants are restricted to either to the edaphically harsh environment of ultramafics or to the climatically harsh environment of more northerly regions. Rune (1953) linked the same pattern to the concept of paleoendemism, speculating that regional endemics such as *Arenaria norvegica*, which is only restricted to serpentine south of 66° N, was outcompeted on other substrates as forest invaded southern Sweden in late postglacial times. In other words, regional indicators such as *A. norvegica* are partial paleoendemics.

An alternative explanation is that, regardless of competition, serpentine provides a microclimate that is warmer and drier than its surroundings. Whittaker (1960) argued that the open conditions on serpentine make it equivalent in terms of moisture balance to a site of lower rainfall, so that the range disjunctions he observed are simply a matter of maintaining constancy in the overall climatic envelope of a species. Borhidi (1996) explained elevational range shifts in the tropics similarly, noting that within the rainforest zone, dry and nutrient-poor serpentine soils support sclerophyllous shrub forest, and that many of the plants that can tolerate such xeric conditions come from the elfin forests at higher elevations.

Whittaker (1960) also documented what he called "shifts to the mesic" even within the ranges of species that do not show large-scale range disjunctions. Where a given species exists on multiple substrates and there is a moisture gradient created by elevation or topography, the species' peak abundances are at more mesic sites on serpentine and more xeric sites on diorite or gabbro. Again, Whittaker attributed this pattern to the warm and open microclimate created by serpentine.

10 Serpentine Plant Assemblages: A Global Overview

Serpentine substrates are found in many parts of the world, but there is considerable variation in the structure, composition, and diversity of the flora they support. To place western North America in a worldwide context, this chapter provides a brief sketch of global patterns in serpentine plant life, drawing on the reviews by Brooks (1987), Baker et al. (1992), and Roberts and Proctor (1992), as well as other sources. Following this is an overview of some of the main physical factors known to cause variation in the vegetation on serpentine both at the regional and local levels. Finally, we discuss what is known about the roles of competition, fire, herbivory, and other ecological processes in shaping plant assemblages on serpentine.

10.1 Global Patterns in the Vegetation and Flora

The availability of botanical information varies considerably around the world. In most countries where serpentine occurs, it is possible to name at least some of the plant species and vegetation types found on it. But in countries where surveys are incomplete, or where information has not been synthesized at a national or larger level, it is generally not possible to estimate the number of serpentine-endemic taxa or to describe patterns of variation within the serpentine vegetation. Indonesia, Malaysia, the Phillippines, and Brazil are particularly notable as countries with serpentine floras that are potentially rich but in need of more study. With this caveat, however, some of the major global trends can be described based on available knowledge. Flora and vegetation of selected parts of the world are summarized in table 10-1, and global contrasts between the vegetation of serpentine and other soils are summarized in table 10-2.

Table 10-1 Serpentine flora and vegetation of selected regions of the world.

Cuba (Brooks 1987, Borhidi 1996)

Well-represented families: Euphorbiaceae, Melastomataceae, Myrtaceae, Rubiaceae

Major vegetation types: (1) Xerophytic shrublands; central to eastern Cuba, <1000 m elevation, 1000–1600 mm annual rainfall. (2) Montane sclerophyllous shrubwoods; eastern Cuba, 600–1000 m, >1000 m annual rainfall. (3) Montane pine forests; eastern Cuba, 600–900 m on latosols. (4) Montane sclerophyllous rainforest; eastern Cuba, 400–900 m elevation, 1800–3200 mm annual rainfall.

New Caledonia (Muller-Dombois and Fosberg 1998, Jaffré 1992, Proctor and Nagy 1992)

Well-represented families: Euphorbiaceae, Myrtaceae, Orchidaceae, Rubiaceae

Major vegetation types: (1) Evergreen rainforest, 15–25 m stature, 500–1000 m elevation, 1500–3500 mm annual rainfall. (2) Montane evergreen forest, 8–15 m stature, >1000 m elevation, >3500 mm annual rainfall with cloud cover. (3) Sclerophyllous scrublands (maquis), 0–1600 m elevation, 900–4000 mm annual rainfall.

Malay Archipelago (Brooks 1987, Proctor and Nagy 1992, Aiba and Kitayama 1999)

Well-represented families: Myrtaceae, Rubiaceae, many others

Major vegetation types: (1) Lowland evergreen dipterocarp rainforest, 40–50 m stature. (2) Sclerophyllous scrub and woodland, >100 m elevation in several locations. (3) Alpine scrub derived from higher-elevation nonserpentine vegetation, >2400 m elevation on Mt. Kinabalu (Malaysia).

Turkey (Brooks 1987, Reeves et al. 2000)

Well-represented families: Brassicaceae, Fabaceae

Major vegetation types: (1) Xerophytic "heathlands" dominated by perennial herbs, <1000 m elevation. (2) *Pinus brutia*–oak woodland, <1000 m elevation. (3) *Pinus nigra* woodland, >1000 m elevation.

Japan (Brooks 1987, Mizuno and Nosaku 1992)

Well-represented families: Rosaceae, Cyperaceae, Asteraceae, Ericaceae

Major vegetation types: (1) *Pinus densiflora* woodland with xerophyllous shrub understory. (2) Conifer forest (spruce, cedar, cypress, hemlock, pine) with shrub understory. (3) Spruce forest. (4) *Pinus pentaphylla* forest. (5) Alpine meadow and heath.

Eastern North America (Brooks 1987, Tyndall and Hull 1999)

Well-represented families: Caryophyllaceae

Major vegetation types: (1) Pine woodland, pine-oak savanna, and open rocky grasslands, succeeding to closed pine forest in the absence of fire (Maryland, Pennsylvania). (2) Birch-juniper scrub and open barrens (Quebec, Newfoundland).

New Caledonia and Cuba lead the world in known serpentine endemic diversity with 900+ species each, >90% of which are also endemics to these islands. Depending on elevation, rainfall, and fire history, the serpentine vegetation on both islands varies from sclerophyllous scrubland that contrasts visibly with the neighboring vegetation, to medium-stature rainforest that is not strikingly different in appearance from the vegetation growing in other soils. The extensive but little-known serpentines of the Malay Archipelago

Table 10-2 Characteristic contrasts between serpentine and neighboring nonserpentine vegetation.

Location	Nonserpentine to serpentine change	Source
Northwestern U.S.	Conifer forest to open woodland or barrens	Kruckeberg 1992
Southwest Oregon	Conifer forest to conifer woodland and scrub	Whittaker 1960
Central California	Oak woodland to chaparral or grassland	Kruckeberg 1992
Eastern Canada	Conifer forest to open conifer barrens	Roberts and Proctor 1992
Eastern U.S.	Broadleaf–conifer forest to conifer savanna	Tyndall and Hull 1999
Cuba		Borhidi 1996
West–central	Semideciduous forest to evergreen thicket	
East, lowland	Evergreen forest to evergreen thicket and pines	
East, montane	Rainforest to sclerophyllous rainforest	
Goias, Brazil	Dense forest to scrub	
Scandinavia	Conifer forest to open barrens	Brooks 1987
Northern Europe	Depends on soil depth and development	Slingsby 1992
Greece	Beech forest to oak woodland and shrubland	
Urals	Taiga to stunted pine–larch or larch–birch heath	
India	Evergreen rainforest (sal) to scrub	
Malaysia (Mt. Kinabalu)		Brooks 1987
Lowlands 1400 m	Evergreen rainforest on both substrates dipterocarp forest to *Agathis–Buxus–Borneodendron* forest	
>2400 m	Low forest to sclerophyllous scrub	
Phillippines (Mt. Bloomfield)	Evergreen rainforest to sclerophyllous scrub	Proctor et al. 1997
Malaysia (Mt. Gunung Silam)	Evergreen rainforest on both substrates	Proctor et al. 1988
New Caledonia		Jaffré 1992
<1000 m	Savanna or open scrub to closed scrub	

(*continued*)

178 Plant Life on Serpentine

Table 10-2 (continued)

Location	Nonserpentine to serpentine change	Source
>1000 m	Low evergreen rainforest on both substrates	
West Australia	Eucalypt/acacia woodland on both substrates	Brooks 1987
East Australia	Eucalypt forest to stunted eucalypt savanna	Davie and Benson 1997
New Zealand		Reeves 1992
<1000 m	Beech forest to microphyllous scrub	
>1000 m	Alpine shrubland to grassland or barrens	
South Africa	Grassland on both substrates	Balkwill et al. 1997
Zimbabwe	Woodland and shrubland to grassland	Reeves 1992

may belong in this top tier of richness as well, although those islands have been much more recently interconnected through the Sunda Shelf and the rest of Southeast Asia than either New Caledonia or Cuba have to their respective mainland neighbors.

Malaysian and New Caledonian serpentine vegetation ranges from lowland rainforest to scrubland and tends to contrast weakly with the nonserpentine vegetation except at montane elevations. Recent investigations on shifts between serpentine shrubland and forest in both New Caledonia and in the Malay Archipelago (McCoy et al. 1999, Proctor 2003) suggest that fire history plays a strong role in the persistence and the extent of the maquis (tropical scrub) in both areas. Aiba and Kitayama (1999) suggest that on Borneo at elevations below 700 m the species composition and structure between serpentine (ultramafic) and nonserpentine vegetation is similar, but as elevation increases, more dissimilarities appear. The decline in forest stature with altitude is steeper on the ultramafic substrates than on the non-ultramafic substrates, and upper montane tree species diversity is generally lower on ultramafic substrates than on non-ultramafic substrates. Proctor et al. (1999) concluded that on Mt. Bloomfield in the Phillipines the difference in stature of forests on greywacke and on serpentine was not related to soil chemistry but to soil moisture availability. The authors concluded that soil water retention, perhaps in combination with fire, is the major cause of stature differences among the vegetation types and, together with soil chemistry, it is an important determinant of floristic composition.

California, Japan, the Balkan Peninsula, and Turkey form a second tier in floristic richness. California has 215 known endemics and perhaps 400 total species that are moderately to strongly associated with serpentine (Kruckeberg 1984, Harrison et al. 2000b).

Stevanovi et al. (2003) listed 123 obligate serpentine taxa from the Balkan Peninsula and 335 taxa (species and subspecies) growing on serpentine. Studies in Japan and Turkey have not been fully synthesized at the national level, but endemic diversity appears to be comparably high. For example, in Turkey the genus *Alyssum* has 33 and the genus *Centaurea* has 15 serpentine-endemic species, respectively. Reeves et al. (2000) and Reeves and Adiguzel (2004) conjecture that of about 500 Turkish plant species that are known from only a single specimen, >100 are serpentine endemics. Two mountains in Japan (Mt. Yupari and Mt. Apoi) have 16 and 22 serpentine endemics, respectively, of which 6 and 14 are also restricted to that single mountain.

Greece, northern Italy, the Iberian Peninsula, New Zealand, South Africa (Changwe and Balkwill 2003), North Africa (Morocco, Ater et al. 2000), eastern Australia, and Zimbabwe follow with 10–40 serpentine endemics each. Regions with appreciable amounts of serpentine, but fewer than 10 endemics each, include northwestern North America (from central Oregon to British Columbia and Alaska), eastern North America (from Alabama to northern Quebec), northern and central Europe, and Russia. Within Europe, Brooks (1987) notes a northwest to southeast increase in the richness of serpentine endemics from Shetland (1), to central Europe (2–5), to Albania (25–30) and Bosnia (7), to the Euboa region of Greece (10–20).

Taxonomic similarity is relatively low among the regional serpentine floras of the world. In general, serpentine endemics tend to have narrow geographic ranges and to be most closely related to nonendemic taxa of their own regions, rather than to the serpentine endemics of other regions (although a handful of endemic species are shared between Northeastern America and Northwestern Europe; Roberts and Proctor 1992). Most serpentine endemics are drawn from a narrow subset of the families present in a given region (see table 10.1). For example, most of the endemics in eastern North America and northern Europe are in the family Caryophyllaceae, whereas southern European endemics are predominantly in Fabaceae and Brassicaceae. In California, families with many endemics include the Apiaceae, Asteraceae, Brassicaceae, Liliaceae, and Polygonaceae. In the wet tropics, not surprisingly, endemics come from a wide range of families; however, the Euphorbiaceae, Myrtaceae, and Rubiaceae appear to be important in multiple regions.

From this global sketch, incomplete as it is, we may infer the broad trends in the richness of the serpentine floras of the world. The highest numbers of endemics are associated with extensive areas (>5000 km^2) of serpentine in wet tropical regions with extremely rich source floras from which the endemics have presumably evolved, in the most extreme cases over millions of years of isolation (New Caledonia and Cuba). Within the temperate zone, the richest serpentine floras are associated with similarly extensive serpentine and with warm and wet climates, or at least with a great enough range of elevations and/or latitudes that the climate is not uniformly cold or arid.

Absence of a history of Pleistocene glaciation also appears to be significant. Brooks (1987) notes the near absence of serpentine endemics from areas that fall within the Wisconsinian glacial maximum. The influence of glaciation is highlighted by western

North America and northern Japan. The richness of the serpentine flora increases steadily northward through California until it reaches 90–100 species in the Klamath-Siskiyou region of northwest California and southwest Oregon, and then drops to fewer than 10 species from the rest of Oregon north to Alaska, almost certainly reflecting the influence of Pleistocene climate history. Unglaciated northern Japan is far higher in serpentine endemism than are any regions of comparable climate and latitude in formerly glaciated western Europe.

Of course, both contemporary climate and past glaciation are major influences on global plant diversity in general, and it is interesting to consider whether the effects of these factors on the richness of serpentine endemic floras have been especially strong. One might speculate that stressful soils and cold or arid climates act as filters through which few species in a source flora can pass, leading to depauperate floras when both conditions are present. In support of this idea, the diversity of serpentine endemic plants in California responds to the same factors as does plant diversity in general, including rainfall, latitude, and topographic variation, but it does so more strongly (Harrison et al. 2000b). Similarly, it may be that recolonization following glaciation is particularly hard for plants confined to scattered outcrops of serpentine, thus making serpentine soils even more species poor than other soils in areas with histories of glaciation.

Almost everywhere in the world, serpentine vegetation is described as being shorter in stature, lower in biomass, and more drought-adapted in foliar structure (e.g., small thick leaves with spines, hairs, or resins) compared with adjacent nonserpentine vegetation. Exceptions to this seem to occur at climatic extremes (table 10-2). As already described, the diverse serpentine vegetation of the wet tropics does not always show a markedly reduced stature; neither does that of hot, arid climates such as western Australia and southern Africa, where savannas are found both on and off of serpentine, and where the diversity of serpentine endemics is low. In between these extremes, serpentine vegetation in most regions of the world is usually marked by striking contrasts in structure compared with adjacent nonserpentine vegetation. However, there are local exceptions that are attributable to variation in soil and other properties. The regions richest in serpentine endemics are not necessarily those in which the serpentine vegetation is the most distinctive in its structure, nor vice versa.

10.2 Causes of Regional and Local Variation

Serpentine plant life varies at a regional-to-local scale in response to the same factors that influence plant life in general. But some factors, including moisture, elevation, soil depth, and other soil properties, may take on greater or lesser significance on serpentine. There are further sources of variation that are more unique to serpentine, such as the degree of serpentinization or the tectonic shearing of the parent material. In this section we briefly review some known causes of regional and local variation in the serpentine vegetation and flora, attempting to highlight those that seem to be significant on serpentine compared with other substrates.

10.2.1 Moisture and Temperature

In the Siskiyou region of southwest Oregon (Whittaker 1954, 1960) and in the Wenatchee Mountains of Washington (del Moral 1982), serpentine floras showed heightened changes in species composition and vegetation structure along topographic gradients in moisture and temperature compared with adjacent nonserpentine floras. In addition, the Siskiyou serpentine flora was more distinctive from the nonserpentine flora on drier than on wetter sites and was more southern in its floristic affinities than the nonserpentine flora, possibly as a result of the greater solar exposure caused by low canopy cover on serpentine (Whittaker 1954, 1960; Wilson 1988). Similarly, Kruckeberg (1969) noted a more distinctive serpentine flora and vegetation in dry inland than in mesic coastal northwestern North America; he suggested that the less stressful coastal climate in some way ameliorated the physiochemical effects of ultramafic substrates on plants.

In contrast to the above findings, species composition in grasslands in central California was highly responsive to temperature and moisture gradients on nonserpentine soils, but not on serpentine soils (McNaughton 1968). Similarly, in northern coastal California, woody species diversity and composition changed markedly along a coastal-to-inland climatic gradient in nonserpentine forests and woodlands, but not in the adjacent serpentine chaparral (Harrison 1997). Thus, we can conclude that the serpentine substrate interacts with geographic and topographic moisture gradients in complex ways, but that the direction of such interactive effects (i.e., the greater or lesser effect of moisture variation on serpentine than on nonserpentine soils) is not consistent across studies.

For California as a whole, the diversity of serpentine endemics is strongly positively correlated with rainfall, as is the diversity of the whole flora. The proportional diversity of endemics shows the same pattern, suggesting that the diversity of serpentine endemic plants responds even more strongly to gradients in moisture than does total plant diversity (Harrison et al. 2000b, 2004, 2006c). This relationship is manifested at the regional scale rather than the field-plot scale, indicating that the explanation must be evolutionary and biogeographic rather than ecological (Harrison et al. 2006c). One aspect of this pattern is that the floristic dissimilarity of north and south slopes increases along a gradient of increasing rainfall and productivity (Harrison et al. 2006b).

10.2.2 Elevation

Serpentine vegetation displays a compression or lowering of altitudinal zones on mountains in Malaysia (Proctor et al. 1988), Cuba (Borhidi 1996), and western North America (Whittaker 1960, Kruckeberg 1969), as well as in other places (Proctor and Woodell 1975). That is, a given vegetation type occurs at a lower altitude on serpentine than it does on other substrates. Proctor et al. (1988) consider that this may be a pure climatic effect unrelated to serpentine, while Borhidi (1996) speculates that it may result from an increasing harshness of the serpentine soil due to a reduction in biological activity at higher elevations. However, the altitudinal pattern could also be a special case of two related phenomena that are sometimes observed on serpentine: range shifts of individual species into more favorable climates (chapter 9) and heightened sensitivity of plant assemblages to gradients in moisture and temperature (see above).

10.2.3 Soils

Transitions from one type of serpentine vegetation to another, such as shrubland to grassland, are often found in association with variation in soil properties including calcium, magnesium, Ca/Mg ratio, nickel and other metals, depth, texture, and/or water-holding capacity (e.g., Lyons et al. 1974, Slingsby and Brown 1977, Koenigs et al. 1982, Del Moral 1982, McCarten 1992, Robinson et al. 1997, also see Brooks 1987). Sometimes such soil variation is attributed to age or topographic position, the degree of weathering and leaching soils of minerals, or varying mixtures of nonserpentine and serpentine parent material; in most cases, however, the causes are unidentified. These patterns are discussed in more detail in chapters 11 and 12 with respect to western North American serpentine vegetation.

Probably the most consistent pattern that can be drawn from the worldwide literature is that deeper serpentine soils often support a less distinctive serpentine vegetation and flora than shallower or rockier serpentine soils within the same region. In the United Kingdom, characteristic vegetation of serpentine barrens is found on exposed rocky debris, while vegetation differing little from that of other soils is found in sites where there are better-developed organic layers or wind deposits (loess) overlying the serpentine rock (Slingsby and Brown 1977, Brooks 1987). In New Caledonia and Cuba, young and shallow serpentine soils support a more distinctive serpentine vegetation than do old and deeply weathered red lateritic soils (Jaffré 1992, Borhidi 1996). In the Pacific Northwest of North America, steep, rocky slopes support a more characteristically distinctive serpentine vegetation than do gentler slopes with deeper alluvial soils, possibly as a result of differences in moisture availability (del Moral 1982).

10.2.4 Geology

Mafic rocks such as gabbro may support a similar but less distinctive flora than ultramafic rocks such as peridotite, dunite, and serpentinite (e.g., Whittaker 1960, Oberbauer 1993). But almost no studies have examined the botanical effects of different ultramafic parent materials. Ultramafic rocks vary in their primary mineral composition: such as the relative amounts of olivine and pyroxene and the levels of trace metals; the degree to which they are serpentinized; and the kind and degree of subsequent mineral alteration they have undergone (e.g., by hydrothermal action). Yet such geologic variation has not been examined in detail as a potential cause of variation in the vegetation and flora.

Ultramafic rock barrens are one case in which geologic processes are responsible for producing special plant habitats within serpentine. In Scotland, the existence of talus with a distinctive flora was attributed to a particular form of serpentinization, "cross-fiber serpentinization," that produced especially friable rock (Proctor and Nagy 1992). In

California, barrens are found on unusual "detrital" serpentines that are thought to have intruded sedimentary formations as masses of shattered rock (Coleman 1996).

10.2.5 Age of Exposure of Serpentine

The scarcity of serpentine endemics in areas glaciated during the Pleistocene, discussed earlier, is one example of the importance of the availability of time for speciation to occur. At the opposite extreme, New Caledonia has been isolated in an equitable climate and continuously above the surface of the South Pacific for more than 80 Ma (Muller-Dombois and Fosberg 1998). It is a complex mountainous island, and its peridotite was emplaced between 38 and 29 Ma. It is no wonder then that there are so many serpentine endemics (900+) there.

The evolutionary time factor may also explain variation in serpentine endemic richness within Cuba. Eastern and western and Cuban serpentines, which are thought to have been exposed above sea level for the past 30 Ma, support many more endemics (750 species and 22 genera) than central Cuba (128 species and 0 genera), where serpentine is thought to have been exposed for 1 Ma (Borhidi 1996).

Within California, Raven and Axelrod (1978) argue that much of the evolution of serpentine endemics in the Sierra Nevada and Coast Ranges must have taken place since the late Pliocene (2–3 Ma) because oceans, sediments, or volcanic deposits generally covered the serpentines before that time. As much of the Klamath region is considered to be an old (Tertiary) land surface (Diller 1902), it is conceivable that a longer time of exposure may have contributed to the high diversity of serpentine endemics in this region. Harrison et al. (2004) used geologic map interpretation to estimate the minimum time that serpentine has been subaerially exposed in different regions of California. They categorized these ages as recent (<1 Ma), Pliocene (2–5 Ma), or Miocene and older (>5 Ma). They found that this measure of age was a significant predictor of the number of serpentine endemic species, in a multiple regression that also included the area of serpentine and the average precipitation (also see Harrison et al. 2006c).

10.2.6 Area and Spatial Distribution of Serpentine

Both the total amount of serpentine in a region and its distribution into islandlike outcrops of varying sizes and degrees of isolation may affect the diversity of the serpentine flora (Kruckeberg 1991). In California, there are more serpentine endemics in regions with a greater total area of serpentine (Harrison et al. 2000b, 2004). Within a single region of California, there were fewer endemic plant species per site on 24 small (<5 ha) outcrops than on 24 equivalently sized and spaced study plots within four large (>1 km^2) outcrops. The small outcrops had greater among-site differences in species composition, so that the total numbers of endemic species were equal on the 24 small-outcrop and the 24 large-outcrop sites (Harrison 1997, 1999a). Harrison and Inouye (2002) found that across California there was greater turnover (dissimilarity) in the serpentine flora than in the total flora, for a given degree of geographic or climatic distance, suggesting the importance of spatial discontinuity in causing differences in species composition. However, there was little indication that the plant community was affected by other aspects of the spatial distribution of serpentine, such as perimeter to area ratios or the isolation of individual outcrops (Harrison et al. 2006c).

10.3 Roles of Ecological Processes

Many studies of serpentine plant life have considered the roles of geographic and physical gradients, but relatively few have addressed the ecological interactions among species that may structure plant communities. Even fewer studies have compared the roles of ecological processes in serpentine versus adjacent nonserpentine communities or along gradients of biomass or soil properties within serpentine. Thus, we know relatively little about the extent to which ecological processes play particularly important roles on serpentine, compared with other soils.

10.3.1 Competition

Plant communities on low-fertility substrates such as serpentine are sometimes thought to be relatively uninfluenced by interspecific competition. Two recent studies from California grasslands are consistent with this idea. Removal of competitors did not enhance the emergence, survival, growth or reproduction of a native grass (*Vulpia microstachys*) in sparse grasslands on rocky serpentine outcrops, although removing competitors did enhance these outcomes in more productive grasslands on deeper serpentine and nonserpentine soils (fig. 10-1; Jurjavcic et al. 2002). In nonserpentine grasslands, the success of an experimentally planted exotic grass (*Dactylis glomeratus*) was lower on plots with higher species richness, but only if the plots were uncleared, suggesting that competition reduced the success of the invader. In serpentine grasslands, success of the same grass was higher in more species-rich plots, but only if the plots were cleared of competitors, suggesting that soil variation was the underlying cause (Williamson and Harrison 2002). As described below, fire has less effect on diversity in serpentine than in sandstone chaparral, which is also consistent with the idea that plant competition is lower on serpentine due to its infertility.

10.3.2 Herbivory

Casual observation suggests that herbivory is relatively low in serpentine plant communities. Hobbs and Mooney (1991) found that excluding above-ground herbivores had no effect on the plant species composition of a serpentine grassland. Similarly to plant communities on other unproductive soils, serpentine vegetation may often be less attractive to herbivores because of its tendency to have small, tough leaves defended with resins, spines and/or hairs. These foliar traits may reflect the fact that tissue loss to herbivory is more damaging to plant fitness on unproductive soils (Coley et al. 1985), or they may be adaptations to nutrient and water stress whose effects on herbivory are incidental.

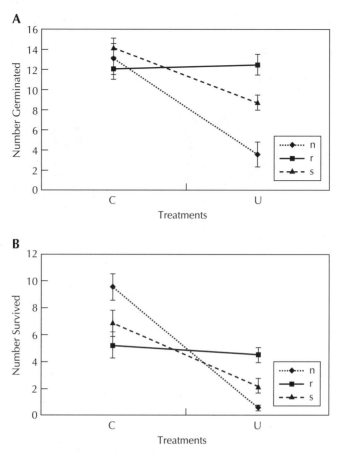

Figure 10-1 Competition had less effect on germination (A) and survival (B) of an experimentally planted native annual grass, *Vulpia microstachys*, in rocky serpentine outcrops (r) than in serpentine grassland (s) or nonserpentine grassland (n). Competition was assessed by planting *Vulpia* in the field, in a treatment in which all other plants were cleared (C) compared with uncleared controls (U). From Jurjavcic et al. (2001).

In addition, herbivory might be expected to have different impacts in serpentine than nonserpentine plant communities simply because of differences in initial plant community structure and species composition. Grazing by livestock tended to increase the diversity of native annual forbs in grasslands on serpentine soils by releasing them from competition with exotic annual grasses. However, the same level of livestock grazing tended to increase the diversity of exotic annual forbs on the already exotic-dominated grasslands on adjacent nonserpentine soils (fig. 10-2). The most likely reason for this different response is that low-statured native forbs—the most likely group of plants to benefit from moderate grazing—were more abundant on serpentine (Safford and Harrison 2001; Gelbard and Harrison 2003; Harrison et al. 2003).

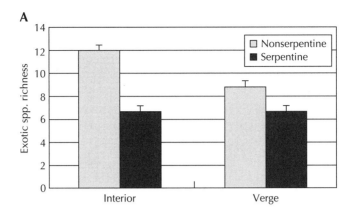

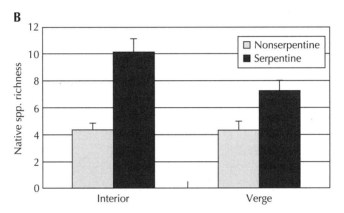

Figure 10-2 Livestock grazing promotes native species in serpentine grasslands, but promotes exotic species in nonserpentine grasslands. Exotic species richness (A) is higher in grazed "interiors" than ungrazed roadside "verges," but only in nonserpentine grasslands. Native species richness (B) is higher in grazed interiors than ungrazed verges, but only in serpentine grasslands. Richness = species/m^2, sampled at 92 sites near fence lines on 5 ranches in northern California. From Safford and Harrison (2001).

10.3.3 Fire

In eastern North America, periodic fire appears to be necessary to the maintenance of characteristic serpentine barrens vegetation, and fire suppression leads to the conversion of these barrens to pine forest (Tyndall and Hull 1999). In contrast, Cuban serpentine vegetation is so poorly fire-adapted that only about 50% of the species in a central Cuban evergreen thicket (cuabal) regenerated successfully after a fire (Matos Mederos and Torres Bilbao 2000).

Fire had effects similar to livestock grazing on grasslands in northern California, in that it enhanced total and exotic species richness more on nonserpentine soils and increased native species richness more on serpentine soils (Harrison et al. 2003). In chaparral in the same region, fire caused a greater increase in total, native, and exotic species richness on nonserpentine (sandstone) than on serpentine soils (Safford and Harrison 2004). Species richness in 50×5 m plots increased by 140% in sandstone chaparral, and by only 45% in serpentine chaparral (fig. 10-3). In contrast to sandstone chaparral, where many shrubs and herbs appeared to be obligately dependent on fire, there were few species in serpentine chaparral that depended on fire for germination or reproduction. Fires were also more frequent, more severe, and more uniform in their severity on sandstone than on serpentine chaparral (table 10-3). Because shrubs on serpentine may be more susceptible to fire-caused mortality and may recruit more slowly, they could be especially sensitive to excessively frequent or badly timed fires (Parker 1990).

10.3.4 Pollination

Plants restricted to serpentine outcrops might be expected to show different suites of pollination and seed dispersal syndromes than plants occupying more widespread or more productive habitats, but these hypotheses have never been tested at a community level. On small isolated outcrops compared with large areas of serpentine, the serpentine morning glory (*Calystegia collina*) received just as many pollinators but produced far fewer viable seeds per flower, probably because plants confined to small outcrops tended to receive incompatible pollen from closely related conspecifics (Wolf et al. 2000, Wolf and Harrison 2001). However, Wolf et al. (1999) did not find differences in the pollination or reproductive success of serpentine sunflower (*Helianthus exilis*) depending on its degree of spatial isolation.

10.3.5 Local Extinction and Colonization

Some of the patterns in species presence and absence within local sites are simply due to the random processes of local extinction and limited colonization. By creating small and isolated populations, the patchy nature of serpentine ought to enhance the role of these chance factors. The smaller numbers of endemic species on small serpentine outcrops compared with sites within large continuous serpentine outcrops, as discussed above, is consistent with this idea (Harrison 1997, 1999a). More evidence for local extinction and recolonization came from a study of 140 populations of five plant species confined to spring-seep habitats within a large serpentine outcrop. Over a 17-year period between two surveys, plant populations that were more isolated from other conspecific populations were more likely to disappear, and new populations were less likely to appear, compared with less isolated populations (Harrison et al. 2000a). One of these spring-seep species, *Helianthus exilis*, was absent from small isolated outcrops of serpentine (Wolf et al. 1999).

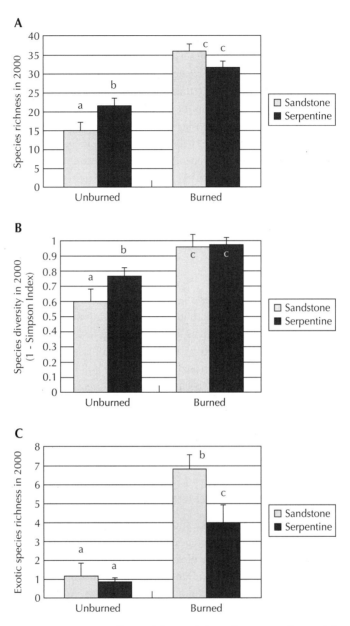

Figure 10-3 Lesser effects of fire on (A) total species richness, (B) species diversity, and (C) exotic species richness, in serpentine compared with sandstone chaparral. Means (± standard error) for burned and unburned sites in sandstone and serpentine chaparral. Bars with different letters are significantly different at $P < .05$. From Safford and Harrison (2004).

Table 10-3 Comparison of measures of fire severity and frequency on serpentine and sandstone chaparral (from Safford and Harrison 2004).

	Sandstone	Serpentine	t	df	P
Mean fire severity (mm)	5.1+1.9	3.3+1.3	3.9	38	<.005
Heterogeneity in severity (%)	34.8+15	50.4+20	2.8	38	<.05
Maximum severity (mm)	5.8+2.1	3.8+1.5	3.8	38	<.005
Heterogeneity in maximum severity (%)	39.9+21	51.2+20	1.8	38	n.s.
Mean time since last burn (years)	18.6+3.1	73.7+39	6.3	38	<.001

Values are calculated from five measurements along each of 40, 50-m transects. Fire severity is measured as the diameters of burned stem tips. Values given are mean + standard deviation. Mean and maximum severity were log transformed, heterogeneity measures were arcsin square root transformed. P values are based on Bonferroni-adjusted t-tests.

10.3.6 Invasion

Serpentine is renowned for its resistance to the invasive exotic species that have adversely affected so many plant communities around the world. The chaparral, barrens, and woodland found on most serpentines in California are largely free of exotic species, in sharp contrast to the majority of other plant communities in the state. Grasslands on serpentine are somewhat less intact but are still much more native-dominated than nonserpentine grasslands (Huenneke et al. 1990, Armstrong and Huenneke 1992, Harrison 1999b). The formation of soils from a mixture of serpentine and nonserpentine parent materials may lead to a gradation from native-dominated to exotic-dominated grasslands (McCarten 1992). Non-native species are more common on small serpentine outcrops and within 50 m of the edges of large outcrops than in the interiors of large outcrops (Harrison 1999a, Harrison et al. 2001).

In some areas the invasion of serpentine soils by exotics appears to be increasing. Of the 194 plants that are partly or wholly restricted to serpentine in California that are officially considered rare (Safford et al. 2005), exotics are considered to be among the principal threats for roughly 20 of them, mostly in the San Francisco Bay area (FWS 1998). Contributing factors include natural disturbance and rainfall variability (Hobbs and Mooney 1991), atmospheric nitrogen deposition (Weiss 1999), the evolution of increased serpentine tolerance in some exotic species, changes in microbial communities (Batten et al. 2006), and the spread of unusually serpentine-tolerant exotics such as barbed goatgrass (*Aegilops triuncialis*) (Meimberg et al. 2005).

Nonetheless, a recent statewide study of serpentine sites in California concluded that there is no evidence that sites with higher cover by exotic species support fewer native, serpentine-endemic, or rare species (Harrison et al. 2006a). Thus, with some localized exceptions, California serpentine appears to still be a comparatively safe refuge for its flora in the face of continued biological invasions.

11 Serpentine Plant Life of Western North America

Serpentine plant life varies dramatically across western North America from north to south and, to a lesser extent, from the coast inland. At the latitudinal extremes in Alaska and Baja California, it follows patterns seen in other climatically harsh parts of the world (as discussed in chapter 10), but the species composition is not very distinctive and there are few endemics. In between, in Washington, Oregon, and especially in the California Floristic Province, lies a great diversity of distinctive serpentine vegetation types and endemic species. This chapter outlines the coarse patterns of variation in vegetation structure and endemic species richness across this region, as a prelude to chapter 12, which describes specific serpentine vegetation types in detail.

11.1 Broad Patterns in the Flora and Vegetation

Little has been published about the serpentine vegetation of Alaska and the Yukon. The Serpentine Slide Research Natural Area in central Alaska was described by Juday (1992) as having a mixture of white spruce (*Picea glauca*) and paper birch (*Betula papyrifera*) with *Rosa acicularis*, *Juniperus communis*, and *Vaccinium uliginosum* in the understory. Several herbs are shared with Swedish serpentines (e.g., *Campanula rotundifolia*, *Minuartia rubella*, *Rumex acetosa*, *Saxifraga oppositifolia*, *Silene acaulis*); several others showed northern range extensions on serpentine (see chapter 9). The ultramafic vegetation of Golden Mountain in southeast Alaska was described by Alexander et al. (1989) as alpine meadow containing forbs, graminoids, and low shrubs, with a transition through low shrubs and stunted lodgepole pines (*Pinus contorta*) down to spruce (Picea sitchensis)–hemlock (mainly *Tsuga mertensiana*) forest, with shrubs and some cedar (*Chamaecyparis nootkatensis*). The forest to alpine transition was lower on serpentine than on other soils. No serpentine endemic species are known from this region.

From south–central British Columbia to central Oregon, serpentine has been described by Kruckeberg (1969, 1992) as supporting open stands of various conifers, which are either a subset of the species occurring in adjacent denser forests on other soils or represent elevational or geographic range shifts. Understories are sparse and may include graminoids, perennial forbs, and shrubs. On rocky ridgetops and at higher elevations and latitudes, these conifer plant communities give way gradually to alpine tundra. Five herb species are endemic to serpentine in this region, including the widespread ferns *Cheilanthes siliquosa* and *Polystichum lemmoni*.

Lewis and Bradfield (2003) note 35 local serpentine indicator species in their study area at Grasshopper Mountain in south-central British Columbia (North Cascades–Fraser River domain). Of these, 24 are dry-habitat herbs that are found elsewhere off serpentine. Lewis and Bradfield also found 37 species entirely or largely excluded from serpentine; 18 were deciduous broad-leaved trees and shrubs, and 11 were herbs from mesic to moist habitats. Thus, the pattern of xeric-habitat species occurring on serpentine within mesic climate zones (chapter 9) appears to hold true in this part of the continent as well.

From southwestern Oregon to the Transverse Ranges of southern California, in the botanical region known as the California Floristic Province, serpentine supports conifer forests and woodlands, chaparral, and grasslands. Scattered among these major vegetation types are special habitats including rock outcrops, barrens, seeps and springs, major riparian zones, and others. This diverse mosaic of serpentine environments is home to an estimated 246 strict endemics and about 400 more species that are positively associated with serpentine (Safford et al. 2005). Families and genera well represented among these endemics or near endemics include Apiaceae (e.g., *Lomatium*, *Perideridia*), Asteraceae (e.g., *Hemizonia*, *Hieracium*, *Layia*, *Lessingia*, *Madia*, *Senecio*), Brassicaceae (e.g., *Streptanthus*), Ericaceae (e.g., *Arctostaphylos*), Hydrophyllacae (e.g., *Phacelia*), Liliaceae (e.g., *Allium*, *Calochortus*, *Erythronium*, *Fritillaria*), Linaceae (*Hesperolinon*), Polygonaceae (e.g., *Chorizanthe*, *Eriogonum*), Scrophulariaceae (e.g., *Collinsia*, *Cordylanthus*, *Mimulus*), and Polemoniaceae (e.g., *Collomia*, *Navarretia*).

Within the California Floristic Province, there are an estimated 90–100 serpentine endemics in the Klamath–Siskiyou region, 90 in the Northern California Coast Ranges, 30 in the Bay Area, 35 in the Southern California Coast Ranges, and 16 in the Sierra Nevada foothills (Kruckeberg 1984). The richness of the Californian serpentine flora thus declines sharply both from north to south and from the California Coast Ranges inland to the Sierra Nevada. As noted in the chapter 10, these patterns in richness broadly correlate positively with rainfall and the area of serpentine (Harrison et al. 2000), and with the length of time that serpentine has been available for plant colonization (Harrison et al. 2004b).

The most ubiquitous serpentine shrubs, occurring from California to Alaska, are low juniper (*Juniperus communis*) in the more open woodland or forest stands and subalpine areas, and *Amelanchier alnifolia* in denser forest stands. Low juniper is a serpentine indicator in many areas, but *A. alnifolia* is more common on nonserpentine soils. Common names for *Amelanchier alnifolia* are Saskatoon to the north and service berry in California.

South of the California Floristic Province, the serpentines of the Baja Californian desert support a sparse cover of trees, shrub and cacti species that are also found on other soils. On the Calmalli ophiolite east of Guerrero Negro, trees include torote (*Bursera* spp.), shrubs include *Encelia farinosa*, *Fouquieria digueti*, and *Larrea divaricata*; cacti include *Machaerocereus gumosus*, *Pachycereus pringlei*, *Ferocactus* sp., and *Opuntia* sp. (locality 1-3 in chapter 13). Serpentine in the Vizcaino Mountains west of Guerrero Negro is characterized by sparse copalquin (*Pachycormus discolor*), some of the same "indifferent" taxa that occur in the Calmalli locality, and a possible serpentine endemic, the leguminous shrub *Errazurizia benthamii*. It occurs on serpentine and is reported as being restricted to the same area of the Baja California Sur, from Cedros Island to the Vizcaino Peninsula, where serpentine is abundant.

11.2 Major Vegetation Types

11.2.1 Alpine Tundra and Fell-Fields

Alpine tundra and fell-fields are treeless areas above forested mountain slopes. They lack the permafrost characteristic of Arctic tundra. Low and cespitose shrubs and subshrubs (plants that look like forbs, but have woody stems) provide good ground cover in the tundra, but poor ground cover in the fell-fields. Alpine tundra and fell-fields are above about 2000 m in the Blue Mountains, slightly lower coastward, slightly higher on the Stillwater complex, and above 500–600 m on Caribou Mountain in north–central Alaska.

Alpine tundra was observed on serpentinized peridotite in north–central Alaska (locality 10-4), south–central Alaska (locality 9-4b), northwestern British Columbia (locality 9-8), and on pyroxenite in the Stillwater complex of the Rocky Mountains. Plants common to all of these localities are cespitose dwarf willow (*Salix* sp.) and mountain-avens (*Dryas dummondii* at locality 9-8 and *D. octapetala* at the other localities). Alpine fescue (*Festuca altaica*) and *Carex bigelovii* are common graminoids in domains 9 and 10, and a low form of tufted hairgrass (*Deschampsia cespitosa*) is common on Chrome Mountain in the Stillwater complex. Mountain heather (commonly *Cassiope tetragona*) and one or more species of sandwort (*Minuartia* spp.) and lupine (*Lupinus* spp.) are generally present, as is commonly silene (*S. acaulis*).

Serpentine alpine fell-fields are represented by the Baldy Mountain area of locality 6-1 in the Blue Mountains. The only notable shrubs are prostrate buckwheat (*Eriogonum* spp.). The cover of forbs is greater than that of buckwheat, until the forbs are desiccated during summer, and there are some grasses. Trees occur at slightly higher elevations than Baldy Mountain on nonserpentine soils in the Strawberry Range. Baldy Mountain is on the north side of the Strawberry Range.

11.2.2 Northern Conifer Woodland and Forest

From British Columbia and southeast Alaska to Oregon, most serpentine vegetation consists of open conifer stands with understories of graminoids, forbs, and shrubs. These conifer plant communities commonly contrast strikingly with the adjacent rich coniferous forests of *Picea*, *Abies*, *Pseudotsuga*, or *Thuja* on nonserpentine soils. At high elevations and latitudes, this serpentine vegetation grades into treeless alpine tundra or fell-fields. The endemic fern *Aspidotis densa* is found throughout this region.

Dominant trees in conifer woodland and forests of serpentine in Alaska, Yukon Territory, and northern British Columbia (domains 9, 10, and the northern part of 8) and locality 8-1) are white spruce (*Picea glauca*) and paper birch (*Betula papifera*), with subalpine fir (*Abies lasiocarpa*) appearing in the southern, inland part of this area. In southeast Alaska (domain 8), the characteristic trees are Sitka spruce (*Picea sitchensis*), hemlock (*Tsuga mertensiana* and *T. heterophylla*), and sparse yellow cedar (*Chamaeciparis nootkatensis*). The hemlocks are common serpentine species along the coast down to Washington (locality 7-5).

From southern British Columbia to northwestern Washington (domain 7) and northeastern Oregon (domain 6), the forests of serpentine habitats are commonly Douglas-fir (*Pseudotsuga menziesii*) at lower elevations and on south-facing slopes and subalpine fir (Abies lasiocarpa) on north-facing slopes and at higher elevations. These forests grade upward through whitebark pine (*Pinus albicaulis*) to alpine subshrubs and herbs above about 2000 m, depending on the slope aspect, or lower coastward. Ponderosa pine (*Pinus ponderosa*) is common on south-facing slopes at lower elevations on the eastern side of the Cascade Mountains and in the Blue Mountains. Rocky Mountain juniper (*Juniperus scopulorum*) woodlands occur at low elevations on serpentine soils in the Blue Mountains (domain 6), with wheatgrass (*Pseudoroegneria spicata*) in the understory, and on Fidalgo Island (locality 7-4), but they are not extensive.

The only serpentine endemics found in British Columbia are *Aspidotis densa* (*Cheilanthes siliquosa* as discussed in Kruckeberg 1969) and *Polystichum lemmoni*, and near-endemics include the ferns *Polystichum scopulinum* and *Adiantum pedatum aleuticum* (Kruckeberg 1969). Endemics in northwestern Washington include *Aspidotis densa*, *Polystichum lemmoni*, *Poa curtifolia*, *Lomatium cuspidatum*, and *Chaenactis thompsoni*, the last three of which are restricted to the Wenatchee Mountains. Nonendemic herbs found on serpentine include many species of Polygonaceae and Caryophyllaceae. Endemic in the Blue Mountains are *Polystichum lemmoni* and *Aspidotis densa*; indicators include *Arenaria obtusiloba*, *Cerastium arvense*, and *Thlaspi alpestre* (Kruckeberg 1969).

Lodgepole pine and aspen are common successional species, but mostly pine grows on serpentine in this area. There are some dense lodgepole pine stands on serpentine soils (e.g., in locality 8-2, Shulaps). In general, most serpentine vegetation does not depend on fire to maintain an open structure as do the serpentine barrens of the eastern United States (Tyndall and Hull 1999).

11.2.3 California Conifer Woodland and Forest

In the Klamath–Siskiyou region of southwest Oregon and northwest California (domains 4 and 5), serpentine generally supports mixed conifer woodlands with *Pinus jeffreyi*, *Calocedrus decurrens*, *Pinus monticola*, *Pseudotsuga menziesii*, *Abies concolor*, and *Pinus lambertiana* among the dominant trees. Tree cover is generally 20%–50% but can reach >70% along streamsides and on lower slopes, which support rich, mixed-conifer forests that contain *Cupressus lawsoniana* in the most mesic locations. *Pseudotsuga menziesii* becomes more dominant in slightly drier sites, where it may be accompanied by a mixture of broadleaved evergreen trees in shrubby form (e.g., *Arbutus menziesii*, *Lithocarpus densiflorus*, *Umbellularia californica*). Moving upslope, *Pinus lambertiana* is common in slightly warmer upland locations, while *Pinus monticola* is prevalent on cooler coastward sites and above 1400 m. The most xeric upper slopes and ridgetops support *Pinus jeffreyi* woodland. Open barrens with sparse *Pinus jeffreyi* and a *Festuca idahoensis* understory are found on unstable rocky slopes (Whittaker 1960, Sawyer and Thornburgh 1988, Jimerson et al. 1995).

Soil depth, tree cover, tree productivity (basal area), and modal stand age generally decline along the mesic to xeric gradient (Whittaker 1960, Jimerson et al. 1995). The understory changes from a continuous shrub layer on mesic sites, through a shrub and grassland mosaic, to grassland in the most xeric *Pinus jeffreyi* woodlands. *Quercus vaccinifolia* is the dominant understory shrub below 1400 m, but others include *Arctostaphylos*, *Amelanchier*, *Ceanothus*, *Garrya*, *Rhamnus*, and *Vaccinium* spp. Above 1400 m, *Arctostaphylos nevadensis* replaces *Quercus vaccinifolia* (Whittaker 1960). The richness of the herb layer increases along the mesic to xeric gradient, as do the numbers of serpentine endemic and/or rare species (Whittaker 1960, Atzet and Wheeler 1984, Jimerson et al. 1995).

This serpentine vegetation is considered to be fire adapted and is probably somewhat more open than it would be in the absence of periodic fires, although its species composition and open structure are primarily due to the effects of the soil (Whittaker 1960). Fire frequency and severity evidently vary with topographic position, because modal tree ages range from 326 to 375 years for mesic *Chamaecyparis lawsoniana* stands to <125 years for xeric *Pinus jeffreyi* woodlands. Localized monodominant stands of *Pinus attenuata* or *Pinus contorta* are considered to be successional stages associated with past fires. Fire may be important for preventing shrub *Quercus vaccinifolia* dominance and for maintaining the open conditions needed by some rare herbs (Jimerson et al. 1995).

Serpentine in the southern, inland part of the Klamath–Siskiyou region (Rattlesnake Creek Terrane) supports sparse woodlands of *Pinus jeffreyi* and *Calocedrus decurrens*, with *Pseudotsuga menziesii* on rocky soils. Coastal forest species such as *Pinus monticola* and *Lithocarpus densiflorus* are absent, and *Quercus durata* presages the appearance of chaparral communities farther south.

Conifer woodlands and forests are also found on serpentines of the northern Sierra Nevada (fig. 11-1, domain 2). Lower elevation stands are dominated by *Pinus ponderosa*, *Abies concolor*, *Pseudotsuga menziesii* and *Pinus contorta*, with *Quercus vaccinifolia* and a

Figure 11-1 Jeffrey pine woodland on serpentine at Frenchman Hill, in the northern Sierra Nevada of California. Photo by H. Safford.

sparse herb layer. Higher elevations support *Pinus jeffreyi* woodlands with a sparse understory. Endemics are much fewer than in the Klamath-Siskiyou region, but include *Polystichum lemmoni, Arabis constancei, Lomatium marginatum,* and *Sedum albomarginatum,* found mostly on rocky outcrops (Clifton 2001).

In addition to these dominant coniferous forests and woodlands, there are localized stands of closed-cone conifers and junipers on serpentine throughout the California Floristic Province. *Pinus attenuata* forms dense single-aged stands, usually on serpentine or other shallow rocky soils (e.g., sandy, granitic, or volcanic substrates), and often on hilltops that receive moisture from clouds or fog. This tree is strictly dependent on fire for reproduction, and trees may senesce at around 50 years old. *Cupressus sargentii,* a serpentine endemic, forms monodominant stands along streamsides or on mountaintops that receive cloud or fog moisture, throughout the California Coast Ranges from Mendocino to Santa Barbara County. *Cupressus macnabiana* is found on serpentine and other rocky substrates in the inner Northern California Coast Ranges, Sierra Nevada foothills, and Cascade Range. *Cupressus bakeri* is restricted to serpentine in the Siskiyous but is found elsewhere on volcanics. All three cypresses tend to be killed by fire and to recruit immediately after fire, but there is some survival of fire by adult trees and also some recruitment during fire-free intervals (Vogl et al. 1988). *Juniperus californicus* is widespread on nonserpentine soils in the southern Californian deserts but ranges north through the California Coast Ranges and Sierra Nevada foothills as isolated stands on dry rocky outcrops, including in some cases serpentine (Vasek and Thorne 1988).

11.2.4 Chaparral

In much of the California Floristic Province below 1500 m, serpentine supports chaparral (evergreen shrub) vegetation, with shrubs that are typically <2 m high and have xeromorphic features such as small, thick leaves (domains 2, 3, and 4). Serpentine chaparral generally has lower cover (30%–60%) and lower species dominance than other types of chaparral in the Floristic Province. Its structure often resembles desert vegetation, with large intershrub gaps that contain small grasses and annual herbs but little litter (Hanes 1988). Common shrubs include such endemics or near-endemics as *Quercus durata, Ceanothus jepsoni, Arctostaphylos viscida* ssp. *pulchella*, and *Garrya congdonii*, as well as nonendemics such as *Adenostoma fasciculatum, Heteromeles arbutifolia*, and a shrubby form of *Umbellularia californica*. Emergent *Pinus sabiniana* are common, especially on north slopes that have not burned recently (McCarten and Rogers 1991).

The most species-rich mixed serpentine chaparral is found in the northern Coast Ranges (fig. 11-2). In this region, there is an outstanding diversity of rare and endemic species in the herb layer of serpentine chaparral (Kruckeberg 1984, McCarten and Rogers 1991, Callizo 1992, Sawyer and Keeler-Wolf 1995). Many of the endemic herbs are narrowly distributed species of *Hesperolinon, Streptanthus, Navarretia*, and *Phacelia* inhabiting the rocky gaps between shrubs. The diversity of endemic species is greatest on the most calcium-poor serpentine soils (Harrison 1999a).

Mixed stands of serpentine chaparral in the Northern California Coast Ranges give way to nearly pure stands of *Arctostaphylos viscida* or *Cupressus macnabiana* on deep, heavily

Figure 11-2 Leather oak chaparral on serpentine at the Knoxville locality in the North Coast Ranges of California. Dominant shrubs include the serpentine endemics *Quercus durata, Arctostaphylos viscida, Ceanothus jepsoni, Garrya congdonii*, and *Cupressus macnabiana*. Photo by S. Harrison.

weathered red clay soils, and to sparse stands dominated by *Quercus durata* on shallow rocky south-facing slopes (Koenigs et al. 1982, McCarten and Rogers 1991). *Adenostoma fasciculatum*, usually sparse on true ultramafic soils, becomes dominant where calcium levels are higher (Koenigs et al. 1982), such as on gabbro- or greenstone-derived soils. Stands of *Cupressus sargentii* or *Pinus attenuata* may replace serpentine chaparral on mountaintops where cloud or fog moisture is available. The herb layer is much less diverse in monodominant *Adenostoma fasciculatum*, *Arctostaphylos viscida*, *Cupressus* spp., or *Pinus attenuata* stands than in mixed stands. In the Northern California Coast Ranges, northern Sierra Nevada, and locally in the Southern California Coast Ranges (e.g., Mount San Benito) serpentine chaparral becomes the understory of montane conifer forest (*Pinus jeffreyi*, *Pinus ponderosa*, *Calocedrus decurrens*, etc.) at around 1500 m.

In the northern and central Sierra Nevada, the woody vegetation on serpentine is less species-rich, often consisting mainly of *Ceanothus cuneatus* or *Arctostaphylos viscida* with scattered emergent *Pinus sabiniana*. The herb layer is often outstandingly rich in native species but includes very few endemics (Medeiros 1984, Stebbins 1984, Farve 1987). In the southern Sierra Nevada, *Pinus sabiniana* and *Ceanothus cuneatus* give way to grasslands and contrasts between the serpentine and nonserpentine vegetation become relatively weak (Latimer 1984). In the Southern California Coast Ranges and the Transverse Ranges, varied forms of serpentine chaparral and cypress woodland ultimately give way to endemic-poor grasslands.

Serpentine chaparral is clearly adapted to fire, as are other types of chaparral. However, fire frequency and severity may be considerably lower because of the sparse and patchy fuel load, and the regeneration of the woody vegetation may be slower because of the infertility of the soil. In serpentine compared with adjacent sandstone chaparral in the inner Northern California Coast Ranges, the diversity of the herb layer increased less in response to a fire, but it also declined more slowly to its prefire values; these changes paralleled a slower recovery of the woody biomass and cover (Safford and Harrison 2004).

11.2.5 Grasslands

Grasslands are most prevalent on serpentine in the San Francisco Bay Area and adjacent parts of the California Coast Ranges (fig. 11-3, domains 3 and 4). In the Bay Area, more than 60% of the serpentine vegetation consists of a mixture of native perennial bunchgrasses, native forbs (mostly annuals), and exotics (mostly annual grasses) (McCarten 1993). Elsewhere in the state, grasslands may appear on small islands of alluvial or colluvial soil within more widespread chaparral or forest vegetation. However, they are absent from steep, rocky, highly erodible serpentine slopes (McCarten and Rogers 1991). *Nassella pulchra* may appear in almost pure stands on serpentine ridges in the coastal prairie belt (Heady 1988), and *Nassella lepida* grasslands are found on serpentine in the foothill zone of the Sierra Nevada and interior California Coast Ranges (Sawyer and Keeler-Wolf 1995).

Within the Bay Area, bunchgrass grasslands are found on the deepest serpentine soils with high moisture content and a low Ca/Mg ratio. These contain native grasses such as

Figure 11-3 Serpentine grassland in Pope Valley, Napa County, California. Dominant species include the native annual forbs *Lasthenia californica* and *Castilleja exserta*. Photo by J. Gelbard.

Calamagrostis ophiditis, Elymus glaucus, Festuca idahoensis, Elymus elymoides, Koeleria macrantha, Poa secunda, and *Nassella pulchra*. Soils that are shallower and rockier, though not as much so as chaparral soils, support "wildflower fields" with native annual forbs such as *Plantago erecta, Lasthenia californica, Minuartia californica, Microseris californica, Layia platyglossa* and *Hemizonia congesta* and scattered bunchgrasses (McCarten 1993). Most of the native annuals found in serpentine grasslands are not endemics, although one exception is *Hesperevax sparsiflora* (Rodriguez-Rojo et al. 2001).

Serpentine grasslands also contain exotic annual grasses such as *Bromus hordeaceus, Avena fatua,* and *Lolium multiflorum*. Although serpentine grasslands are considerably less-invaded by exotic Mediterranean annuals than other Californian grasslands, they are more invaded than other types of western North American serpentine vegetation. The prevalence of exotic species within serpentine grassland correlates positively with soil depth, organic content, nitrogen, phosphorous, and/or Ca/Mg ratio (McNaughton 1968, Huenneke et al. 1990, McCarten 1992, Harrison 1999b, Safford and Harrison 2001). The invasion of serpentine grassland by annual grasses may be increasing in the Bay Area due to atmospheric nitrogen deposition (Weiss et al. 1999).

Moisture and disturbance are also important in creating local variation. At the well-studied grassland of Jasper Ridge (Santa Clara County), several years of above-average rainfall led to an increase in the exotic annual grass *Bromus hordeaceus* and shifted the dominance among native annuals from *Plantago erecta* to *Lasthenia californica* (Hobbs and Mooney 1991). However, several years of drought disproportionately reduced the abundance of *Bromus hordeaceus* on serpentine soil, probably because on these soils it could not grow deep enough roots to survive drought (Armstrong and Huenneke 1992). Gopher disturbance increased the local abundance of several natives (e.g., *Lotus wrangelianus*) and exotics (e.g., *Bromus hordeaceus*) that are good colonizers, but it adversely affected bulb plants (e.g., *Brodiaea* spp.) and other native annuals (e.g., *Lasthenia californica*) (Hobbs and Mooney 1991). Local adaptation by certain serpentine-tolerant annual species such as *Lasthenia californica* (Rajakaruna et al. 2003) may also enable individual native taxa to respond differently under the variable rainfall patterns in these serpentine grasslands. The native annual grass *Vulpia microstachys*, an indifferent species, produced the highest biomass in openings in serpentine chaparral, rather than in moist serpentine meadows where competition from other species may be high (Jurjavcic et al. 2002). In this case an interaction between seed source and habitat, affecting emergence and survival, indicated ecotypic adaptation to the rocky serpentine slope habitat.

Local microsite variability within a given serpentine grassland may also strongly affect the microdistribution of individual native and non-native species. In a study in the Southern California Coast Ranges (Gram et al. 2004), many native species appeared to be restricted to isolated outcrops of shallow serpentine soil, or "hummocks." Of the 27 most common plant species sampled along hummock-grassland transects, eight were hummock specialists, seven edge specialists, eight matrix specialists, and four generalists. Despite their hypothesis that hummocks would be refugia for native species and the deeper soils would contain mostly non-native species, Gram et al. (2004) found both the hummock and matrix specialist groups included native species. This suggests that a high proportion of non-native species in a stand of serpentine grassland does not preclude its value as an important habitat for some native species.

Fire and grazing may benefit native species in serpentine grassland by reducing the dead material (thatch) produced by exotics. In a northern Californian serpentine grassland (Lake, Napa and Yolo Counties), an autumn wildfire burned a mosaic of serpentine and nonserpentine, cattle-grazed and ungrazed grasslands. Serpentine grasslands were more species-rich than nonserpentine grasslands before the fire, but species richness increased more in response to fire on nonserpentine soils. Two years after the fire, richness returned to approximately the same levels as before the fire. Both native and exotic species increased in response to fire on both soils. However, fire mainly increased native species on serpentine soils and exotics on nonserpentine soils. Grazing by cattle similarly increased the richness of native species on serpentine soils, and decreased the richness of natives on nonserpentine soils (Harrison et al. 2003).

11.2.6 Deserts

Deserts are defined as arid areas with sparse vegetation. The only notable serpentine deserts in western North America are in Baja California (domain 1). The physiognomy of the serpentine vegetation is not noticeably different from that in adjacent desert on other kinds of rocks and soils. They are dominated by shrubs and cacti. Trees are quite sparse; the forbs are ephemeral, depending on the sporadic rainfall, and grazing has practically eliminated the grasses. Rain that invigorates the plants occurs mostly in late summer to early autumn in the more tropical (actually subtropical) Magdelena–Margarita locality (1-4) and both in winter and in late summer to early autumn in localities 1-1, 1-2, and 1-3 that are farther north (fig. 13-2). Coastal fog is an important factor in locality 1-4, and the trees and shrubs are festooned with lichen (*Niebla* sp.). There are no satisfactory descriptions of serpentine alliances and plant associations in the deserts of Baja California, but preliminary assessments are given in chapter 13.

11.3 Special Serpentine Habitats

11.3.1 Barrens and Rock Outcrops

Interspersed in all of the major serpentine vegetation types is a variety of special habitats supporting unique floras. Serpentine barrens are areas of finely to coarsely fractured serpentine rock on which plant cover is generally much less than 1% (fig. 11-4). They may occur in a variety of physical settings: on streambanks where erosion has produced steep slopes, on mountaintops where adverse growth conditions have inhibited plant colonization, or in localized areas where a combination of tectonic shearing and rapid uplift have occurred. These barrens represent one of the most extreme serpentine environments, in which the majority of species are serpentine endemics with restricted geographic distributions. A variety of endemic *Arabis, Allium, Eriogonum,* and *Streptanthus* species typify the special flora of serpentine barrens in the California Floristic Province. The shallow, rocky, and dry soils of serpentine barrens make them extreme environments for serpentine plants (McCarten 1993, Hughes et al. 2001).

Figure 11-4 Serpentine talus barrens at The Cedars, Sonoma County, California. Photo by R. Raiche.

Similar to barrens in their lack of plant and soil cover, but differing in their greater physical stability, are the outcrops of exposed bedrock that often form bare hilltops or ridgetops within larger serpentine landscapes. Plants on these outcrops are confined to crevices that trap water and allow some development of soil. Many of the same plant genera found on talus barrens are also found on outcrops (e.g., *Arabis, Allium, Erogonum,* and *Streptanthus*). In addition, the outcrop habitat supports numerous species of ferns and succulent taxa. Ferns include the northernmost serpentine endemics, *Polystichum lemmoni* and *Aspidotis densa*. Succulents include several serpentine-endemic *Sedum* spp. and nonendemic *Dudleya, Lewisia,* and *Claytonia* spp. (McCarten 1993).

11.3.2 Seeps and Riparian Zones

Serpentine areas are often rich in seeps, or small spring-fed streams, for reasons that are not fully understood (see chapter 4). The northernmost serpentine seep that has been described, in central British Columbia (domain 7), supports a dense stand of *Carex sitchensis* (Kruckeberg 1969, p. 87). Some of the most striking seep floras are found in the Klamath–Siskiyou region (domain 5), where abundant fens (nitrogen-poor bogs with slowly flowing water) on serpentine are characterized by conspicuous stands of the carnivorous pitcher plant *Darlingtonia californica*, under a coniferous overstory. These *Darlingtonia* fens also support a variety of other endemic or near-endemic species, including *Narthecium californicum, Hastingsia alba,* and *Cypripedium californicum* (Jimerson et al. 1995, Sawyer and Keeler-Wolf 1995).

Northern California Coast Ranges serpentine seeps (fig. 11-5) also support a rich characteristic flora, including the endemics *Senecio clevelandi, Mimulus nudatus, Astragalus clevelandii, Helianthus exilis, Delphinium uliginosum,* and *Carex serratodens,* and nonendemics such as *Epipactis gigantea, Mimulus guttatus,* and *Stachys albens* (McCarten and Rogers 1991, Callizo 1992, Wolf et al. 1999, Harrison et al. 2000a, Freestone and Harrison 2006). Seeps in the San Francisco Bay area harbor the rare thistles *Cirsium fontinale* and *Cirsium campylon,* as well as *Mimulus guttatus* and *Calamagrostis ophiditis* (McCarten 1993, Evens and San 2004). The Southern California Coast Ranges seeps are less botanically distinctive, but support a few of the endemics also found in the north, such as *Delphinium uliginosum, Castilleja minor* and *Aquilegia eximia*. *Senecio clevelandii* is found in Sierran seeps.

Larger riparian areas on serpentine also form a distinctive habitat. In the Klamath-Siskiyou region, as has already been discussed, riparian zones often support a rich mixed conifer forest including *Cupressus lawsoniana*, interspersed with seeps and streamside thickets with *Darlingtonia californica, Hastingsia alba,* and many other specialized herbaceous species (Whittaker 1960). In the Northern California Coast Ranges, serpentine riparian woody vegetation may include *Cupressus sargentii, Rhododendron occidentale, Rhamnus tomentella crassifolia, Calycanthus occidentalis,* and *Salix breweri*; the first and last of these are also found in the Southern California Coast Ranges serpentine riparian zones (Vogl 1988, McCarten and Rogers 1991, Sawyer and Keeler Wolf 1995).

Figure 11-5 Spring-seep on serpentine at the Knoxville locality in the North Coast Ranges of California. Photo by S. Harrison.

11.3.3 Cracking-Clay Soils (Vertisols)

Localized areas of clay-rich soils on serpentine are distinctive for their dark color, high moisture capacity, and tendency to crack when they dry. These soils support elements of both seep and grassland floras, plus some species unique to this habitat. In the San Francisco Bay area, the rare herb *Acanthomintha obovata* ssp *duttoni* is found on heavy serpentine clays with an extremely low Ca/Mg and high moisture capacity (McCarten 1993). In the Northern California Coast Ranges, typical species of cracking-clay soils include *Fritillaria pluriflora*, *Trifolium fucatum*, *Castilleja rubicundula* ssp. *lithospermoides*, *Lasthenia glaberrima* and the exotic grass *Lolium multiflorum*. In both the Northern California Coast Ranges and the Klamath Mountains, *Perideridia* spp., *Danthonia* spp., and

Lotus purshianus are characteristic of these soils. Composition varies depending on the slope of the terrain. Flat or concave sites tend to have more mesophytic species or even vernal pool species, whereas sites with more slope support early spring annual species including some non-native grasses such as wild oats (*Avena* sp.), soft brome (*Bromus hordaceous*), red brome (*Bromus madratensis*), nitgrass (*Gastridium ventricosum*), and foxtail fescue (*Vulpia megalura*).

12 Serpentine Vegetation of Western North America

As discussed in chapter 11, the general patterning of vegetation on serpentine up and down the western North American continent is relatively straightforward. However, many of the distinctive nuances relating to the structure and composition of the vegetation, particularly in comparison to adjacent nonserpentine vegetation have yet to be described. In this chapter we use vegetation as a tool to describe the variation of biotic diversity on serpentine throughout western North America. Vegetation is valuable in this regard because, by describing it, one assembles the information on all plants growing in different patterns in a landscape.

This chapter expands on some of the concepts mentioned in chapter 11 and addresses some of the specific questions of interest to ecologists and biologists regarding the influence of serpentine on groups of plant species, using examples from western North America. Western North America provides an excellent template for understanding general questions about serpentine effects on species and vegetation. The broad latitudinal distribution and the local topographic and geologic diversity of serpentine exposures throughout this area produce an array of gradients of temperature, moisture, soil development, disturbance patterns, and day length to produce multiple ecological gradients operating at multiple scales. Also, within western North America a wide number of species from many different genera and families are influenced by serpentine.

12.1 Vegetation Classification

Vegetation classification is a tool used for several purposes, including efficient communication, data reduction and synthesis, interpretation, and land management and planning. Classifications provide one way of summarizing our knowledge of vegetation patterns. Although there are many different classification concepts, all classifications require the identification of a set of discrete vegetation classes. The fundamental unit of these discrete classes that is identifiable in the field is the stand. A stand is defined by two main unifying characteristics (CNPS 2003):

1. It has compositional integrity. Throughout the site, the combination of plant species is similar. The stand is differentiated from adjacent stands by a shift in plant species composition that may be abrupt or indistinct. That shift relates to a concomitant shift in certain ecological features such as temperature, moisture, or soil fertility that maintain control over the plant species composition.
2. It has structural integrity. It has a similar history or environmental setting that affords relatively similar horizontal and vertical spacing of plant species. For example, a hillside forest originally dominated by the same species that burned on the upper part of the slopes, but not the lower, would be divided into two stands. Likewise, a sparse woodland occupying a slope with very shallow rocky soils would be considered a different stand from an adjacent slope with deeper, moister soil and a denser woodland or forest of the same species.

Vegetation may show distinct physiognomic patterns that translate to relatively minor differences in floristic composition. For example, a very distinct-looking stand of serpentine vegetation by virtue of its openly spaced trees and shrubs may actually share those tree and shrub species largely with other adjacent stands of nonserpentine vegetation. Thus it is important to have a classification that describes the structure of the stand as well as its composition. Height, vertical layering of vegetation strata, and spacing of individual plants within these strata are all important features that may distinguish serpentine vegetation.

The structural and compositional features of a stand are often combined into the concept of homogeneity. For an area of vegetated ground to meet the requirements of a stand, it must be homogeneous.

Several ideas are central to the conceptual basis for classification (Mueller-Dombois and Ellenberg 1974):

1. Similar combinations of species recur from stand to stand under similar habitat conditions, though similarity declines with geographic distance.
2. No two stands are exactly alike, due to chance events of dispersal, disturbance, extinction, and history.
3. Species assemblages change more or less continuously if one samples a geographically widespread community throughout its range.
4. Stand similarity depends on the spatial and temporal scale of analysis.

Vegetation classifications can be either qualitative or quantitative. Both have their advantages. Both may be hierarchical and have progressively more detailed levels nested within the coarser levels, a facet that is very useful for scoping the most appropriate scale of the classification based on needs and objectives. Qualitative vegetation classifications are easy to develop and usually are based on some intuitive life-form approach such as describing woodlands versus forests and scrub versus grasslands. This approach is appealing because most people can readily understand such strong patterns. However, the more detail we try to draw from such classifications, the more we rely on incomplete existing knowledge to do so, and the less certainty we have.

Vegetation classification is an emerging science in western North America. The historical record of vegetation classification in the west and recent developments in classification standards show an evolution toward quantification of vegetation units (Phister and Arno 1980, Allen 1987, Franklin and Dyrness 1973, Sawyer and Keeler-Wolf 1995). The quantitative approach to classification sets certain rules, usually based on the percent cover or other biomass measure of species and/or life-forms present in a stand of vegetation. In an area such as this, where vegetation classification is still a dynamic activity, some of these rules must be general and must be applied at a coarser level of the classification. We have not quantitatively defined all forests, scrubs, or grasslands in western North America, but we can first make rules to differentiate what we consider vegetation characterized by trees, vegetation characterized by shrubs, and vegetation characterized by grasses and herbs. Then we can apply our growing knowledge of more specific vegetation types to this system. This kind of combined top-down and bottom-up classification system is mirrored in the current National Vegetation Classification System (Grossman et al. 1998).

In this chapter we use the language of the National Vegetation Classification system, or NVCS, to describe vegetation (Grossman et al. 1998). The NVCS uses specifically defined terms to identify different levels in a quantifiable vegetation hierarchy. This hierarchical system uses a blend of floristic and structural terms to describe vegetation from the very broad scale to the fine individual-stand level.

A classification of vegetation that includes both a floristic and a life-form component has the capacity to synthesize the influence of the major plant species and their concomitant structure and density. The fundamental floristic units in the NVCS are the alliance and the association. These are the principal units we use in this discussion of serpentine vegetation because they reflect the importance of the species composition in stands of vegetation. The standard definitions are:

> Association: A recurring plant community with a characteristic range in species composition, specific diagnostic species, and a defined range in habitat conditions and physiognomy or structure.
>
> Alliance: A grouping of associations with a characteristic physiognomy, and which shares one or more diagnostic species. These species, as a rule, are found in the uppermost or dominant stratum of the vegetation.

Above the floristic level there are purely physiognomic levels, which enable less detailed, but still useful, perspectives on vegetation. Figure 12-1 shows an example of this hierarchy.

Physiognomic Categories
Category   Example
Class . . . . Open Tree Canopy
Subclass . . . . Evergreen Open Tree Canopy
Group . . . . . . Temperate or Subpolar Needle-leaved Evergreen Open Tree Canopy
Subgroup . . . . . . Natural/Seminatural
Formation . . . . . . Rounded-crowned temperate or subpolar needle-leaved evergreen open tree canopy
Floristic Categories
Alliance . . . . *Pinus jeffreyi* Woodland Alliance   Association . . . . *Pinus jeffreyi*/*Quercus vaccinifolia-Arctostaphylos nevadensis*/*Festuca idahoensis* Association

Figure 12-1 An example of the National Vegetation Classification hierarchy of a serpentine vegetation type found on the Josephine Peridotite body of the Klamath Mountains. The order of listing of the species and the uses of the dashes and slashes have specific meaning in the classification: Tallest strata are always listed first; a dash separates characteristic species in the same layer; a forward slash separates characteristic species in different layers.

When observing stands of vegetation, it is usually possible for the uninitiated to classify vegetation at the alliance level by identifying the most abundant or conspicuous species in the tallest layer of plants in the stand and assuming that that species will confer the name to the vegetation alliance that the stand represents. For example, stands strongly dominated by whiteleaf manzanita (*Arctostsaphylos viscida*) would clearly be members of the whiteleaf manzanita alliance. More subtle distinctions have to be made only when there are stands with several species that share dominance or are equally conspicuous. In some cases alliances are named by a pair of regularly co-occurring species in the same strata, such as the chamise–bigberry manzanita (*Adenostema fasciculatum–Arctostaphylos glauca*) alliance, or the Douglas-fir–ponderosa pine (*Pseudotsuga menziesii–Pinus ponderosa*) alliance. In these cases, these characteristic species usually fall within certain bounds of relative cover, such as between 30%–60% relative cover of all plants in the tallest strata of vegetation. A useful guide to identification of individual vegetation alliances and associations as treated in the NVCS may be found at the NatureServe (2004) website.

Vegetation classification is approached through multivariate statistical analysis of individual samples where species composition and percent cover of each species is compared. The floristic similarity of all of the vegetation samples can be compared and arrayed in cluster analyses (McCune and Grace 2002). Cluster analysis ranks the similarity between vegetation sample plots based on species composition and a measure of each species' relative biomass. Degrees of similarity between the samples may be shown as dendrograms. The ecologically closely related samples cluster tightly together, and the most distant relationships split off at the most general divergence of the dendrogram. The patterns of floristic similarity can be graphically compared. As an example, these comparisons can graphically show the degrees of similarity between serpentine and nonserpentine vegetation and suggest certain trends and ecological patterns. For example, in a cooperative project between California Native Plant Society and California Department of Fish and Game (CNPS-DFG. 2006) a quantitative classification analysis of 565 samples was conducted for a contiguous area of serpentine and nonserpentine in the Southern California Coast Ranges of San Benito and Fresno County.

12.2 Organization of the Discussion of Serpentine Vegetation

In chapter 9 we made the distinction between plant species that are serpentine obligates, serpentine tolerators, and serpentine avoiders. There are also analogous terms for vegetation. These characteristics, as with individual species of plants, also appear at different levels of the taxonomic hierarchy. In some cases there are largely endemic alliances (analogous to genera of plants) that are largely endemic to serpentine (*Hesperolinon, Streptanthus, Harmonia*, etc.). However, in many cases there are alliances with only a portion of them represented by serpentine associations. For example, most of the widespread Douglas-fir (*Pseudotsuga menziesii*) alliance is represented by associations found largely off serpentine—some 65 associations are known from British Columbia to central California (NatureServe 2004); only seven are known largely from serpentine soils (Jimerson et al. 1995). Some alliances are known to have an even mixture of serpentine and nonserpentine associations. A good example is the Port Orford-cedar (*Chamaecyparis lawsoniana*) alliance. It has 24 associations with seven known only from serpentine, eight are indifferent, known from either serpentine or nonserpentine, and nine are only known from nonserpentine substrates (Jimerson 1994, Jimerson et al. 1999). At the ultimate extent of the continuum of serpentine-specialist vegetation are the serpentine specialist alliances, which are wholly or almost entirely represented only on serpentine. These are broken into unique stands and endemic alliances. Another group at the polar opposite of this is the alliances that predictably avoid serpentine such as the western hemlock (*Tsuga heterophylla*) alliance.

We organize the following discussion by types of alliances. First are the serpentine endemic alliances and unique stands, followed by alliances that have some of their component associations known solely from serpentine. Next are the alliances that may occur on serpentine but that have differentiated into serpentine associations. Finally, we discuss the vegetation composed of non-native species that have adapted to serpentine. Table 12-1 displays

Table 12-1 Summary of vegetation types on serpentine including information on endemic alliance or association, locations of each alliance within the domains discussed in this book.

Group	Vegetation alliance	Serpentine endemic association	Serpentine endemic alliance	Domains represented
Broadleaf evergreen shrubland	Leather oak (*Quercus durata*)	Yes	Yes	2, 3, 4, 5
	Coast whiteleaf manzanita (*Arctostaphylos viscida* ssp. *pulchella*) alliance	Yes	Yes	4
	Thick-leaved coffeeberry (*Frangula californica* ssp. *crassifolia*) alliance	Yes	No	3, 4
	Chamise (*Adenostoma fasciculatum*) alliance	Yes	No	2, 3, 4
	Chamise-eastwood manzanita (*Adenostoma fasciculatum–Arctostaphylos glandulosa*) alliance	Yes	No	1
	White leaf manzanita (*Arctostaphylos viscida* ssp *viscida*)	Yes	No	2, 5
	Bigberry manzanita (*Arctostaphylos glauca*) alliance	Yes	No	3
	Buckbrush (*Ceanothus cuneatus*) alliance	Yes	No	2, 3, 4, 5
	Huckleberry oak (*Quercus vaccinifolia*) alliance	No	No	2, 5
	Pinemat manzanita (*Arctostaphylos nevadensis*) alliance	No	No	2, 5
	Dwarf tanbark oak (*Lithocarpus densiflorus* ssp. *echinoides*) alliance	No	No	5
	Toyon (*Heteromeles arbutifolia*) alliance	No	No	3, 4
	Glandular Labrador-tea (*Ledum glandulosum*) alliance	No	No	5

(*continued*)

Table 12-1 (continued)

Group	Vegetation alliance	Serpentine endemic association	Serpentine endemic alliance	Domains represented
Drought deciduous shrubland	California sagebrush (*Artemisia californica*) alliance	Yes	No	3
	California buckwheat (*Eriogonum fasciculatum*) alliance	No	No	3
	Wright's buckwheat (*Eriogonum wrightii*) alliance	No	No	3, 4
	Black sage (*Salvia mellifera*) alliance	No	No	3
Needleleaf evergreen forest and woodland	Sargent cypress (*Cupressus sargentii*) alliance	Yes	Yes	3, 4
	Knobcone pine (*Pinus attenuata*) alliance	Yes	No	1, 2, 3, 4, 5
	Coulter pine (*Pinus coulteri*) alliance	Yes	No	1, 3
	Jeffrey pine (*Pinus jeffreyi*) alliance	Yes	No	2, 3, 4, 5
	Foothill pine (*Pinus sabiniana*) alliance	Yes	No	2, 3, 4, 5
	Incense-cedar (*Calocedrus decurrens*) alliance	Yes	No	2, 3, 4, 5
	Douglas-fir (*Pseudotsuga menziesii*) alliance	Yes	No	2, 4, 5, 7,
	Western white pine (*Pinus monticola*) alliance	Yes	No	5, 7,
	Foxtail pine (*Pinus balfouriana*) alliance	Yes	No	5
	Lodgepole pine (*Pinus contorta* ssp. *latifolia*)	Yes	No	5, 7, 8
	Shore pine (*Pinus contorta* ssp. *contorta*) alliance	No	No	7, 8
	Sugar pine (*Pinus lambertiana*) alliance	Yes	No	5
	Ponderosa pine–Douglas-fir (*Pinus ponderosa*–*Pseudotsuga menziesii*) Forest alliance	Yes	No	5
	Subalpine fir (*Abies lasiocarpa*) alliance	Yes	No	7, 9

Table 12-1 (continued)

Group	Vegetation alliance	Serpentine endemic association	Serpentine endemic alliance	Domains represented
	McNab cypress (*Cupressus macnabiana*) alliance	Yes	No	2, 4
	Engelmann spruce (*Picea engelmanii*) alliance	No	No	7
	White spruce (*Picea alba*) alliance	No	No	9, 10
	Brewer spruce (*Picea breweriana*) alliance	No	No	5
	Western juniper (*Juniperus occidentalis*) woodland alliance	No	No	5
	White fir (*Abies concolor*) alliance	No	No	5
	Red fir (*Abies magnifica*), including Noble fir (*A. procera*) alliance	No	No	5
	Grand fir (*Abies grandis*) alliance	No	No	7
	Pacific silver fir (*Abies amabilis*) alliance	No	No	7
	Mountain hemlock (*Tsuga mertensiana*) alliance	No	No	5, 7, 8
	Whitebark pine (*Pinus albicaulis*) alliance	No	No	5, 7
Winter deciduous forest and woodland	White alder (*Alnus rhombifolia*) alliance	No	No	5
	Aspen (*Populus tremuloides*) alliance	No	No	8, 9
	Paper birch (*Betula papyrifera*) alliance	No	No	8, 9, 10
	Blue oak (*Quercus douglasii*) alliance	No	No	2, 4
	Oregon oak including Brewer oak (*Quercus garryana* var. *breweri*) alliance	No	No	5

(continued)

Table 12-1 (continued)

Group	Vegetation alliance	Serpentine endemic association	Serpentine endemic alliance	Domains represented
Evergreen broadleaf forest and woodlands	California bay (*Umbellularia californica*) alliance	No	No	3, 4
	Curl-leaf mountain mahogany (*Cercocarpus ledifolius*) alliance	No	No	5, 7
Winter-deciduous shrubland	Brewer's willow (*Salix breweri*) alliance	Yes	Yes	3, 4
	Tea-leaf willow (*Salix planifolia*) alliance	No	No	9, 10
	Shrub birch (*Betula glandulosa*) alliance	No	No	9, 10
Winter-deciduous dwarf shrubland	Snow willow (*Salix reticulata*) alliance	No	No	9, 10
Evergreen microphyllous dwarf shrubland	Mountain heather (*Cassiope tetragona*) alliance	No	No	9, 10
	Creeping crowberry (*Empetrum nigrum*) alliance	No	No	8, 9
Evergreen needleleaf shrubland	Low juniper (*Juniperus communis*) alliance	No	No	7
Evergreen microphyllous shrubland	Big sagebrush (*Artemisia tridentata*) alliance	No	No	7
	Rubber rabbitbrush (*Chrysothamnus nauseosus* var. *mojavensis*) alliance	No	No	3
Temperate perennial herb and grassland	Twotooth sedge (*Carex serratadens*) alliance	Yes	Yes	3, 4
	Tufted hairgrass (*Deschampsia casepitosa*) alliance	No	No	5
	Big squirreltail (*Elymus multisetus*) alliance	Yes	No	3, 4
	Purple needlegrass (*Nassella pulchra*) alliance	Yes	No	2, 3, 4
	Rough fescue (*Festuca altaica*) alliance	No	No	9, 10

Table 12-1 (*Continued*)

Group	Vegetation alliance	Serpentine endemic association	Serpentine endemic alliance	Domains represented
	Bigelow's sedge (*Carex bigelowii*) alliance	No	No	9, 10
	Mountain avens (*Dryas octopetala*) alliance	No	No	10
	California pitcherplant (*Darlingtonia californica*) alliance	Yes	Yes	2, 5
	Mount hamilton thistle (*Cirsium fontinale* var. *campylon*) alliance	Yes	Yes	3
	Irisleaf rush (*Juncus xiphioides*) alliance:	Yes	Yes	3, 4
Temperate annual herb and grass-lands	California goldfields (*Lasthenia californica* and other species) alliance	Yes	No	2, 3, 4
	Plantain (*Plantago erecta*) alliance	Yes	No	2, 3, 4
	Fescue (*Vulpia microstachys*) alliance	Yes	No	3, 4
	Vernal pool vegetation	No	No	1, 2, 3, 4
	Goatgrass (*Aegilops triuncialis*) alliance	Yes	No	3, 4
	Soft chess (*Bromus hordaceous*) alliance:	No	No	2, 3, 4
	Italian ryegrass (*Lolium multiflorum*) alliance	No	No	3, 4

Domains: 1. Baja California; 2. Sierra Motherlode; 3. Southern California Coast Ranges; 4. Northern California Coast Ranges; 5. Klamath Mountains; 6. Blue Mountains, Oregon; 7. Northern Cascade–Fraser River; 8. Gulf of Alaska; 9. Denali-Yukon; 10. Northern Alaska–Kuskokwim Mountains.

some of these patterns, showing their affinities for specialization and their geographic subdivision by serpentine domain.

As with most biological data, there are few absolutes. Given difficulties such as the absence of a full distributional record of serpentine species and not knowing for certain whether species are truly wholly endemic to serpentine substrates, it is even more risky to make such claims for vegetation types that are as yet incompletely known. Thus, we expect that as more becomes known, some of the groups treated below will be revised.

12.3 Vegetation Alliances and Unique Stands Largely Endemic to Serpentine

The following account is a provisional discussion, based on data that are not wholly representative of the vegetation of all areas of serpentine in western North America. As more sampling and vegetation analysis take place, particularly throughout California, a clearer understanding of the unique and endemic alliances and associations will arise. In particular, it is likely that more information will be developed from the Sierra Motherlode and the Northern California Coast Ranges domains, where sampling and analysis of vegetation is just now getting underway.

Because California is at the heart of the diversity and individuality of the serpentine syndrome in western North America, it should come as no surprise that the most distinctive vegetation types endemic to serpentine are all thus far known only from California and adjacent southwestern Oregon. Although the largest variety of California vegetation endemic to serpentine is at the association level, many of the most characteristic vegetation types in the state are largely endemic at the level of alliance or unique stand (table 12-1).

12.4 Unique Stands

One of the interesting features of serpentine vegetation is that there are many highly localized shrubs and herbs that are restricted to certain small areas of serpentine outcrops. Although many of these species are considered rare (criteria in Tibor 2001), they are by no means always inconspicuous. On the contrary, some of these serpentine endemics actually form one to several stands where they dominate, albeit only in their restricted geographic range. Typically, such stands are so distinct that they cannot be categorized as variants of existing vegetation types (e.g., associations of a given alliance), so they must be treated as separate entities. These species also do not form a sufficient number of stands over a broad enough range to be considered as defining their own alliance. They form their own category and have been recognized as "unique stands" (Sawyer and Keeler-Wolf 1995).

Several species of serpentine endemic manzanita (*Arctostaphylos*) are known to form unique stands. These include Mt. Tamalpias manzanita (*Arctostaphylos hookeri* ssp. *montana*), Obispo manzanita (*A. obispoensis*), and Baker manzanita (*A. bakeri* ssp. *bakeri*) (fig. 12-2). Species of *Ceanothus* also form local unique stands. Siskiyou mat (*Ceanothus pumilus*) forms such stands in the Klamath domain, where they may be associated with a number of other rare herbaceous species (Jimerson et al. 1995), while Coyote Ridge ceanothus (*Ceanothus ferrisiae*) forms unique stands at Coyote Ridge, Santa Clara County (Evens and San 2004). In the Baja California domain, there is a unique stand of Cuyamaca cypress (*Cupressus arizonica* ssp. *cuyamacae*), which is only known from the Las Posas soils derived from gabbro on the southwestern flank of Cuyamaca Peak in San Diego County (fig. 12-3). In the same region there is a unique stand of the Vail Lake ceanothus (*Ceanothus ophiochilus*) near the Agua Tibia Wilderness Area. The stand is on a mesic north-facing slope surrounded by other chaparral species. Other unique stands are mon-

Figure 12-2 A unique stand of Baker manzanita (*Arctostaphylos bakeri* ssp. *bakeri*) at the California Department of Fish and Game Harrison Grade Ecological Reserve, Sonoma County, Northern California Coast Range Domain. Trees in the background are Sargent Cypress (*Cupressus sargentii*). Several other endemic serpentine species are present at this locality. Photo by T. Keeler-Wolf.

Figure 12-3 A unique stand of Cuyamaca cypress (*Cupressus arizonica* ssp. *stephensonii*) at the King Creek Research Natural Area, San Diego County, California. Cuyamaca Peak (1980 m) in background. Photo taken May 1990 by T. Keeler-Wolf. Stands have since been burned by fire in 2003.

tane or alpine in nature such as the herbaceous community on the upper ridges of Mt Eddy in the Klamath Mountains domain. There *Penstemon procerus* dominates with a mixture of other herbs, some like *Eriogonum siskiyouense*, largely endemic to this mountain. Other high montane, sparsely vegetated stands may also be considered unique including some outside of California. One likely example is the serpentine barrens mentioned by Kruckeberg (1969) from the Ingalls locality in central Washington (chapter 19), characterized by *Eriogonum pyrolaefolium, Arenaria obtusiloba,* and *Poa curtifolia*. These unique stands will be discussed further within the context of the domain descriptions in part IV.

12.5 Endemic Upland Serpentine Alliances

12.5.1 Leather Oak *(*Quercus durata*) Alliance*

Perhaps the most widespread of the largely endemic alliances is the leather oak (*Quercus durata*) alliance. It is found in the Sierra Motherlode, the Southern California Coast Ranges, the Northern California Coast Ranges, and the Klamath Mountains domains. *Quercus durata* is typically the most common shrub in this alliance, which is typically found on north- and east-facing slopes. Although individuals of leather oak are found on the most xeric exposures on serpentine, they are not usually abundant enough to form their own alliance until the topography shifts to a more mesic setting.

Individuals of leather oak may occasionally be found off serpentine, particularly in the Northern California Coast Ranges, where they occur on sterile Franciscan sandstone metasediments in such places as Vaca Mountain and Hale Ridge (Keeler-Wolf 1987a). However, in these cases leather oak is usually present as isolated individuals associated with other vegetation alliances such as *Quercus wislizeni* var. *fructescens, Adenostoma fasciculatum,* or *Quercus berberidifolia.*

Leather oak is a shrub that re-sprouts readily after fire, and individual root crowns may reach several hundred years of age. Depending on the fire intervals, stands may have most associated woody species composed of either re-sprouting or obligate seeding species (Safford and Harrison 2004). The basis of classification of the leather oak alliance into associations may thus relate to fire history as well as to local geographic conditions.

A number of associated species occur within the leather oak alliance and vary geographically. The alliance occurs in the southwestern portion of the Klamath Mountains in the Hayfork and the Rattlesnake Creek Terrane, southeast Trinity County and northwest Tehama County (see chapter 17). This is the warmest and driest portion of the Klamath Mountains. Here leather oak is often mixed with buckbrush (*Ceanothus cuneatus*), a widespread chaparral species that has stronger affinities for serpentine in the Klamath and Sierra Motherlode domains than in other parts of its range (Taylor and Teare 1979; J. Sawyer, J. Kagan, personal communication, 2004). Other associated species with leather oak in this area include *Garrya congdonii* (mostly in northwest Tehama County) and *G. buxifolia* (mostly in Trinity County).

Some characteristic associated woody species of the leather oak alliance include:

Adenostoma fasciculatum (chamise)

Arctostaphylos bakeri (the cedars manzanita)

Arctostaphylos glauca (bigberry manzanita)

Arctostaphylos stanfordiana (Raiche manzanita)

Arctostaphylos viscida (whiteleaf manzanita)

Arctostaphylos viscida ssp. *pulchella* (coast whiteleaf manzanita)

Ceanothus ferrisae (coyote ceanothus)

Ceanothus jepsonii (musk bush)

Cercocarpus betuloides (birch-leaf mountain-mahogany)

Cupressus macnabiana (McNab cypress)

Cupressus sargentii (Sargent cypress)

Eriodictyon californicum (Yerba santa)

Garrya congdoni, G. flavescens (silktassels)

Heteromeles arbutifolia (toyon)

Juniperus californica (California juniper)

Pinus attenuata (knobcone pine)

Pinus sabiniana (foothill pine)

The alliance occurs from the Middle to the Inner Northern California Coast Ranges. Perhaps the most widespread associates in the North California Coast Ranges are Jepson ceanothus (*Ceanothus jepsonii*) and coast whiteleaf manzanita (*Arctostaphylos viscida* ssp. *pulchella*), with the latter species occurring on more xeric exposures. Although not always found with the alliance, one of the most beautiful shrub species that occurs with it on mesic exposures is the snowdrop bush (*Styrax officinalis* var. *redivivus*).

The leather oak alliance is scattered through the northern portion of the Sierra Motherlode. Here it also tends to occur on relatively mesic exposures often associated with buckbrush (*Ceanothus cuneatus*) and white leaf manzanita (*Arctostaphylos viscida* ssp. *viscida*). In several areas such as Grass Valley and Volcanoville it is associated with McNab cypress (*Cupressus macnabiana*) and may form mixed or discrete stands.

In the Southern California Coast Ranges the alliance is widespread on serpentine from the vicinity of Mount Diablo in Contra Costa County to Figueroa Mountain in Santa Barbara County. Its most common associates are bigberry manzanita (*Arctostaphylos glauca*) and toyon (*Heteromeles arbutifolia*). The leather oak alliance has been better studied here than in other parts of its range. Five associations have been defined within the alliance here, most from detailed studies of the vegetation of Coyote Ridge, Santa Clara County (Evens and San 2004), and San Benito Mountain–New Idria locality, San Benito and Fresno Counties (CNPS-DFG 2006). At Coyote Ridge, the alliance ranges from mesic chaparral, where it blends with adjacent California bay woodland (*Quercus durata–Heteromeles arbutifolia–Umbellularia californica* shrubland association), to more typical chaparral (*Quercus durata–Arctostaphylos glauca–Garrya congdonii/Melica torreyana* Shrubland Association), to relatively xeric chaparral (*Quercus durata–Rhamnus tomentella–Arctostaphylos glauca* shrubland association), and finally to quite xeric mixtures with coastal scrub (*Quercus durata–Arctostaphylos glauca–Artemisia californica*/grass shrubland association).

At Mount San Benito the alliance is also varied with five associations ranging from low xeric to high-elevation mesic with emergent conifers. The following associations have been identified:

Quercus durata/Pinus sabiniana, shrubland

Quercus durata-Adenostoma fasciculatum-Quercus wislizeni shrubland

Quercus durata–Arctostaphylos glauca/Pinus sabiniana, shrubland

Quercus durata–Arctostaphylos pungens–/Pinus sabiniana shrubland

Quercus durata–Arctostaphylos glauca-Garrya congdonii–Melica torreyana, shrubland

Quercus durata–Cercocarpus betuloides, shrubland

12.5.2 Coast Whiteleaf Manzanita (Arctostaphylos viscida ssp. pulchella) *Alliance*

The coast whiteleaf manzanita alliance is conspicuous on xeric serpentine slopes from Napa to Mendocino and western Tehama Counties, although the plant (not representing the alliance) continues northward into the Klamath Province of northwestern California and adjacent Oregon. This alliance has been recently segregated from the broader *A. viscida* alliance now believed to be largely dominated by the varieties *A. v. viscida* and *A. v. mariposa* of the Sierra Nevada and parts of the Klamath domain, and largely nonserpentine localities in the Northern California Coast Ranges (Sawyer and Keeler-Wolf, 2004). As far as is known, this alliance is restricted to serpentine of the Northern California Coast Range domain from the vicinity of Mount Saint Helena to the serpentine chaparral west of Paskenta in Tehama County.

In the Northern California Coastal Ranges this alliance often occupies the opposite warmer slopes from the leather oak alliance, where it may be mixed with chamise, foothill pine, *Garrya congdoni*, and toyon. Coast white-leaf manzanita is an obligate seeder and tends to occur as widely spaced individuals in stands with <50% shrub cover. Throughout much of the serpentine of the Middle and Inner Northern California Coast Ranges it occupies the more xeric south-facing slopes, often with a sparse emergent canopy of foothill pine (*Pinus sabiniana*). Many of the very open barrens of this area are low-density representatives of this alliance. As a result, many of the localities of serpentine endemic "barrens" species are frequently associated with low-density stands of this alliance. These include *Streptanthus morrisonii, S. breweri, Hesperolinon* spp., *Asclepias solanoana, Allium cratericola, Calypridium quadripetalum, Claytonia exigua,* and *Mimulus nudatus*.

12.5.3 Sargent Cypress (Cupressus sargentii) *Alliance*

The Sargent cypress alliance is the only vegetation alliance dominated by a conifer that is largely endemic to serpentine soils. Sargent cypress is typically a short tree rarely >8 m tall. It is considered here an "upland" alliance, although in reality it is generally found in mesic settings, often adjacent to stream channels or in shaded defiles and north-facing slopes. Although some of the most extensive stands occur in upland settings (e.g, Cedar Roughs, Napa County), the Cedars (Sonoma County), and Cuesta Ridge (San Luis Obispo County), these are usually associated with high water tables or climatically situated to receive additional moisture through summer fog-drip. As with other California members of the genus *Cupressus*, Sargent cypress is a serotinous- (closed-) cone conifer, which germinates most successfully after the fused cone scales are heated by fire. A short interval between fires tends to reduce the seed and cone crop stored in the crowns of the trees (Sawyer and Keeler-Wolf 2004).

The Sargent cypress alliance occurs in the South California Coast Ranges regularly on ridges near the coast in Monterey and San Luis Obispo County, where it may associate with leather oak, Obispo manzanita, and chamise alliances on serpentine. It is also found close to the coast in Sonoma County and may occur adjacent to mesic Douglas-fir and redwood forest near Occidental. Further inland, as at Cedar Mountain in Alameda County, it co-occurs with chamise and big-berry manzanita and also with leather oak alliance stands. North in Napa, Lake, Colusa, and Mendocino counties, stands of this alliance regularly occur along narrow creeks and in the bottom of ravines in serpentine landscapes. Some of these stands, if not burned for many years, contain trees >15 m tall and >150 years old.

Occasionally in the Northern California Coast Ranges this species co-occurs with its congener McNab cypress (*C. macnabiana*). The two species are usually separated ecologically with the more xerophytic McNab cypress occurring on upper slopes and ridges while the mesophytic Sargent occurs in canyon bottoms near streams. In a few cases, McNab cypress actually does occur along streams as at the McLaughlin Reserve in Napa County.

12.6 Endemic Bottomland and Wetland Serpentine Alliances

12.6.1 Brewer's Willow (Salix breweri) Alliance

The small, shrubby Brewer's willow forms characteristic riparian stands along permanent and intermittent creeks in both the Northern and the Southern California Coast Ranges. In some cases it may make up small stands in wet meadows. It usually forms scattered cover over active alluvium of sand, gravel, cobbles, or boulders. It is largely associated with large serpentine bodies, but it may move downstream following alluvial outwash from serpentine bodies for several miles. The associated species vary considerably depending on the profile and other physical features of the streambeds.

In the South Coast Ranges Brewer's willow alliance is common in the New Idria area of the Clear Creek Management area of the Bureau of Land Management. Associated species include the shrub *Rhamnus (Frangula) tomentella*, and the herbs *Castilleja miniata, Aquilegia eximia, Stachys pycnantha,* and *Solidago guiradonis*.

In the Northern Coast Ranges stands of Brewer's willow line small to large creeks and may be associated with stands of *Rhamnus californica crassifiolia, Cupressus sargentii,* or *Carex serratadens* alliances. The northernmost stands of this alliance appear to reach into the Inner Northern Coast Ranges of Glenn County both along creeks and in wet meadows. Associated species include *Rhododendron occidentalis, Calycanthus occidentalis, Astragalus clevelandii, Castilleja mineata, Carex serratadens, Aqulegia eximia, Stachys albens, Epipactis gigantea,* and *Senecio clevelandii*.

12.6.2 Twotooth Sedge (Carex serratadens) Alliance

The twotooth sedge alliance is restricted to saturated seeps and swales in serpentine throughout much of the Northern California Coast Ranges, Southern California Coast Ranges, and Sierra Motherlode domains. It usually forms small stands <0.5 ha in size, being limited to small springs and seeps. Stands of this sedge also occur in the beds of small creeks and streams, where it may be proximate to thickleaf coffeeberry, (*Frangula californica* ssp. *crassifolia*), Sargent cypress, and Brewer willow alliances. Associations of this alliance have been defined from Mt. San Benito (twotooth sedge–Mexican rush association) (CNPS-DFG 2006) and from Tuolumne County near Jackson (Evens et al. 2004). Another association with this species and the Coyote Hills thistle was called the Mt. Hamilton thistle–twotooth sedge–meadow barley forbland association by Evens and San (2004).

Although individuals of the two-tooth sedge have been collected from the Klamath Province in the Rattlesnake Creek terrane (locality 5-12, chapter 17), the alliance has not been identified in the Klamath Mountains.

12.6.3 California Pitcherplant (Darlingtonia californica) Alliance

The California pitcherplant alliance is often considered the quintessential serpentine wetland vegetation of California Floristic Province. However, it is not wholly restricted to serpentine. More stands do occur on serpentine, but in many areas it is possible to find the alliance away from serpentine. Although California pitcher plant is largely restricted to the Klamath Mountains, it is also found in the Northern Sierra Nevada, in the Sierra Motherlode domain.

In the Klamath Mountains and the Sierra Motherlode domains this alliance is restricted to seasonally flooded to saturated meadows and fens, which range from flat to moderately steep slopes. Substrates may range from serpentenite and peridotite to gabbro. In the Klamath Mountains the Port Orford-cedar (*Chamaecyparis lawsoniana*) alliance is commonly seen adjacent to it. However, in the Northern Sierra Nevada it is more apt to be found adjacent to drier woodlands of Jeffrey Pine or mixed ponderosa pine, Douglas-fir, and incense-cedar (as at Butterfly Valley, Plumas County (Mohlenbrock 2003).

Despite the generally basic nature of serpentine soils, the accumulation of organic matter particularly in sites within shallow basins tends to make the surface pH relatively acidic in these stands (Jimerson et al. 1995). It is neutral at Howel fen in Josephine County, Oregon.

At one site, 38 species were associated with the California pitcherplant alliance (Keeler-Wolf 1986, LE Horton RNA); at another more than 50 species were enumerated (Keeler-Wolf 1982, Cedar Basin). At this latter site surrounding a shallow montane lake, the *Darlingtonia* alliance occupied slightly lower mucky sites, while raised hummocks a few inches above the permanently saturated fen-bottom were occupied by bog laurel (*Kalmia polifolia*), ledum (*L. glandulosum*), bog blueberry (*Vaccinium uliginosum*), and other ericaceous shrubs. Common associates with California pitcherplant in the Klamath Mountains and the Sierra Motherlode are the primitive lily relatives (*Hastingsia alba*), bog asphodel (*Narthecium californicum*), and western Tolfieldia (*Tolfieldia glutinosa*). Twotooth sedge (*Carex serraradens*) as individual plants has been found in some of the lower elevation stands of *Darlingtonia* in both the Klamath Mountains and the Sierra Motherlode domains. Some of the most interesting species of plants associated with this alliance include other carnivorous species such as the sundew (*Drosera rotundifolia*) and the California butterwort (*Pinguicula macroceras*).

12.6.4 Thick-leaved Coffeeberry (Frangula californica ssp. crassifolia) Alliance

The thick-leaved coffeeberry and its close relatives *Frangula californica* ssp. *cuspidata*, and *F. c.* ssp. *tomentella* are largely restricted to moist serpentine bottomlands in the Klamath Mountains Province, the Northern California Coast Ranges, and the South California

Coast Ranges. There are taxonomic intergrades between the subspecies, and certain stands of riparian *F. californica* ssp. *tomentella*-like plants occur on serpentine in the Southern California Coast Ranges, as well. Currently, the best examples of this alliance come from the Inner North California Coast Ranges of Napa, Lake, and Colusa counties (CNPS Rapid Assessment Database, unpublished, 2004) and from San Benito County (CNPS-DFG 2006) and Santa Clara County (Evens and San 2004) in the South Coast Ranges. These stands are typically narrow riparian stringers, which lie back from the immediate streambed or wet meadow, suggesting that this alliance prefers moist but not long-saturated substrates during the growing season.

12.6.5 Mount Hamilton Thistle (Cirsium fontinale var. campylon) Alliance

Although the Mount Hamilton thistle alliance is limited in extent to a few stands in the Coyote Ridge and adjacent Mount Hamilton and Diablo Ranges of the South California Coast Ranges, it is locally represented by multiple stands that have been sampled extensively by Evens and San (2004). It is therefore included as an alliance.

This alliance is characterized by *Cirsium fontinale* var. *campylon* as the main overstory herbaceous species, while emergent shrubs or trees may be present in low cover. Stands are restricted in range to a few central coast counties, and they vary in size from a hundred square-meters to a few acres. They usually occur on serpentine soils in drainages that appear recharged by seeps, or they occur on sedimentary soils that are influenced by runoff of adjacent serpentine seeps. A variety of mesic serpentine and riparian species occur in the stands, including *Carex* spp., *Juncus xiphioides*, *Mimulus guttatus*, *Polypogon monspeliensis*, and *Frangula* (*Rhamnus*) *tomentella* at the base of the Mount Hamilton Range, below Almaden Calero Canal, near Calero Reservoir, at Coyote Ridge, and the base of Coyote Ridge at Coyote Creek Golf Course.

12.6.6 Irisleaf Rush (Juncus xiphioides) Alliance

Although irisleaf rush is known from California, Oregon, Arizona, and Baja California, it only dominates in repeating stands that constitute vegetation on serpentine in the Sierra Motherlode, the Northern California Coast Ranges, and the Southern California Coast Ranges.

This alliance is characterized by the dominance of *Juncus xiphioides* in the overstory. Other native and non-native species may be present, but in low numbers and in overall lower cover than *J. xiphioides* (e.g., relative cover is <25%), including *Carex* spp., *Juncus balticus*, *Lolium multiflorum*, and *Lythrum* spp. Irisleaf rush has been found dominating in natural, wetland habitats such as streams of serpentine alluvial substrate from the central coast and North Coast California Ranges (unpublished California Native Plant Society data; J. Evens, personal communication, 2004). However, it may also be found in other parts of California, Nevada, Arizona, New Mexico, Idaho, Oregon, or Washington. Stands have been sampled at Coyote Ridge (Evens and San 2004). *J. xiphioides* also dominates stands on serpentine in eastern Napa County and adjacent Colusa County, with *Sisyrinchium bellum, J. balticus,* and other wetland species (T. Keeler-Wolf and J. Evens, personal observation, 2002).

12.7 Upland Alliances with Serpentine Associations

Many more vegetation alliances may be associated with serpentine soils in some parts of their range than are largely or wholly restricted to them. As mentioned in chapters 10 and 11, the notion of the serpentine syndrome as a strong limiting influence on density and distribution of plants but not an exclusive restrictor is also apparent in the distribution of vegetation alliances on serpentine throughout western North America. The following section briefly identifies the alliances with some representation on serpentine and their specific associations known or thought to be endemic to serpentine.

12.7.1 Tree Alliances with Serpentine Associations

Knobcone pine *(Pinus attenuata)* alliance

Stands are commonly found on serpentine in the Northern California Coast Ranges, the Klamath Mountains, and in the Sierra Motherlode. This is a fire-adapted alliance and would be expected to occur where stands of serpentine chaparral are relatively dense (Safford and Harrision 2004). Some of the southernmost stands (Santa Ana Mountains, Orange County, California) also occur on serpentine (see chapter 13). Little vegetation classification work has been carried out on this alliance. It is uncertain how many unique associations are largely restricted to it. Yet, as many of the associated species are also serpentine endemics, it stands to reason that some associations will be defined that are largely exclusive to serpentine. Stands on serpentine commonly occur with a dense understory of chaparral shrubs including chamise, white-leaf or Stanford manzanita, muskbush (*Ceanothus jepsonii*), leather oak, and toyon.

Coulter pine *(Pinus coulteri)* alliance

This alliance is commonly associated with serpentine only in the South California Coast Ranges domain. At San Benito Mountain, six different serpentine associations were identified (CNPS-DFG 2006). Most are found over combinations of chaparral shrubs such as leather oak, bigberry manzanita, and Mexican manzanita. However, two were found in association with other conifers, one with foothill pine with or without incense-cedar with an open understory. Coulter pine is relatively short lived and usually requires fire to regenerate (Borchert et al. 2004).

The following associations have been defined from serpentine at Mt. San Benito:

Pinus coulteri /Arctostaphylos glauca

Pinus coulteri–Pinus sabiniana/Quercus durata–Arctostaphylos pungens

Pinus coulteri–Calocedrus decurrens–Pinus jeffreyi/Quercus durata

Pinus coulteri–Calocedrus decurrens/Rhamnus tomentella/Aquilegia eximia

Pinus coulteri/Quercus durata

Jeffrey pine *(Pinus jeffreyi)* alliance

Although Jeffrey pine occurs many places off of serpentine, this alliance is probably the most ubiquitous of all the montane serpentine alliances south of central Oregon and has been differentiated into several serpentine endemic associations. It occurs on serpentine in the Klamath Mountains, Northern California Coast Ranges, Sierra Motherlode, and Southern California Coast Ranges domains. The most striking representations of this alliance are found in the Klamath Mountains domain, where open, stunted stands may occur over shrubs such as huckleberry oak, or pinemat manzanita, or over grasses such as Idaho fescue *(Festuca idahoensis)* (Jimerson et al. 1995). Similar stands occur in the Southern California Coast Ranges at Mt. San Benito, where a single association has been defined mixed with Coulter pine, incense-cedar, and foothill pine over an open shrub cover of primarily leather oak, and bigberry manzanita (CNPS-DFG 2006). Taller stands may occur mixed with Douglas-fir and incense-cedar over much of the lower to-mid elevations in the Klamath Mountains. In the Rattlesnake Creek terrane (locality 5-13, chapter 17) the Dunning timber site index for old growth stands is commonly III (125 ft in 300 years) on both serpentine and adjacent stands on nonserpentine soils, even though the stands are generally more open on serpentine (E.A. Alexander, personal observation).

Stands of Jeffrey pine also occur mixed with red and white fir on the slopes of Red Hill and Red Mountain in Plumas County. These stands above 1800 m elevation are the highest stands of serpentine vegetation in the Sierra Motherlode domain. In the Klamath Mountains domain, the following associations have been defined by Taylor and Teare (1979): *Pinus jeffreyi/Ericameria ophititis* and *Pinus jeffreyi/Calamagrostis koelerioides*.

Also, the following have been defined by Jimerson et al. (1995):

Pinus jeffreyi/Festuca idahoensis

Pinus jeffreyi/Quercus vaccinifolia–Artctostaphylos nevadensis

Pinus jeffreyi–Pseudotsuga menziesii/Quercus vaccinifolia/Festuca californica

Pinus jeffreyi–Calocedrus decurrens/Ceanothus cuneatus

Pinus jeffreyi–Calocedrus decurrens/Ceanothus pumilus

Pinus jeffreyi–Calocedrus decurrens/Quercus vaccinifolia/Xerophyllum tenax

Pinus jeffreyi–Calocedrus decurrens–Abies concolor/Quercus vaccinifolia

The alliance has also been identified from serpentine on Mt. San Benito in the Southern California Coast Ranges (CNPS-DFG 2006). In southern Oregon *Pinus jeffreyi–Pseudotsuga menziesii/Arctostaphylos viscida* (NatureServe 2004) occurs on serpentine in the Klamath Mountains domain.

Foothill pine *(Pinus sabiniana)* alliance

The foothill pine alliance is widespread both on and off serpentine in California. It occurs in the Klamath Mountains, Northern California Coast Ranges, Southern California Coast Ranges, and Sierra Motherlode domains. In serpentine areas it usually occurs as an open overstory with open to dense shrubs beneath. In the Sierra Motherlode it is commonly associated with buckbrush (*Ceanothus cuneatus*) and with native annual herbs such as *Plantago erecta* and *Vulpia microstachys*. One association has been defined on serpentine: *Pinus sabiniana/Ceanothus cuneatus/Plantago erecta* (Evens et al. 2004). In the Northern California Coast Ranges on south-facing slopes it is commonly associated with leather oak and coast white-leaf manzanita and on north-facing slopes with shrubby California bay. In the Southern California Coast Ranges domain it is often associated with leather oak, bigberry manzanita, and chamise. Recent classification work in the Southern Coast Range (Evens and San 2004) has described some of the variation on serpentine. One association has been defined from Coyote Ridge (*Pinus sabiniana/Artemisia californica–Ceanothus ferrisiae–Heteromeles arbutifolia*) the latter identified by its association with the endemic coyote ceanothus (*C. ferrisiae*).

Incense-cedar *(Calocedrus decurrens)* alliance

The incense-cedar alliance is widespread usually as small stands in convexities and near streams, creeks, and meadows throughout California and southwestern Oregon. On serpentine, stands often occur at lower elevations and in somewhat less mesic conditions. For example, small stands occur in the middle and inner North Coast Ranges on serpentine, surrounded by serpentine chaparral or by low-elevation Douglas-fir forest. In the South California Coast Ranges stands of incense-cedar occur on slopes mixed with Jeffrey pine and Coulter pine in the Mt. San Benito area. No associations have yet been formally described from serpentine.

Douglas-fir *(Pseudotsuga menziesii)* alliance

The widespread Douglas-fir alliance has endemic associations on serpentine in the Klamath Mountains domain (Jimerson et al. 1995). Some 80 associations are known from British Columbia to Colorado and central California (NatureServe 2004). Only seven are known largely from serpentine soils (Jimerson et al. 1995):

Pseudotsuga menziesii–Umbellularia californica/Toxicodendron diversilobum

Pseudotsuga menziesii–Lithocarpus densiflora/Quercus vaccinifolia–Holodiscus discolor

Pseudotsuga menziesii–Calocedrus decurrens/Festuca californica

Pseudotsuga menziesii–Pinus jeffreyi/Festuca californica

Pseudotsuga menziesii/Quercus vaccinifolia

Pseudotsuga menziesii/Quercus vaccinifolia–Lithocarpus densiflora ssp. *echinoides*

Pseudotsuga menziesii/Quercus vaccinifolia–Rhododendron macrophyllum

However, it is likely that many structural differences (based primarily on lower site class ratings of the mature trees and lower average percent cover) are present between stands of *Pseudotsuga menziesii* on and off serpentine in the Blue Mountains and Northern Cascade Mountains–Fraser River domains of Oregon, Washington, and southern British Columbia (see chapters 18 and 19). One likely association that shows a more open structure on serpentine is the *Pseudotsuga menziesii–Arbutus menziesii/Vicia americana* association on the San Juan Islands. Douglas-fir–dominated forests grade into grand fir (*Abies grandis*) forests on serpentine in the Blue Mountains. In the Tulameen area of British Columbia, stands of Douglas-fir, low juniper (*J. communis*), and bluebunch wheatgrass (*Agropyron spicatum*) occur (Kruckeberg 1969). At the same locality on Inceptisols, a denser Douglas-fir stand with sparse ponderosa pine (*Pinus ponderosa*), common kinnikinnick (*Arctostaphylos uva-ursi*), and some saskatoon (*Amelanchier alnifolia*) occurs (locality 7-3b).

At the Ingalls locality (central Washington) a Douglas-fir–subalpine fir–whitebark pine/grouseberry (*V. scoparium*) plant association occurs on a Eutrochrept at 1800 m elevation. About 1650 m in elevation, a Douglas-fir (*Pseudotsuga menziesi*)/juniper (*Juniperus* sp.)/wheatgrass (*Pseuoroegnaria spicatum*) plant community occurs on a shallow Eutrochrept. Also occurring at Ingalls is a Douglas-fir/pinemat manzanita (*Arctostaphylos nevadensis*)/sedge association containing some ponderosa pine (*P. ponderosa*) on drier slopes and some white pine (*P. monticola*) on cooler or moister slopes and on mesic slopes: *Pseudotsuga menziesii/Juniperus communis–Ledum glandulosum*. Although these types have not formally been accepted within the National Vegetation Classification system, they appear to represent local serpentine expressions of the alliance.

Western white pine *(Pinus monticola)* alliance

The western white pine alliance is widespread in western North America from California to Idaho and Montana. It is known from serpentine in the Klamath Mountains domain, where it may occur at elevations as low as 700 m and as high as 2200 m. The classification work done on the serpentine manifestation of this alliance in the Klamath domain suggests at least four associations. Three are low-elevation types from the Josephine Peridotite: western white pine–Del Norte lodgepole pine/dwarf tanbark-Pacific rhododendron (*Pinus monticola–Pinus contorta* ssp *latifolia/Lithocarpus densiflorus* ssp. *echinoides–Rhododendron macrophyllum*), western white pine–sugar pine/huckleberry oak–dwarf tanbark (*Pinus monticola–Pinus lambertiana/Quercus vaccinifolia–Lithocarpus densiflorus* ssp. *echinoides*), and western white pine–Douglas-fir/huckleberry oak-dwarf tanbark (*Pinus monticola–Pseudotsuga menziesii/Quercus vaccinifolia–Lithocarpus densiflorus* ssp. *echinoides*) (Jimerson et al. 1995). One is a high-elevation type from the subalpine woodlands of the Trinity pluton area: western white pine/Angelica (*Pinus monticola/Angelica arguta*) (Whipple and Cole 1979).

Other stands of western white pine are known from exposed upper slopes of the Blue Mountains (domain 6, chapter 18). These have not been differentiated from existing associations of the alliance.

Foxtail pine *(Pinus balfouriana)* alliance

The foxtail pine alliance is a subalpine woodland alliance only known from the high mountains of California. Most stands are found on granitic or other crystalline rocks. Only in the Klamath domain does it occur on serpentine (including gabbro). No serpentine associations have been defined for the alliance (although sample plots have been taken at Crater Creek RNA, Keeler-Wolf 1990).

Lodgepole pine (*Pinus contorta* ssp. *latifolia*) and shore pine (*Pinus contorta* ssp. *contorta*) alliance

The lodgepole and shore pine alliance, in the broad sense as it is treated here, is very widespread and includes a coastal race of shore pine as well as the extensive northern interior and montane races of lodgepole pine. Although a distinctive race from serpentine on the Josephine Peridotite of the Klamath domain has been identified (Keeler-Wolf 1986, Oliphant 1992), it mixes with various shrubs such as scrub tanbark oak, huckleberry oak, and Labrador tea (*Ledum glandulosum*) and with other serpentine tolerant trees such as Port Orford-cedar, western white pine, sugar pine, and knobcone pine. Stands including this race have been included largely in the Sugar Pine alliance (Jimerson et al. 1995, see below). Farther north in Washington the coastal serpentine outcrops of Cypress Island also support the more typical form of shore pine (Kruckeberg 1969). Other stands of *P. contorta* ssp. *latifolia* from serpentine in the North Cascades–Fraser River domain are known, but it is uncertain if this is closely related to the Klamath race found on the Josephine Peridotite (see chapter 17).

Kruckeberg (1969) described a stunted woodland stand which could be called *Pinus contorta latifolia/Juniperus communis–Sheperdia canadensis/Aspidotis densa–Polystichum kruckebergii*. This stand is likely to represent a unique serpentine association. He also defined stands from the Ingalls locality in central Washington as a moderately dry woodland on alluvium: *Pinus contorta* ssp. *latifolia–Pseudotsuga menziesii–Pinus ponderosa/ Arctostaphylos nevadensis*.

Lodgepole pine is the main successional coniferous tree alliance on the serpentine soils of parts of the North Cascades–Fraser River domain. A lodgepole pine (*P. contorta*)/white-flowered rhododendron (*Rhododendron albiflorum*)/low blueberry (*Vaccinium myrtillus*) plant community has been identified at 1550 m at the Tulameen locality (Kruckeberg 1969).

Farther north in the Gulf of Alaska domain, lodgepole pine is found only in southeastern Alaska and is generally on boggy, poorly drained sites. It may form low stands near the coast, which resemble the nonserpentine association *Pinus contorta/Empetrum nigrum* defined by Martin et al. (1985) and Neiland and Viereck (1977).

Sugar pine (*Pinus lambertiana*) alliance

The sugar pine alliance is local in the mountains of California, the nominate species being usually only a component of several other vegetation alliances. However, on serpentine in the Klamath domain, Jimerson et al. (1995) have identified three associations that regularly occur on serpentine. The first two contain the serpentine race of the lodgepole pine which resembles *Pinus contorta latifolia*: sugar pine–lodgepole pine/huckleberry oak–dwarf tanbark oak (*Pinus lambertiana–Pinus contorta/Quercus vaccinifolia–Lithocarpus densiflora* var *echinoides*), Sugar pine–lodgepole pine/huckleberry oak–Pacific rhododendron (*Pinus lambertiana–Pinus contorta/Quercus vaccinifolia–Rhododensron macrophyllum*), and sugar pine–western white pine/huckleberry oak–dwarf silktassel (*Pinus lambertiana–Pinus monticola/Quercus vaccinifolia–Garrya buxifolia*).

Ponderosa pine–Douglas-fir (*Pinus ponderosa-Pseudotsuga menziesii*) forest alliance

Although widespread in western North America, this mixed alliance is known on serpentine only from the Klamath Mountains and small portions of the North Cascades Frazer River domain. North of the Klamath Mountains, this alliance is not known to have distinctive associations on serpentine. Ponderosa pine–Douglas-fir forests with grasses and elk sedge (*Carex geyeri*) dominating the understory are known from the Blue Mountains of northeastern Oregon on serpentine (chapter 18). Stands of mixed Ponderosa pine and Douglas-fir also occur on warmer aspects in the Ingalls locality with open to sparse forests of either ponderosa pine and Douglas-fir with bitterbrush (*Purshia tridentata*) and bluebunch wheatgrass (*Pseudoroegneria spicata*) or ponderosa pine with bluebunch wheatgrass. All of these associations are found elsewhere off of serpentine.

In the Klamath Mountains this mixed alliance is common on peridotite and gabbro in the Trinity and the Rattlesnake localites, where one serpentine endemic association has been identified: *Pseudotsuga menziesii–Pinus ponderosa/Quercus vaccinifolia* (Sawyer et al. 1978). It occurs on warm south-facing slopes between 1000–1500 m elevation.

Subalpine fir (*Abies lasiocarpa*) alliance

Widespread in subalpine western North America, the subalpine fir alliance is known from serpentine in Oregon, Washington, British Columbia, and southeast Alaska. In British Columbia (Tulameen locality) subalpine fir is common on upper north-facing slopes, with some western white pine (*P. monticola*). In locality 7-2 (shulaps) a krumholtz subalpine fir (*Abies lasiocarpa*)–whitebark pine (*P. albicaulis*)/willow (*Salix* spp.)–white mountain heather (*Cassiope mertensiana*)/arctic lupine (*L. arcticus*) plant community occurs at 2150 m in mesic settings while stands of sparse whitebark pine with an understory of common juniper (*J. communis*) occur in more xeric, rocky areas. Semi-dense stands of lodgepole pine and subalpine fir and an understory of white rhododendron and black huckleberry (*Vaccinium membranaceum*) occur on lower sites. None of these types has been identified as solely serpentine associations (D. Meidinger, personal communication, 2004).

Kruckeberg (1969) identified several stands in the Northern Cascades of subalpine fir with lodgepole pine. Two xeric types, depending on their extent and redundancy, are likely to represent serpentine associations, but have been so far not formally defined: xeric moraine upland (*Pinus contorta latifolia–Abies lasiocarpa/Phyllodoce glandulifera/Silene acaulis–Polystichum lemmonii*) and xeric open woodlands (*Abies lasiocarpa–Chamaecyparis nootkatensis/Juniperus communis/ Aspidotis densa–Polystichum lemmonii*). They both have the regional serpentine indicator ferns *Aspidotis densa* and/or *Polystichyum lemmonii*.

Another extremely xeric stand described by Kruckeberg (1969) at the Ingalls locality as *Abies lasiocarpa–Pseudotsuga menziesii/Agropyron spicatum* may also represent an undescribed serpentine association.

In Alaska in the Denali domain, subalpine fir reaches its northern limits of distribution, where on serpentine small patches of forest similar to the nonserpentine association *Abies lasiocarpa/Phyllodoce aleutica–Fauria crista–galli* (Harris 1965, Worley and Jacques 1973) exist.

Subalpine fir occurs in stands with Engelmann spruce and whitebark pine on pyroxenite in the Stillwater complex of the Rocky Mountains.

McNab cypress (*Cupressus macnabiana*) alliance

Along with the Sargent cypress alliance, the McNab cypress alliance is the most widespread of the cypress woodlands in western North America. About half of the known stands occur on serpentine, particularly in the Northern California Coast Ranges where virtually all stands are closely associated with serpentine. The remaining stands, particularly in the northern Sierra and Southern Cascade foothills, often occur on volcanic substrates (Griffin and Critchfield 1972). Stands are typically found surrounded by serpentine chaparral of either the leather oak or the coast whiteleaf manzanita alliance. Stands are often shrubby as a result of low moisture, low fertility, and recurring fire. No associations have been formally defined, but because all associated species in a large portion of its range are serpentophiles, we have included this alliance in this part of the discussion.

12.7.2 Shrub Alliances with Serpentine Associations

California sagebrush (*Artemisia californica*) alliance

Although it not usually associated with serpentine, Evens and San (2004) described two associations from the serpentine soils of Coyote Ridge in the Southern California Coast Ranges domain. In one association, *Artemisia californica* and the rare endemic *Ceanothus ferrisiae* codominate in the shrub layer (average cover 13% and 7%, respectively). *Pinus sabiniana* and *Heteromeles arbutifolia* are constants, but tend to have somewhat lower cover (average 3% and 2%). In the other association, California sagebrush occurs over a grassy and herbaceous understory. *Artemisia californica* is the dominant species in the shrub layer (average 24% cover). *Avena barbata, Bromus madritensis, Eschscholzia californica, Bromus hordeaceus,* and *Nassella pulchra* are often associated with these stands (each average around 2%–4% cover, 66% frequency).

Chamise (*Adenostoma fasciculatum*) alliance (including the related chamise–whiteleaf manzanita alliance)

The chamise alliance is the quintessential alliance of the California chaparral. It is divided into a number of associations, most of which have been defined on nonserpentine substrates. In the Southern California Coast Ranges chamise may strongly dominate or have a minor component of buckbrush (*Ceanothus cuneatus*). It may also have a significant component of leather oak, bigberry manzanita, or toyon. In studies of the vegetation of both Coyote Ridge (Evens and San 2004) and Mt. San Benito (CNPS-DFG 2006), chamise often codominates with bigberry manzanita. Chamise may also occur on serpentine in the Sierra Motherlode (Evens et al. 2004). However, in the Sierra most serpentine chaparral is not dominated by chamise. A chamise–buckbrush–bigberry manzanita association has been defined from serpentine at Mt. San Benito (CNPS-DFG 2006).

Chamise–Eastwood manzanita (*Adenostoma fasciculatum–Arctostaphylos glandulosa*) alliance

The chamise–Eastwood manzanita alliance is only known from serpentine rock on the gabbro of the Peninsular Ranges in the Baja California domain. One association has been defined from the Las Posas soil series (Gordon and White 1994). It is known as the chamise–Eastwood manzanita/mafic association and harbors several rare and endemic species of the San Diego gabbro area.

Whiteleaf manzanita (*Arctostaphylos viscida* ssp *viscida*) alliance

Stands of whiteleaf manzanita occur in the Klamath Mountains Province, often with an open overstory of Jeffrey pine and associated with buckbrush (*Ceanothus cuneatus*). Stands also occur in the Sierra Motherlode. Unlike the serpentine endemic subspecies, coast whiteleaf manzanita (*A. viscida* ssp. *pulchella*), the alliance dominated by the nominate subspecies is very common both on and off serpentine. In the more favorable moisture regimes of the Northern Sierra Motherlode and the Klamath domains, whiteleaf manzanita tends to remain seral to Douglas fir and other coniferous alliances, even on serpentine.

One association has been formally defined on serpentine from the Klamath Province: *Arctostaphylos viscida–Ceanothus cuneatus/Festuca idahoensis–Achnatherum lemmonii* Shrubland (NatureServe 2004).

Bigberry manzanita (*Arctostaphylos glauca*) alliance

The bigberry manzanita alliance is in some ways the Southern California Coast Ranges domain's ecological equivalent of the Northern California Coast Range's coast whiteleaf manzanita alliance. Although not exclusive to serpentine, when it does grow on serpentine, it also dominates on south-facing exposures and frequently mixes with leather oak. Both bigberry and coast whiteleaf manzanitas look quite similar to one another. The principal difference is that although bigberry manzanita is often found on serpentine, particularly in the northern portion of its range (see also chapter 11), it may also occur off of serpentine nearby. Data from Coyote Ridge (Evens and San 2004) and Mt. San Benito (CNPS-DFG 2006) show several associations: on Coyote Ridge are *Arctostaphylos glauca* mixed (*–Artemisia californica–Salvia mellifera*) shrubland (indifferent, on and off serpentine) and *Arctostaphylos glauca*/(*Melica toreyana*) Torrey's Melic grass shrubland (serpentine endemic); and on San Benito Mountain (all on serpentine) are *Arctostaphylos glauca–Quercus durata* shrubland, *Arctostaphylos glauca–Quercus durata–Ceanothus cuneatus* shrubland, *Arctostaphylos glauca–Quercus durata–Rhamnus tomentella–Rhamnus ilicifolia* shrubland, and *Arctostaphylos glauca–Adenostoma fasciculata–Quercus durata–Ceanothus cuneatus* shrubland.

Buckbrush (*Ceanothus cuneatus*) alliance

The buckbrush alliance occurs from Baja California north to southern Oregon. It is one of the characteristic components of the California chaparral formation, and many stands occur off of serpentine on a variety of substrates. Many stands also occur on serpentine, particularly in the Sierra Motherlode and in the Klamath Mountains domains. Buckbrush alliance stands are common below ~1300 m elevation in the drier portions of the Klamath Mountains domain of both Oregon and California (Taylor and Teare 1979; J. Sawyer and J. Kagan, personal communication, 2004). There is some question as to the identity of the Klamath buckbrush. Many individuals growing on serpentine have smaller leaves and a more spreading habit than the typical *C. cuneatus* to the south (J. Sawyer, personal communication, 2004).

Currently a single association has been identified from the serpentine of the Klamath domain: *Ceanothus cuneatus/Elymus elymoides* (Taylor and Teare 1979). In the Sierra Motherlode the alliance is common on serpentine, often with an emergent overstory of foothill pine. Evens et al. (2004) have identified one association (*Ceanothus cuneatus/Plantago erecta*) as endemic to serpentine in the central Sierra Nevada Foothills.

12.7.3 Herb and Grass Alliances That May Occur on Serpentine

California goldfields *(Lasthenia californica* and other *Lasthenia* species) alliance

The California goldfields alliance is dominated by the native annual composite wildflower genus *Lasthenia*. Classification work has just begun for the vegetation dominated by these species and they have not yet been divided into separate alliances. *L. californica* is widespread throughout California and stretches eastward into the western Mojave Desert. *L. glabrata* is more restricted to vernally moist areas on serpentine or alkaline nonserpentine soils. Stands in Lake and Colusa counties have been identified on alluvial bottomland soils derived from serpentine rock. Another common associate of these stands is the blue-eyed grass (*Sisyrinchium bellum*). On serpentine it is usually restricted to open grassy areas that become quite dry in the summer. Many of the upland serpentine goldfields stands have *Lasthenia californica* codominant with *Plantago erecta* and those that have been sampled and analyzed have been treated under the *P. erecta* alliance (see below).

Foothill plantain (*Plantago erecta*) alliance

The foothill plantain alliance, characterized by a diminutive annual plantain, is largely restricted to clay-rich soils of California and may or may not occur on serpentine. It has been described from serpentine soils at Coyote Ridge, Santa Clara County (Evens and San 2004), and Mt. San Benito (CNPS-DFG 2006) in the Southern California Coast Ranges domain. Evens et al. (2004) also described it from serpentine in the Sierra Motherlode domain. In central California, it is integral to the survival of the endangered Bay Checkerspot butterfly because it is the sole larval food of this species.

Small fescue (*Vulpia microstachys*) alliance

The small fescue alliance is characterized by the dominance of *Vulpia microstachys* in the overstory, herbaceous layer. A variety of native and non-native species may also be present, but in overall lower cover than the *Vulpia* (e.g., relative cover is <25%). It has been observed in the Central Coast Range of California on northeast- to southwest-facing slopes, including well-developed soils (e.g., clay loams) on serpentine and sedimentary substrates that are vernally mesic. It may occur in other locations in western North America, such as the Central Valley and North California Coast Ranges. One serpentine association has been defined from the Sierra Motherlode domain, *Vulpia microstachys–Plantago erecta–Calycadenia truncata* (Evens et al. 2004), and one from the Southern California Coast Ranges domain, *Vulpia microstachys-Plantago erecta* (Evens and San 2004). Until more data are analyzed from a wider range of localities, those stands that share dominance with *Plantago erecta* are placed in this alliance (see above).

12.8 Vegetation Alliances Largely Undifferentiated on Serpentine

12.8.1 Undiffferentiated Tree Alliances

Engelmann spruce (*Picea engelmanii*) alliance

The Engelmann spruce alliance is widespread in the mountains of western North America. On serpentine it is known only from the Northern Cascade–Fraser River domain (chapter 19), where it can be confused with white spruce, and it occurs on pyroxenite of the Stillwater Complex in Montana. No specific information is available on its species composition, but it is assumed to be undifferentiated from other similar local nonserpentine stands of the alliance.

White spruce (*Picea glauca*) alliance

This is an extensive alliance in the northern part of North America. It represents the classic climatic timberline in many areas of Canada and Alaska. White spruce alliance stands are common on serpentine soils near the lower (north) end of Dease Lake, in northwestern British Columbia. In the Gulf of Alaska domain, white spruce (*Picea glauca*) and paper birch (*Betula papyrifera*) prevail on some serpentine north-facing slopes, with much bunchberry (*Cornus canadensis*) and mosses in the understory. Although white spruce-paper birch stands generally are generally dominated by large white spruce and paper birch that reach maximum diameters of about 30–35 cm (12–14 in diameter at breast height) and maximum heights of 18–23 m (60–75 ft), stands on serpentine in Alaska are generally much smaller in stature. Viereck et al. (1992) do not differentiate these from nonserpentine stands of the same association, which they call *Picea glauca–Betula papyrifera/Alnus crispa/Calamagrostis canadensis*.

In the Northern Alaska–Kuskokwim Mountains (chapter 22), stands with low white spruce, alder (*Alnus crispa*), crowberry (*Empetrum nigrum*), alpine blueberry (*Vaccinium uliginosum*), Labrador tea (*Ledum palustre*), mosses, and lichens (mostly *Cladina* sp. and *Flavocetraria* sp.) form a timberline scrub on serpentine which resembles two nonserpentine associations defined in Viereck et al. (1992): *Picea glauca/Alnus crispa–Salix* spp./*Equisetum arvense* (Craighead et al. 1988) and *Picea glauca/Vaccinium spp.–Empetrum nigrum* (Craighead et al. 1988).

Brewer spruce (*Picea breweriana*) alliance

This local alliance in only found in the Klamath Mountains, where it occasionally occurs on serpentine (including gabbro). Usually, it is associated with Douglas-fir, white fir, red fir, and/or western white pine and frequents steep, north-facing slopes. No classification work has been done on the serpentine stands.

Rocky Mountain juniper *(Juniperus scopulorum)* woodland alliance

The Rocky Mountain juniper alliance occupies serpentine outcrops in the Blue Mountains (domain 6). No detailed classification work has been done on serpentine, but the vegetation appears to resemble other previously defined associations such as the *Juniperus occidentalis/Pseudoroegneria spicata* association (NatureServe 2004). A small area of serpentine on Fidalgo Island (locality 7-4) is dominated by this alliance.

Western juniper *(Juniperus occidentalis)* woodland alliance

The western juniper alliance occupies serpentine outcrops in the eastern Klamath Mountains, where it abuts Great Basin scrub vegetation dominated by *Artemisia tridentata* and *Purshia tridentata*.

White fir *(Abies concolor)* alliance

The white fir alliance is common throughout the Rocky Mountains and in the California Mountains. However, it is only known from serpentine in the Klamath domain, where it occasionally forms stands at higher elevations in the eastern portion of the domain as in the Trinity locality. It is usually limited in extent and exists as small stands on north-facing aspects. It may co-occur with Douglas-fir, sugar pine, and Jeffrey pine.

Red fir *(Abies magnifica,* including noble fir–*A. procera)* alliance

This alliance is indicative of the high snow accumulation belt of the mountains of California and southern Oregon. It is widespread off of serpentine, but does occur on serpentine and gabbro in the Trinity pluton and Josephine peridotite areas, usually above 1550 m on north-facing slopes. As with many of the montane forest alliances, those stands on serpentine tend to be relatively open. No specific red-fir serpentine associations have been identified yet. Common associated trees may be western white pine and mountain hemlock; associated shrubs include *Quercus vaccinifolia* and *Arctostaphylos nevadensis*.

Grand fir (*Abies grandis*) alliance

The grand fir alliance ranges from northwestern California to British Columbia but is known on serpentine only from the Blue Mountains domain (chapter 18), where it occurs with Douglas-fir and some Pacific silver fir (*Abies amabilis*), on warm slopes. The understory is dominated either by grouseberry (*Vaccinium scoparium*) or pinemat manzanita (*Arctostaphylos nevadensis*) (Kruckeberg 1969).

Pacific silver fir (*Abies amabilis*) alliance

The Pacific silver fir alliance is largely limited to Oregon, Washington, and British Columbia and occurs on serpentine in British Columbia's Hozameen locality (chapter 19) with silver fir, western hemlock (*Tsuga heterophylla*), and yellow cedar (*Chamaecyparis nootkatensis*) with an understory of black huckleberry (*Vaccinium membranaceum*), mosses, and star flower (*Trientalis latifolia*). This alliance is not thought to be restricted to serpentine.

Mountain hemlock (*Tsuga mertensiana*) alliance

The mountain hemlock alliance is widespread from California to southeastern Alaska. Some of its stands occur on serpentine in the Klamath Mountains domain. Stands also occur in the coastal ranges of southeast Alaska. Most of these are sparse and occur on recently glaciated rock. The stands resemble the *Tsuga mertensiana–Abies lasiocarpa–Chamaecyparis nootkatensis*/*Cassiope mertensiana–Luetkea pectinata* association defined by Viereck et al. (1992). At the Twin Sisters locality in British Columbia, a mountain hemlock– subalpine fir (*Abies lasiocarpa*) with a semi-dense understory of pink and white mountain heather (*Phyllodoce empetriformis* and *Cassiope mertensiana*) and black huckleberry (*Vaccinium membranaceum*) occurs on deep Cryepts.

Whitebark pine (*Pinus albicaulis*) alliance

The whitebark pine alliance is one of the classical timberline vegetation types of western North America. It ranges from California to British Columbia, Montana, and Colorado. It is known from serpentine in the Klamath Mountains and the Blue Mountains domains. No specific associations have been identified from serpentine. However, in the eastern part of the Klamath Mountains domain, whitebark pine seems to mix with *Cercocarpus ledifolius* on gabbro, although it does not appear to occur on peridotite. In the

Blue Mountains of eastern Oregon whitebark pine mixes with low juniper (*Juniperus communis*).

At the Ingalls locality in Washington, Kruckeberg (1969) identified stands on xeric southwestern serpentine exposures as *Pinus albicaulis/Juniperus communis/Poa curtifolia*, which are probably similar to other dry nonserpentine whitebark pine timberline stands in the region.

White alder (*Alnus rhombifolia*) alliance

The white alder is a widespread species in the foothills, lower mountains, and coastal ranges of California and Oregon. It is restricted to riparian settings along creeks and rivers. In the Klamath Mountains domain it forms scattered stands along streams and creeks with rocky and cobbly beds composed of serpentine. No specific serpentine associations have been defined.

Aspen (*Populus tremuloides*) alliance

The aspen is the most widespread tree in North America (Little 1978) and one of the most widespread vegetation alliances as well. Where it occurs on serpentine it is typically successional and is largely known from British Columbia (chapter 19) and domains in Alaska (chapters 20–22). In the Gulf of Alaska domain, the alliance is composed of semi-dense to open vegetation on steep, southwest-facing slopes. It is predominantly aspen (*Populus tremuloides*) and paper birch (*Betula payrifera*), with much buffaloberry (*S. canadensis*), rose (*Rosa acicularis*), fireweed (*Epilobium angustifolium*), and other forbs in the understory, and with sparse white spruce and mosses. According to Viereck et al. (1992), quaking aspen occurs widely on extremely dry sites on steep southern slopes in interior and south-central Alaska.

Paper birch–quaking aspen is found on moderately warm sites in interior and south-central Alaska. The following plant associations are likely to occur on serpentine: *Populus tremuloides-Betula papyifera/Rosa acicularis/Arctostaphyos uva-ursi/lichens* (Yarie 1983), *Populus tremuloides/Elaeagnus commutata–Shepherdia canadensis/Arctostaphylos* spp./lichens (Neiland and Viereck 1977), and *Populus tremuloides/Shepherdia canadensis/Calamagrostis purpurascens* (Viereck et al. 1983).

Balsam poplar (*Populus balsamifera*) may be present in forests of aspen on the lower slopes on both northern and southwestern exposures on serpentine in the Gulf of Alaska domain. Such stands resemble *Populus tremuloides–P. balsamifera/Rosa acicularis* (Yarie 1983). Quaking aspen–balsam poplar occurs on flood plains in interior Alaska on and off serpentine.

Paper birch (*Betula papyrifera*) alliance

The paper birch alliance is widespread in the boreal regions of North America. In the Denali domain, paper birch occurs on many upland sites, both with and without permafrost, in interior and south-central Alaska. Some stands on serpentine resemble *Betula papyrifera/Alnus crispa/Calamagrostis* spp. (Viereck et al. 1992). An alder (*Alnus crispa*)–paper birch (*Betula paperifera*)–willow (*Salix* sp.)–scrub birch (*Betula nana*) scrub thicket covers the slopes. It is probably analogous to the widespread *Betula papyrifera/Alnus* spp.–*Salix* spp. association (Racine 1976) or the *Betula papyrifera/Ledum groenlandicum/Pleurozium schreberi–Polytrichum juniperinum* association (Jorgenson et al. 1986).

Blue oak (*Quercus douglasii*) alliance

This widespread woodland alliance is one of the signature vegetation types of the foothills of the California region. However, blue oak typically avoids serpentine except in the Sierra Motherlode, where it is occasionally found on gently sloping serpentine soils often associated with buckbrush and non-native annual grasses. Rare stands of blue oak are also found on serpentine in the Inner Northern California Coast Ranges.

Oregon oak (including Brewer oak, *Quercus garryana* var. *breweri*) alliance

Oregon oak is common in Oregon and in the Northern California Coast Ranges and Klamath Mountains domain. It only occasionally occurs on serpentine. Most of the serpentine stands are composed of small, shrubby trees and resemble var. *breweri*. One population of the federally listed endangered *Phlox hirsuta* is listed as occurring with *Q. garryana* at the Soap Creek Ridge population, according to a U.S. Fish and Wildlife Service recovery plan. (FWS 2004.)

California bay (*Umbellularia californica*) alliance

Bay is a good tolerator of serpentine soils, commonly forming shrubby understories to foothill pine stands and occasionally forming its own dominant stands (often as shrubs) on north-facing slopes in serpentine landscapes in the Southern California Coast Ranges and Northern California Coast Range domains. No classification work has been conducted on the serpentine stands.

12.8.2 Undifferentiated Shrub Alliances

Low juniper *(Juniperus communis)* alliance

Low juniper alliance stands are known from a variety of substrates throughout much of northern and montane North America. Stands are found on exposed rocky hilltops, ridges, and slopes. Soils are thin and contain sand and gravel. Exposed bedrock of several origins (basalt, granitic, serpentine) is common. High winds move over the ridges and slopes and are an important factor in maintaining the structure of stands of this alliance. They are common and characteristic of shallow serpentine soil, particularly in the North Cascades-Fraser River domain. At the Ingalls locality a low juniper/bluebunch wheatgrass (*Agropyron spicatum*) plant community develops on a Lithic Cryorthent at about 1800 m elevation in association with a sparse forb plant community. No strictly serpentine associations have been defined for this alliance.

Curl-leaf mountain mahogany (*Cercocarpus ledifolius*) alliance

The curl-leaf mountain mahogany alliance is widespread in the Great Basin and in the drier parts of the west, where it usually does not come into contact with serpentine. However, in the Klamath Mountains and in the Blue Mountains domains, it does occasionally grow on serpentine. Keeler-Wolf (1987b) described stands in the eastern Klamath Mountains growing on both serpentinite and gabbro between 1700 and 2365 m. The lower elevation types tend to be more open and have many herbs and grasses, while the upper elevation types are mixed with emergent subalpine conifers such as foxtail, whitebark, lodgepole, and western white pines. The curl-leaf mountain mahogany stands in the Blue Mountains domain have not been described in detail (see chapter 18).

Big sagebrush (*Artemisia tridentata*) alliance

Big sagebrush is a widespread scrub of the drier and cooler portions of western North America and is known from upper slopes in the Blue Mountains domain. The south-facing concave slopes are dominated by sagebrush (*Artemesia tridentata*) and mountain sorrel (*Rumex acetosella*), with patches of buckwheat (*Eriogonum flavum* and *E. heracleoides*), both widespread, dry montane species of western North America. No specific associations of big sagebrush alliance have been defined for serpentine.

Rubber rabbitbrush (*Chrysothamnus nauseosus* var. *mojavensis*) alliance

Although typically considered a nonserpentine Great Basin vegetation alliance, rubber rabbitbrush does occur on disturbed serpentine sites in the Southern California Coast Ranges at Mt. San Benito (CNPS-DFG 2006). It is usually a short-lived plant in that area, rapidly colonizing recently flooded alluvium along creeks and also tailings from recently closed asbestos mines. It is frequently associated with California buckwheat (*Eriogonum fasciculatum* var. *fasciculatum*).

California buckwheat (*Eriogonum fasciculatum*) alliance

Occasional stands of this widespread Southern California and Southern California Coast Ranges alliance are seen on serpentine in the Southern California Coast Ranges domain. In general, the California buckwheat alliance appears to avoid serpentine.

Wright's buckwheat (*Eriogonum wrightii*) alliance

Wright's buckwheat is a widespread but never common alliance. When interpreted broadly as including all three subspecies of *E. wrightii*, it occurs from extreme southern California up thorough the southern Sierra Nevada and the California Coast Ranges as far north as Tehama County. However, only a few stands have been found on serpentine. This alliance often occupies harsh rocky substrates and has been seen on serpentine alluvial outwash in Bear Valley and in openings at the upper elevations of Mt. San Benito in the Southern California Coast Ranges domain.

Huckleberry oak (*Quercus vaccinifolia*) alliance

The huckleberry oak alliance is very common on serpentine in the Klamath Mountains, but otherwise it appears only to occur on serpentine in the extreme northern portion of the Sierra Motherlode (Fiedler et al.1986). It is typically a montane alliance, frequently occurring above 1500 m, but it may occur as low as 500 m in the western Klamath Mountains. No detailed classification work has been done with serpentine stands of this alliance. In the eastern Klamath Mountains it frequently associates with montane coniferous alliances such as red fir, Jeffrey pine, western white pine, and pinemat manzanita on coarse gabbro. It also often occurs on peridotite in the western Klamath Mountains, as on the Josephine peridotite (chapter 17), where associates are usually lower elevation species.

Pinemat manzanita (*Arctostaphylos nevadensis*) alliance

In the Klamath Mountains, the pinemat manzanita alliance occurs on serpentine in the western part of the domain (Siskiyou Mountains, Josephine peridotite), where it may associate with endemic serpentine species (Keeler-Wolf 1986). It is also known from gabbro in the eastern Klamath Mountains. No serpentine associations have been defined.

Dwarf tanbark oak (*Lithocarpus densiflorus* ssp. *echinoides*) alliance

The recently described dwarf tanbark oak alliance (Sawyer and Keeler-Wolf, 2004) is largely endemic to the Klamath Mountains domain, where it may occur both on and off serpentine. Concerning the serpentine portion of its extent, it is most conspicuous on gabbro and occasionally on peridotite in the Trinity pluton area. It also occurs off of serpentine on granitic substrate in the Whiskeytown National Recreation Area (J. Stuart personal communication, 2002). Stands on serpentine are frequently seral and appear to be related to fire or tree harvest and may be short-lived (T. Keeler-Wolf, personal observation). No serpentine associations have been described.

Glandular Labrador-tea (*Ledum glandulosum*) alliance

Glandular Labrador tea is a wetland alliance found in bogs and fens in montane and coastal northern California and Oregon. Small stands in the Klamath Mountains domain are occasionally found on serpentine (Keeler-Wolf 1982, 1986), where they are usually associated with fens or bogs. Stands of *Ledum* are often associated with *Kalmia polifolia*, *Spiraea densiflora*, *Vaccinium uliginisum*, and other acid-loving shrubs and herbs. Stands are often associated with pitcher plant (*Darlingtonia californica*) and Port Orford-cedar alliance stands; however, no specific serpentine associations have been defined.

Black sage (*Salvia mellifera*) alliance

The black sage alliance is a common component of the coastal sage scrub formation of the Southern California Coast Ranges and the Baja California domains. It is more widespread off serpentine, with only a few known occurrences on serpentine. At Coyote Ridge

Toyon (*Heteromeles arbutifolia*) alliance

Toyon is a relatively widespread, but sporadic chaparral alliance throughout much of California west of the deserts. It is known from serpentine on xeric slopes in the Northern California Coast Range and Southern California Coast Ranges domains, where it is grows as open scrub with a scattered rocky, herbaceous understory. It usually occurs on steep south-facing exposures. No detailed classification work has been done on serpentine stands of this alliance.

Tea-leaf willow (*Salix planifolia*) alliance

The tea-leaf willow alliance forms thickets in subalpine settings from California to Alaska. In the Denali domain it is known from serpentine, where stands of *S. planifolia* associate with patches of scrub birch (*Betula nana*). Similar shrub birch-willow stands occur in poorly drained lowlands and on moist slopes in northern, interior, south-central, and southwestern Alaska. These resemble the *Salix planifolia/Betula nana–Vaccinium uliginosum* association (Brock and Burke 1980).

Shrub birch (*Betula nana*) alliance

This northwestern boreal alliance is known from the Denali domain, where stands resembling *Betula nana–Salix glauca–S. planifolia/Festuca altaica, Vaccinium vitis-idaea–Arctostaphylos alpina/Hylocomium splendens* (Viereck et al. 1992) and *Betula nana/Ledum decumbens–Vaccinium* spp. (Jorgenson et al. 1986) have been reported from serpentine. None of these associations are considered to be serpentine types. (Note that *Betula nana* and *B. glandulosa* are no longer differentiated from one another.)

Snow willow (*Salix reticulata*) alliance

Snow willow, a widespread boreal-alpine dwarf willow alliance, has been reported from Alaskan serpentine in the Denali domain, where stands resembling the *Salix reticulata–Dryas integrifolia–Carex bigelowii–Tomenthypnum nitens* association in northeastern Alaska occur (Hettinger and Janz 1974).

Mountain heather (*Cassiope tetragona*) alliance

This boreal tundra alliance has been reported from the Denali domain: *Cassiope* tundra is widespread on moist alpine sites throughout Alaska. The serpentine stands are not significantly different from nonserpentine stands of *Cassiope tetragona* (Viereck et al. 1992).

Creeping crowberry (*Empetrum nigrum*) alliance

The boreal alliance of creeping crowberry is known from serpentine in the Gulf of Alaska domain. There crowberry (*Empetrum nigrum*), alpine blueberry (*Vaccinium uliginosum*), dwarf willow (*Salix* sp.), Labrador tea (*Ledum palustre decumbens*), mountain-avens (*Dryas* sp.), white mountain heather (*Cassiope tetragona*), mosses, and lichens (especially *Cladina* sp.) form a moist alpine tundra. The vegetation is similar to the widespread *Empetrum nigrum–Vaccinium uliginosum* association of (Hultén 1962).

Tufted hairgrass (*Deschampsia casepitosa*) alliance

Tufted hairgrass is a widespread alliance often found in moist meadow and wetland settings from sea level to alpine zones. It is known from a few serpentine seeps in the Klamath domain (Taylor and Teare 1979). No serpentine associations have been defined.

Vernal pool vegetation

Several vernal pool complexes are associated with serpentine in the Northern California Coast Ranges and the Southern California Coast Ranges domains. These have not been analyzed thoroughly in vegetation classification, although there is ongoing work (Barbour and Witham 2004). As with all vernal pools, vegetation within them varies at a fine scale, and several alliances and associations may occur within one pool. One example of a likely vernal pool alliance that occurs at least in part on serpentine is characterized by Vasey's coyote thistle (*Eryngium vaseyi*). Stands are known from the Spanish Lakes vernal pools on the summit ridge of San Benito Mountain, San Benito County. Certain species of goldfields (*Lasthenia* spp.) often associate with vernal pools. *L. glabrata* has been mentioned under the *Lasthenia* alliance already. Spikerushes (*Eleocharis* spp.) are also common in vernal pools both on and off serpentine. Some of these have already had their own alliances defined (*E. pauciflora, E. macrostachys*) but are not clearly described from serpentine pools yet. So far it appears that there are no serpentine-endemic vernal pool plants. Thus, hydrology appears to trump soil chemistry for determining endemism of species—and likely vegetation as well—in vernal pools.

Big squirreltail (*Elymus multisetus*) alliance

This xerophytic and typically open grassland may occur on serpentine in California in the Northern California Coast Ranges, Southern California Coast Ranges, and Sierra Motherlode domains. It is typically sparse grassland and may occupy recently disturbed sites such as burns and roadcuts. This alliance is characterized by *Elymus multisetus* as an indicator, perennial grass species. Other grass species may be present, including *Lolium multiflorum, Nassella pulchra, Poa secunda,* and *Bromus hordeaceus*. A variety of other native herb species are usually present in low cover, including *Plantago erecta, Achillea millefolium,* and *Eschscholzia californica*. *Lolium multiflorum* may be codominant to dominant in the stands, especially in serpentine areas with air pollution and resulting nitrogen deposition (Weiss 1999). It occurs on sedimentary and serpentine slopes in the central Coast Ranges of California, and it may extend into other western North American regions.

Purple needlegrass *(Nassella pulchra)* alliance

This largely Californian alliance is common on clay soils in California west of the Sierra-Cascade divide. It was probably more widespread off of serpentine, but since the invasion of non-native exotic grasses to California it has many of its best remaining stands on serpentine in the Southern California Coast Ranges and the Northern California Coast Ranges domains. This alliance is characterized by *Nassella pulchra*, which consistently occurs in herbaceous stands that have deep soils with high clay content. *N. pulchra* is usually dominant. However, some locations are type-converting to annual non-native grasslands with minor components of native bunchgrass and forbs. Here *N. pulchra* is subdominant, but indicative, with native forbs and non-native grass species dominant. In the Northern California Coast Ranges in Marin County, Fiedler and Leidy (1987) have described this alliance as serpentine bunchgrass, occurring on upper slopes and ridge tops that are flat to moderately steep (0%–25%). Characteristic species include *Lolium multiflorum, Nassella pulchra,* and *Chlorogalum pomeridianum*. Other associations have been defined from northern to southern California (cf. Sawyer and Keeler-Wolf 1995), with the alliance extending into Baja California. At Coyote Ridge (Evens and San 2004) the following associations have been described, both with the non-native *Lolium multiflorum* as the most abundant species: *Lolium multiflorum–Nassella pulchra–Astragalus gambelianus–Lepidium nitidum*, and *Lolium multiflorum–Nassella pulchra–Calystegia collina*.

Rough fescue *(Festuca altaica)* alliance

The rough fescue alliance, which constitues arctic–alpine tundra grassland, is common on localized, steep, south-facing bluffs in interior and south-central Alaska. In the Gulf of Alaska domain, it may occur on serpentine along with scattered plants of kinnikinnick (*Arctostaphylos uva-ursi*), and prickly saxifrage (*Saxifraga tricuspinata*). In the Denali–Yukon domain the rough fescue associates with *Vaccinium* species in serpentine tundra, resembling the *Festuca altaica–Vaccinium vitis–idaea–V. uliginosum–Empetrum nigrum–Dryas octopetala* association described by Hanson (1951). Also in Denali are simpler rough fescue meadows with crowberry (*Empetrum nigrum*) and cespitose willows resembling the *Festuca altaica–Empetrum nigrum–Salix reticulata* association (Scott 1974).

Mountain avens *(Dryas octopetala)* alliance

On serpentine mountain or ridge summits in the Gulf of Alaska domain, mountain-avens (*Dryas octapetala*), cushion silene (*Silene acaulis*), and lichens are dominant plants, along with arctic sandwort (*Minuartia arctica*) and arctic lupine (*Lupinus arcticus*). This association resembles the *Dryas octopetala*–lichens vegetation type (Viereck et al. 1992).

Bigelow's sedge *(Carex bigelowii)* alliance

In the Gulf of Alaska domain several serpentine localities have *Carex bigelowii* dominating in wet hollows. This resembles previously defined nonserpentine associations such as the *Carex bigelowii–C. rariflora–C. saxatilis* association (Hettinger and Janz 1974). Bigelow's sedge is also common in the Northern Alaska domain.

12.9 Non-Native Herb and Grass Alliances on Serpentine

One of the most disturbing aspects of serpentine vegetation is the fact that over recent decades more evidence is showing that the "resistance" that serpentine landscapes have enjoyed against invasive alien plants has not been as insurmountable as was once thought. In several areas, vegetation dominated and characterized by exotic grasses has taken hold on serpentine. The following alliance types are among those that appear to be the most threatening to serpentine ecosystems.

12.9.1 Goatgrass (Aegilops triuncialis) *Alliance*

The goatgrass alliance is one of the most worrisome of the non-native grasslands because of its ability to tolerate and proliferate on serpentine soils. It is characterized by the dominance of *Aegilops triuncialis* in the overstory. A variety of other native and non-native species may be present, but in overall lower cover than *Aegilops* (e.g., relative cover <25%). Goatgrass has been observed in the Northern California central Coast Ranges and Southern California Coast Ranges on serpentine soils with clay loam below 1500 ft elevation, but *Aegilops* could dominate in other locations such as the Cascade Range foothills, northern and central Sierra Nevada foothills, Sacramento Valley, and San Francisco Bay region. Its occurrence and distribution is increasing, and it is known to survive grazing and controlled fires. Further, grazing animals should be removed from invaded areas before plants mature (where seeds can be easily carried by animals), and *Aegilops* should be prevented from occurring and spreading because controlling it after establishment is very difficult.

12.9.2 Soft Chess (Bromus hordeaceus) *Alliance*

Soft chess is the most ubiquitous of the non-native annual grass alliances in the California Floristic Province (Sawyer and Keeler-Wolf, 2004). Although typically inhabiting nonserpentine substrates, soft chess is known from serpentine in the Southern California Coast Ranges and in the Sierra Motherlode domains. Evens et al. (2004) define a single association, *Bromus hordeaceus–Amsinckia menziesii–Hordeum murinum* ssp. *leporinum*. Another association defined by Evens et al. (2004) that may sometimes occur on serpentine, the *Bromus hordeaceus–Holocarpha virgata–Taeniatherum caput–medusae*, also includes the notoriusly invasive medusa-head grass *(Taeniatherum caput–medusae)*, which otherwise has not been reported as a major component on serpentine grassland.

12.2.3 Italian Ryegrass (Lolium multiflorum) *Alliance*

The Italian ryegrass alliance is characterized by strong dominance and high cover of *Lolium multiflorum*. Other non-native annual grasses (e.g., *Bromus hordeaceus, Hordeum* spp.) may be present but in relatively low cover. Italian ryegrass is commonly found in California within lowlands that have periodic flooding such as Suisun Marsh and Elkhorn Slough, within disked fields and managed uplands, and within coastal serpentine slopes that have well-developed clay soils with nitrogen deposition. The alliance may extend into Alaska and eastern North America; it is native to Europe. Italian ryegrass has been identified from serpentine on Coyote Ridge (Evens and San 2004), where its presence has been attributed to the non-natural fertilization due to exhaust emissions from nitrogen fallout (Weiss 1999).

12.10 Undefined Types

Several stands of vegetation described from serpentine are not adequately understood to be placed within the NVCS system. These include a widely scattered array of types from Alaska to Baja.

Stands of Rocky Mountain juniper (*J. scopulorum*) grassland, with much tufted phlox (*Phlox diffusa*) and pod fern (*Aspidotis densa*) in the North Cascades–Fraser River domain at the Fildago Island locality do not fit well into the existing vegetation classification system for Washington state.

Stands of Sitka sedge (*Carex stichensis*) in seeps on serpentine at Tulameen in British Columbia and stands of *Rubus leucodermis* defined in openings on serpentine also at Tulameen, both described by Kruckeberg (1969), do not mesh with the existing British Columbia vegetation classification (R. del Meidinger, personal communication, 2004).

No detailed surveys of the vegetation on the Baja California serpentine exposures have been conducted. Most of the Baja California serpentine exposures occur in the Vizcaino Subdivision of the Sonoran Desert (Shreeve and Wiggins 1964) within a few kilometers of the ocean. This subdivision located on the Pacific side of the Baja California peninsula consists of intriguing plants such as the strangler figs (*Ficus petiolaris*), blue palm trees (*Brahea armata*), Baja elephant trees (*Pachycormus discolor*), cardones (*Pachycereus pringlei*), tree yucca (*Yucca valida*), and boojums (*Fouquieria columnaris*). However, most of these species are restricted inland to the core of the Vizcaino Desert and are not found on serpentine along the coast. Certain succulent species such as the coastal agave (*Agave shawii*) are common, but otherwise the coastal Vizcaino Plain is a flat, cool, desert of shrubs barely a foot tall, with occasional mass blooms of annual species. It is assumed that much of the serpentine vegetation of this area is dominated by the widespread brittlebush (*Encelia farinosa*) alliance. Any divergence at the association level remains to be discovered.

In the Northern Alaska–Kuskokwim Mountains domain, Juday's (1992) serpentine barrens do not mesh well with the current Alaska vegetation classification (Viereck et al. 1992). The closest existing match is to a sparse herbaceous type called *Bromus pumpellianus–Trisetum spicatum–Bupleurum triradiatum* defined by Koranda (1960). However, Juday's photographs suggest it should be called a *Minuartia arctica–Bupleurum triradiatum* type, currently undescribed from the Alaska Vegetation Classification. According to Viereck et al. (1992), alpine herbs occur as sparse vegetation on talus and blockfields and in some well-vegetated herbaceous meadows in alpine valleys throughout the state. *Minuartia arctica* would be considered as part of this physiognomic type considered as a level IV in the Alaska vegetation hierarchy (Viereck et al. 1992).

12.11 Conclusions

The main purpose of this chapter was to describe the range of vegetation patterns on serpentine in western North America. By so doing we can make some comparisons about the effect of serpentine on vegetation in a wide range of climatic and other physical conditions. Below are some of the highlights of this comparison, many of which can be gleaned from table 12-1.

There are 75 alliances that currently have some representation on serpentine in western North America. Twenty-four alliances are only found in a single domain. Eight alliances are found in 4 or more of the 10 serpentine domains: Douglas-fir (5), leather oak (4), incense-cedar (4), foothill pine (4), Jeffrey pine (4), buckbrush (4), knobcone pine (4), and vernal pool (4). Of all 75 alliances with some representation on serpentine, the numbers of alliances of each type are:

Needleleaf evergreen forest and woodland (25)

Broadleaf evergreen shrubland (14)

Temperate perennial grasslands and herblands (11)

Temperate annual grasslands and herblands (6)

Winter deciduous forests and woodlands (4)

Drought deciduous shrublands (4)

Winter deciduous shrublands (4)

Evergreen microphyllous dwarf shrublands (2)

Evergreen microphyllous shrublands (2)

Broadleaf evergreen dwarf shrubland (1)

Evergreen broadleaf forest and woodland (1)

Evergreen needleleaf shrubland (1)

There are nine endemic alliances. The endemic alliances all occur in the Northern and the Southern California Coast Ranges domains, though some also occur in the Klamath Mountains. There are no serpentine endemic alliances north of the Klamath Mountains.

There are 34 alliances that have endemic serpentine associations. The number occurring in each of the following domains is:

Southern California Coast Ranges domain (22)

Northern California Coast Ranges domain (21)

Klamath Mountains domain (14)

Sierra Motherlode domain (14)

Blue Mountains domain (4)

North Cascades–Fraser River domain (3)

Baja California domain (3; all are restricted to the California Peninsular Ranges portion of this domain)

No serpentine endemic alliances occur in any of the other domains. There are apparently no endemic serpentine associations north of the Ingalls locality in central Washington. The endemic associations represent a wide array of physiognomic types:

Needleleaf evergreen woodlands or forests (24)

Temperate perennial herb and grassland (11)

Temperate annual herb and grasslands (6)

Winter-deciduous forests and woodlands (5)

Winter-deciduous shrublands (3)

Winter-deciduous dwarf shrubland (1)

Because there is no complete inventory of vegetation yet for western North American serpentine, unique stands are not represented in this analysis. However, it appears that these stands may occur in some settings even as far north as Alaska. Most unique serpentine stands recognized so far appear to occur in the five California domains. This observation, along with many of the other statistics, may be more an artifact of the completeness of the vegetation classification than anything else.

It is difficult to conclude much about the moisture relationships of serpentine vegetation because most moisture descriptors relate to regional moisture regimes. For example, Alaskan aspen stands are generally considered to be very dry by Alaskan standards, but very moist by Californian standards. Regardless, it does appear that serpentine vegetation does occur over a range from relatively wet to relatively dry conditions. However, the only endemic alliances tied to higher than ambient moisture occur in California, whereas most associations restricted to serpentine are clearly more xeric than mesic or hydric.

In conclusion, based on diversity of alliances, associations, and their distribution, it appears that the latitudinal range between 32° and 43° contains the majority of the diversity of serpentine vegetation, whether defined by the diversity of physiognomic types, floristic alliances, endemic alliances, or associations.

Part III References

Alexander, E.B., P. Cullen, and P.J. Zinke. 1989. Soils and plant communities of ultramafic terrain on Golden Mountain, Cleveland Peninsula. Pages 47–56 *in* E.B. Alexander (ed.), Proceedings of Watershed '89: a Conference on the Stewardship of Soil, Air and Water Resources. USDA Forest Service, Alaska Region, Juneau, AK.

Allen, B.H. 1987. Ecological type classification for California. USDA, Forest Service, Pacific Southwest Research Station, General Technical Report PSW-98.

Anderson, R.C., J.S. Fralish, and J.M. Baskin (eds.). 1999. Savannas, Barrens and Rock Outcrop Communities of North America. Cambridge University Press, New York.

Armstrong, J. and L.F. Huenneke. 1992. Spatial and temporal variation in species composition in California grasslands: the interaction of drought and substratum. Pages 207–212 in A.J.M. Baker, J. Proctor, and R.D. Reeves (eds.), The Vegetation of Ultramafic (Serpentine) Soils. Intercept, Andover, UK.

Ater, M., M.C. Lefèbvre, W.Gruber, and P. Meerts. 2000. A phytogeochemical survey of the flora of ultramafic and adjacent normal soils in North Morocco. Plant and Soil 218: 127–135.

Atzet, T. and D.L. Wheeler. 1984. Preliminary plant associations of the Siskiyou Mountain Province. U.S. Forest Service, Pacific Northwest Region, Portland, OR.

Baker, A.J.M., J. Proctor, and R.D. Reeves (eds.), 1992. The Vegetation of Ultramafic (Serpentine) Soils. Intercept, Andover, UK.

Baldwin, B. 1999. New combinations and new genera in the North American tarweeds (Compositae – Madiinae). Novon 9: 462–471.

Balkwill, K., S.D. Williamson, C.L. Kidger, E.R. Robinson, M. Stalmans, and M.-J. Balkwill. 1997. Diversity and conservation of serpentine sites in southern Mpumalanga (eastern Transvaal), South Africa. Pages 133–138 in T. Jaffre, R.D. Reeves, and T. Becquer (eds.), The Ecology of Ultramafic and Metalliferous Areas. French Institute of Research for Cooperation and Development (ORSTOM), Noumea, New Caledonia.

Barbour, M.G. and C. Witham. 2004. Islands within islands: viewing vernal pools differently. Fremontia 32(2): 3–9.

Batten, K.M., K.M. Scow, K.F. Davies, and S. Harrison. 2006. Two invasive plants alter soil microbial community composition in serpentine grasslands. Biological Invasions 8: 217–230.

Borchert, M., A. Lopez, C. Bauer, and T. Knowd. 2004. Field guide to coastal sage scrub and chaparral alliances of Los Padres National Forest, California. USDA Forest Service, Pacific Southwest Region, Technical Paper R5-TP-019.

Borhidi, A. 1996. Phytogeography and Vegetation Ecology of Cuba. Akademiai Kiado, Budapest, Hungary.

Bradshaw, H.D. 2005. Mutations in CAX-1 produce phenotypes characteristic of plants tolerant to serpentine soils. New Phytologist 167: 81–88.

Brady, K.U, Kruckeberg, A.R., and Bradshaw Jr., H.D. 2005. Evolutionary ecology of plant adaptation to serpentine soils. Annual Review of Ecology, Evolution, and Systematics 36: 243–266.

Brock, S. and I. Burke. 1980. Vegetation. Pages 147–202 in N. Farquhar, and J. Schubert (eds.), Ray Mountains, Central Alaska: Environmental Analysis and Resources Statement. Northern Studies Program, Middlebury College, Middlebury, VT.

Brooks, R.R. 1987. Serpentine and Its Vegetation: A Multidisciplinary Approach. Dioscorides Press, Portland, OR.

Callizo, J. 1992. Serpentine habitats for the rare plants of Lake, Napa and Yolo Counties. Pages 35–52 in A.J.M. Baker, J. Proctor, and R.D. Reeves (eds.), The Vegetation of Ultramafic (Serpentine) Soils. Intercept, Andover, UK.

Changwe, K. and K. Balkwill. 2003. Floristics of the Dunbar Valley serpentinite site, Songimvelo Game Reserve, South Africa. Botanical Journal of the Linnean Society 143: 271–285.

Chiarucci, A., D. Rocchini, C. Leonzio, and V. De Dominicis. 2001. A test of serpentine soils of Tuscany, Italy. Ecological Research 16: 627.

Clifton, G. 2001. Plumas County Flora. Unpublished manuscript.CNPS. 2003. California Native Plant Society relevé protocol, CNPS Vegetation Committee. Unpublished document. California Native Plant Society, Sacramento, CA. Available: *http://www.cnps.org/archives/forms/releve.pdf*.

CNPS-CDFG. 2006. Vegetation Classification, Descriptions, and Mapping of the Clear Creek Management Area, Joaquin Ridge, Monocline Ridge, and Environs in San Benito and Western Fresno Counties, California. Unpublished cooperative report by California Native Plant Society and California Department of Fish and Game. On file at California Native Plant Society, Sacramento, CA.

Coleman, R.G. 1996. New Idria serpentinite: a land management dilemma. Environmental and Engineering Geoscience 2: 9–22.

Coley, P.D., J.P. Bryant, and F.S. Chapin 1985. Resource availability and plant anti-herbivore defense. Science 230: 895–899.

Craighead, J.J., F.L. Craighead, D.J. Craighead, and R.L. Redmond. 1988. Mapping arctic vegetation in northwest Alaska using Landsat MSS imagery. National Geographic Research 4: 496–527.

Davie, H. and J.S. Benson 1997. The serpentine vegetation of the Woko-Glenrock region, New South Wales, Australia. Pages 155–162 in T. Jaffre, R.D. Reeves, and T. Becquer (eds.), The Ecology of Ultramafic and Metalliferous Areas. ORSTOM Noumea Doc. Sci. Tech. III 2.

del Moral, R. 1982. Control of vegetation on contrasting substrates: herb patterns on serpentine and sandstone. American Journal of Botany 69: 227–238.

Diller, J.S. 1902. Topographic development of the Klamath Mountains. U.S. Geological Survey Bulletin 196.

Evens, J. and S. San. 2004 Vegetation associations of a serpentine area: Coyote Ridge, Santa Clara County, California. Unpublished report. California Native Plant Society, Sacramento CA.

Evens, J., S. San, and J. Taylor. 2004. Vegetation classification and mapping of Peoria Wildlife Area, South of New Melones Lake, Tuolumne County, CA. Unpublished report. California Native Plant Society, Sacramento, CA.

Farve, R.M. 1987. A management plan for rare plants in the Red Hills of Tuolumne County, California. Pages 425–427 in T.S. Elias (ed.), Conservation and Management of Rare and Endangered Plants. California Native Plant Society, Sacramento, CA.

Fiedler, P.L. 1992. Cladistic test of the adaptational hypothesis for serpentine endemism. Pages 421–434 in A.J.M. Baker, J. Proctor, and R.D. Reeves (eds.), The Vegetation of Ultramafic (Serpentine) Soils. Intercept, Andover, UK.

Fiedler, P.L. 1987. Ecological survey of the Antelope Creek Lakes proposed Research Natural Area. Unpublished report. USDA Forest Service, Pacific Southwest Research Station, Berkeley, CA.

Fiedler, P.L. and R.A. Leidy. 1987. Plant communities of Ring Mountain Preserve, Marin County, California. Madroño 34: 173–192.

Fiedler, P.L., N. Carnal, and R. Leidy. 1986. Ecological survey of the Green Island Lake proposed Research Natural Area. Unpublished report. USDA Forest Service, Pacific Southwest Research Station, Berkeley, CA.

Franklin, J.F. and C.T. Dyrness. 1973. Natural vegetation of Oregon and Washington. General Technical Report PNW-8. USDA Forest Service, Pacific Northwest Research Station.

Freestone, A.L. and S. Harrison. 2006. Regional enrichment of local assemblages is robust to variation in local productivity, abiotic gradients, and heterogeneity. Ecology Letters (in press).

FWS 1998. Recovery plan for serpentine soil species of the San Francisco Bay Area. U.S. Fish and Wildlife Service, Portland, OR.

FWS. 2004. U.S. Fish and Wildlife Service Draft Recovery Plan for *Phlox hirsuta* (Yreka Phlox). Federal Register 69(137).

Gabbrielli, R. and T. Pandolfini 1984. Effect of Mg and Ca on the response to nickel toxicity in a serpentine endemic and nickel-accumulating species. Physiologia Plantarum 62: 540–544.

Gabbrielli, R., T. Pandolfini, O. Vergnano, and M.R. Palandri. 1990. Comparison of two serpentine species with different metal tolerance strategies. Plant and Soil 122: 271–277.

Gardner, M. and M.R. MacNair. 2000. Factors affecting the coexistence of the serpentine endemic *Mimulus nudatus* Curran and its presumed progenitor *Mimulus guttatus* Fischer ex DC. Biological Journal of the Linnean Society 69: 443–459.

Gelbard, J.L. and S. Harrison 2003. Roadless habitats as refuges for native plant diversity in California grassland landscapes. Ecological Applications 13: 404–415.

Gordon H. and T. White. 1994. Ecological guide to southern California chaparral plant series: Transverse and Peninsular Ranges: Angeles, Cleveland, and San Bernardino National Forests. Publication R5-ECOL-TP-005. USDA Forest Service, Pacific Southwest Region.

Gram, W.K., E.T. Borer, K.L. Cottingham, E.W. Seabloom, V.L. Boucher, L. Goldwasser, F. Micheli, B.E. Kendall, and R.S. Burton. 2004. Distribution of plants in a California serpentine grassland: are rocky hummocks spatial refuges for native species? Plant Ecology 172: 159–171.

Griffin, J.R. and W.B. Critchfield. 1972. The distribution of forest trees in California. Research Paper PSW-82. USDA Forest Service, Pacific Southwest Research Station.

Grossman, D.H., D. Faber-Langendoen, A.S. Weakley, M. Anderson, P.S. Bourgeron, R. Crawford, K. Goodin, S. Landaal, K. Metzler, K. Patterson, M. Pyne, M. Reid, and L. Sneddon. 1998. International Classification of Ecological Communities: Terrestrial Vegetation of the United States. Volume I. The National Vegetation Classification System: Development, Status, and Applications. The Nature Conservancy, Arlington, VA.

Hanes, T.L., 1988. Chaparral. Pages 417–470 *in* M.G. Barbour and J. Major (eds.), Terrestrial Vegetation of California. California Native Plant Society, Sacramento, CA.

Hanson, H.C. 1951. Characteristics of some grassland, marsh, and other plant communities in western Alaska. Ecological Monographs 21: 317–378.

Harris, A.S. 1965. Subalpine fir on Harris Ridge near Hollis, Prince of Wales Island, Alaska. Northwest Science 39: 123–128.

Harrison, S. 1997. How natural habitat patchiness affects the distribution of diversity in Californian serpentine chaparral. Ecology 78: 1898–1906.

Harrison, S. 1999a. Local and regional diversity in a patchy landscape: native, alien and endemic herbs on serpentine soils. Ecology 80: 70–80.

Harrison, S. 1999b. Native and alien species diversity at the local and regional scales in a grazed Californian grassland. Oecologia 121: 99–106

Harrison, S., K.F. Davies, J.B. Grace, H.D. Safford, and J.H. Viers. 2006a. Exotic invasion in a diversity hotspot: disentangling the direct and indirect relationships of exotic cover to native richness in the Californian serpentine flora. Ecology 87: 695–703.

Harrison, S., K.F. Davies, H.D. Safford, and J.H. Viers. 2006b. Beta diversity and the scale-dependence of the productivity-diversity relationship: a test in the Californian serpentine flora. Journal of Ecology 94: 110–117.

Harrison, S. and B.D. Inouye. 2002. High beta diversity in the flora of Californian serpentine "islands." Biodiversity and Conservation 11: 1869–1876.

Harrison, S., B.D. Inouye, and H.D. Safford. 2003. Ecological heterogeneity in the effects of grazing and fire on grassland diversity. Conservation Biology 17: 837–845.

Harrison, S., J. Maron, and G. Huxel. 2000a. Regional turnover and fluctuation in populations of five plants confined to serpentine seeps. Conservation Biology 14: 769–779.

Harrison, S., K. Rice, and J. Maron. 2001. Habitat patchiness promotes invasions by alien grasses on serpentine soil. Biological Conservation 100: 45–53.

Harrison, S., H.D. Safford, J.B. Grace, J.H. Viers, and K.F. Davies. 2006. Regional and local species richness in an insular environment: serpentine plants in California. Ecological Monographs 76: 41–56).

Harrison, S., H.D. Safford, and J. Wakabayashi. 2004. Does the age of exposure of serpentine explain variation in endemic plant diversity in California? International Geology Review 46: 235–242.

Harrison, S., J.L. Viers, and J.F. Quinn. 2000b. Climatic and spatial patterns of diversity in the serpentine plants of California. Diversity and Distributions 6: 153–161.

Heady, H.F. 1988. Valley grassland. Pages 491–514 in M.G. Barbour and J. Major (eds.), Terrestrial Vegetation of California. California Native Plant Society, Sacramento, CA.

Hettinger, L.R. and A.J. Janz. 1974. Vegetation and soils of northeastern Alaska. Arctic Gas Biological Report Series, volume 21.

Hobbs, R.J. and H.A. Mooney. 1991. Effects of rainfall variability and gopher disturbance on serpentine grassland dynamics. Ecology 72: 59–68.

Holdridge, L.R. 1967. Life Zone Ecology. Tropical Science Center, San Jose, Costa Rica.

Huenneke, L., S. Hamburg, R. Koide, H. Mooney, and P. Vitousek 1990. Effects of soil resources on plant invasion and community structure in Californian serpentine grassland. Ecology 71: 478–491.

Huggett, R. 1995. Geoecology: An Evolutionary Approach. Routledge, London.

Hughes, R., K. Bachmann, N. Smirnoff, and M.R. Macnair. 2001. The role of drought tolerance in serpentine tolerance in the *Mimulus guttatus* Fischer ex DC. complex. South African Journal of Science 97: 581–586.

Hultén, E. 1962. Flora and vegetation of Scammon Bay, Bering Sea Coast, Alaska. Svensk Botanisk Tidskrift 56: 36–54.

Jaffré, T. 1992. Floristic and ecological diversity of the vegetation on ultramafic rocks in New Caledonia. Pages 101–108 in A.J.M. Baker, J. Proctor, and R.D. Reeves (eds.), The Vegetation of Ultramafic (Serpentine) Soils. Intercept, Andover, UK.

Jennings, M., O. Loucks, D. Glenn-Lewin, R. Peet, D. Faber-Langendoen, D. Grossman, A. Damman, M. Barbour, R. Pfister, M. Walker, et al. 2003. Guidelines for describing associations and alliances of the U.S. National Vegetation Classification. Vegetation Classification Panel, version 2.0. Ecological Society of America, Washington, DC.

Jimerson, T.M. 1993. Preliminary plant associations of the Klamath province, Six Rivers and Klamath National Forests. Unpublished report. USDA Forest Service, Six Rivers National Forest, Eureka, CA.

Jimerson, T.M. 1994. A field guide to Port Orford cedar plant associations in northwest California. Technical Publication R5-ECOL-TP-002. USDA Forest Service, Pacific Southwest Region.

Jimerson, T.M. and S. Daniel. 1999. Supplement to A field guide to Port Orford cedar plant associations in northwest California. Technical Publication R5-ECOL-TP-002. USDA Forest Service, Pacific Southwest Region.

Jimerson, T.M., L.D. Hoover, E.A. McGee, G. DeNitto, and R.M. Creasy. 1995. A field guide to serpentine plant associations and sensitive plants in northwestern California. Technical Publication R5-ECOL-TP-006. USDA Forest Service, Pacific Southwest Region.

Johnston, W.R. and J. Proctor. 1981. Growth of serpentine and nonserpentine races of *Festuca rubra* in solutions simulating the chemical conditions in a toxic serpentine soil. Journal of Ecology 69: 855–869.

Johnston, W.R. and J. Proctor. 1984. The effects of magnesium, nickel, calcium, and micronutrients on the root surface phosphatase activity of a serpentine and nonserpentine clone of *Festuca rubra* L. New Phytologist 96: 95–101.

Jorgenson, M.T., C.W. Slaughter, and L.A. Viereck. 1986. Relation of vegetation and terrain in the Caribou-Poker Creek research watershed, central Alaska. Unpublished document. Institute of Northern Forestry, Fairbanks, AK

Juday, G.P. 1992. Alaska's Research Natural Areas. 3. Serpentine Slide. General Technical Report PNW-GTR-271. USDA Forest Service.

Jurjavcic, N.L., S. Harrison, and A.T. Wolf. 2002. Abiotic stress, competition, and the distribution of the native annual grass *Vulpia microstachys* in a mosaic environment. Oecologia 130: 555–562.

Keeler-Wolf, T. 1982. An ecological survey of the proposed Cedar Basin Research Natural Area, Shasta-Trinity National Forest, California. Unpublished report. USDA, Forest Service Pacific Southwest Research Station.

Keeler-Wolf, T. 1986. An ecological survey of the proposed Stone Corral–Josephine Peridotite Research Natural Area (L.E. Horton-*Darlingtonia* Bog RNA) on the Six Rivers National Forest, Del Norte County, California. Unpublished report. USDA Forest Service, Pacific Southwest Research Station.

Keeler-Wolf, T. 1987a. An ecological survey of the proposed Hale Ridge Research Natural Area, Mendocino National Forest, Lake County, California. Unpublished report. USDA Forest Service, Pacific Southwest Research Station, Berkeley, CA.

Keeler-Wolf, T. 1987b. An ecological survey of the proposed Crater Creek Research Natural Area, Klamath National Forest, Siskiyou County, California. Unpublished report. USDA, Forest Service, Pacific Southwest Research Station, Berkeley, CA.

Keeler-Wolf, T. 1990. Ecological surveys of forest service research natural areas in California. General Technical Report PSW-125. USDA Forest Service, Pacific Southwest Research Station.

Keeler-Wolf, T., J. Evens, A. Klein, and D. Hickson. 2004. Preliminary classification of the Mount San Benito-Joaquin Rocks Area of Western Fresno and Eastern San Benito counties, California. Report on file at Wildlife Habitat and Data Analysis Branch, California Department of Fish and Game, Sacramento, CA.

Koenigs, R.L., W.A. Williams, and M.B. Jones. 1982. Factors affecting vegetation on a serpentine soil. I. Principal components analysis of vegetation data. Hilgardia 50: 1–14.

Koranda, J.J. 1960. The plant ecology of the Franklin Bluffs area, Alaska. Ph.D. thesis, University of Tennessee, Knoxville.

Kruckeberg, A.R. 1951. Intraspecific variability in the response of certain native plant species to serpentine soil. American Journal of Botany 38: 408–419.

Kruckeberg, A.R. 1954. The ecology of serpentine soils: a symposium. III. Plant species in relation to serpentine soils. Ecology 35: 267–274.

Kruckeberg, A.R. 1957. Variation in the fertility of hybrids between isolated populations of the serpentine species, *Streptanthus glandulosus* Hook. Evolution 11: 185–211.

Kruckeberg, A.R. 1967. Ecotypic responses to ultramafic soils by some plant species of northwestern United States. Brittonia 19: 133–151.

Kruckeberg, A.R. 1969. Plant life on serpentine and other ferromagnesian rocks in northwestern North America. Syesis 2: 15–114.

Kruckeberg, A.R. 1984. California Serpentines: Flora, Vegetation, Geology, Soils, and Management Problems. University of California Press, Berkeley.

Kruckeberg, A.R. 1991. An essay: geoedaphics and island biogeography for vascular plants. Aliso 13: 225–238.

Kruckeberg, A.R. 1992. Plant life of western North American ultramafics. Pages 31–74 *in* B.A. Roberts and J. Proctor (eds.), The Ecology of Areas with Serpentinized Rocks: A World View. Kluwer, Dordrecht, the Netherlands.

Latimer, H. 1984. Eastern Fresno County. Fremontia 11(2): 29–30.

Lewis, G.L. and G.E. Bradfield. 2003. A floristic and ecological analysis at the Tulameen ultramafic (serpentine) complex, southern British Columbia, Canada. Davidsonia 14: 121–144.

Little Jr., E.L. 1978. Checklist of United States Trees (Native and Naturalized). Agriculture Handbook 541. USDA Forest Service, Washington, DC.

Lyons, M.T., R.R. Brooks, and D.C. Craig. 1974. The influence of soil composition on the vegetation of the Coolac serpentinite belt in New South Wales. Proceedings of the Royal Society of New South Wales 107: 67–75.

MacNair, M.R. 1992. Preliminary studies on the genetics and evolution of the serpentine endemic *Mimulus nudatus* Curran. Pages 409–420 in A.J.M. Baker, J. Proctor, and R.D. Reeves (eds.), The Vegetation of Ultramafic (Serpentine) Soils. Intercept, Andover, UK.

MacNair, M.R., and M. Gardner. 1998. The evolution of edaphic endemics. Pages 157–171 in D.J. Howard and S.H. Berlocher (eds.), Endless Forms: Species and Speciation. Oxford University Press, New York.

Madhok, O.P. and R.B. Walker. 1969. Magnesium nutrition of two species of sunflower. Plant Physiology 44: 1016–1022.

Martin, J.R., W.W. Brady, and J.M. Downs. 1985. Preliminary forest plant associations (habitat types) of southeast Alaska: Chatham Area, Tongass National Forest. USDA Forest Service, Alaska Region, Juneau, AK.

Matos Mederos, J. and A. Torres Bilbao. 2000. Primeros estadios sucesionales del cuabal en las serpentinas de Santa Clara. Revista del Jardin Botanico Nacional, 21: 167–184.

Mayer, M.S., and P.S. Soltis. 1999. Intraspecific phylogeny analysis using ITS sequences: insights from studies of the *Streptanthus glandulosus* complex (Cruciferae). Systematic Botany 24: 47–61.

Mayer, M.S., P.S. Soltis, and D.E. Soltis. 1994. The evolution of the *Streptanthus glandulosus* complex (Cruciferae): genetic divergence and gene flow in serpentine endemics. American Journal of Botany 81: 1288–1299.

McCarten, N. 1988. Systematics and ecology of the *Hesperolinon disjunctum* complex. M.A. thesis, San Francisco State University, San Francisco, CA.

McCarten, N. 1992. Community structure and habitat relations in a serpentine grassland in California. Pages 207–211 in A.J.M. Baker, J. Proctor, and R.D. Reeves (eds.), The Vegetation of Ultramafic (Serpentine) Soils. Intercept, Andover, UK.

McCarten, N.F. 1993. Serpentines of the San Francisco Bay region: vegetation, floristics, distribution and soils. Report to Endangered Plant Program, California Department of Fish and Game, Sacramento.

McCarten, N.F. and C. Rogers 1991. Habitat management study of rare plants and communities associated with serpentine soil habitats in the Mendocino National Forest. USDA Forest Service, Mendocino National Forest, Willows, CA.

McCoy, S., T. Jaffré, F. Rigault, and J.E. Ash. 1999. Fire and succession in the ultramafic maquis of New Caledonia. Journal of Biogeography 26: 579–585.

McCune, B., and J.B. Grace. 2002. Analysis of Ecological Communities. MjM Software, Gleneden Beach, OR.

McNaughton, S.J. 1968. Structure and function in California grasslands. Ecology 49: 962–972.

Medeiros, J. 1984. The Red Hills. Fremontia 11(2): 28–29.

Meimberg, H., J.I. Hammond, C.M. Jorgensen, T.W. Park, J.D. Gerlach, K.J. Rice, and J.K. McKay. 2006. Molecular evidence for an extreme bottleneck during the introduction of barbed goatgrass, Aegilops triuncialis, to California. Biological Invasions (in press).

Mizuno, N. and S. Nosaku 1992. The distribution and extent of serpentinized rocks in Japan. Pages 249–270 in B.A. Roberts and J. Proctor (eds.), The Ecology of Areas with Serpentinized Rocks: A World View. Kluwer, Dordrecht, the Netherlands.

Mohlenbrock, R.H. 2003. Valley high: a California forest harbors cobra plants and other treats for plant lovers willing to get their feet wet. This Land Natural History Magazine, July–August: 47–48.

Mueller-Dombois, D. and H. Ellenberg. 1974. Aims and Methods of Vegetation Ecology. Wiley, New York.

Mueller-Dombois, D. and R. Fosberg. 1998. Vegetation of the Tropical Pacific Islands. Springer-Verlag, New York.

Nagy, L. and J. Proctor. 1997a. Soil Mg and Ni as causal factors of plant occurrence and distribution at the Meikle Kilrannoch ultramafic site in Scotland. New Phytologist 135: 561–566.

Nagy, L. and J. Proctor. 1997b. Plant growth and reproduction on a toxic alpine ultramafic soil: adaptation to nutrient limitation. New Phytologist 137: 267–274.

NatureServe. 2004. Online catalog of vegetation types. Available: http://www.natureserve.org/explorer/servlet/NatureServe?init=Ecol.

Neiland, B.J. and L.A. Viereck. 1977. Forest types and ecosystems. Pages 109–136 in North America Forest Lands at Latitudes North of 60 Degrees. University of Alaska, School of Agriculture and Land Resources Management, Agricultural Experiment Station, Cooperative Extension Service, Fairbanks, AK.

Oberbauer, T. 1993. Soils and plants of limited distribution in the Peninsular Ranges. Fremontia 21(4): 3–7.

O'Dell, R.E., J.J. James, and J.H. Richards. 2006. Congeneric serpentine and nonserpentine shrubs differ more in leaf Ca:Mg than in tolerance of low N, low P, or heavy metals. Plant and Soil 280: 49–64.

Ohmann, J.L., and T.A. Spies. 1998. Regional gradient analysis and spatial pattern of woody plant communities of Oregon forests. Ecological Monographs 68: 151–182.

Oliphant, J.M. 1992. Geographic variation of lodgepole pine in Northern California. M.S. thesis, Humboldt State University, Arcata, CA.

Parker, V.T. 1990. Problems encountered while mimicking nature in vegetation management: an example from fire-prone vegetation. Pages 231–234 in R.S. Mitchell, C.J. Sheviak, and D.J. Leopold (eds.), Proceedings of the 15th Annual Natural Areas Conference, Ecosystem Management: Rare Species and Significant Habitats. New York State Museum, Bulletin 471.

Pepper, A.E. and L.E. Norwood. 2001. Evolution of *Caulanthus amplexicaulus var. barbarae* (Brassicaceae), a rare serpentine endemic plant: a molecular phylogenetic perspective. American Journal of Botany 88: 1479–1489.

Pfister, R.D. and S.F. Arno. 1980. Classifying forest habitat types based on potential climax vegetation. Forest Science 26:52–70.

Proctor, J. 2003. Vegetation and soil and plant chemistry on ultramafic rocks in the tropical Far East. Perspectives in Plant Ecology, Evolution and Systematics 6: 105–124.

Proctor, J., A.J.M. Baker, M.M.J. Van Balgooy, S.H. Jones, L.A. Bruijnzeel, and D.A. Madulid 1997. Mount Bloomfield, Palawan, The Phillippines: the scrub and Gymnostoma woodland on ultramafic rocks. Pages 123–130 in Jaffre, T., R.D. Reeves, and T. Becquer (eds.), The Ecology of Ultramafic and Metalliferous Areas. ORSTOM Noumea Doc. Sci. Tech. III 2.

Proctor, J., L.A. Bruijnzeel, and A.J.M. Baker. 1999. What causes the vegetation types on Mount Bloomfield, a coastal tropical mountain of the western Philippines? Global Ecology and Biogeography 8: 347–354.

Proctor, J., Y.F. Lee, A.M. Langley, W.R.C. Munro, and T. Nelson 1988. Ecological studies on Gunung Silam, a small ultrabasic mountain in Sabah, Malaysia. I. Environment, forest structure and floristics. Journal of Ecology 76: 320–340.

Proctor, J. and L. Nagy 1992. Ultramafic rocks and their vegetation: an overview. Pages 469–494) in A.J.M. Baker, J. Proctor, and R.D. Reeves (eds.), The Vegetation of Ultramafic (Serpentine) Soils. Intercept, Andover, UK.

Proctor, J., and S.R.J. Woodell. 1975. The ecology of serpentine soils. Advances in Ecological Research 9: 255–366.

Racine, C.H. 1976. Flora and vegetation. Pages 39–139 in H.R. Melchior (ed.), Biological Survey of the Proposed Kobuk Valley National Monument. Cooperative Park Studies Unit, Biological and Resource Management Program, University of Alaska, Fairbanks.

Rajakaruna, N. and B.A. Bohm 1999. The edaphic factor and patterns of variation in *Lasthenia californica*. American Journal of Botany 86: 1576–1596.

Rajakaruna N., M.Y. Siddiqi, and J. Whitton. 2003. Differential responses to Na^+/K^+ and Ca^{2+}/Mg^{2+} in two edaphic races of the Lasthenia californica (Asteraceae) complex: A case for parallel evolution of physiological traits. New Phytologist 157: 93–103.

Raven, P. J. and D. Axelrod 1978. Origin and relationships of the California flora. University of California Publications in Botany, no. 72. Berkeley, CA.

Reeves, R.D. 1992. The hyperaccumulation of nickel by serpentine plants. Pages 253–278 in A.J.M. Baker, J. Proctor, and R.D. Reeves (eds.), The Vegetation of Ultramafic (Serpentine) Soils. Intercept, Andover, UK.

Reeves, R.D., and N. Adiguzel. 2004. Rare plants and nickel accumulators from Turkish serpentine Soils, with special reference to *Centaurea* species. Turkish Journal of Botany 28: 147–153.

Reeves, R.D., and A.J.M. Baker 1984. Studies on metal uptake by plants from serpentine and non-serpentine populations of *Thlaspi goesingense* Halacsy (Cruciferae). New Phytologist 98: 191–204.

Reeves, R.D., A.J.M. Baker, A. Borhidi, and R. Berazain. 1996. Nickel-accumulating plants from the ancient serpentine soils of Cuba. New Phytologist 133: 217–224.

Reeves, R.D., A.R. Kruckeberg, N. Adiguzel, and U. Kramer. 2000. Studies on the flora of serpentine and other metalliferous areas in western Turkey. South African Journal of Science 97: 513–517.

Roberts, B.A. and J. Proctor 1992. The Ecology of Areas with Serpentinized Rocks: A World View. Kluwer, Dordrecht, the Netherlands.

Robinson, B.H., R.R. Brooks, J.H. Kirkman, P.E.H. Gregg, and H. Varela Alvarez 1997. Edaphuic influences on a New Zealand ultramafic ("serpentine") flora: a statistical approach. Plant and Soil 188: 11–20.

Rodriguez-Rojo, M.P., D. Sanchez-Mata, R.G. Gavilan, S. Rivas-Martinez, and M.G. Barbour. 2001. Typology and ecology of Californian serpentine annual grasslands. Journal of Vegetation Science 12: 687–698.

Rune, O. 1953. Plant life on serpentine and related rocks in the north of Sweden. Acta Phytogeographica Suecica 31, Uppsala, Sweden.

Safford, H. and S. Harrison. 2001. Ungrazed roadside verges in a grazed landscape: interactive effects of grazing, invasion and substrate on grassland diversity. Ecological Applications 11: 1112–1122.

Safford, H.D. and S. Harrison. 2004. Fire effects on plant diversity in serpentine versus sandstone chaparral. Ecology 85: 539–548.

Safford, H.D, J.H. Viers, and S. Harrison. 2005. Serpentine endemism in the Calfornia flora: a database of serpentine affinity. Madroño 52: 222–257.

Sawyer, J.O. and T. Keeler-Wolf. 1995. A Manual of California Vegetation. California Native Plant Society, Sacramento, CA.

Sawyer, J.O. and Keeler-Wolf. 2004. A Manual of California Vegetation, 2nd ed. Unpublished manuscript.

Sawyer, J.O., and D.A. Thornburgh. 1988. Montane and subalpine vegetation of the Klamath Mountains. Pages 699–732 in M.G. Barbour and J. Major (eds.), Terrestrial Vegetation of California. California Native Plant Society, Sacramento, CA.

Sawyer, J.O., J. Palmer, and E. Cope. 1978. An ecological survey of the proposed Preacher Meadows Research Natural Area, Trinity County, California. Unpublished report. USDA, Forest Service, Pacific Southwest Research Station, Berkeley, CA.

Scott, R.W. 1974. Alpine plant communities of the southeastern Wrangell Mountains, Alaska. Pages 283–306 in V.C. Bushnell and M.G. Marcus (eds.), Icefield Ranges Research Project Scientific Results, vol. 4. American Geographical Society, New York.

Shreve, F., and I.L. Wiggins. 1964. Vegetation and Flora of the Sonoran Desert. Stanford University Press, Stanford, CA.

Skinner, M.W., and B.A. Pavlik 1994. California Native Plant Society's inventory of rare and endangered plants in California. Special Publication no. 1, 5th ed. California Native Plant Society, Sacramento, CA.

Skinner, M.W., D.P. Tibor, R.L. Bittman, B. Ertter, T.S. Ross, S. Boyd, A. C. Sanders, J.R. Shevock, and D.M. Taylor. 1995. Research needs for conserving California's rare plants. Madroño 42: 211–241.

Slingsby, D.R. 1992. The Keen of Hamar, Shetland–a long-term site-specific study of classic serpentine soils. Pages 235–242 in A.J.M. Baker, J. Proctor, and R.D. Reeves (eds.), The Vegetation of Ultramafic (Serpentine) Soils. Intercept, Andover, UK.

Slingsby, D.R. and D.H. Brown 1977. Nickel in British serpentine soils. Journal of Ecology 65: 597–618.

Stebbins, G.L. 1942. The genetic approach to problems of rare and endemic species. Madroño 6: 241–272.

Stebbins, G.L. 1984. The Northern Sierra Nevada. Fremontia 11(2): 26–28.

Stebbins, G.L. and J. Major 1965. Endemism and speciation in the California flora. Ecological Monographs 35: 1–35.

Stevanovi, V., K. Tan, and G. Iatrou. 2003. Distribution of the endemic Balkan flora on serpentine. I. – Obligate serpentine endemics. Plant Systematics and Evolution 242: 149–170.

Taylor, D.W. and K.A. Teare. 1979. Ecological survey of the vegetation of the proposed Smokey Creek Research Natural Area, Shasta-Trinity National Forest, Trinity County, CA. Unpublished report. USDA, Forest Service, Pacific Southwest Research Station, Berkeley, CA.

Tibor, D. (ed.), 2001. California Native Plant Society's Inventory of Rare and Endangered Plants of California. California Native Plant Society, Sacramento, CA.

Tilstone, G.H. and M.R. Macnair 1997. Nickel tolerance and copper-nickel co-tolerance in *Mimulus guttatus* from copper mine and serpentine habitats. Plant and Soil 191: 173–180.

Tyndall, R.W. and J.C. Hull. 1999. Vegetation, flora and plant physiological ecology of serpentine barrens of eastern North America. Pages 67–82 in R.C. Anderson, J.S. Fralish, and J.M. Baskin (eds.), Savannas, Barrens and Rock Outcrop Communities of North America. Cambridge University Press, New York.

Vasek, F.C. and R.F. Thorne. 1988. Transmontane coniferous vegetation. Pages 797–834 in M.G. Barbour and J. Major (eds.), Terrestrial Vegetation of California. Special Publication 9. California Native Plant Society, Sacramento, CA.

Viereck, L.A., C.T. Dyrness, A.R. Batten, and K.J. Wenzlick. 1992. The Alaska Vegetation Classification. General Technical Report PNW-GTR-286. USDA Forest Service. Pacific Northwest Research Station.

Viereck, L.A., C.T. Dyrness, K. Van Cleve, and M.J. Foote. 1983. Vegetation, soils, and forest productivity in related forest types in interior Alaska. Canadian Journal of Forest Research 13: 703–720.

Vogl, R.J., W.P. Armstrong, K.L. White, and K.L. Cole 1988. The closed-cone pines and cypresses. Pages 295–358 in M.G. Barbour and J. Major (eds.), Terrestrial Vegetation of California. Special Publication 9. California Native Plant Society, Sacramento, CA.

Walker, R.B. 1954. The ecology of serpentine soils: a symposium. III. Factors affecting plant growth on serpentine soils. Ecology 35: 259–266.

Walker, R.B., H.M. Walker, and P.R. Ashworth 1955. Calcium-magnesium nutrition with special reference to serpentine soils. Plant Physiology 30: 214–221.

Wallace, A., M.B. Jones, and G.V. Alexander. 1982. Mineral composition of native woody plants growing on a serpentine soil in California. Soil Science 134: 42–44.

Walter, H. 1994. The Vegetation of the Earth and Ecological Systems of the Geobiosphere. Springer-Verlag, Berlin.

Weiss, S.B. 1999. Cars, cows, and checkerspot butterflies: nitrogen deposition and management of nutrient-poor grasslands for a threatened species. Conservation Biology 13: 1476–1486.

Whipple, J., and E. Cole. 1979. An ecological survey of the proposed Mount Eddy Research Natural Area. USDA Forest Service, Pacific Southwest Research Station, Berkeley, CA.

Whittaker, R.H. 1954. The ecology of serpentine soils. IV. The vegetation response to serpentine soils. Ecology 35: 275–288.

Whittaker, R.H. 1960. Vegetation of the Siskiyou Mountains, Oregon and California. Ecological Monographs 30: 279–338.

Whittaker, R.H. 1975. Communities and Ecosystems. MacMillan, London.

Willett, I.R. and T. Batey 1977. The effects of metal ions on the root surface phosphatase activity of grasses differing in tolerance to serpentine soil. Plant and Soil 48: 213–221.

Williamson, J.N. and S. Harrison. 2002. Biotic and abiotic limits to the spread of exotic revegetation species in oak woodland and serpentine habitats. Ecological Applications 12: 40–51.

Wilson, M.V. 1988. Within-community vegetation structure in the conifer woodlands of the Siskiyou Mountains, Oregon. Vegetatio 78: 61–72.

Wolf, A.T., P.A. Brodmann, and S. Harrison. 1999. Distribution of the rare serpentine sunflower (*Helianthus exilis* Gray, Asteraceae): the roles of habitat availability, dispersal limitation and species interactions. Oikos 84: 69–76.

Wolf, A.T., and S. Harrison. 2001. Natural habitat patchiness affects reproductive success of serpentine morning glory (*Calystegia collina*, Convolvulaceae) in northern California. Conservation Biology 15: 111–121.

Wolf, A.T., S. Harrison, and J.L. Hamrick, 2000. The influence of habitat patchiness on genetic diversity and spatial structure of a serpentine endemic plant. Conservation Biology 14: 454–463.

Worley, I.A. and D. Jacques. 1973. Subalpine fir *(Abies lasiocarpa)* in coastal western North America. Northwest Science 47: 265–273.

Yang, X., V.C. Baligar, D.C. Martens, and R.B. Clark. 1996a. Plant tolerance to nickel toxicity: I. Influx, transport and accumulation of nickel in four species. Journal of Plant Nutrition 19: 73–85.

Yang, X., V.C. Baligar, D.C. Martens, and R.B. Clark. 1996b. Plant tolerance to nickel toxicity: II. Nickel effects on influx and transport of mineral nutrients in four species. Journal of Plant Nutrition 19: 265–279.

Yarie, J. 1983. Forest community classification of the Porcupine River drainage, interior Alaska, and its application to forest management. General Technical Report PNW-154. USDA Forest Service, Pacific Northwest Forest and Range Experiment Station.

Part IV Serpentine Domains of Western North America

Most ultramafic rocks of North America are set in allochthonous terranes that formed beyond the continental margin and subsequently have been annexed to the core of the continent, which is called a craton. During the early Paleozoic, sandstone, shale, and limestone accumulated in shallow seas along the western edge of the North American continent. There was little volcanic activity. As Pangea, a supercontinent, began to breakup during the Triassic (245–206 Ma), reorganization occurred on a global scale. From the Triassic to the Tertiary, allochthonous terranes, including those formed along the western edge of the continent, have drifted to North America and been welded onto it. These processes were accompanied by tectonic events and volcanism along the western margin of the continent. All of the larger bodies of ultramafic rocks in western North America, west of the Canadian Shield and the Rocky Mountains, are in these accreted terranes, with the exception of the Stillwater complex in the Rocky Mountains.

The ultramafic bodies in the allochthonous terranes of western North America have been grouped into 10 domains (see fig. 13-1). A *domain* is a geographical area with somewhat similar geologic origins and climate. In some cases a domain corresponds to a major physiographic region (e.g., the Klamath Mountains), and in other cases it does not (e.g., Baja California). Each domain is described in a separate chapter (chapters 13–22). In each chapter, a brief domain description, including definitive domain characteristics, topography, and climate, is followed by separate sections on geology, soils, and vegetation. Climate is represented for two stations in each domain by plotting mean monthly precipitation and temperature on scales of 20 mm/10°C. This gives a fair indication of the length of summer drought, as explained in appendix E.

Discrete ultramafic bodies or ophiolitic suites within the domains are called *localities*. Most of the ultramafic bodies in western North America are included in these localities, but the coverage is not complete. Each locality has a brief description of the geology followed in most cases by descriptions of soils and vegetation. We have much more information about the soils, plants, and plant communities, or vegetation, in some localities than in others. The uneven coverage of the environmental attributes in localities generally reflects the different levels of knowledge about different localities.

Stillwater Complex

The Stillwater complex is in the Beartooth Mountains, which are just northeast of the Yellowstone volcanic plateau (see fig. 13-1). The Stillwater complex is one of the world's largest layered ultramafic-gabbroic intrusions. The layers dip steeply to the north and strike west-northwesterly along the northern edge of the Beartooth Mountains. The topmost layers have been lost, mostly by glacial erosion, but a vertical thickness of 6 km is now exposed. The complex is divided into a basal chill zone (mostly gabbro) conformably overlain by a layered ultramafic zone (layered peridotite, olivine+pyroxene) approximately 1000 m thick that in turn is overlain by a bronzitite layer ~4000 m thick, with alternating layers of norite, gabbro, and anorthosite. The striking feature of this complex is the regular and persistent layering, which is remarkably similar to sedimentary bedding. This complex is considered the product of fractional crystallization of liquid basaltic magma trapped in a magma chamber with a horizontal floor. Ultramafic rocks are present in minor amounts. The Stillwater complex contains large reserves of chromite and associated platinum and palladium. There are large areas of barren rock, boulder or block fields, and talus.

Near the Ben Bow Mine (~2500 m above sea level [asl]), the soils over bronzite (an orthopyroxene with more iron than most enstatite and less than most hypersthene) are Alfisols (Haplocryalfs) and Inceptisols (Cryepts) dominated by young stands of lodgepole pine (*Pinus contorta*) with sparse subalpine fir (*Abies lasiocarpa*) and white pine (*Pinus* sp.) and with grouseberry (*Vaccinium scoparium*) and serviceberry (*Amelanchier* sp.) in the understory. Low juniper (*Juniperus communis*) and bearberry (*Arctostaphylos uva-ursi*) are present on the more shallow soils.

The main plants around the summit of Iron Mountain (3075 m asl) are mountain avens (*Dryas octapetala*), cespitose phlox (*Phlox pulvinata*), bistort (*Polygonum bistortoides*), miniature lupine (*Lupinus bicolor*), cinquefoil (*Potentilla* sp.), and sedge (*Carex* sp.). Although quartzite is exposed at the summit of Iron Mountain, the same plants occur over pyroxenite around the summit. Most of the same plants dominate alpine slopes below the summit, except that mountain-avens is missing, and a cespitose sandwort (*Minuartia obtusiloba*) is present. Whitebark pine trees are scattered sparsely across the alpine, or subalpine, slopes. Forests with spruce (*Picea engelmannii*), whitebark pine (*Pinus albicaulis*), and subalpine fir (*Abies lasiocarpa*) in the overstory and grouseberry (*Vaccinium scoparium*) in the understory prevail on the north side of Iron Mountain. A serpentine (pyroxenite) Mollic Haplocryalf at 3060 m asl on Chrome Mountain had a cover of about 35% forbs, 20% grass (*Deschampsia cespitosa*), 20% sedge (*Carex* sp.), and scattered whitebark pine trees. The dominant forbs were miniature lupine, bistort, and sandwort.

13 Baja California, Domain 1

Ophiolites occur in Baja California (fig. 13-1) along the outer coast from San Benito and Cedros Islands through the Vizcaíno Peninsula to Magdalena and Santa Margarita Islands. This is a mountainous region with altitudes up to 920 m (3018 ft) on the Vizcaíno Peninsula, >300 m (~1000 ft.) on Magdalena Island, and about 550 m (~1800 ft) on Santa Margarita Island. The ophiolite of Calmalli, which is geologically distinct from ophiolites on the outer coast, is in low hills (mostly <500 m, or 1640 ft) near El Arco, about midway from Guerrero Negro to the Gulf of California (fig. 13-2). Ophiolites of the outer coast are in the Cochimí terrane, whereas the ophiolite of Calmalli is in the Alisitos terrane (Sedlock et al. 1993, Sedlock 2003).

Mafic rocks of the Peninsular Ranges batholith that extends from California into Baja California are included in this domain. A major feature of the Peninsular Ranges is this batholith with plutons that range in composition from granite to gabbro, with tonalite the most common composition. Also, gabbro is common in the "western zone" of the batholith (Sedlock 2003). This zone is mostly southwest of the Elsinore fault zone in the California and north of the Agua Blanca fault in Baja California.

All the ophiolites are in desert areas. Mean annual temperatures are about 20°C, and mean annual precipitation is about 10 cm on Cedros Island and along the outer coast of Baja California Sur and about 15 cm in the ophiolite of Calmalli locality (Hastings and Turner 1965). The precipitation falls mostly in winter in the Cedros Island and Puerto Nuevo localities, in September in the Magdalena–Margarita locality, and in both September and in winter in the Calmalli locality. Fog and dew are common along the outer coast around Santa Margarita and Magdalena Islands. Drought persists for most of each year at all the localities (Hastings and Humphrey 1969; fig 13-3).

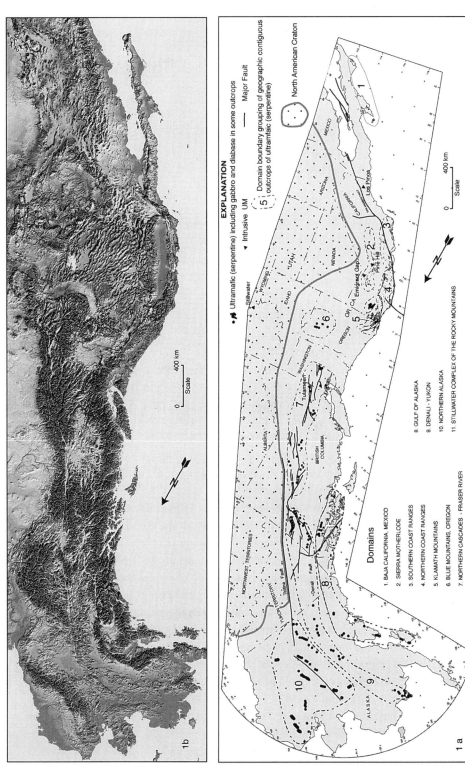

Figure 13-1 Western North America. (A) Serpentine domains from Baja California (right) to Alaska (left). (B) A shaded relief map of the same area.

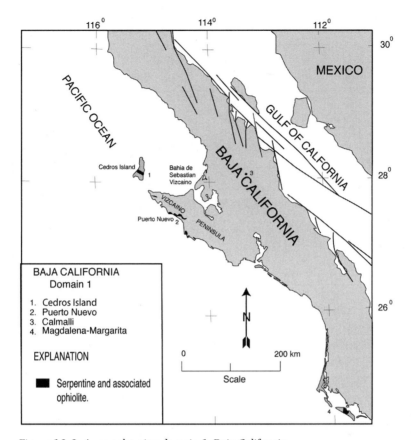

Figure 13-2 A map showing domain 1, Baja California.

The gabbro belt in the northern part of the Peninsular Ranges has been added to this domain. Descriptions of the geology, climate, soils, and vegetation of the gabbroic plutons are given in section 13.8, describing the Los Pinos locality.

13.1 Geology

Basement rocks in the Cedros, Puerto Nuevo, and Magdalena-Margarita localities are made up of accreted terranes of ophioite (ocean crust) and island arc complexes (Blake et al. 1984, Moore 1986, Sedlock 1993). The ophiolites in these localities are associated with Jurassic–Cretaceous ocean crust that was accreted to North America at the same time as many of the ophiolites from California to Alaska. Detailed reconstruction of the ophiolite sequences is difficult because of tectonic dismembering. Serpentine acts as a lubricant along low angle faults and is displaced from its original setting within the ocean crust. Serpentine rocks are found as serpentinized peridotite and as melange matrix.

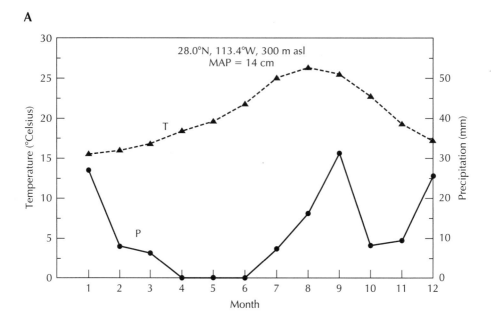

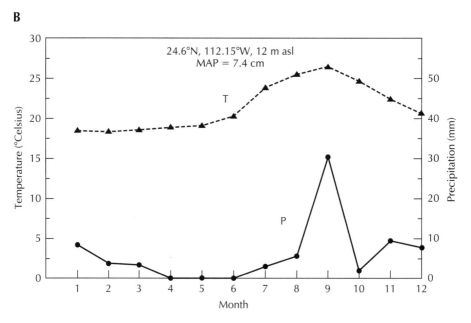

Figure 13-3 Mean monthly temperatures and precipitation at two sites in the Baja California domain. (A) El Arco in Baja California (Norte, BCN). (B) Bahia Magdalena in Baja California Sur (BCS).

13.2 Soils

Serpentine soils in this domain are mostly shallow Entisols (Lithic Torriorthents) on steep and very steep slopes and Aridisols (Haplargids) on steep and gentler slopes. The soil temperature regime is hyperthermic in all of the soils. Although the soils are commonly calcareous, the serpentine soils do not have calcic horizons. Free magnesium-carbonates may be more pervasive than calcium carbonates in them. Layers of magnesite are common in the ophiolite suites, and large chunks of it are scattered across many soils. Some of the alluvium in washes, or arroyos, in the ophiolite of Calmalli, and in the Sierra el Placer, is cemented with silica.

13.3 Vegetation

The serpentine localities of Baja California are considered to be in the Sonoran Desert (Shreve 1951, Wiggins 1980). This is a hot desert with mild winters containing many cacti and other succulent plants. Trees are sparse, and shrub cover is open, seldom dense.

13.4 Cedros Island, Locality 1-1

In the Cedros Island locality, basal tectonic melange with harzburgite, wherlite, gabbro, and blueschist in a serpentine matrix is overlain by cumulative gabbro (Sedlock 1993). The age of the ophiolite is 104–148 Ma. We have no information on the soils of Cedros Island, but there is no reason to believe that they are any different from those in the Puerto Nuevo locality on the Vizcaíno Peninsula.

13.5 Puerto Nuevo, Locality 1-2

On the outer, or coastal, side of the Vizcaíno Peninsula, there are several exposures of ultramafic rocks associated with accreted ophiolites. The Puerto Nuevo complex consists of a basal serpentinite-matrix melange with blocks of metamorphic rocks, including blue schist. The serpentinite is mainly lizardite–chrysotile, with no recognizable peridotite protolith. Above the melange, a serpentinized peridotite forms the base of the ophiolite section. It contains relics of olivine and pyroxene altered to lizardite and chrysotile. The Puerto Nuevo ophiolite is considered to be Jurassic (Moore 1986).

The serpentine soils in this locality, or in the ophiolite of San Andrés, are predominantly Entisols (Lithic Torriorthents) and Aridisols (Haplargids). Most of them are shallow, with highly serpentinized peridotite soil parent materials. Pedon PN1 is representative of the majority of serpentine soils on mountain sides in the Sierra el Placer area of the Puerto Nuevo locality.

Pedon PN1–114°35'W, 27°31'N, south-facing 42% backslope

A (0–5 cm) dark brown (7.5YR 3/4, 5/5 dry) very gravelly sandy loam, pH 6.8

Bt (5–12 cm) yellowish red (5YR 4/6, 5/6 dry) very gravelly sandy clay loam, red (2.5YR 4/6) on subangular blocky ped faces, pH 7.0

R (12–15+ cm) hard, highly fractured, weathered serpentinized peridotite bedrock

This soil is an Entisol that would be an Aridisol if the argillic horizon were 0.5 cm thicker. Some soils on footslopes have Btk horizons, but not enough lime for calcic horizons (fig. 13-4).

Copalquin (*Pachycormus discolor*) appears to be the only tree of serpentine slopes in the locality, and it is generally sparse. *Errazurizia benthami* and endemic buckwheats (*Eriogonum* spp.) are common shrubs on the serpentine soils. These shrubs are endemic to the Viscaíno Peninsula and Cedros Island areas, where they occur on both serpentine and nonserpentine soils. Some of the other shrubs found on the serpentine soils are spurge (*Euphorbia magdaleneae*), lentisco (*Rhus lentii*), snapdragon (*Galvesia juncea*), jojoba (*Simmondsia chinensis*), bladderpod (*Isomeris arborea*), and sweetbush (*Bebbia juncea* var. *juncea*). Cenizo (*Atriplex canescens*) is common on some serpentine footslopes. Pitaya agria *(Machaerocereus gummosus)* and other cacti are present, and datylillo (*Agave* sp.) occurs on some serpentine soils. No forbs were seen on serpentine soils in any of three visits to the Puerto Nuevo locality. Lichens were not evident on the ultramafic rocks, although they are present on some of the more silicic rocks in the area.

13.6 Calmalli, Locality 1-3

The Calmalli ophiolite contains imbricated slices of serpentine consisting of foliated antigorite that surround peridotite, pyroxenite, and amphibolite. The ophiolite and serpentine are strongly folded and metamorphosed to greenschist facies. The ophiolite is probably Jurassic and was emplaced during the Cretaceous (Radelli 1989).

Peridotite and serpentinite are so intricately folded and faulted along with metabasalt and other rocks in the ophiolite of Calmalli (Radelli 1989) that it is difficult to find a large block of ultramafic rock to describe a serpentine soil landscape (fig 13-5). Aridisols (Haplargids) dominate the serpentine landscapes. Shallow Haplargids have Bt horizons and deeper ones on colluvial footslopes have Btk horizons, but not enough lime for calcic horizons. Sparse torote colorado (*Bursera microphylla*) is present on the shallow soils, and torote prieto (*B. hindsiana*) and palo verde (*Cercidium microphyllum*) occur on serpentine or adjacent soils with mafic parent materials. Brittlebrush (*Encelia farinosa*) is the most abundant shrub on the shallow soils. It is more abundant, and burrobrush (*Ambrosia dumosa*) is less abundant, on serpentine than on surrounding nonserpentine soils. Other shrubs on the shallow soils are palo Adán (*Fouquieria digueti*), *Ephedra* sp., creosotebush (*Larrea divaricata*), cliff spurge (*Euphorbia misera*), candelillo (*Pedilanthus macrocarpus*), and box-thorn (*Lycium andersoni*). There are other shrubs

Figure 13-4 A serpentine landscape in the Sierra de Placer, locality 1-2, Puerto Nuevo. The spade on the right is in a shallow to moderately deep Haplargid. White rocks on the ground surface are magnesite from veins in bedrock, although some magnesite may form in the soils.

such as mariola (*Solanum hindsianum*) and fairy duster (*Calliandra californica*) on mixed ophiolite colluvium and alluvium in washes, or arroyos. Common succulents are *Agave* sp., cardon (*Pachycereus pringlei*), pitaya agria (*Machaerocereus gummosus*), cholla (*Opuntia* sp.), and *Ferocactus* sp. Herbaceous species are sparse or absent from most sites.

Figure 13-5 A layer of serpentine (green) and magnesite (white) in metabasalt of the ophiolite of Calmalli, locality 1-3. The shrub on the upper right is brittlebush (*Encelia farinosa*).

13.7 Magdalena–Margarita, Locality 1-4

Outcrops of ophiolite on Magdalena and Santa Margarita Islands have been correlated with those of the Cedros Island and Puerto Nuevo localities and are probably the same age (Blake et al. 1984).

The serpentinite is in melange, mixed with gabbro and other ophiolitic rocks, making it difficult to find a block of serpentinite large enough to describe a serpentine soil landscape. Entisols (Torriorthents) and Aridisols (Haplargids) are the common soils on ophiolites of the locality. Pedon MI1, a Lithic Torriorthent, is representative of the serpentine soils:

Pedon MI1–112°08.5'W, 24°33'N, NNE-facing 42% linear mountain slope; noncalcareous white coatings on stones on ground surface

A (0–2 cm) dark brown (7.5YR 3/4, 5/4 dry) extremely gravelly sandy loam; pH 7.4

AB (2–9 cm) dark reddish brown (5YR 3/4, 7.5YR 5/4 dry) very gravelly loam; weak, fine subangular blocky structure; pH 7.6

Bw (9–17 cm) dark reddish brown (5YR 3/4, 7.5YR 5/4 dry) very gravelly loam; moderate, very fine subangular blocky structure; pH 7.6

R (17+ cm) hard, fractured serpentinite; thin, white, calcareous coatings on bedrock surfaces

The only obviously living shrub at the site of pedon MI1 when it was described in early January was *Schoepfia californica*. A subshrub, saladillo (*Atriplex barclayana*), was common.

There were several forb species at the site and one species of cactus, but the most striking occurrence was the abundance of fruiticose lichens (*Niebla* sp.) on rocks and on branches of shrubs. An abundance of lichens at the Magdalena–Margarita locality may be related to the heavy morning dew observed in this area.

13.8 Los Pinos, Locality 1-5

The Peninsular Ranges batholith of southern California extends south from Mt. San Jacinto and the Santa Ana Mountains into Baja California. A gabbroic belt, from the Santa Ana Mountains in Orange and western Riverside Counties through San Diego County into Baja California, is represented by small bodies <10 km in length that occupy about 15% of the land surface in this area; for example, the Los Pinos pluton (Walawender 1976). The outcrops are distinctive, with dark red soils and prominent peaks such as Cuyamaca Peak, McGinty Mountain, San Marcos Mountains, Viejas Mountain, and Los Pinos Mountain towering above the light-colored boulders of granite on their lower slopes. The main rock type is an amphibole-bearing gabbro consisting of plagioclase and black amphibole. The Los Pinos concentrically zoned intrusive has a central core of peridotite, olivine-gabbro, and anorthosite enclosed by amphibole-gabbro surrounded by granitic rocks of the main batholith. These mafic and ultramafic rocks show layering and banding that indicates formation from a mafic magma chamber within the crust, with the peridotite and olivine-gabbro representing the earliest differentiates of cooling. This occurrence is similar to the Alaskan-type zoned mafic-ultramafic intrusions described in chapter 20 and the Emigrant Gap mafic–ultramafic intrusion in the Sierra Nevada in chapter 14.

Gabbro of the Peninsular Ranges is in moderately steep to steep terrain with elevations from about 100 m asl up to 1985 m on Cuyamaca Peak. Mean annual precipitation ranges from 30 to 60 cm, or more on Cuyamaca Peak, with dry summers. The frost free period is 240–320 days, or less at the highest elevation.

Most of the soils on the gabbro are Alfisols in the Las Posas Series of fine, smectitic, thermic Typic Rhodoxeralfs. They are predominantly moderately deep to bedrock that has been weathered soft, although the ground surface is commonly stony. The Las Posas soils are slightly acid to neutral in surface horizons and neutral to slightly or moderately alkaline in the subsoil. These soils are moderately fertile. About 30,000 ha have been mapped in San Diego County (Bowman 1973). No serpentine soils have been mapped in the gabbroic belt (Bowman 1973, Wachtell 1978); therefore it may be assumed that any ultramafic rock exposures are small and not extensive.

There is a small patch of slightly serpentinized peridotite on the northwest side of Los Pinos. Soils there look similar to the Las Posas soils, except that those on peridotite have slightly redder hues than the surrounding Las Posas soils. The vegetation on the the peridotie soils is essentially the same oak–manzanita–chamise–sage plant community that is common on surrounding gabbro soils. Coulter pine, the only conifer on the gabbro of Los Pinos, is sparse and mostly on the north-facing slopes.

Gabbro generally has relatively high calcium concentrations, higher than other plutonic rocks, but some olivine-gabbros have relatively low calcium concentrations and more magnesium than calcium. It is the odd gabbros with more olivine than plagioclase and amphibole that attract the attention of plant ecologist, although gabbros with relatively high calcium concentrations are much more common and extensive in this locality.

The gabbroic bodies in this area contain a diversity of plant species and vegetation types. However, the overall vegetation physiognomy is chaparral with emergent conifers such as Coulter pine (*Pinus coulteri*) or the endemic Cuyamaca cypress (*Cupressus arizonica* ssp. *stephensoni*) and Tecate cypress (*Cupressus forbesii*). Gordon and White (1994) describe an endemic chamise-eastwood manzanita mafic association that is characteristic of the chaparral in the gabbroic belt. There are also several rare and endemic vascular plant species in the area.

An integrated survey of soils and plant communities in the Santa Ana Mountains (EA Engineering, Science, and Technology 1995) found scrub oak chaparral on north-facing Las Posas soils and mixed chaparral with inclusions of coastal sage scrub on south-facing Las Posas soils. The predominant plants are scrub oak (*Quercus berberidifolia*) and chamise (*Adenostoma fasciculatum*) in the scrub oak chaparral and chamise, ceanothus (*Ceanothus crassifolius*), and laurel sumac (*Malosma laurina*) in the mixed chaparral plant community. The main plants in the coastal sage scrub plant community, which is more common on some other soils in the Santa Ana Mountains, are coastal sagebrush (*Artemesia californica*), black sage (*Salvia mellifera*), and white sage (*Salvia apiana*).

There are a few very small areas of serpentinite along faults in the Santa Ana Mountains. They are too small to map in a detail soil survey, so we have no soils information for them other than that provided by Vogl (1973) in investigations of the serpentine vegetation. The main vegetative feature of the serpentine and associated hydrothermally altered rocks is knobcone pine (*Pinus attenuata*). Vogl (1973) estimated that the total area in all of the pine stands is 194 ha. He found that manzanita (*Arctostaphylos glandulosa*) is more abundant than pine in the open stands, and the cover area of ceanothus (*Ceanothus crassifolius*) is about the same as that of pine. In the Santa Ana Mountains, small stands of this pine occur in otherwise chaparral-dominated areas on the slopes of Sugarloaf, Pleasants, and Santiago Peaks (Thorne 1977; F. Roberts, personal communication, 2004).

Keeler-Wolf (1990a, summarized in Cheng 2004) described the vegetation of the King Creek Research Natural Area, which ranges from moist coast-live-oak (*Quercus agrifolia*) forests and meadows dominated by San Diego sedge (*Carex spissa*) and deer grass (*Muhlenbergia rigens*) to extensive chaparral, which varies from mesic north-slope types in the scrub oak (*Quercus berberidifolia*) alliance, to relatively open xeric chamise–eastwood manzanitia (*Adenostema fasciculatum–Arctostaphylos glandulosa*) mafic association, as described by Gordon and White (1994). Stands of Coulter pine and the endemic Cuyamaca cypress emerge from this sea of chaparral. Burke (1985) describes similar chaparral from the Organ Valley Research Natural Area about 40 km north of King Creek.

13.8.1 Endemic and Localized Plants

Some of the most striking (and rare) endemic plants in San Diego and adjacent western Riverside County occur on gabbro soils (Beauchamp 1986, Oberbauer 1993, Reiser 1994). The following species are particularly noteworthy rare endemics on the Las Posas soils:

Vail Lake ceanothus (*Ceanothus ophiochilus*) is a rounded, rigidly branched shrub in the buckthorn family (Rhamnaceae) with pale blue to pinkish-lavender flowers. This species was first discovered during a spring 1989 botanical survey of the property surrounding Vail Lake in southwestern Riverside County. Two additional populations of Vail Lake ceanothus were discovered in 1993 within the Agua Tibia Wilderness of the Cleveland National Forest, also in southwestern Riverside County. Both populations include hybrids between Vail Lake ceanothus and the common hoaryleaf ceanothus (*Ceanothus crassifolius*). Its specific name refers to its affinity for ophiolitic rock. All occurrences of the species are on north-facing slopes and on soils derived from an unusual pyroxene-rich gabbro oucrop.

Dunn's Mariposa lily (*Calochortus dunnii*) is scattered throughout the chamise–eastwood manzanita mafic association on the slopes of Cuyamaca and Laguna Mountains, as well as Guatay Mountain, San Miguel Mountain, and Otay Mountain in central San Diego County. Here it grows along with two other widespread species that locally only occur on gabbro, chaparral pea (*Pickeringia montana*), and creeping sage (*Salvia sonomensis*). *Calochortus dunnii* has a few basal leaves, which wither early, and a slender, branched flower stalk reaching to 0.5 m tall. The flowers are erect and bell-shaped with broad white petals, sometimes flushed with pink. Each of the petals has a reddish-brown yellow-hairy blotch above the round nectar gland.

Cuyamaca cypress (*Cupressus arizonica* ssp. *stephensoni*) is known only from Cuyamaca Peak and perhaps one other locality in northern Baja California at the edge of an arroyo at El Agua Colorado in the Sierra Juarez (Reiser 1994). It is threatened by high fire frequency but is currently protected in a Research Natural Area set aside, as are other rare gabbro species (Keeler-Wolf 1990b). Southern mountain misery (*Chamaebatia australis*) is an endemic sprawling shrub in the rose family, which may form impenetrable thickets. It is apparently restricted to gabbroic or metavolcanic derived soils. It may be found in conjunction with disjunct stands of the shrub *Pickeringia montana*, and is typically surrounded by chamise (*Adenostoma fasciculatum*) chaparral. This species occurs on several isolated mountains in San Diego County, including Tecate Peak and Otay Mountain. At the San Marcos Mountains it is definitely known on the Las Posas soil series.

Dehesa beargrass (*Nolina interrata*) typically occurs in clay soils derived from gabbro or metavolcanic bedrock (FWS 1998). *Nolina interrata* occurs in restricted and localized populations from the interior foothills of San Diego County to northwestern Baja California, Mexico. The habitat (gabbroic chaparral) for this species is relatively restricted and probably includes fewer than 5000 acres, of which <500 is occupied habitat (FWS 1998). Open southern mixed chaparral and chamise chaparral are the preferred habitats of this distinctive shrub. Near the Dehesa School, a series of fires have left stands of this *Nolina* in a disturbed annual grassland. Most populations apparently occur on Las Posas stony, fine sandy loams. Repeated subjection to fire may be an important factor in this species' life history (Reiser 1994).

Other rare associates of San Diego gabbros include Gander's butterweed (*Senecio ganderi*), tufted pine-grass (*Calamagrostis koelerioides*), felt-leaved monardella (*Monardella hypoleuca* ssp. *lanata*), and a dioecious shrub in the spurge family (Euphorbiaceae), *Tetracoccus dioicus*, which typically grows on Las Posas soils in San Diego County.

14   Sierra Motherlode, Domain 2

The Sierra Motherlode domain is in a series of allochthonous terranes, sometimes called the "Foothill Belt," along the western edge of the north-northwest–south-southeast trending Sierra Nevada, adjacent to the Great Valley of California (fig. 14-1). It is a discontinuous belt from the southern Sierra Nevada, in Tulare and Fresno counties, to Butte County in the northern Sierra Nevada, but a branch within the belt is practically continuous from El Dorado County about 140 km north to Plumas County at the north end of the range.

Cenozoic block faulting has lifted the Sierra Nevada and tilted the mountain range toward the west; therefore the highest elevations are on the east side of the range. Uplift is more pronounced in the southern than in the northern Sierra Nevada. Altitudes range from <200 m adjacent to the Great Valley to more than 4000 m along the crest of the central to southern part of the mountain range. The highest altitudes in the Sierra Motherlode domain are 1939 m (6360 feet) on Red Mountain and 1935 m (6335 feet) on Red Hill in Plumas County, and even higher on some of the granitic plutons that are within the outer limits of the serpentine domain. These plutons were intruded into the allochthonous terranes after the terranes had been accreted onto the continent. Much of the western slope of the northern Sierra Nevada is an undulating to rolling plateau. This plateau is a remnant from the early Tertiary when its surface was deeply weathered to produce lateritic serpentine soils with silica deposited in the subsoils and in fractures in the bedrock (Rice and Cleveland 1955, Rice 1957). The ancient plateau was capped by volcanic flows that produced a practically continuous cover in the northern Sierra Nevada (Durrell 1966). Uplift along the eastern side of the northern part of the Sierra Nevada to initiate its current relief commenced 4 or 5 Ma ago (Wakabayashi and Sawyer 2001). Since the range began to rise a few million years ago, the larger streams flowing across it have cut deep canyons up to about 600 m below the plateau. Erosion has removed much

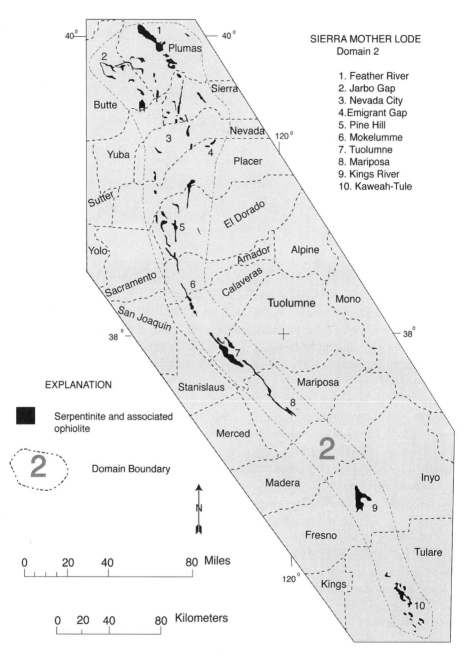

Figure 14-1 A map showing domain 2, Sierra Motherlode.

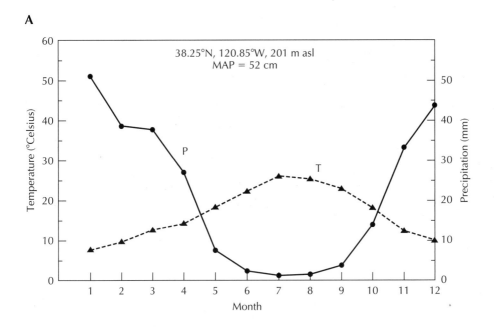

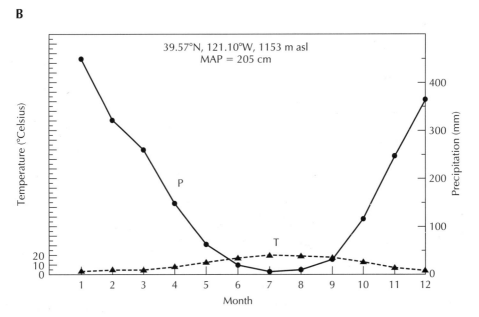

Figure 14-2 Mean monthly temperature and precipitation at two sites in the Sierra Motherlode domain. (A) Camp Pardee, California. (B) Strawberry Valley, California.

of the volcanic cover to expose ultramafic rocks in allochthonous terranes of the Sierra Motherlode from Mariposa County north only within the last 4 or 5 Ma. Geologic erosion rates have been particularly high on the Bear Mountain block, which is an oval shaped area outlined by the west and east Bear Mountain faults (Springer 1980).

Climates range from warm with hot summers at lower elevations, adjacent to the Great Valley, to cool with cold winters at the higher elevations. The mean annual precipitation in the Sierra Motherlode domain ranges from about 30 cm at lower elevations on the south to over 225 cm at higher elevations on the north. Most it falls as winter rain at lower elevations and much of it falls as snow at the higher elevations. Summer soil water deficits are moderate at higher elevations that are wetter and severe at lower elevations (fig. 14-2). The mean freeze-free period ranges from about 50 days at high elevations to 300 days at low elevations on the south.

14.1 Geology

From the American River north, the Foothill Belt, or Western Sierra Nevada Metamorphic Belt, has been divided into a "Western Belt" of Jurassic rocks, a "Central Belt" of mostly Paleozoic rocks, and an "Eastern Belt" with mixtures of both Paleozoic and Mesozoic rocks (Schweickert and Cowan 1975). North–south fault zones are marked by lenses of ultramafic rocks (mostly serpentinite) forming melange, which is commonly enclosed in melange of metavolcanic and metasedimentary rocks.

The Western Belt, which is west of the Big Bend–Wolf Creek fault zone, is an ancient volcanic arc consisting of basalt and andesitic rocks interbedded with volcanoclastic rocks. The rocks have been transformed to greenstones and schists by invading Jurassic–Cretaceous plutons and are strongly deformed by folding and transpressional movements.

The Central Belt is between the Big Bend–Wolf Creek fault zone on the west and the Rich Bar–Goodyears Creek fault zone on the east. It represents an earlier arc sequence with lithologies similar to those of the Western Belt, except that the rocks are older (Carboniferous to Permian) based on fossils in limestone.

The Eastern Belt is east of the Melones fault zone. It contains early Paleozoic silicic pyroclastic rocks that are overlain by Paleozoic to early Mesozoic arc volcanic rocks interbedded with marine and nonmarine sandstone and mudstone. This belt has more continental affinity than the Western and Central Belts, which are decidedly oceanic in origin (Mayfield and Day 2002).

Each of the three belts is represented by volcanic activity related to subduction along the edge of the North American continent. A strip of ultramafic rocks called the Feather River complex lies between the Central and Eastern Belts, from Plumas County south. South of the American River, from Amador to Mariposa counties, the Foothill Belt is 20–30 km wide. From Madera County south through Fresno and Tulare counties, it occurs as isolated patches, rather than as a continuous belt.

Volcanic arc rocks were strongly deformed by folding, faulting, and low-grade metamorphism during accretion to the North American continent. After accretion, they were invaded by Sierra Nevada granitic rocks. Intrusions of granitic plutons intensified the metamorphism, particularly adjacent to boundaries of the intrusions (Day et al. 1988). Ultramafic rocks in the three belts of the Motherlode domain represent fragments of ocean crust that were basement for volcanic arcs. During accretion, small slices of ocean crust, including ultramafic basement, were incorporated in the volcanic arc assemblages (Irwin 1979, Saleeby 1982).

14.2 Soils

The soils are deep in materials weathered from bedrock on gentle slopes that have not been eroded greatly in the last few million years. They are shallow to deep on steep slopes, and shallow on more gentle slopes that have been eroded intensively. Floodplains and stream terraces are generally narrow and are not extensive. The soils are well drained, with few minor exceptions. Salts and carbonates are leached from the well-drained soils. Clay is leached from surface soils and has accumulates in subsoils to form argillic horizons on relatively stable landforms. Ochric epipedons predominate, although mollic epipedons are common in soils under chamise chaparral. Cambic and argillic horizons are the common subsurface diagnostic horizons.

Clay in subsoils is commonly dominated by minerals inherited from soil parent materials. Kaolinite accumulates in older nonserpentine soils in wetter climates and smectites accumulate in soils in drier areas. Smectites and iron oxides are more prevalent in serpentine soils than in most soils with other kinds of parent materials.

Serpentine Mollisols and Alfisols are common in warm grassland, woodland, and chaparral at low elevations. Alfisols predominate in cool conifer forest at intermediate elevations and Lithic Xerorthents or Inceptisols at the higher elevations. There are some Ultisols in the wetter areas.

Soil temperature regimes are mostly mesic, and thermic at lower elevations. Soil moisture regimes are xeric, and possibly udic in some of the cooler and wetter areas.

14.3 Vegetation

The vegetation of the Motherlode domain contains elements of the California Cismontane Foothill and Sierra–Cascades floristic regions. The majority of the vegetation in the north tends to exhibit more montane characteristics due to the higher elevation expression of the serpentine bodies there and the more northerly latitude. According to Fiedler et al. (in Keeler-Wolf 1990b), some of the most northerly and highest elevation serpentine vegetation in the Sierra is made up of sparse stands of shrubs considered part of the *Quercus vaccinifolia* alliance and *Arctostaphylos nevadensis* alliance. These occur on open rocky ridges. Open coniferous forest of *Pinus jeffreyi* alliance also occurs at higher elevations in the northern portion of the domain. At slightly lower elevations vegetation may be dominated by open stands of *P. jeffreyi* and *Calocedrus decurrens*. Local areas of seeps and springs may have *Darlingtonia californica* alliances and associated wetland alliances. Farther south, starting in Nevada County, the serpentine belt trends away from the mid and upper elevations of the range and tends to occupy narrow exposures at lower elevations. Most of the serpentine vegetation in this area is below the winter snow accumulation zone and is within the general zone of chaparral, annual grassland, and foothill woodlands. Predominant vegetation includes stands of *Ceanothus cuneatus, Adenostoma fasciculatum, Pinus sabiniana, Bromus diandrus–B. hordeaceous,* and *Quercus douglasii* alliances. Local stands of *Cupressus macnabiana* and *Pinus attenuata* alliances also occur on serpentine in the northern Sierra Motherlode.

Much of the vegetation on serpentine in the Sierra Motherlode domain is only marginally different from adjacent stands off of serpentine. This is the only serpentine domain where *Quercus douglasii* woodlands occur commonly on serpentine. It is also known for its relatively high proportion of non-native annual species. Kruckeberg (1984) and Stebbins (1984) have also commented on the relatively low diversity of serpentine endemic plants in the Sierra Motherlode. No endemic *Arctostaphylos* species are known to form dominant stands on serpentine here. A few distinctive species of herbs or shrubs occur and some of the most distinctive are more tied to gabbro, rather than strictly serpentinized rocks (Hunter and Horenstein 1992).

14.4 Feather River, Locality 2-1

The Feather River complex is a long (54 km) and narrow (3–6 km wide) body between the Eastern and Central Belts, mostly in Plumas County (Ehrenberg 1975, Mayfield and Day 2002). Serpentine soils and vegetation extending for another 80 km south along the Melones fault are included in the Feather River locality. As such, the locality stretches from the Georgetown area in El Dorado County across the Forest Hill divide in Placer County north through Nevada and Sierra Counties to the head of Canyon Creek, a tributary of the North Yuba River, and from there northwest across Plumas County to Red Hill and the North Fork of the Feather River.

The Feather River complex is mostly peridotite (predominantly harzburgite and dunite) that has undergone two different metamorphic events. The main minerals in the peridotite (forsterite and orthopyroxene) are commonly accompanied by tremolite, anthophyllite, diopside, and talc that were produced during amphibolite facies metamorphism.

A sample of peridotite from the Feather River complex contained 23 g/kg CaO (Hietanen 1973). This is close to an estimate of Ehrenberg (1975) based on analyses of individual minerals in the ultramafic rocks. Most of the calcium is in tremolite, which tends to be more concentrated in rocks near the margin of the Feather River complex than in rocks nearer the center of the body. Depleted harzburgite in the California Coast Ranges and Great Valley ultramafics commonly contains only 3–7 g/kg CaO, while many altered ultramafic rocks in the Sierra Nevada contain CaO in the 20–39 g/kg range that is characteristic of high-temperature peridotite intrusions (Ehrenberg 1975). If higher calcium is characteristic of ultramafic rocks in the Sierra Nevada, it is not known if it is high enough to have discernible effects on the plant communities. In peridotite with 20–40 g/kg Ca, the molar Ca/Mg ratio is still <0.1.

Tectonic slices of amphibolite are concordant with the northwest trend of the Feather River ultramafic body indicating that these foreign bodies are inclusions of mafic rock introduced during accretion. There is no field evidence to show that this ultramafic body was basement rock for oceanic crust. The Feather River ultramafic occupies the northeasterly extension of the Melones Fault zone and its tectonic position suggests that it was emplaced during the accretion of the Central Metamorphic Belt.

Altitudes range from about 800 m (~2500 ft) along some of the rivers that cross the Feather River complex to 1935 m (6335 ft) on Red Hill. Slopes range from undulating on broad divides between deeply incised streams to steep or very steep in deep canyons and on Red Hill. Mean annual precipitation ranges from about 225 cm on the north to 150 cm on the south.

The serpentine soils are mostly moderately deep Mollic Haploxeralfs (Dubakella series) with open Jeffrey pine–incense-cedar–Douglas-fir forest cover, and some Lithic Mollic Haploxeralfs with sparse Jeffrey pine–incense-cedar forest and shrub cover. In wetter areas to the north, in Plumas County, there are some Lithic Xerorthents on narrow ridge crests and very steep slopes, moderately deep Ultic Haploxeralfs (Cerp series) on moderately steep to steep mountain slopes, and deep to very deep Ultisols (tentative, not confirmed with laboratory data) on mountain benches. These Ultic Haploxeralfs and Ultisols may have a cover of dense mixed conifer forest. Practically all of these soils have mesic soil temperature regimens (STRs) and xeric soil moisture regimens (SMRs).

Shallow serpentine soils on the summit of Red Hill are dominated by buckbrush and buckwheat (*Eriogonum umbellatum*). Deeper serpentine soils, Alfisols, east of the summit of Red Hill have predominantly Jeffrey pine–incense-cedar–sugar pine forests with pine-mat manzanita (*Arctostaphylos nevadensis*) and buckbrush (*Ceanothus cuneatus*) in the understory. White fir (*Abies concolor*) is another conifer that is common where there is gabbro. *Phlox diffusa* and *Senicio* sp. are common and conspicuous when in bloom, and there is sparse coffeeberry *Rhamnus rubra* on the serpentine soils.

There is dense mixed conifer cover on a few thousand acres of Oxisols (Forbes series), which have been described and sampled on the Forest Hill Divide in Placer County (Alexander 1988).

Stebbins (1984) describes alder (*Alnus rhombifolia*) and bay (*Umbellularia californica*) along Traverse Creek, a few miles southeast of Georgetown, and adjacent serpentine slopes with *Rhamnus tomentella*, *Ptelea crenulata*, and *Quercus durata*. Xeric serpentine slopes in this part of El Dorado County have gray pine (*P. sabiniana*), buckbrush, and whiteleaf manzanita (*A. viscida*) alliances. Stebbins (1984) noted some distinctive herbs on serpentine in the El Dorado County area: *Allium sanbornii, A. membrenaceum, Cryptantha intermedia, Githopsis pulchella, Arenaria californica, Heliathus bolanderi, Senecio layaneae, Calystegia fulcrata, Parvisedum congdonii, Phacelia purpusii, Streptanthus polygaloides, Trichostoma simulatum, Calochortus superbus, Sidalcea hirsuta, Eriogonum tripodum, Lewisia rediviva, Lomatium marginatum*, and *Viola douglasii*. Jeffrey pine grows at lower elevations on serpentine than on other substrates. An extensive stand of *P. jeffreyi* alliance northeast of Forest Hill, at

1050–1200 m elevation, is one of the few below 1800 m on the west side of the Sierra (Stebbins 1984). Constance rockcress (*Arabis constancei*) is a sensitive plant that grows on the serpentine soils of this locality.

14.5 Jarbo Gap, Locality 2-2

The ultramafic rocks of the Jarbo Gap locality are well exposed along the North Fork of the Feather River around Pulga (Day et al. 1988, Dilek 1989). It is a belt of serpentinites that extend east from Butte County into Plumas County and from there south to the North Yuba River in Yuba and Sierra counties. It occupies the northern part the Central Metamorphic Belt. The serpentinites crop out as discontinuous, elongate, fault-bounded lenses. They are within a melange zone that consists of metavolcanic and metasedimentary rocks. Mesozoic granitic rocks have intruded this belt and metamorphosed serpentinites on their intrusive contacts. The protoliths of these serpentinites are harzburgite and dunite; however, metamorphism has produced schistosity that has obscured the original textures. Antigorite, chrysoltile, tremolite, chlorite, talc, and carbonate minerals are common phases in these recrystallized serpentinites.

Altitudes range from about 400 m (~1250 feet) along the North Fork of the Feather River to 1939 m (6360 feet) on Red Mountain. Mean annual precipitation ranges from about 125 cm on the west to >200 cm on the east.

The common serpentine soils of this locality are Alfisols. More specifically, they are moderately deep-to-deep Mollic and Ultic Haploxeralfs (Dubakella, Woodleaf, and Cerpone series) with open forest cover and Lithic Haploxeralfs with sparse forest and shrub cover. Most of the soils have mesic STRs, but some Lithic Haploxeralfs may have thermic STRs where the cover is gray pine and blue oak with wedgeleaf ceanothus, California bay, toyon, and whiteleaf manzanita in the understory. The Cerpone series sampled at the type location was found to have Ca/Mg ratios from 1.6 in the upper part of the argillic horizon (9–24 cm) to 0.1 in the lower part, below 1 m. It has been suggested that relatively high surface soil Ca/Mg ratios are characteristic of serpentine soils in the Sierra Nevada.

Vegetation at lower elevations, below about 800 m, includes stands of gray pine, blue oak (*Quercus douglasii*), *Arctostaphylos viscida*, and *Ceanothus cuneatus* alliances, and mixed stands of *P. sabiniana* over *A. viscida*, *C. cuneatus*, and shrubby *Umbellularia californica*, or herbaceous species. Small inclusions or separate stands of incense-cedar (*Calocedrus decurrens*) and Douglas-fir (*Pseudotsuga menziesii*) alliances occur on mesic soils. There are local stands of *Quercus durata* and *Cupressus macnabiana* alliances in mesic (moisture) areas at the lower elevations. Stands of *Chrysothamnus nauseosus* ssp. *albicaulis*, isolated from similar stands at higher elevations several miles to the east, occur in the Magalia area. Also, *Streptanthus polygaloides*, *Phacelia corymbosa*, *Senecio eurycephalus* ssp. *lewisrosei*, and the local serpentine indicators *Allium cratericola* and *A. sanbornii* occur in the area. Schlising (1984) characterized the flora of the Magalia area as being composed of 119 species; 56 (47% of the flora) are annuals, most of which (46 species) are natives. The 10 species of non-native annual grasses on serpentine soils were mostly restricted to disturbed areas such as quarries and borrow-pits. Many of the native annuals are not serpentine indicators but locally appear to prefer serpentine. Included in these are *Mimulus layneae* and *M. torreyi*, which produce spectacular displays in spring. *Erythronium multiscapoideum*, *Senecio eurycephalus* ssp. *lewisrosei*, and *Calamagrostis koeleroides* are notable herbs in the Pulga area (Stebbins 1984).

14.6 Nevada City, Locality 2-3

The Nevada City locality is in Nevada County, along the western edge of the Central Metamorphic Belt, between the Weimer fault on the east and the Wolf Creek fault zone on the west. It contains narrow north–northwest trending bands of serpentinite (Day and Moores 1985) from Grass Valley north about 15 km. Jurassic mafic igneous and sedimentary rocks structurally overlie a foliated and serpentinized mixture of harzburgite, pyroxenite, and dunite. The ultramafic unit is intruded by gabbro and quartz diorite that in turn are intruded by mafic dike swarms. Volcanic flows and pyroclastics have covered

---

A Possible Source of Extra Calcium

Although not evident from the data in table 14-1, it has been suggested that exchangeable Ca/Mg ratios are higher in serpentine soils of the Sierra Nevada than in those of the California Coast Ranges. If true, could this be explained by atmospheric fallout, or deposition? Both dry and wet deposition from the atmosphere have higher molar Ca/Mg ratios in the Sierra Nevada (3.0 dry and 1.6 wet, $n=3$) than in the Coast Ranges (1.6 dry and 0.75 wet, $n=5$) (McColl et al. 1982). Lower ratios in the Coast Ranges are attributable to its location near the Pacific Ocean and greater concentrations of magnesium than calcium in the oceans. The higher ratios for dry deposition are assumed to be related to more local sources for the materials in it than in wet deposition. Although calculating the cation balance in a soil requires knowledge of rates of cation release from weathering, recycling by plants, and leaching losses, a hypothetical case is easy to quantify by ignoring these transfers and losses.. Assume that the cations on all exchange sites in the Dubakella-like soil (table 14-1) are replaced with calcium and magnesium from wet deposition, which has a sum of $1.6 \text{ mmol}(+)/\text{m}^2$ annually (McColl et al. 1982). It would take 6 ka to replace all cations in the upper 15 cm of soil and 50 ka to replace them in the upper 50 cm of soil, resulting in an exchangeable Ca/Mg ratio of 1.6 mol/mol. Much less than 6000 years might be required for the exchangeable Ca/Mg ratios in Sierra Nevada serpentine surface soils to become significantly higher than in those of the Coast Ranges. This time might be reduced substantially if dry deposition is taken into account. Much of the material in dry deposition in the Sierra Motherlode domain may be from the Sacramento and San Joaquin valleys, rather than from local sources.

Table 14-1 Selected data from Natural Resources Conservation Service pedon S68CA-29-36, Dubakella series or a similar soil that is less stony, in Nevada County.

Horizon	Depth (cm)	Texture	Organic C (g/kg)	N (g/kg)	Exchangeable cations (mmol/kg)				CEC (pH 7)	Acidity (pH) (1:1)
					Ca	Mg	Na	K		
A1	0–5	GrSL	15.8	0.73	36	81	—	2	153	5.8
A2	5–15	SL	6.2	0.31	16	97	—	1	132	5.9
Bt	15–40	C	3.3	0.31	13	431	1	—	379	6.1
C	40–66	GrC	2.0	—	15	448	1	—	417	6.3

CEC, cation-exchange capacity; C, clay; GC, gravelly clay; GrSL, gravelly sandy loam; SL, sandy loam.

all of these units. The serpentinite is considered to be oceanic basement for an ancient volcanic arc complex.

Altitudes are mostly 600–800 m (2000–2500 feet), but down to about 450 m (~1500 feet) along the South Yuba River. Mean annual precipitation is 100–125 cm.

Serpentine soils in this area are predominantly moderately deep Mollic Haploxeralfs (Dubakella series) with open forest cover, and presumably Lithic or Lithic Mollic Haploxeralfs with sparse forest and shrub cover. The shallow soils were not mentioned in the soil survey of the Nevada County area (Britton 1975) but are presumed to be inclusions in soil map units with moderately deep Mollic Haploxeralfs. A serpentine soil sampled in Nevada County has a typical range of exchangeable Ca/Mg values from <0.1 in the subsoil to 0.44 in the surface (table 14-1). Plants recycle more calcium than magnesium, which is returned to the surface when plants shed leaves or die, and dust may also add more calcium.

Vegetation in the vicinity of Nevada City and Grass Valley is commonly buckbrush chaparral with scattered emergent gray pine, occasionally in stands dense enough to form its own alliance. Chaparral pea (*Pickeringia montana*) is often conspicuous in these stands, and stands of whiteleaf manzanita alliance are common. The most notable serpentine vegetation alliances are stands of McNab cypress (*Cupressus macnabiana*) and locally leather oak (*Quercus durata*).

14.7 Emigrant Gap, Locality 2-4

The ultramafic bodies of Emigrant Gap are part of a mafic complex that is the core of a large granitic pluton at the crest of the Sierra Nevada. The mafic complex is roughly zoned with ultramafic rocks in the core, and gabbro forms an intermediate medial zone surrounded by various kinds of granitic rock. Peridotite (wherlitic) is dominant, accompanied by dunite and websterite (olivine and pyroxene), mostly unserpentinized. This complex is similar to the zoned ultramafic complexes of southeast Alaska and the Los Pinos pluton. The Monumental Ridge exposure in locality 2-4 is a differentiated ultramafic body ranging in composition from dunite through wehrlite (James 1971).

The largest ultramafic rock exposure in this locality is wehrlite and clinopyoxenite on Monumental Ridge and the south slope from the summit at 2092 m asl down to about 1650 m. This is "serpentine rock" that lacks serpentine; a sample of wehrlite from near the summit of the ridge has a density of 3.30 Mg/m^3. Bedrock exposures dominate Monumental Ridge, with dense huckleberry oak (*Quercus vaccinifolia*)–silktassel (*Garrya fremontii*)–greenleaf manzanita (*Arctostaphylos patula*) chaparral between exposures of bedrock. A few small patches of shallow soils along the ridge have a cover of pinemat manzanita and prostrate ceanothus (*Ceanothus prostratus*). The ceanothus has small leaves (~0.5 cm wide), but it does not appear to be *Ceanothus pumilis*. Talus that is barren, except for abundant crustose lichens, extends downward from just below the ridge, with some dense patches of huckleberry oak–silktassel chaparral that grade downward to huckleberry oak–coffeeberry chaparral. Soils below the talus are steeply sloping, cold (frigid STRs) Typic Haploxerepts with semidense mixed conifer forest and an understory of huckleberry oak and coffeeberry. Trees in the mixed conifer forest are Jeffrey pine, incense-cedar, sugar pine, Douglas-fir, and white fir. Stony, shallow soils on broad spur ridges have open Jeffrey pine–incense-cedar forests with huckleberry oak, pinemat manzanita, prostrate ceanothus, and sparse Idaho fescue (*Festuca idahoensis*) in the understory. The ceanothus in these plant communities has leaves that are more characteristic of prostrate ceanothus, rather than the small leaves of prostrate ceanothus along the summit of Monumental Ridge. Prostrate ceanothus avoids serpentine soils in the Klamath Mountains. The higher calcium contents of wehrlite and clinopyroxenite may be the factor that allows it to grow on serpentine soils in the Monumental Ridge locality.

The largest exposure of dunite in this locality is just east of Onion Valley at 1500–1650 m asl. Cool (mesic STRs) Mollic Haploxeralfs (Dubakella Series) and Lithic Mollic Haploxeralfs (possibly the tentative Wildmad Series) have been mapped in this area (Hanes 1994). Most of the dunite exposure is on a steep, rocky, west-facing slope. It is covered by a Jeffrey pine, incense-cedar, Douglas-fir, bay forest with deerbrush (*Ceanothus integerrimus*) in the understory. It is uncertain why deerbrush appears on the dunite soils, because it avoids serpentine in other areas. There are no oak trees, but sparse huckleberry oak occurs on colluvial footslopes. The main forb is rough bedstraw (*Gallium* sp.), and spreading phlox (*Phlox diffusa*), milkwort (*Polygala* sp.), monardella (*Monardella* sp.), and prostrate buckwheat are common, along with serpentine fern (*Aspidota densa*) and squirreltail (*Elymus elymoides*). Bird's-beak (*Cordylanthus* sp.) is common on disturbed soils on a summit at the top of the west-facing slope.

An interesting feature of the talus on the south side of Monumental Ridge is thick (1 or 2 mm) white coatings on stones beneath some of the boulders. Crushed coatings effervesce strongly in dilute HCl and appear similar to those on overhanging cliff faces of Cement Bluff in the Klamath Mountains that are composed of nesquehonite, calcite, magnesite, and hydromagnesite (Alexander 1995).

14.8 Pine Hill, Locality 2-5

Pine Hill is on the Bear Mountain block, which is an oval-shaped block wedged between two parts of the Bear Mountain fault zone. It has been raised above the surrounding terrain and all traces of the Tertiary volcanic cover have been eroded away. The core of the block is occupied by gabbro that has intruded and metamorphosed the surrounding serpentine belts (Springer 1980). The intrusive temperature of gabbro melts is higher than for granitic melts, and the evidence of this is clearly seen in the metamorphosed serpentinite around the Pine Hill intrusive. Veins of tremolite asbestos have formed near the Pine Hill intrusion, and semi-nephrite, a compact form of tremolite, are found within the contact zone. Much cross-fiber asbestos in the serpentine has been modified to picrolite, a form of chrysotile with the fibers welded into a dense nonfibrous mass. The Pine Hill intrusion is similar to the Smartsville complex, which is farther north in the Western Metamorphic Belt.

The topography is characterized by interconnecting, north–northwest trending ridges that are broad and rounded, with gentle summit slopes and moderately steep side slopes. Altitudes range from 300 m (~1000 feet) along the South Fork of the American River to 628 m (2060 feet) on Pine Hill. The mean annual precipitation is 60–90 cm, and the frost-free period is 170–270 days.

Shallow Haploxeralfs of the Delpiedra series (Mollic Haploxeralfs, or Lithic Mollic Haploxralfs) were the only serpentine soils mapped in the Pine Hills area (Rogers 1974). The Delpiedra map units contain inclusions of very shallow Entisols (Lithic Xerorthents) and moderately deep Mollic Haploxeralfs, however. Gray pine–chamise chaparral or mixed chaparral plant communities are common on the serpentine soils. They have thermic STRs.

Plant communities on gabbro soils of the Rescue series have many endemic species, including some that are commonly assumed to be restricted to serpentine soils (Hunter and Horenstein 1992). A representative Rescue soil has Ca/Mg ratios >3 in the surface and nearly 2 in the subsoil (Rogers 1974) and chromium, cobalt, and nickel concentrations in the soils are not particularly high (Hunter and Horenstein 1992). It is not known why the plant communities on the Rescue soils have such unique species compositions.

The Pine Hill Rare Plants are a collection of eight rare plant species that share the unusual growing conditions of a small area of western El Dorado County. They grow in a roughly oval area centering around Green Valley Road and stretching from Folsom Lake in the north to Highway 50 in the south. Three of these rare plants are endemic to the Pine Hill region. Another two species are nearly endemic, with only a few small colonies of the plants found elsewhere. This assemblage of rare species is part of a unique plant community that has been termed the Northern Gabbroic Chaparral (Holland 1986). Hunter and Horenstein (1992) sampled vegetation there and noted a more fine grain matrix of vegetation ranging from *Quercus wislizenii* dwarf woodland and *Q. kelloggii* tall woodland in more sheltered mesic areas to open or dense stands of chaparral of either *Adenostoma fasciculatum* and *Arctostaphylos viscida* or other mixtures of endemic and widespread shrubs. The endemic species are distributed variably depending on fire history and microclimate. Several of the endemics including *Ceanothus roderickii, Fremontodendron californicum* ssp. *decumbens, Helianthemum suffrutescens, Chlorogalum grandiflorum, Senecio layneae,* and *Calystegia stebbinsii* occur in the drier southwest-facing slopes of the area, while others such as *Galium californicum* ssp. *sierrae* are more common in the understory of the *Q. wislizenii* or *Q. kelloggii* alliance stands on the north-facing slopes. Others such as *Wyethia reticulata* may occur in either mesic or xeric settings. Several species appear to be most prolific following fire, including *Ceanothus roderickii, Fremontodendron californicum* ssp. *decumbens, Helianthemum suffrutescens,* and *Calystegia stebbinsii*.

Fire management is a major issue in the area. Current management plans for the Pine Hill Preserve call for periodic fire to maintain the diversity of fire following endemics (California Department of Fish and Game [CDFG] and Bureau of Land Management [BLM] management plans). Stands of chaparral with longer fire intervals tend to have denser stands of chaparral dominated by *Adenostoma fasciculatum* and *Arctostaphylos viscida*. These widespread species have been considered less desirable than the endemic shrubs, subshrubs, and herbs that may show increased density following fire and other disturbance. However, the temptation to apply fire as a regular management tool to the area should be carefully weighed against its potential detriments. Many of the endemic species are capable of maintaining long-persisting seedbanks or survive vegetatively for many years without regular fire.

14.9 Mokelumne, Locality 2-6

The Mokelumne locality is confined to a narrow band of highly sheared serpentinite in the Bear Mountain Fault Zone, in Amador and Calaveras Counties, south of the Pine Hill locality (Duffield and Sharp 1975). On the west the serpentinite is in fault contact with the lower Jurassic Copper Hill volcanics, and on the east it is faulted against upper Paleozoic metamorphosed melange of the Central Metamorphic Belt. The serpentinite is so highly sheared that it is difficult to establish the nature of the original peridotite protolith. East of Valley Springs the serpentinite is covered by Tertiary volcanics, including the burial of some silicified serpentinite from the base of a lateritic soil (Rice and Cleveland 1955). Remnants of silica "boxwork" from the lateritic soils cap many of the serpentine hills in this locality and some farther south.

The topography is hilly, and altitudes range from about 150 m (500 feet) where the Mokelumne and Calaveras rivers run from east to west across the locality to about 300 m (1000 feet). Mean annual precipitation is 50–60 cm.

Serpentine soils that have been mapped in this area are Alfisols–shallow and moderately deep Mollic Haploxeralfs of the Delpiedra and Fancher series. Characteristically, the Delpiedra soils have cover of chamise or mixed chaparral, and the Fancher soils have cover of blue oak woodland or grass. Gray pine trees are sparse in these vegetation types, and blue oak trees are generally absent from shallow serpentine soils. Stands of buckbrush alliance ofeeten occupy recently burned areas, and gray pine is common.

Figure 14-3 A typical south-facing slope in the Red Hills (locality 2-6) with a gray pine (*Pinus sabiniana*)/buckbrush (*Ceanothus cuneatus*)/grassland plant community on Haploxeralfs.

14.10 Tuolumne, Locality 2-7

Serpentine of the Tuolumne locality is in a band 6 or 7 km wide and about 25 km long from Calaveras County across the Stanislaus River southeast to the Tuolumne River in Tuolumne County (Eric et al. 1955, Morgan 1976). It has been drawn into a faintly sigmoid shape by deformation between the Melones Fault Zone on the northeast and the Bear Mountain Fault Zone on the southwest. The rocks west of the Bear Mountain fault are deformed metavolcanics and metasediments of Jurassic age. The ultramafic rocks east of the fault are within the Central Metamorphic Belt consisting of metamorphosed Upper Paleozoic metavolcanic and metasedimentary rocks. The peridotite is nearly completely serpentinized and consists mainly of lizardite and chrysotile. Abundant cross-fiber chrysotile veins near Sugarloaf Mountain have been mined for asbestos (Leney and Loeb 1971). The serpentinized peridotite is intensively sheared along the Bear Mountain fault. In this sheared zone are large blocks of amphibolite.

The topography is hilly, and altitudes range from about 150 m (500 feet) where the Stanislaus and Tuolumne rivers run from east to west across the locality to about 500 m (1640 feet), but exceed 700 m at the extreme southeast end near Moccasin Peak. Mean annual precipitation is 50–75 cm.

The Miocene Table Mountain latite volcanic flow forms prominent bluffs where it rests on the Mehrten formation consisting of volcanic diamicton interbedded with conglomerate, sandstone, and siltstone. The Mehrten formation was deposited on an old Tertiary erosion surface over serpentinite.

The serpentine soils are mainly shallow Mollisols with chamise chaparral cover and shallow and moderately deep Alfisols with blue oak woodland and gray pine/mixed chaparral cover. The shallow Mollisols have been mapped as Lithic Argixerolls of the Henneke Series (Gowans and Hinkley 1964, Soil-Vegetation Survey). The shallow Alfisols have shallow lithic contacts on serpentinized peridotite (Lithic Mollic Haploxeralfs), but may have only paralithic contacts within 50 cm depth on sheared serpentinite (shallow Mollic Haploxeralfs, Delpiedra series). Below is a brief description of a serpentine Lithic Mollic Haploxeralf in the Red Hills of Tuolumne County, about 6 km southeast of Chinese Camp:

> loamy-skeletal, magnesic, thermic, Lithic Mollic Haploxeralf
>
> Latitude 37.8353°N, longitude 120.4687°W, 430 m altitude, convex 36% SW slope
>
> Sparse gray pine, buckbrush (40%), wild oat, annual fescue, soft chess, onion grass, and many forbs
>
> Oa <1 cm of herb litter
>
> A1 (0–3 cm) dark reddish brown (5YR 3/3, 4/4 dry) very gravelly loam; weak, very fine, subangular blocky; slightly hard; few very fine roots; pH 6.8; abrupt, smooth boundary
>
> A2 (3–8 cm) dark reddish brown (5YR 3/3, 4/4 dry) very cobbly loam; moderate, very fine subangular blocky; soft, friable; common very fine roots; pH 6.9; clear, smooth boundary
>
> AB (8–18 cm) dark reddish brown (5YR 3/4, 5/5 dry) very cobbly loam; moderate, very fine subangular blocky; friable; common very fine, few fine, and very few medium and coarse roots; pH 7.0; clear, wavy boundary
>
> Bt (18–39 cm) dark reddish brown (5YR 3/4, 5/6 dry) very cobbly clay loam; moderate, medium, subangular blocky; firm; slightly sticky and slightly plastic; continuous thin coatings on ped faces; few fine and very few medium roots; pH 7.0; abrupt, irregular boundary to moderately fractured serpentinized dunite (R horizon)

Medeiros (1984) has described the vegetation of the Red Hills area. Evens et al. (2004) have identified several serpentine plant associations in the area: *Pinus sabiniana/Ceanothus cuneatus–Rhamnus illicifolia, Quercus douglasii/Bromus–Vulpia microstachys*, and herbaceous associations of *Plantago erecta–Lasthenia californica, Vulpia macrostachys*, and *Coreopsis stillmanii*. Widespread or local serpentine indicators include *Steptanthus polygaloides, Allium cratericola, Lewisia rediviva, Lupinus spectabilis, Fritilaria agrestis*, and *Brodiaea pallida*. Endemic and rare serpentine species include *Chlorogalum grandiflorum, Lomatium congdonii, Allium sanbornii* var. *tuolumnense*, and *Verbena californica* (Medeiros 1984).

14.11 Mariposa, Locality 2-8

The Mariposa serpentine belt in Mariposa County is entirely within the northwest–southeast trending Melones fault zone. On the west it is thrust over slates and graywackes of the Mariposa formation, and on the east Upper Jurassic Peñon Blanco and Gopher Ridge arc volcanics of Middle Jurassic age are thrust over the serpentinite. The serpentinite is broadly folded, and along its margins a matrix melange has developed (Schweickert et al.

Figure 14-4 A Jeffrey pine (*Pinus jeffreyi*)–incense-cedar (*Calocedrus decurrens*)/buckbrush (*Ceanothus cuneatus*)/Idaho fescue (*Festuca idahoensis*) plant community on serpentinite with a moderately deep–shallow Haploxeralfs complex of soils. It is 1190 m asl, about 2.5 km south of Washington in Nevada County, California.

1988). Originally dunite and harzburgite, it has been completely serpentinized and highly sheared. Antigorite and chrysotile are the main serpentine minerals in the serpentinite, with only sparse relicts of the original olivine and pyroxene. Numerous leucocratic dikes have invaded the serpentinite in the Bagby area and have produced extensive lenses of nephrite jade along the contacts (Evans 1966). Talc and tremolite have formed along the faulted margins of the Mariposa serpentine belt.

The band of serpentinite in the southernmost part of the Melones fault zone, which ends in Mariposa County, is generally <1 km wide. It is considerably broader only at a kink in the fault zone from the Merced River at Bagby, 8 or 10 km northwest toward Coulterville. This is a steep, hilly area. Altitudes range from 250 m (800 feet) along the Merced River to about 800 m (2600 feet) on the summits. Mean annual precipitation is 60–75 cm.

Only Mollisols (Lithic Haploxeralfs, Henneke series) have been mapped on serpentine in Mariposa County (Butler and Griffith 1974), with no indication of what the deeper associated soils might be like. These soils have chaparral (*Adenostoma fasciculatum*, *Ceanothus cuneatus*, *Arctostaphylos viscida* alliances) and blue oak woodland cover, with some gray pine (*P. sabiniana*/*Ceanothus cuneatus* association).

14.12 Kings River, Locality 2-9

The Kings River serpentine of the Sierra Foothill Belt is in Fresno County, adjacent to the Kings River. It consists of blocks of serpentinized peridotite that have been disrupted by tectonic activity (Saleeby 1978). These blocks are within a serpentinite melange of sheared antigorite, accompanied by metagabbros and metabasalts. Invading Mesozoic granitic plutons of the Sierra Nevada have metamorphosed the Kings River melange.

Peridotite and serpentine melange of the Kings River ophiolite cover 75 or 80 square km of high hills and low mountains. Altitudes range from 200 m (600 feet) along the Kings River to 729 m (2390 feet) on Red Mountain and 957 m (3140 feet) on Hog Mountain. Mean annual precipitation is 40–60 cm.

Mollic Haploxeralfs, both shallow (Delpiedra series) and moderately deep (Fancher series), were mapped in this locality (Huntington 1971). The Delpidra soils are on sheared serpentinite. Lithic Mollic Haploxeralfs are expected to be prevalent on partially serpentinized peridotite. Grassland is prevalent on the shallow Alfisols, with some buckbrush and sparse gray pine. There is also some blue oak woodland on the moderately deep Alfisols, particularly on north-facing slopes. These soils have thermic STRs; any serpentine soils with mesic STRs on north-facing slopes at the higher elevations are not extensive.

Latimer (1984) described serpentine vegetation in this locality. Woody vegetation is prominent only on north- and east-facing slopes, primarily the blue oak alliance. Gray pine, buckbrush, and Mariposa manzanita (*A. mariposa*) are principal species within this alliance, although much of it may be dominated by herbaceous species or scattered low shrubs. Stands of *Lupinus albifrons* alliance occur on steep southwest exposures. Native perennial bunch grass stands include both *Nassella pulchra* and *Aristida hamulosa* alliances. The common non-native annual grasses are *Bromus hordaceous* and *Avena barbata*, with both forming their own alliances. Extensive stands of *Selaginella hanseni* alliance occur on rocky skeletal soils. They form terracettes and mats over exposed rocks. These provide good substrates for *Allium hyalinum*, *Dichelostemma pulchella*, and *Triteleia lutea* var. *scabra*. The open serpentine habitats often have spectacular floral displays in the spring, in contrast to the more strongly grass dominated granitic habitats.

14.13 Kaweah-Tule, Locality 2-10

Serpentinite can be traced south from the Kings River to the Kaweah–Tule serpentine melange belt in Tulare County (Saleeby 1979). The melange is in a shear zone where it consists of antigorite containing tectonic blocks of serpentinized peridotite, metagabbro, metabasalt, and metachert. These rocks have undergone several periods of deformation and metamorphism related to their emplacement and later intrusion by Sierra Nevada granitic plutons. Some of the Mesozoic plutonism has altered the serpentinites to silica–carbonate rock.

The serpentine in this locality crops out on a series of low hills along the edge of the Great Valley from the Kaweah River south–southeast about 50 km to the Tule River. Altitudes range from about 140 m (~460 feet) adjacent to the valley up to 400 or 500 m (~1500 feet) on the higher hills. The mean annual precipitation is in the range from 30 to 40 cm.

Vegetative cover in this locality is commonly blue oak and annual grassland.

15 Southern California Coast Ranges, Domain 3

The Southern California Coast Range domain is a mountainous region with subparallel ridges aligned north–south, or more precisely north, northwest–south, southeast, and with intervening valleys that are controlled by strike-slip faulting. It extends about 400 km from the Golden Gate at the entrance to San Francisco Bay south to the Transverse Ranges that have east–west trending ridges. The domain corresponds to a physiographic region about 400 km long and 100 km wide that is bound by the Pacific Ocean on the west, the Great Valley of California on the east, on the north by the drainage outlet of the Sacramento and San Joaquin Rivers through the Carquinas Straight and San Pablo Bay, and on the south by the Transverse Ranges (fig. 15-1).

Ridges in the Southern California Coast Ranges generally have nearly level crests (Page et al. 1997), but they range considerably in height up to about 1500 m on some of the higher peaks. No streams from the Great Valley cross the Southern California Coast Ranges to the Ocean; the Great Valley drains through the Carquinez Straight and Golden Gate at the north end of these ranges. The larger streams in the Southern California Coast Ranges drain from the Santa Clara Valley, Salinas Valley, and Cuyama Valley to the San Francisco, Monterey, and San Luis Obispo bays. Only relatively small streams drain to the Great Valley, but some of them have large alluvial fans in the valley. There are many Tertiary-fault-bound valleys and basins among the mountain ranges. Some of the more prominent basins are the Santa Maria basin, Carrizo Plains, Paso Robles basin, and Watsonville basin. Serpentine is scattered in relatively small bodies throughout the domain and is concentrated along some of the major faults and in the New Idria area (locality 3-12).

Climates range from cool and foggy along the coast to warm inland, with hot and dry summers inland from the fog belt. Mean annual precipitation ranges from 50 cm, or less, in the interior ranges and in the south, around the Santa Maria basin, up to 150 cm in

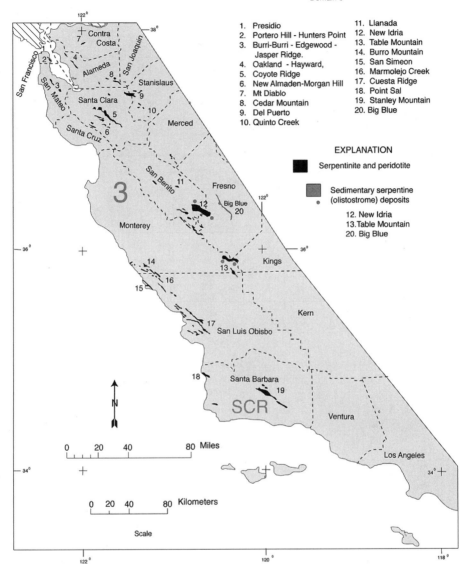

Figure 15-1 A map showing domain 3, Southern California Coast Ranges.

the Santa Cruz Mountains on the northwest and in the Santa Lucia Mountains, south of Monterey Bay. Precipitation is as low as 15 cm/year in the Cuyama Valley. Summer drought ranges from moderate along the coast northwest from San Luis Obispo Bay to severe in the drier areas (fig. 15-2). Some snow falls at the higher elevations, but it does not persist through winter. The frost-free period ranges from about 120 days at the higher elevations to >300 days along the coast.

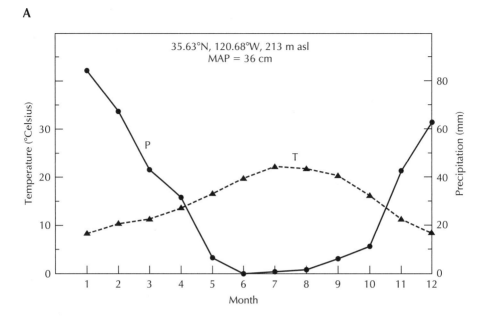
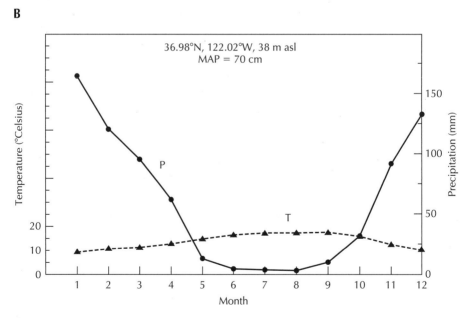

Figure 15-2 Mean monthly temperature and precipitation at two sites in the Southern California Coast Ranges domain. (A) Paso Robles inland from the coast. (B) Santa Cruz on the Pacific coast.

15.1 Geology

The Franciscan complex is the predominant basement rock in domain 3. It consists largely of graywacke sandstone, gray-colored siltstone and black shales. Erratically distributed within these sediments are serpentine and ophiolitic blocks associated with blueschist metamorphic blocks. These rocks range in age from Late Jurassic to Cretaceous. This association is often called a melange that formed within an oceanic trench before being subducted under the edge of the North American continent. The serpentines within this melange have experienced many tectonic events and most often are present as diapirs or along the base of thrust sheets. The extreme eastern part of this area is underlain by Mesozoic Great Valley forearc sediments. At the base of the Great Valley sediment is the Late Jurassic Coast Range ophiolite containing serpentine bodies. These bodies are distinct from those in the Franciscan complex because they have not been metamorphosed and commonly display preserved stratigraphic sequences of oceanic crust.

The San Andreas fault is a prominent feature of the Southern California Coast Ranges. It extends diagonally across the entire length of the ranges, from the Big Pine fault on the northern side of the Transverse Ranges, near the southern end on the Great Valley, up through the Santa Cruz Range and off into the Pacific Ocean about 16 km south of the Golden Gate.

West of the San Andreas fault is the Salinian block, which has been transported northward >300 km, relative to the Franciscan complex and Great Valley sequence on the northeast side of the fault. This block is composed of granitic and metamorphic rocks that represent a southern extension of the Sierra Nevada plutonic belt and contain no serpentinites.

There is more serpentine-bearing Franciscan complex west of the Salinian block where the western edge of the block is defined by the left–lateral Sur–Nacimiento fault and others. The land southwest of the Salinian block has been referred to as the Nacimiento block (Page et al. 1997). It is composed of rocks of the Franciscan complex and the Great Valley sequence. Within the Nacimiento block, peridotite–serpentinite bodies have been tectonically imbricated and dismembered.

15.2 Soils

Serpentine soils in the California Coast Ranges are in the mountains, where they are commonly in shallow to very deep colluvium on steep mountain slopes. Some serpentine soils are in melange, where they occur as shallow to very deep to blocks of serpentinite. These blocks that are floating in melange are barriers to plant roots just as effective as bedrock. Even the matrix of melange containing sufficient clay can set-up like concrete and be a nonlithic barrier to roots that pedologists call a "paralithic" contact to distinguish it from a "lithic" contact between soil and hard rock. Salts are leached from well-drained serpentine soils. and carbonates accumulate in few of them, although calcium-carbonate accumulations are common in nonserpentine soils in the drier parts of the California coast ranges. Magnesium-carbonate does appear to accumulate in some of the serpentine soils, but fragments of veins from dissintegrated bedrock may be the common source of magnesite in the soils. Clay is leached from surface soils and accumulates in subsoils on relatively stable landforms.

Ochric epipedons predominate, with mollic epipedons in the drier areas and some umbric epipedons on nonserpentine soils in the wetter climates. Poorly drained soils with histic epipedons are sparse. Cambic and argillic horizons are the common subsurface diagnostic horizons.

Clay in subsoils is commonly dominated by minerals inherited from soil parent materials. Kaolinite accumulates in older nonserpentine soils in wetter climates and smectites accumulate in soils in drier areas and in basins where silica and cations are washed in from soils in higher landscape positions. Cracking-clay soils are common in these basins and on serpentine in the drier areas. Smectites and iron oxides are more prevalent in serpentine soils than in soils with other kinds of parent materials.

Inceptisols and Alfisols are most extensive in the wetter areas and Mollisols are most extensive in drier areas. Ultisols are sparse in the Southern California Coast Ranges and no ultisols are in serpentine materials. Vertisols occur in basins where the soils are not well drained. Most of the soils with serpentine parent materials are in the same orders and suborders as those with nonserpentine parent materials. There are no serpentine soils on the coastal plains and terraces, where there are small areas of Spodosols in addition to the orders mentioned already.

Soil temperature regimes are isomesic to isothermic along the coast and thermic, mesic, and possibly frigid at increasing elevations inland. Any occurrences of frigid soils are not extensive. Soil moisture regimes are predominantly xeric, with relatively small areas with udic soil moisture regimens (SMRs) along the coast. Also, there are some small areas of poorly drained soils with aquic SMRs.

15.3 Vegetation

Grassland, wooldland, chaparral, and forest are all common on serpentine soils of the Southern California Coast Ranges domain. The vegetation and species show some strong effects of serpentine, with numerous examples of serpentine endemic plants and extensive stands of serpentine vegetation on the larger exposures of serpentine rocks and soils; for example, Coyote Ridge, New Almenden, New Idria, Los Burros Mountain, and Figerora Mountain. Several recent studies of vegetation in this domain (McCarten 1992, Keeler-Wolf et al. 2003, Evens and San 2004, CNPS-CDFG 2006) enable more detailed description than in some other serpentine domains.

15.4 Presidio, Locality 3-1

The Presidio locality is at the south end of the Golden Gate Bridge. Serpentinites of the Franciscan complex that were transported westward along the Coast Range fault were then displaced along strike-slip faults of the San Andreas fault zone in the San Francisco Bay area (Page 1992). Serpentinite melange is exposed in small areas on steep cliffs facing the ocean and on sloping to moderately steep hill slopes inland from the shoreline (Schlocker 1974). The south pier of the Golden Gate Bridge rests on serpentinite melange. Small exposures of serpentinite are scattered through the city of San Francisco, where they have been covered by buildings and streets. These serpentinites are in the Franciscan complex or at the base of the Great Valley sequence.

All exposures of serpentinite are below 100 m altitude. Mean annual precipitation ranges from 40 to 50 cm, possibly slightly lower at the south end of the Golden Gate Bridge. Snowfall is rare; it does not occur every year, nor even every decade.

The serpentine soils have mostly been covered by eolian sand or disturbed in urban operations. Those that are exposed along the coast are Mollisols. Soil temperature regimes are isomesic, and moisture regimes may be xeric, although udic SMRs have been considered for areas with persistent summer fog.

Serpentine plant communities are being protected in the Presidio, which is now in the Golden Gate National Park. On the steep slope of serpentine melange facing the Pacific Ocean, they are chaparral dominated by blue blossom (*Ceanothus thyrsiflorus*) and seaside woolly sunflower (*Eriophyllum staechadifolium*) and grassland communities. The only known native populations of the endangered Presidio manzanita (*Arctostaphylos hookeri* ssp. *ravenii*) and of the rare and endangered annual Presidio Clarkia (*Clarkia fransiscana*) occur on serpentine outcrops. Most of the serpentine vegetation away from the immediate coast has been disturbed by long-term recreational use and has been converted to non-native plantings of Monterey cypress (*Cupressus macrocarpa*) and Monterey pine (*Pinus radiata*).

15.5 Potrero Hill-Hunters Point, Locality 3-2

Potrero Hill-Hunters Point is a locality of undulating to rolling hills in the southeastern part of San Francisco. Tracts of serpentinite melange larger than those in the Presidio locality occur in the Potrero–Hunters Point locality (Bonilla 1971). Major quarrying of the melange for landfill and building sites has exposed large bodies of serpentinite rock.

All of the serpentinites in San Francisco (localities 3-1 and 3-2) are highly sheared and contain tectonic blocks of graywacke, chert, greenstone, rodingite, and amphibolite. Lizardite, chrysotile, brucite, and magnetite are the main minerals in the serpentinite, and relicts in massive blocks of serpentinite reveal that the harzburgite and dunite are the most important protolith peridotites. These serpentinites are in the Franciscan complex or at the base of the Great Valley sequence.

All exposures of serpentinite are below 100 m altitude. The mean annual precipitation is about 50 cm. Snowfall is rare; it does not occur every year, nor even every decade.

The serpentine soils have been disturbed in urban operations. They are man-made, or anthromorphic, soils that have isomesic or thermic soil temperature regimens (STRs) and xeric SMRs.

15.6 Burri Burri–Edgewood–Jasper Ridge, Locality 3-3

Locality 3-3 consists of a series of outcrops of serpentinite melange along the east side of the San Andreas fault on the peninsula south of San Francisco (Coleman 2004). Here thin sheets of fault-bounded serpentinite melange are thrust into and over the larger Franciscan complex melange. These thin sheets are exposed on low, rounded hilltop surfaces (Page and Tabor 1967, Brabb and Pampeyan 1983, Pampeyan 1993) in an area of undulating to rolling hills. The San Andreas fault truncates the melange sheets on their western boundary, but serpentinite is not incorporated into the fault zone. The sheets consist of a highly sheared matrix of lizardite, chrysotile, brucite, and magnetite enclosing tectonic blocks of greenstone, graywacke, gabbro, chert, and glaucophane schist. Blocks of massive serpentinized peridotite up to 1 m in nominal diameter contain relict olivine and pyroxene preserving a tectonic fabric typical of oceanic peridotites.

There is no clear evidence to relate serpentinite melange on the peninsula south of San Francisco to the Coast Range ophiolite. The serpentinites were probably introduced into the Franciscan complex during the late Jurassic. Elevations in this locality range up to about 200 m altitude. Mean annual precipitation ranges from 50 to 75 cm. Snowfall is rare; it does not occur every year.

The soils are mostly shallow Mollisols—very dark grayish brown to black soils both with and without clay accumulation in argillic horizons. They have thermic STRs and xeric SMRs. Lithic Haploxerolls of the Montara series with loamy subsoils and the Obispo series with clayey subsoils, both mapped by J. Kashiwagi for the Jasper Ridge Biological Preserve, are representative of the serpentine soils in this area. Associated Argixerolls are similar to these soils, but they contain more illuviated clay in their B horizons. E. Alexander described some soils on Jasper Ridge for a Western Society of Soil Science field trip, and R. Southard and his students at U.C. Davis provided laboratory data (table 15-1). Most of the serpentine soils are in grassland, but there is some leather oak–chamise chaparral on a serpentine Lithic Argixeroll on Jasper Ridge. Some chaparral shrubs, such as California coffeeberry, are more abundant on the serpentine soils than on others and some shrubs, such as chaparral pea, are absent from the chaparral on serpentine soils of Jasper Ridge.

The Jasper Ridge Biological Reserve is adjacent to the Stanford University and is managed by Stanford staff. They have conducted many studies on the organisms and vegetation of the serpentine on Jasper Ridge. These include long-term studies of the endemic Bay checkerspot butterfly (*Euphydryas editha bayensis*), whose larval host plant dwarf

Table 15-1 Some field and laboratory data for two soils (Montara and Obispo series) on Jasper Ridge, San Mateo County.

Horizon	Depth (cm)	Texture	Organic carbon (g/kg)	N (g/kg)	Exchangeable cations (mmol/kg)				CEC (pH 7)	Acidity (pH) (1:1)
					Ca	Mg	Na	K		
Montara										
A	0–7	GrL	23.1	1.97	29	198	0.6	2.2	263	6.5
Bt1	7–25	GrCL	14.0	1.22	17	233	0.6	1.0	300	6.8
Bt2	25–35	GrCL	A discontinuous soil horizon, bromthymol blue pH 7.2							
R	Highly fractured serpentinized peridotite									
Obispo										
A1	0–6	CL	19.3	1.56	19	230	0.6	2.9	268	6.6
A2	6–22	CL	8.1	0.76	16	273	0.6	1.2	311	6.9
C	22–30	GrSL	3.3	0.41	7	168	0.5	0.5	189	7.1
R	Green, highly fractured serpentinized peridotite									

CEC, cation-exchange capacity; CL, clay loam; GrCL, gravelly clay loam; GrL, gravelly loam; GrSL, gravelly sandy loam. Laboratory analyses of fine earth (soil fraction <2 mm) by students of R. Southard, University of California-Davis.

plantain (*Plantago erecta*) is well represented in the serpentine grassland at this locality, by Paul Ehrlich and many others (Ehrlich and Hanski 2004); long-term studies of phenology and plant composition in serpentine grasslands by Harold Mooney and others (e.g., Hobbs and Mooney 1995); and studies of the effects of carbon dioxide (Dukes 2001, 2002).

15.7 Oakland-Hayward, Locality 3-4

Oakland-Hayward is a steep, hilly locality along the east side of the north-northwest trending Hayward fault in Alameda County. Narrow, elongate outcrops of serpentinite parallel to the Hayward fault zone are associated with both Franciscan complex rocks and the Great Valley sequence (Graymer et al. 1995). Serpentine occurrences are mainly in highly sheared serpentinite melange containing a variety of tectonic blocks. The presence of rhyolite indicates that these serpentinites may be dismembered portions of the Coast Ranges ophiolite. On the other hand, the presence of blue-schist in serpentinite melange of the Oakland-Hayward domain indicates that their source is the Franciscan complex. It is possible that serpentinites from the Coast Range ophiolite and the Franciscan complex have been tectonically mixed and imbricated within the Hayward fault zone.

Elevations range from about 100 m along the Hayward fault up to 500 m in the Oakland Hills. The mean annual precipitation is about 50 cm. Snowfall is rare at the lower elevations, but it is an occasional occurrence on the ridge summits.

The soils are Mollisols with grassland and shrub cover and introduced trees, such as Monterey pine, that have been planted in the area. Only Lithic Haploxerolls of the Montara series have been mapped in the area and inclusions of other serpentine soils were not mentioned in the soil survey report of the NRCS (Welch 1981). The soils have thermic STRs and xeric SMRs.

The vegetation of this locality has been heavily altered by urbanization. Numerous homes have been constructed even on the steeper slopes. A portion has been conserved in the Redwood Regional Park of the East Bay Regional Parks District. However, much of the serpentine in this park has been planted to Monterey cypress. Small remnant stands of purple needlegrass (*Nassella pulchra*) grassland exist along road banks and a few of these areas have small populations of the rare and endangered serpentine endemic Presidio Clarkia.

15.8 Coyote Ridge, Locality 3-5

At the south end of the Hayward fault zone, where it merges with the Evergreen thrust fault, it contains a sheet of serpentinite that marks the eastern boundary of the Santa Clara Valley (Page et al. 1999). The serpentinite is exposed on the steep sides and gently sloping summits of Coyote Ridge, which has a northwest to southeast orientation.

Elevations range from about 100 m on serpentinite near the base of Coyote Ridge to 400 m on the summit. The mean annual precipitation is about 50 cm. Snowfall is rare; it does not occur every year.

The soils in the grassland of Coyote Ridge are Mollisols, mostly Lithic Haploxerolls of the Montara series and associated Argixerolls. They have thermic STRs and xeric SMRs. Some Argixerolls, particularly Lithic Argixerolls of the Henneke series, occur in chamise chaparral on steep side slopes. Smectites are the dominant clay minerals in Mollisols sampled on Coyote Ridge (R. Southard, unpublished data, June 23, 1999). The serpentine grassland on Coyote Ridge is a core habitat for the Bay checkerspot butterfly (Harrison et al. 1988).

The vegetation of Coyote Ridge has been thoroughly studied by McCarten (1992) and Evens and San (2004). It includes extensive grassland and herbaceous communities dominated by both native and nonnative species. Despite the resistance some serpentine areas have shown to invasion by exotic plant species elsewhere, the Coyote Ridge locality shows significant incursion by non-native species. Apparently native herbaceous stands have been recently type-converted to non-native grasslands, with Italian ryegrass (*Lolium multiflorum*) as one of the dominant species. Research results suggest that air pollution and nitrogen deposition from the adjacent heavily urbanized Santa Clara Valley have contributed significantly to the success of non-native species (Weiss 1999). Nevertheless, extensive areas of largely native herbaceous vegetation still exist at Coyote Ridge. Some of the characteristic native herbaceous species are the serpentine indifferent dwarf plantain (*Plantago erecta*), small fescue (*Vulpia microstachys*), California goldfields (*Lasthenia californica*), erect dwarf-cudweed (*Hesperevax sparsiflora*), and hayfield tarweed (*Hemizonia congesta* var. *luzulifolia*). There are also wetland and seep communities dominated by native herbs; one of the most distinctive is the endemic Mt. Hamilton thistle (*Cirsium fontinale* var. *campylon*), a rare and endangered species. Scrub and chaparral also occur on serpentine at this locality. Leather oak (*Quercus durata)* dominates and associates with a mix of serpentine chaparral species. Bigberry manzanita (*Arctostaphylos glauca*) also mixes with a variety of shrubs, including coastal sage scrub species such as California sagebrush (*Artemisia californica*) and black sage (*Salvia mellifera*).

15.9 New Almaden–Morgan Hill, Locality 3-6

The New Almaden–Morgan Hill locality is on west-flanking hills of the Santa Clara Valley, south of San Jose. It is dominated by Franciscan complex with numerous small, elongate bodies of serpentine (Bailey and Everhart 1964, Coleman 1991). All of these serpentinites are pervasively sheared, developing an irregular foliation. Elongated blocks of massive serpentinite in the melange contain relict olivine and pyroxene that define a relict tectonic fabric. Lizardite, chrysotile, brucite, and magnetite are the main minerals, and in certain areas there is widespread alteration of the serpentinized peridotites to silica–carbonate rock that contains the rich cinnebar deposits of the New Almaden mine. Intense Tertiary deformation of these serpentine bodies by thrust faulting, slip-strike faulting, and folding between the Hayward–Calaveras fault system on the east and the San Andreas fault system on the west has imbricated and mixed the serpentinites such that it is now difficult to distinguish which sheets were derived from the Franciscan complex and which from the Coast Range ophiolite (Page et al. 1999).

This is an area of hills and low mountains, with elevations from about 100 m up to a maximum of 1155 m altitude on Loma Prieta. The mean annual precipitation ranges from 40 to 80 cm. Snowfall is rare at the lower elevations, but it is an occasional occurrence at the higher elevations, where the snow may remain on the ground for a few days.

The serpentine soils are mostly Mollisols. They have thermic STRs and xeric SMRs, and possibly there are some Alfisols or Mollisols with mesic STRs at the higher elevations. Release of chrysotile asbestos fibers from the local rocks and soils during urban construction have led to strict guidelines for dust control during subdivision construction. The vegetation and flora of this locality are similar to those of Coyote Ridge, locality 3-5.

15.10 Mt. Diablo, Locality 3-7

Mt. Diablo is at the north end of the Diablo Range. A strip of serpentinite about 8 km long and 0.5 km wide is exposed on the north flank of Mt. Diablo in Contra Costa County. It separates a Franciscan assemblage of greenstone and graywacke on the south from a thick section of sheeted dikes on the north (Pampeyan 1963). The serpentinite is highly sheared, but it contains small blocks of massive peridotite with relict olivine and orthopyroxene.

Although the summit of Mt. Diablo is 1173 m high, altitudes on the serpentinite range from 210 to 760 m. Slopes are mostly steep, but locally gentle, particularly on Long Ridge. The mean annual precipitation ranges from 40 to 80 cm. Snowfall is rare on the serpentinite, but it is an occasional occurrence on the summit of Mt. Diablo.

The serpentine soils are Mollisols, predominantly Lithic Argixerolls of the Henneke series. Its distribution is shown in figure 6 in Ertter and Bowerman (2002). It occurs in both grassland and chaparral on Mt. Diablo. There are minor amounts of serpentinized peridotite and pyroxenite rock outcrop in the area. Below is a brief description of a serpentine Lithic Argixeroll in mixed chaparral on Mt. Diablo:

Clayey-skeletal, magnesic, thermic, Lithic Argixeroll

Latitude 37.900°N, longitude 121.919°W, 450 m altitude, linear 56% NW slope

Sparse gray pine, toyon (40%), whiteleaf manzanita (*A. manzanita*, 30%), leather oak (*Q. durata*, 10%), Jim brush (*C. oliganthus* var. *sorediatus*, 10%), buckbrush (*C. cuneatus*, 5%), and sparse California fescue (*F. californica*), onion grass (*M. torreyana*), rough bedstraw (*Galium* sp.), and goldenback fern (*Pentagramma triangularis*)

Oi (2–0 cm) loose shrub leaves and moss between shrubs where leaves are absent

A (0–8 cm) dark brown (7.5YR 3/2 moist) very gravelly loam; moderate, very fine, subangular blocky; friable, common very fine and few fine roots; neutral; clear, smooth boundary

AB (8–22 cm) dark brown (7.5YR 3/2 moist) very gravelly clay loam; weak, fine subangular blocky; firm; few very fine, fine, and medium roots; neutral; clear, wavy boundary

Bt (22–45 cm) dark brown (7.5YR 3/4 moist) very gravelly clay; moderate, fine subangular blocky; very firm; very sticky and very plastic; many thin and common moderately thick coatings on ped faces; few fine and medium roots; neutral; abrupt, irregular boundary

Cr (45–50 cm) bedrock weathered soft

R (50 cm) hard, moderately fractured, serpentinized peridotite

The serpentine plant communities on Mt. Diablo are shrub and grassland, with sparse gray pine (*Pinus sabiniana*) in many areas. Plants recorded at the pedon (soil) site above are common ones in serpentine chaparral on Mt. Diablo.

15.11 Cedar Mountain, Locality 3-8

A small patch of peridotite and serpentinite is exposed on Cedar Mountain, south of Livermore, in Alameda County, and in an east–west strip about 8 km long just south of Cedar Mountain, between Arroyo Mocho and Arroyo del Valle. The peridotite of this locality is offset northward from that of the Del Puerto locality 5 or 10 km along the Greenville fault.

This is a mountainous area with altitudes from 450 m near Arroyo del Valle to 1120 m on Cedar Mountain. The mean annual precipitation is 40–50 cm. Snowfall is rare; it does not occur every year.

The serpentine soils are predominantly Mollisols (Welch 1981). Lithic Argixerolls of the Henneke series are common, and they generally are covered with shrubs. The soils have thermic STRs and xeric SMRs.

Vegetation on Cedar Mountain is characterized by a stand of low Sargent cypress (*Cupressus sargentii*) and mixed to pure chaparral composed of chamise, bigberry manzanita, leather oak, and buckbrush. Much of the land is privately owned and is not readily accessible to botanists.

15.12 Del Puerto, Locality 3-9

Del Puerto serpentine is a body of peridotite and dunite that extends from the Greenville fault in Santa Clara County, where it is truncated by the fault, across Red Mountain into Stanislaus County and several kilometers down along Del Puerto Canyon.

A nearly complete, intact sequence of oceanic crust with large areas of only partly serpentinized peridotite and dunite is preserved in this area. The base of the Great Valley sedimentary sequence is marked by an andesitic pyroclastic tuff deposited on basalt and rhyolitic (keratophyre) flows of Del Puerto ophiolite that grades upward into Upper Jurassic mudstones. Highly sheared serpentinite is present along the basal contact where the Del Puerto ophiolite has been thrust westward over the Franciscan complex consisting of metagraywacke and metagreenstone (Evarts 1977).

This is a mountainous area with altitudes from about 760 m near the mouth of Deer Park Canyon to 1115 m on Red Mountain. The mean annual precipitation is 40–50 cm. Snowfall is rare; it does not occur every year.

The serpentine soils are predominantly Mollisols. Lithic Argixerolls of the Henneke have been mapped on Red Mountain in Santa Clara County (Lindsay 1974) and are common in western Stanislaus County (Ferrari and McElhiney 2002). Magnesite concentrations are common in the serpentine soils. The soils have thermic STRs and xeric SMRs.

The vegetative cover is generally gray pine and shrubs. Leather oak and manzanita (*Arctostaphylos glauca*) are two of the more common shrubs. Oniongrass and blue pinegrass (*Poa secunda*) are common on north-facing slopes.

15.13 Quinto Creek and Llanada, Localities 3-10 and 3-11

South of Del Puerto Canyon, from Stanislaus County through Merced County to San Benito County, the fundamental boundary between the Great Valley sedimentary sequence and the Franciscan complex is marked by discontinuous, narrow belts of highly sheared serpentinite occupying a vertical fault zone (Coast Range thrust?) between these two units. The serpentinite belts are exposed in the Quinto Creek locality in western Stanislaus County (Hopson et al. 1981) and in the Llanada locality in San Benito County (Giaramita et al. 1998) where fragments of ophiolite rocks are immersed in a serpentinite melange. The serpentinite of the Llanada locality has been altered to silica–carbonate rock with very little of the original serpentinized peridotite preserved in the area. Llanada is at the south end of the Diablo Range where the range disappears under the Panoche Valley.

15.14 New Idria, Locality 3-12

South from Santa Clara County through San Benito County, only small slivers of serpentinite are present in the San Andreas fault zone. Late Tertiary folding as a result of crustal shortening related to movement on the San Andreas fault has formed a series of northwest-trending anticlines and synclines (Page et al. 1997). The New Idria antiform exposes a core of highly sheared serpentinite approximately 15 km long and 5 km wide, parallel to the San Andreas fault. This serpentinite body is in fault contact with rocks of the Franciscan complex, Great Valley sediments, and Tertiary marine and nonmarine sediments. The displacement of these rocks is the result of diapiric emplacement and continuing rise of the serpentinite core since the Miocene. Numerous tectonic inclusions of Franciscan complex rocks are found within the New Idria dome, and some have been recrystallized under high pressure to glaucophane schist, indicating that these rocks were exhumed from depths >15–20 km in the Franciscan subduction zone.

The New Idria serpentinite contains highly sheared and crushed material that consists of soft crumbly aggregates and sheets of asbestos (Coleman 1996). Chrysotile is the predominant mineral, with some lizardite, brucite, and magnetite. Within the wet weathering zone, many secondary minerals, such as artinite, coalingite, hydromagnesite, and pyroaurite, are present. These minerals do not persist in well-drained soils. Cinnabar (mercury) deposits were mined from 1856 to 1970, and asbestos is still mined in the New Idria locality (Eckel and Myers 1946). Government regulations concerning the introduction of toxic mercury into local streams terminated the cinnabar mining (Coleman 1996).

The New Idria locality, which straddles the boundary between San Benito and Fresno counties, is a mountainous area with sloping to moderately steep summits and steep side slopes. Elevations range from about 800 m adjacent to the San Benito River up to a maximum of 1579 m on San Benito Mountain. Mean annual precipitation ranges from 40 to 60 cm. Snowfall is rare at the lower elevations, but it is an occasional occurrence at the higher elevations, where snow may remain on the ground for a few days.

Lithic Argixerolls of the Henneke and Hentine series, Typic Argixerolls of the Atravesada Series, and other Mollisols, all with a cover of chaparral, prevail on steep slopes, except where they are severely eroded (Isgrig 1969, Arroues 2005). Gray pine trees are common, but sparse. A striking feature of the New Idria locality is large areas of severely

Figure 15-3 A barren serpentine landscape in western Fresno County, locality 3-12, New Idria. The soil is a very shallow Entisol in strongly sheared serpentinite.

eroded, barren land with very shallow Entisols lacking vegetative cover (fig. 15-3). These soils consist of a thin layer of loose serpentine detritus, commonly <10 cm thick, over bedrock that has been weathered soft enough to dig into with a spade. Generally, they are not quite thick enough to be called soils in soil taxonomy (Soil Survey Staff 1999) and have been mapped as rock outcrop in the soil survey of San Benito County and as asbestos in the soil survey of western Fresno County. Moderately deep-to-deep Mollisols are common on the summit slopes, along with shallow Mollisols. A moderately deep clayey–skeletal, magnesic, mesic Ultic Argixeroll with an open stand of Coulter pine (*Pinus coulteri*) and an understory of leather oak, buckbrush, and Mexican manzanita (*A. pungens*) was described near the summit of San Benito Mountain. The soils in the New Idria locality have thermic and mesic STRs and xeric SMRs.

The serpentine vegetation of the area has been recently studied by Evens et al. (2004). It includes pure and mixed serpentine chaparral dominated by leather oak, bigberry manzanita, riparian scrub dominated by Brewer's willow (*Salix breweri*) with associated serpentine endemic herbs such as *Aquilegia eximia* and *Solidago guiradonis*, as well as the unique barrens with occasional scattered endemic herbs such as *Streptanthus breweri*, *Camissonia benetensis*, and possibly *Layia discoidea*. Scattered moist meadows and seeps contain examples of the two-toothed sedge vegetation alliance (*Carex seratodens*), and the higher peaks have pure or mixed stands of Coulter pine, Jeffrey pine (*Pinus jeffreyi*), and incense-cedar (*Calocedrus decurrens*). Much of the serpentine body has been open to off-road vehicle access through the Bureau of Land Management, but there is much concern that several of these unique botanical resources of the area are compromised by this activity.

15.15 Table Mountain, Locality 3-13

Southward from the New Idria body, the Great Valley sequence and Franciscan complex are folded into a series of northwest-trending antiforms and synforms that are truncated along the San Andreas fault. Just east of Parkfield, Table Mountain is an anti-formal ridge capped by a sheet of highly sheared serpentinite (Dickinson 1966). At this locality, the Great Valley sedimentary sequence was thrust over melange of the Franciscan complex during the early Tertiary, and shortening of the crust related to the San Andreas fault produced the folding that lifted light serpentinite upward by diapiric movement. There is no depositional contact of the basal Great Valley sediments upon the serpentinite of the Coast Ranges ophiolite sequence at this locality. The serpentinite has a sheared and foliated matrix that encloses massive blocks of serpentinized peridotite. Lizardite, chrysotile, brucite, and magnetite are the main minerals within the serpentinite. The serpentinite sheet was formed by extrusion of the serpentinite along a fissure (fault) within melange of the Franciscan complex as a diapir during late Tertiary folding. The incompetent serpentinite formed a sheet along the crest of Table Mountain that is now sliding and washing downslope by mass wasting and stream erosion (Dickinson 1966). Massive Pleistocene flows of tectonized serpentinite have moved downslope to form ramps and fill stream valleys, coalescing into grand sheets of serpentinite. With continuing uplift, recent streams have dissected the sheets of serpentinite and triggered modern landslides covering the Pleistocene sheets of debris.

Table Mountain and its side slopes are mostly moderately steep to steep. The soils are too unstable to maintain very steep slopes. Elevations of the serpentinite along Table mountain range from 900 to 1200 m. Elevations of serpentine soils on the sides of Table Mountain range down to about 450 m. The mean annual precipitation is 40–50 cm. Snowfall is rare; it does not occur every year.

Serpentine soils in the area are Mollisols and Vertisols. The soils that have been mapped across Table Mountain in Monterey (Cook 1978), Kings (Arroues and Anderson 1986), and western Fresno (Arroues 2005) counties, are Montara (Lithic Haploxerolls) with grassland, Henneke (Lithic Argixerolls) with shrubs, and Climara (Aridic Haploxererts) with grass. Soils in serpentinite melange have paralithic contacts to incompetent melange matrix, or lithic contacts to bedrock, or to blocks of hard rock within melange (fig. 15-4).

A range site was inventoried on the Climara soil (Arroues 2005). Its cover was annual grassland. Wild oat (*Avena fatua*) and soft chess (*Bromus hordaceus* ssp. *hordaceus*) were the dominant grass, with common annual fescue (*Vulpia myuros*), red brome (*B. rubens*), sparse pine grass (*Poa secunda*), and purple needlegrass. The estimated annual forage production is 2800 ± 900 lbs/acre (3185 kg/ha).

15.16 Burro Mountain, Locality 3-14

Burro Mountain is a small mountainous locality, about 2 km in nominal diameter, of partially serpentinized peridotite in southern Monterey County. The peridotite is tectonically

Figure 15-4 A serpentine landscape on the Parkfield Grade in southeastern Monterey County, locality 3-13, Table Mountain. Rocks visible in the foreground are chert and silica–carbonate, which are in a melange, along with serpentinite and other kinds of rocks.

embedded in melange of the Franciscan complex that consists of metagraywacke with chaotic blocks of metachert, glaucophane schist, metagreenstone, and amphibolite. It is truncated on the northwest by the Sur–Nacimiento fault. Plastic deformation in the foliated, partially serpentinized peridotite (harzburgite and dunite) reveals a deep mantle history. Olivine and orthopyroxene are the main primary minerals, accompanied by clinopyroxene and chrome spinel. Serpentinization is concentrated along the faulted boundaries of the peridotite body. Lizardite, chrysotile, brucite, and magnetite are the main minerals formed during serpentinization (Coleman and Keith 1971). There is no associated gabbro or basalt within the peridotite, nor is there any sediment of the Great Valley sequence. The Burro Mountain body was tectonically emplaced into the Franciscan complex during subduction of the Pacific Plate (Burch 1968). Numerous smaller and elongate bodies of serpentinite are immersed in the Franciscan complex south and west of the Nacimiento fault.

Slopes are steep to very steep on Burro Mountain. Elevations range from about 450 m along Los Burros Creek up to 862 m on the top of Burro Mountain. The mean annual precipitation is 70–80 cm. Snowfall is rare; it does not occur every year.

Rock outcrop and shallow Mollisols were mapped on Burro Mountain (Cook 1978). The main Mollisol is a Lithic Argixeroll of the Henneke series and is covered by chaparral. The soils have thermic STRs and xeric SMRs.

Vegetation and flora of Burro Mountain has been studied (B. Paynter and D. Taylor, personal communication 2004) and botanical collections have been made. Several serpentine endemics occur here, ranging from widespread endemics such as leather oak and Sargent cypress to more localized specialties such as Obispo manzanita (*Arctostaphylos obispoensis*) and possibly *Chlorogalum purpureum*.

15.17 San Simeon, Locality 3-15

From southern Monterey County, west of Burro Mountain, elongated slices of fault-bound serpentinite extend south along the coast of the Pacific Ocean into San Luis Obispo County (Hopson et al. 1981). The rocks include dismembered blocks of gabbro and sheeted diabase dikes that have a cover of tuffaceous radiolarian chert interbedded with Great Valley upper Jurassic shale. The serpentinites are highly sheared and consist mainly of lizardite, chrysotile, brucite, and magnetite.

Mollisols of the Obispo series were mapped at 600–900 m elevation in the Burnett Mountain area north of San Simeon (Ernstrom 1984). Grassland is the predominant cover on these soils. All serpentine soils in this locality have thermic STRs and xeric SMRs.

A substantial Sargent cypress stand is known from the serpentine on the upper slopes of Zaca Peak within this locality.

15.18 Marmolejo Creek, Locality 3-16

East of the Hearst Castle, northeast of San Simeon, fault-bounded and highly sheared serpentinite is present, but poorly exposed, along Marmolejo Creek (Hopson et al. 1981). The serpentinite occupies the footwall of a west-directed thrust fault (Coast Range fault?). Disrupted sheeted dikes and dioritic sills are present, but there is no clear depositional contact of Great Valley sedimentary rocks on these dismembered ophiolite rocks.

Mollisols of the Henneke series were mapped at 400–624 m elevation on Red Mountain east of San Simeon (Ernstrom 1984). Chaparral is the predominant cover on these soils. All serpentine soils in this locality have thermic STRs and xeric SMRs.

15.19 Cuesta Ridge, Locality 3-17

Serpentinites in a synclinal zone just northeast of San Luis Obispo are exposed for about 25 km along the northwest trend of the zone. Although the Coast Range ophiolite is dismembered, the Coast Range thrust has placed the serpentinite above melange of the Franciscan complex in this locality. The serpentinites are strongly sheared, and only remnants of the original peridotite remain. Minerals in the serpentinite are lizardite, chrysotile, and magnetite. Original plastic deformation can be ascertained by banding of the altered orthopyroxene. The serpentinites are cut by diabase dikes that have rodingitized contacts with the serpentinite. The sheet of serpentinite on Cuesta Ridge has been modified by extensive landsliding along the southwestern edge of the ridge.

Table 15-2 Some field and laboratory data for surface and subsoil (argillic horizon) samples from two soils (Henneke and Cuesta series).

Depth	Texture	Exchangeable cations (mmol/kg)				Acidity (pH) (1:1)	Common plants
		Ca	Mg	Na	K		
Henneke Series, 285 m elevation, mean (n = 3)							Grass, *Yucca whipplei*
Surface	GrCL	75	109	4	9	6.6	
Subsoil	vCbC	70	151	3	4	6.7	
Henneke Series, 752 m elevation, mean (n = 3)							Buckbrush, manzanita, chamise
Surface	GrCL	118	74	2	5	6.5	
Subsoil	vCbC	45	140	3	8	6.6	
Cuesta Series, 219 m elevation, mean (n = 3)							Grass, including *Nassella pulchra*
Surface	CL	70	87	4	11	6.6	
Subsoil	vCbC	52	111	3	3	6.7	
Cuesta Series, 687 m elevation, mean (n = 3)							Buckbrush, Obispo manzanita
Surface	L	132	60	3	7	6.5	
Subsoil	vCbCL	47	136	3	12	6.7	

Results here are means for three pedons in each soil sampled on Cuesta Ridge and for three each in Brizziolari Canyon on the lower southwest slope from Cuesta Ridge, near San Luis Obispo. CL, clay loam; GrCL, gravelly clay loam; L, loam; vCbC, very cobbly clay; vCbCL, very cobbly clay loam. Soils were sampled and analyzed by M.A. Menetrey under the direction of T.J. Rice, California Polytechnical State University, San Luis Obispo.

Slopes are steep, and elevations are mostly in the range from 400 to 800 m. The mean annual precipitation is 50–60 cm. Snowfall is rare; it does not occur every year.

Mollisols of the shallow Henneke and Obispo series and moderately deep Cuesta series were mapped in this locality (O'Hare and Hallock 1980). All of these soils have argillic horizons, making them Argixerolls. Chaparral is the predominant cover of these soils on Cuesta Ridge, and grassland is predominant on the lower southwest slope from Cuesta Ridge toward San Luis Obispo (table 15-2). All serpentine soils in this locality have thermic STRs.

The Cuesta Ridge Botanical Area is in this locality. It has been established to preserve an example of the Sargent cypress woodland and associated rare serpentine plants such as Obispo manzanita (*Arctostaphylos obispoensis*). This area burned in the 1995 Cuesta fire, and regeneration from the fire-stimulated closed-cone cypress and the population of knobcone pine (*Pinus attenuata*) has been very good (D. Hillyard, personal communication, 2004). The grove at Cuesta Ridge in the southern Santa Lucias is the southernmost large occurrence and one of several isolated groves ranging from northern Mendocino County south to Zaca Peak in Santa Barbara County.

15.20 Point Sal, Locality 3-18

At Point Sal near Santa Maria, in the northwestern corner of Santa Barbara County, an unusually complete ophiolite section is exposed in the sea cliffs (Hopson et al. 1981). Its nearly complete exposure has lead to extensive studies of this locality. Only fragmentary blocks of serpentinized peridotite are exposed, and serpentinite exposures are minimal. The significant aspect of this exposure is that careful stratigraphic studies indicate that the ophiolite is exotic to North America, having formed as oceanic crust well south of its present position.

15.21 Figueroa and Stanley Mountains, Locality 3-19

An exposure of serpentinite is found in a part of the San Raphael Range between Stanley and Figueroa Mountains, about 25 km northeast of Santa Maria (Hopson et al. 1981). It crosses the boundary between San Luis Obispo and Santa Barbara counties. In this locality, two narrow lenses of serpentinite are contained within melange of the Franciscan complex. The melange consists of metagraywacke, metaconglomerate, metachert, greenstone, glaucophane schist, and amphibolite. A volcanic section grades upward into radiolarian tuffs of Oxfordian to Tithonian age, suggesting a connection with the Great Valley sequence. There is no evidence that the serpentinites are part of this volcanic sequence.

Olistostrome Serpentinites

The characteristic weakness of sheared and fragmental tectonized serpentinites has given rise to significant sedimentary deposits as a result of mass flow and stream erosion. These accumulations can usually be traced to a serpentinite source that has been tectonized or brecciated before its dispersal. Depositional modes are manifold and range from stream transport of clastic serpentinite debris to gravity transport of large volumes of serpentinite or olistostrome (Phipps 1984, Raymond 1984). Many of the California serpentinite localities within the Coast Ranges are found within active tectonic regimes. Here the weak and sheared serpentinite is situated along antiformal axes, where folding has forced the light and weak material to migrate upward in a manner similar to squeezing toothpaste from a container or pushing a grape from its skin. When the associated ophiolite mafic components and surrounding country rock become mixed into the matrix of sheared serpentinite, it is referred to as melange. When mass movement such as landslides or sheetflow modifies these melanges, it becomes difficult to distinguish between in situ tectonic serpentinite melange and the newly deposited mass wasting of a serpentinite melange.

Olistostromes are dominant in locality 3-20 and common in localities 3-13, 4-17, and 4-18.

The Figueroa–Stanley Mountains locality is well known (Smith 1998) as a botanical hot spot due to the reliable and often spectacular spring wildflower displays and the endemic serpentine species. Vegetation includes grassland with a high native component: chaparral, including the southernmost stands of leather oak; and woodland, including the southernmost stands of Sargent cypress. Santa Barbara jewelflower (*Caulanthus amplexicaulis* var. *amplexicaulis*) is a notable serpentine endemic to this region. This species appears to increase in abundance in the years immediately after fire (Smith 1998). The proximity to the Sedgwick Ranch Reserve of the University of California has spurred numerous studies of the vegetation and flora in this locality (see Seabloom et al. 2003, Gram et al. 2004).

15.22 Big Blue, Locality 3-20

The Big Blue locality is on the Big Blue formation in low hills on the southwest edge of the San Joaquin Valley. It is 230–550 m in elevation, and the mean annual precipitation is in the 20–25 cm range. The frost-free period is 240–270 days.

The uplift and mass wasting of the New Idria serpentinite (locality 3-12) in Fresno and San Benito counties, initiated during the Miocene (15 Ma), contributed serpentinite debris to the Big Blue sedimentary sequence, which is called an olistostrome (box 15-1). These deposits consist of debris flows that grade into serpentinite–clast conglomerate and breccia that in turn grade into serpentine–flake mudstones and laminated serpentinite grain sands. It is estimated that the initial erosion of the protruding New Idria serpentinite continued through middle to late Miocene, depositing >5 cubic miles of serpentinite debris. A lesser volume of serpentine debris continued to be shed from the New Idria dome as landslides around its flanks.

The dominant soil on the weakly consolidated Big Blue formation is a moderately deep Mollic Haploxeralf (Arroues 2005). Its cover is annual grassland, with common shrubs. The shrubs are mostly allscale (*Atriplex polycarpa*) and sparse California buckwheat (*Eriogonum fasciculatum*). Annual fescue is the dominant grass, with common soft chess, mouse barley (*Hordeum murinum*), and sparse red brome. The estimated annual forage production is 900 ± 300 lbs/acre (1020 kg/ha).

16 Northern California Coast Ranges, Domain 4

The Northern California Coast Ranges domain is in a mountainous region in which most of the mountain ranges are aligned north–south, or more precisely north, northwest–south southeast, curving around the Klamath Mountains into Oregon where the domain branches to north–south and northeast–southwest trends on the northwest side of the Klamath Mountains. It extends about 600 km from the Golden Gate at the entrance to San Francisco Bay north to about the Coquille River in Coos County and nearly to the North Umqua River in Douglas County, Oregon. The domain corresponds to a physiographic region that is bounded by the Pacific Ocean on the west, the Coast Range of Oregon and Washington (Orr and Orr 1996) on the north, the Klamath Mountains on the northeast, the Great Valley of California on the southeast, and on the south by the drainage outlet of the Sacramento and San Joaquin rivers through the Carquinas Straight and San Pablo Bay (fig. 16-1). Serpentine is scattered in relatively small ultramafic bodies throughout the Northern California Coast Ranges and is concentrated along some of the major faults.

For 200 or 300 km south from the Klamath Mountains, the Northern California Coast Ranges region is a rectangular strip 90–110 km wide between the Ocean and the Great Valley of California. The Klamath Mountains crowd the region to a narrow strip only 10 or 12 km wide in Del Norte County. Most of the mountain ranges have approximately concordant summits that are tilted up toward the east–northeast. Therefore, the highest altitudes are on the east, just south of the Klamath Mountains. North Yolla Bolly at 2397 m (7865 feet) and South Yolla Bolly at 2466 m (8092 feet) have the highest summits. Both of these and some neighboring mountains have cirques and moraines indicative of glaciation on their north slopes. There is no evidence of glaciation in any areas with serpentine rocks.

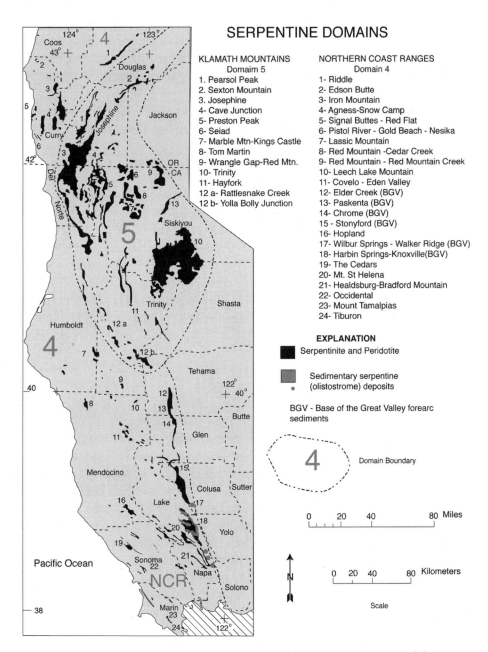

Figure 16-1 A map showing domain 4, Northern California Coast Ranges, and domain 5, Klamath Mountains.

Only the Rogue and Klamath rivers cut from east to west all of the way across the Northern California Coast Ranges, except for a few smaller streams such as the Chetco and Smith rivers that have headwaters in the Klamath Mountains. South of the Klamath Mountains, no streams, other than the Sacramento–San Joaquin system, cross the California Coast Ranges. The larger streams flow either north or south between the elongated mountain ranges until they reach gaps through which they can flow either toward the sea or toward the Sacramento River Valley.

Climates range from cool and foggy along the coast to warm inland, with hot and dry summers inland from the fog belt. Fog moves long distances up the Eel (>100 km), Van Duzen, and other rivers that open toward the northwest, but much shorter distances up the Mattole, Russian, and other river valleys that are not oriented toward the northwest. Mean annual precipitation ranges from about 100 cm adjacent to the Sacramento Valley and along the drier parts of the coast to 250–300 cm at some of the wetter locations, mostly at higher elevations. Summer drought is hardly evident along the coast (represented by Gold Beach in fig. 16-2) but pronounced inland (represented by Ukiah in fig. 16-2). Winter snowfall is common above 600 m, and snow remains on the ground through the winter above about 1200 m in the north and a little higher toward the south. Because the southern end of the northern California Coast Ranges region is lower, the ridge from Goat Mountain (1866 m altitude) southeastward along the Lake-Colusa County line is the southern limit of sustained winter snow cover. The frost-free period ranges from about 120 days in the Yolla Bolly Mountains to >300 days along the coast.

16.1 Geology

Graywacke is the dominant kind of rock in the Northern California Coast Ranges. It is fine-grained sandstone consisting of poorly sorted feldspar and quartz grains and basalt, chert, and shale rock fragments in a matrix of mud. The graywacke sandstones, mudstones, and chert formed in deep trenches along the edge of the North American Plate, above a subduction zone. These sedimentary and associated rocks comprise the Franciscan complex. Associated with the dominant graywackes are various types of oceanic crust formed at oceanic spreading centers. New oceanic crust forms crudely layered sequences at these spreading centers. Submarine eruptions of basalt form pillow lavas at the top. These pillow basalts are underlain by sheeted dikes of diabase through which magma was fed to the ocean floor from chambers of gabbro, which represents the magmatic precursor of the diabase and basalt. This sequence of gabbro, sheeted dikes, and basalt is called ocean crust. Underlying the ocean crust is peridotite, the precursor of serpentinite, which formed within the mantle and is the basement upon which ocean crust forms at oceanic spreading centers.

New oceanic crust is commonly subducted under the continental margins along a trench where graywacke sediments are being deposited. Changes in plate motion may lead to the break up of oceanic crust, and fragments of crust may underplate the continental

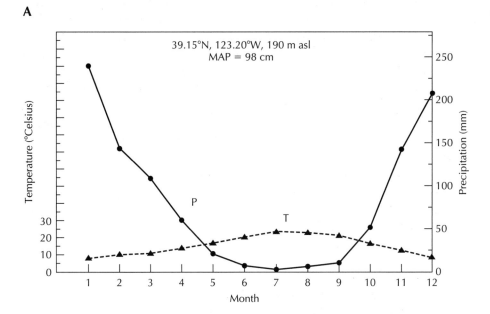

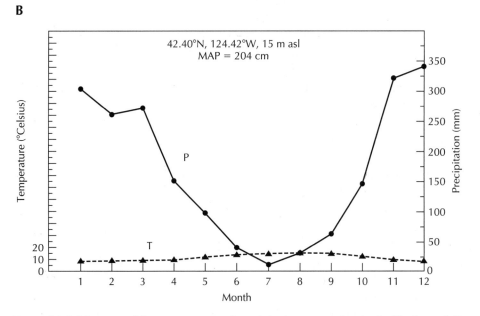

Figure 16-2 Mean monthly temperature and precipitation at two sites in the Northern California Coast Ranges domain. (A) Ukiah, California, inland from the coast. (B) Gold Beach, Oregon, on the Pacific coast.

margin or be obducted into the trench sediments. Fragments of oceanic crust may be altered by low-grade metamorphism deep within the trench. Gabbro, diabase, and basalt in the fragments are altered to greenstone containing metamorphic chlorite, epidote, and actinolite that impart greenish colors to the rocks. Serpentine is formed from olivine (and pryoxenes) in peridotite when water is added at temperatures < 500°C. Red and green thin-bedded cherts are ubiquitous, representing slow accumulations of pelagic sediments. Limestone is uncommon in the Northern California Coast Ranges, as are silicic plutonic and volcanic rocks.

No basement of oceanic crust has been exposed beneath the Franciscan complex. Some of the ultramafic rocks in the Northern California Coast Ranges have been resurrected from subduction trenches along the continental margin and others are from oceanic crust beneath sedimentary sequences in the Klamath Mountains and Great Valley that have been thrust over the Franciscan complex. The most prominent thrusting has been along the Coast Range fault zone extending from the Klamath Mountains >200 km southward into Napa County. This thrust fault zone is considered the fundamental boundary between the Franciscan complex and the Great Valley sequence. The Great Valley sequence consists of ocean crust and overlying sediments. Great Valley sediments are contemporaneous with Jurassic to Cretaceous sediments in the Franciscan complex but were deposited in a forearc basin rather than in a deep trench. Water introduced into the basal peridotite during faulting produced large masses of serpentine that acted as an effective lubricant for thrusting of the Great Valley ophiolite westward over the Franciscan complex. Tertiary folding and strike–slip faulting of the Coast Range thrust sheet west of the Coast Range thrust fault zone have produced somewhat chaotic terrain that contrasts markedly with long north–south trending ridges, or hogbacks, of upturned Great Valley sedimentary rocks that dip eastward from the Coast Ranges thrust fault. Geologic mapping has outlined the distorted and sinuous trace of the Coast Range thrust and has clearly shown that large horizontal westward displacement of the Great Valley sequence and its basal serpentinite took place before Tertiary folding and strike–slip faulting (Dickinson et al. 1996).

Numerous bodies of peridotite and serpentinite are scattered through the Northern California Coast Ranges. Many small serpentinite bodies are associated with fault zones. Where these occupy vertical faults their three-dimensional shape appears to be discoidal. In areas of low relief, where slumping and landslide obscure contacts, it is difficult to ascertain the shape of the serpentinite, particularly when the surrounding country rock is in unstable melange. Areomagnetic surveys of serpentinite in graywacke-greenstone terranes can lead to accurate three-dimensional modeling of their shape. In contrast, partially serpentinized peridotite bodies such as those at Red Mountain–Cedar Creek (locality 4-8), Red Mountain in southern Trinity County (locality 4-9), and the Cedars (locality 4-19) have narrow zones of sheared serpentinite at their boundaries and usually form rugged high-relief outcrops that can easily be mapped. Geophysical studies of these peridotite bodies reveal a slablike shape with sheared serpentinite at their base (Thompson and Robinson 1975). The close association of these peridotites along regional thrust faults indicates their final tectonic emplacement. Recent studies suggest that

Blue Schists, Indicators of Subduction Deep into the Mantle

Blue schists are indicators of the great depths to which some rocks in the Franciscan complex have been subducted below trenches. The main mineral in these schists is a blue amphibole called *glaucophane*. It forms at depths where the pressure is very high, but the temperature is moderate compared with many metamorphic environments. Large tectonic blocks of blueschist found completely immersed in serpentine melange demonstrate that light diapiric serpentinites can act as tectonic elevators transporting these blocks upward through great vertical distances in subduction zone melanges. Blueschists are associated in and around serpentinite from north to south throughout the California Coast Ranges (Coleman and Lamphere 1971). The tectonic inclusions within serpentine melanges have a great diversity, including graywacke, greenstone, chert, and amphibolite.

---

these peridotite bodies may be fragments of oceanic fracture zones underplated into the subduction wedges. Large bodies of peridotite are present at Riddle (locality 4-1; Hotz 1964, Ramp 1978) and Agness-Snow Camp (locality 4-4; Ramp 1978), also.

Sedimentary serpentine in olistostromes near the base of the Great Valley forearc sediments records a period of destruction and incorporation of the tectonized serpentinites in the Late Jurassic–Early Cretaceous subduction regime. The low strength of sheared serpentinites permits initial plastic tectonic flow in the crust and on the earth's surface. Late Tertiary folding and thrusting has mobilized these serpentinite melanges into anticlinal axis producing diapiric protrusions of serpentinite melange. This instability promotes mass movement by gravity as landslides, sheet flows or as olistostromes (see inset above).

16.2 Soils

Most serpentine soils in the Northern California Coast Ranges are shallow or moderately deep to bedrock or shallow to very deep in colluvium on steep mountain slopes. Floodplains and stream terraces are generally narrow and lake and coastal plains are not extensive. Salts are leached from well-drained soils and carbonates accumulate in few of them. Clay is leached from surface soils and accumulates in subsoils on relatively stable landforms.

Ochric epipedons predominate, with mollic epipedons in the drier areas and some umbric epipedons on nonserpentine soils in the wetter climates. Poorly drained soils with histic epipedons are sparse. Cambic and argillic horizons are the common subsurface diagnostic horizons.

Clay in subsoils is commonly dominated by minerals inherited from soil parent materials. Kaolinite accumulates in older nonserpentine soils in wetter climates and smectites accumulate in soils in drier areas and in basins where silica and cations are washed in from soils in higher landscape positions. Cracking-clay soils are common in these basins. Smectites and iron oxides are more prevalent in serpentine soils than in soils with other kinds of parent materials.

Inceptisols and Alfisols are most extensive, followed by Mollisols in drier areas. Ultisols are present in the wetter climates but less common with serpentine than with nonserpentine parent materials. Vertisols occur in basins where the soils are not well drained. At the order and suborder levels, most of the soils with serpentine parent materials are classified the same as those with nonserpentine parent materials. There are no serpentine soils on the coastal plains and terraces, where there are small areas of Spodosols in addition to the orders mentioned already.

Soil temperature regimes are isomesic along the coast and thermic, mesic, frigid, and cryic at increasing elevations inland. Soil moisture regimes are predominantly xeric, with relatively small areas with udic soil temperature regimens (SMRs) along the coast. Also, there are some areas of poorly drained soils with aquic soil moisture regimes.

16.3 Vegetation

The vegetative cover is mostly open to semidense forest, and in the warmer and drier areas, chaparral. Grassland prevails in ephemerally wet basins with Vertisols. The forest includes Douglas-fir (*Pseutotsuga menziesii*) and mixed coniferous forests of Douglas-fir, ponderosa pine, and incense-cedar at the higher elevations. Oak woodlands typically avoid serpentine and often form distinct boundaries with open gray pine and leather oak scrub on serpentine. In the inner and middle portions of the coast ranges, the majority of the serpentine vegetation occurs below 1500 m elevation and is predominantly chaparral with chamise (*Adenostoma fasciculatum*), various manzanita (*Arctostaphylos*) species, ceanothus species, and leather oak (*Quercus durata*) as the main shrub components. Forests of four serotinous coned conifers occur in the domain; these are knobcone pine (*Pinus attenuata*), lodgepole pine (*P. contorta*), and local stands of Sargent cypress and McNab cypress. Stands of these species are usually dense. In Sonoma and Mendocino counties, several species of endemic manzanitas and ceanothus may form localized and

Figure 16-3 Chaparral on a serpentine soil of the Henneke Series (Lithic Argixerolls) southeast of Morgan Valley in Lake County, California. The shrubs are mainly whiteleaf manzanita (*Arctostaphylos viscida*), leather oak (*Quercus durata*), chamise (*Adenostoma fasiculatum*), Jepson ceanothus (*Ceanothus jepsonii*), and valley silktassel (*Garrya congdonii*).

relatively unique stands. Other than the Klamath Mountains (domain 5), this domain has the greatest number of endemic serpentine plants in western North America (Kruckeberg 1984, Harrison et al. 2000).

16.4 Riddle, Locality 4-1

At Riddle in Douglas County, a sheet of peridotite (harzburgite and dunite) that lacks appreciable serpentinization has been thrust over sheared serpentinite and sedimentary and volcanic rocks of the Dothan formation (Ramp 1978). This sheet of peridotite occupies the summit and upper slopes, above ~850 m, of Nickel Mountain (elevation 1202 m). The geology of this area has been studied intensively for more than a century, because weathering and leaching of the peridotite over millions of years has concentrated nickel to make the saprolite and lateritic soil commercially viable ores. This regolith has been called "nickel-laterite" by geologists. Mining was discontinued in the latter part of the twentieth century.

Ramp (1978) has described a typical lateritic soil profile and given approximate nickel concentrations at Nickel Mountain: (1) dark, reddish brown surface soil 0.3–1 m thick with 120 g Fe/kg and 1–8 g Ni/kg; (2) yellowish brown soil with 160 g Fe/kg and 6–15 g Ni/kg; (3) yellowish brown soft saprolite with 10–25 g Ni/kg; (4) greenish brown, slightly weathered peridotite or hard saprolite with 5–15 g Ni/kg; and (5) fresh peridotite with 2–3 g Ni/kg. A boxwork of criss-crossing silica (chacedony) veins is common in the lower part of the saprolite on Nickel Mountain. The soil or regolith thickness ranges ~3 m, with a maximum ~12 m, or 60 m, depending on the reference cited by Ramp (1978). The deeper sites may be in landslide deposits.

16.5 Edson Butte and Iron Mountain, Localities 4-2 and 4-3

Thin thrust sheets of highly sheared serpentinite are sandwiched between the overlying Colebrooke schist and the underlying Dothan graywacke trench deposits at these localities in northern Curry County (Coleman 1972). Some of the best examples of blueschist are found at Edson Butte (Coleman and Lamphere 1971). Elevations are 850 m on Edson Butte and 1227 m at the top of Iron Mountain. Mean annual precipitation is up to about 275 cm in these localities.

16.6 Agness–Snow Camp, Locality 4-4

A long, narrow, fault-bound ultramafic body of slightly serpentinized peridotite extends north about 30 km into Coos County and south about 25 km from Agness, which is on the Rogue River, in Curry County. This locality is broader in the Saddle Mountain to Pistol River area on the south, 4 or 5 km wide. It is 25–30 km inland from the Pacific Ocean.

Elevations range from about 50 m along the Rogue River to 1339 m on Saddle Mountain. This is a mountainous area with steep slopes. Mean annual precipitation ranges from 157 cm at Powers to about 325 cm at higher elevations. Snow accumulates and persists through winters at the higher elevations.

Serpentine soils around Agness and north nearly to Powers have been mapped as Dystric (Mislatnah and Serpentano series) and Lithic (Greggo series) Eutrudepts with mesic soil temperature regimens (STRs). They generally have open forest cover of Douglas-fir, white pine (*Pinus monticola*), tanoak (*Lithocarpus densiflora*), and California bay (*Umbellularia californica*), with an understory of shrubs and beargrass (*Xerophyllum tenax*). The common shrubs are huckleberry oak (*Quercus vaccinifolia*), California coffeeberry (*Rhamnus californica*), whiteleaf or hoary manzanita (*Arctostaphylos viscida* and *A. canescens*), and sparse red huckleberry (*Vaccinium parvifolium*). The shallow soils have more open forest with more Jeffrey pine (*P. jeffreyi*) and incense-cedar (*Calocedrus decurrens*).

Serpentine soils south of Agness, in the Saddle Mountain to Snow Camp Mountain area, have been mapped as Dystric (Cedarcamp and Snowcamp series) and Lithic (Flycatcher series) Eutrudepts with frigid STRs. Some of these soils may have argillic horizons and be Hapudalfs rather than Eutrudepts (Burt et al. 2001), and they may be in andic subgroups. They generally have open forest cover of Douglas-fir and white pine with an understory of shrubs and bear grass. The common shrubs are huckleberry oak and California coffeeberry and sparse tanoak, California bay, and red huckleberry. Also, greenleaf manzanita (*A. patula*) and Sadler oak (*Q. sadleriana*) are common at the higher elevations. The shallow soils have more open forest with Jeffrey pine and incense-cedar and with most of the same shrubs as on deeper soils plus pimemat manzanita (*A. nevadensis*) and common juniper (*Juniperus communis*). Lawson cypress (*Chamaecyparis lawsoniana*) and azalea (*Rhododendron occidentalis*) are common around seeps and on footslopes.

16.7 Signal Buttes-Red Flat, Locality 4-5

A sheet of slightly serpentinized peridotite grading downward to sheared serpentinite and Colebrooke schist has been thrust over the Dothan and Otter Point formations in locality 4-5 in Curry County (Ramp 1978). The locality stretches from Signal Butte on the north about 15 km south to the Pistol River, parallel to the Pacific coast and about 10 km inland from the ocean. Elevations range from about 400 m along Hunter Creek and the Pistol River to 1070 m on Signal Buttes. This is a mountainous area with steep slopes, except on a gently sloping, north–south trending ridge from Red Flat to the vicinity of Hunter Creek Bog. Mean annual precipitation is on the order of 275 cm. Snowfall is common, but it may not remain on the ground throughout winters.

Serpentine soils in this locality have been mapped as Dystric (Mislatnah, Serpentano, and Redflat series) and Lithic (Greggo series) Eutrudepts with mesic STRs. They generally have open forest cover of Douglas-fir, white pine, tanoak , and California bay, with an understory of shrubs and bear grass. The common shrubs are huckleberry oak, California coffeeberry, whiteleaf or hoary manzanita, and sparse red huckleberry. The shallow soils have more open forest with more Jeffrey pine and incense-cedar.

The soil on Redflat has been designated a lateritic nickel source containing about 8 g Ni/kg, 1.5 g Co/kg, 8 g Cr/kg, and 180 g Fe/kg, and the average depth of soil and saprolite was estimated to be 2.5 m (Ramp 1978). Open forests of white pine and lodgepole pine, with an understory of tanoak, California bay, huckleberry (*V. ovatum* and *V. parvifolium*), hoary manzanita, and azalea are common on the Redflats soils. Knobcone pine is very common following fires and pinemat manzanita, common juniper, and beargrass are common in forest openings.

16.8 Pistol River–Gold Beach–Nesika, Locality 4-6

There are many small exposures of serpentinized peridotite and serpentinite along the coast of the Pacific Ocean from Nesika to the Pistol River. This a hilly area in which steep slopes predominate. Elevations are mostly < 300 m. The mean annual precipitation is 200–225 cm.

Hapludolls, both shallow (Sebastian series) and deeper (Rustybutte series), have been mapped in this area. They have isomesic STRs. Because of human disturbance and grazing, it is difficult to discern whether the natural cover was forest or grassland, or both. Forest may encroach on grasslands in the absence of fire and grazing. The common conifers are Douglas-fir and grand fir (*Abies grandis*) and the common understory or successional trees are tanoak and California bay, and in some cases madrone (*Arbutus menziesi*). Sitka spruce (*Picea sitchensis*) occurs where salt spray is blown in from the ocean. Evergreen huckleberry (*Vaccinium ovatum*) is abundant in forest openings.

16.9 Lassic Mountain, Locality 4-7

Ultramafic rocks at the Lassic Mountian locality are apparently in an outlier of the Coast Range ophiolite embedded in sheared serpentinite (Kaplan 1983). This locality lies across the boundary between Humboldt and Trinity counties, California. Elevations are up to 1791 m on Red Lassic. This is a mountainous area with moderately steep to very steep slopes. The mean annual precipitation is about 200 cm, with much of it falling as snow, some of which persists through winters.

The summit of Red Lassic is nonserpentine rock outcrop. Black Lassic is a metabasalt barren with very shallow soils. Serpentine soils mapped around Red and Black Lassic for the Forest Service are mostly deep-to-very deep Inceptisols (Typic Xerumbrepts of the Hungry series) on steep-to-very steep slopes and moderately deep Alfisols (Mollic Haploxeralfs of the Madden series) on moderately steep-to-steep slopes. Also, shallow Haploxeralfs were recognized in the soil map units. These soils have frigid and mesic STRs, depending mainly on elevation and slope aspect. Open Jeffrey pine–incense-cedar forests are the predominant cover on the serpentine soils. White fir (*Abies concolor*) is prevalent on nonserpentine forest soils. There are some patches of serpentine barrens on serpentinite of the Lassic locality.

The locality is characterized by Jeffrey pine–incense-cedar woodland, chaparral vegetation, serpentine barrens, and seasonal wetland habitats. Several endemic herbs occur in the serpentine barrens and the open forests. A forest service botanical area has been established to conserve the unique combination of species, which include *Allium hoffmanii* (Hoffman's onion), *Gentiana calycosa* (mountain bog gentian), *Lathyrus biflorus* (two-flowered pea), Lassics lupine (*Lupinus constancei*), *Minuartia decumbens*, *Raillardiopsis scabrida*, *Sanicula tracyi*, and *Sedum laxum* ssp. *flavidum*. Forest Service botanists (Six Rivers National Forest, Eureka, CA) have compiled an extensive plant spcesies list for the Lassics Botanical and Geological Area.

16.10 Red Mountain–Cedar Creek, Locality 4-8

Large amounts of only slightly serpentinized peridotite are exposed on Little Red Mountain and on Red Mountain in Mendocino County. Elevations range from about 500 m along Cedar Creek, between Red Mountain and Little Red Mountain, to 981 m (3718 ft) on Little Red Mountain and about 1250 m on Red Mountain. The mean annual precipitation is 150–175 cm.

Shallow Entisols (Hiltabidel Series), moderately deep Inceptisols (Dann Series), and very deep Ultisols (Littlered Series) have been mapped in the locality. These are reddish (2.5–5 YR hues) soils that are slightly acid to neutral. A sample from the Littlered subsoil at the type location contained 33% citrate-dithionite extractable iron, which would represent 55% goethite (FeOOH) or 47% hematite (Fe_2O_3). The soil probably contains appreciably more goethite than hematite, although it is hematite that imparts the reddish hues to the soils.More complete analyses of a Littlered soil on Red Mountain indicate that it is a Xeric Kanhapudult.

The plant communities on serpentine soils of the locality are mostly open to semidense Jeffrey pine, incense-cedar, and sugar pine forests, with extensive shrub understory. Huckleberry oak is the predominant shrub on north-facing slopes and hoary manzanita on south-facing slopes. Locally, knobcone pine, ceanothus (*Ceanothus jep*-

Figure 16-4 A view eastward from near the summit of Red Mountain in locality 4-8. The open forest with a shrub understory in the foreground is on serpentine and the dense forest in the middleground and the grassland in the background are on other strata of the Franciscan complex. Photo by I. Noell, Bureau of Land Management, U.S. Department of Interior.

sonii), beargrass, and lousewort (*Pedicularis* sp.) are common. Grasses (*Festuca* spp. and *Elymus elymoides*) are sparse. Several rare plant species occur here, including the southernmost localities of the McDonald's rock cress (*Arabis macdonaldiana*). The U.S. Bureau of Land Management has designated Red Mountain as an Area of Critical Environmental Concern and has conducted studies on the population dynamics and reproductive biology of McDonald's rock cress there. Other rare taxa known to occur on the serpentine soils of Red Mountain include the California State-listed endangered Kellogg's buckwheat (*Eriogonum kelloggii*) and Red Mountain catchfly (*Silene campanulata* ssp. *campanulata*), as well as the unlisted but rare Red Mountain stonecrop (*Sedum laxum* ssp. *eastwoodiae*).

16.11 Red Mountain–Red Mountain Creek, Locality 4-9

Slightly serpentinized peridotite is exposed on the Red Mountain in southern Trinity County. Another Red Mountain just 26 km to the north–northeast is also in Trinity County, but it is in the Klamath Mountains, instead of in the California Coast Ranges. The summit of the Red Mountain in the southern part of Trinity County is at 1272 m elevation, and the mean annual precipitation is about 150 cm.

16.12 Leech Lake Mountain, Locality 4-10

Ultramafic rocks at the Leech Mountain locality are in an outlier of the Coast Range ophiolite embedded in sheared serpentinite (Chesterman 1963). This locality is in a mountainous area with elevations 1500–2000 m. It is near the northeast corner of Mendocino County. The mean annual precipitation is about 150 cm.

Vegetation from the area is not well known. The locality is apparently the northernmost location of the rare and localized Snow Mountain willowherb (*Epilobium nivium*) (Jepson Herbarium records JEPS74983, JEPS74984, JEPS74985), which is often found on serpentine (Kruckeberg 1984).

16.13 Covelo–Eden Valley, Locality 4-11

Small to moderately large, disconnected exposures of serpentinite occur in fault contact with the Franciscan complex south of Covello in northern Mendocino County. This is a low mountainous area with predominantly moderately steep and steep slopes. Elevations range from about 427 m at Covelo in Round Valley up to about 1090 m. The mean annual precipitation is 100–125 cm. Much of the precipitation falls as snow, but it does not accumulate or persist though winters.

The soils of this area have been mapped in some detail (Howard and Bowman 1991). Predominant serpentine soils in the locality are in the moderately deep Dingman and shallow Beaughton series. Both are Mollisols with prominent argillic horizons. Beaughton is a Lithic Argixeroll, and Dingman is a Pachic Ultic Argixeroll. Both soils have mesic STRs. There are minor areas of alluvial fill with serpentine Vertisols of the Maxwell series (Typic Haploxererts) with a thermic STR. The main cover is mostly chaparral on the Mollisols and grassland on the Vertisol.

The main chaparral species are leather oak and whiteleaf manzanita, plus Sargent cypress (*Cupressus sargenti*) on the cooler slopes and much chamise and sparse gray pine (*Pinus sabiniana*) on warmer slopes (Cuff et al. 1951). Knobcone pine follows severe fires, and fireweed (*Eriodictyon californicum*) is common on disturbed soils.

16.14 Coast Range Fault Zone, Localities 4-12, -13, -14, and -15

Elder Creek, Paskenta, Chrome, and Stonyford are along the Coast Range fault zone, from Tehama County through Glenn into Colusa County, where the fault zone separates the Great Valley sediments on the east from the California Coast Ranges on the west.

The northern segment of the Coast Range serpentinite forms a nearly continuous belt along the California Coast Ranges eastern mountain front. In Tehama County the South Fork of Elder Creek exposes a fault bounded sheet of serpentine melange (locality 4-12, Hopson et al. 1981). To the south of Elder Creek the serpentinite belt broadens to the west at Paskenta (locality 4-13, Hopson et al. 1981). Here the basal serpentinite melange marks the trace of the Coast Range thrust and is overlain by broken slabs of intact ocean crust. To the south, the Coast Range thrust is marked by a sheared serpentinite matrix melange enclosing tectonic blocks of serpentinized harzburgite, massive greenstone, and pillowed greenstone. South of Thomes Creek, a sinuous zone of serpentinite melange that stretches into Glen County marks the Coast Range thrust. These thin serpentine melanges have sharp contacts with the South Fork Mountain schist that is derived from the Franciscan complex by regional metamorphism. At Stonyford (locality 4-15) in Colusa County and from the very southern part of Glenn County, the serpentine at the base of the Coast Range thrust broadens and becomes distorted by thick piles of volcanic rock formed as sea mounts that formed on the oceanic crust before they were subducted (Brown 1964). These seamounts, several kilometers high, were not easily accommodated during subduction, producing major distortion in the subduction zone. The Great Valley basal rocks are pillow lavas and tuffs instead of marine shales and are faulted against the serpentinized peridotite.

Most of the serpentine soils mapped in these localities are Lithic Argixerolls in the Henneke series with chaparral cover (Gowans 1967, Begg 1968). There are minor areas of moderately deep Mollic Haploxeralfs in the Dubakella series with open forest cover and Lithic Haploxerolls in the Montara series with grassland.

A detailed inventory of vegetation in the mostly serpentine Frenzel Creek drainage in Colusa County by Keeler-Wolf (1983) provides a very good perspective of the serpentine vegetation in these localities. Frenzel Creek flows into Little Stony Creek about 10 km south of Stonyford at an elevation about 425 m. The head of the Frenzel Creek drainage is about 945 m above sea level (asl). Of ~290 ha of serpentine in the drainage 81% was chaparral, 11% barrens, and 8% valley bottom and riparian. The valley bottom was dominated by Sargent cypress adjacent to the riparian areas and bay in upper areas. Riparian vegetation was characterized by azalea, Brewer's willow (*Salix breweri*), and spicebush (*Calycanthus occidentalis*). Chaparral was dominated by whiteleaf manzanita on upper and west-facing slopes, leather oak on north- and east-facing slopes, and by McNab cypress (*Cupressus macnabiana*) in many places. Other common shrubs were Jepson ceanothus (*Ceanothus jepsonii* var. *albiflorus*), valley silktassel (*Garrya congdoni*), and toyon (*Heteromeles arbutifolia*), plus yerba santa (*Eriodictyon californicum*) in disturbed areas and chamise along with McNab cypress. Gray pine was scattered sparsely through the chaparral. Only sparse forbs were generally present on the barrens. The most ubiquitous were jewelflower (*Streptanthus morrisonii*), wild buckwheats (*Eriogonum vimineum*, *E. nudum*, and *E. dasyanthemum*), wild onion (*Allium* sp.), and milkweeds (*Asclepias cordifolia* and *A. solanoana*).

16.15 Hopland, Locality 4-16

Small, disconnected exposures of serpentinite occur in fault contact with the Franciscan complex near Hopland in southeastern Mendocino County. This is a low mountainous area with predominantly moderately steep and steep slopes; elevations are between 200 and 900 m. The mean annual precipitation is about 100 cm. Some of the precipitation falls as snow, but it does not accumulate or persist through winters.

The soils of this area have been mapped in some detail (Howard and Bowman 1991). Predominant serpentine soils in the locality are in the shallow Henneke and Montara series, with inclusions of deeper soils. Both Henneke and Montara are Mollisols, but only soils in the Henneke series have argillic horizons. Henneke is a Lithic Argixeroll and Montara is a Lithic Haploxeroll. There are minor areas of alluvial fill with serpentine Vertisols of the Maxwell series (Typic Haploxererts). All of these soils have thermic STRs. The vegetative cover is mostly chaparral on the Henneke soil and grassland on the Montara and Maxwell soils.

The main chaparral species are leather oak and whiteleaf manzanita, plus Sargent cypress on the cooler slopes and much chamise and sparse gray pine on the warmer slopes (Neuns 1950, Swensen 1950). Also, Jepson ceanothus, toyon, and coffeeberry (*Rhamnus californica* and *R. tomentella*) are common. Grassland also occurs locally on the serpentine at Hopland and species such as squirreltail grass (*Elymus multisetus*) and purple needlegrass (*Nassella pulchra*) may form small but conspicuous stands.

16.16 Wilbur Springs–Walker Ridge, Locality 4-17

The serpentinite belt of the Coast Range fault zone extending southward from Stonyford (locality 15) pinches out just north of Wilbur Springs, along the boundary between Colusa and Lake Counties (McLaughlin et al. 1985). This segment from Stonyford up along Red and Walker Ridges is a serpentinite matrix melange that encloses blocks of massive serpentinized peridotite and exotic blocks of metasedimentary and metavolvamic rocks. Near Wilbur Springs the basal sequence of the Great Valley consists of detrital serpentine derived from the underlying Coast Range serpentinite.

The basal sedimentary sequences of the Great Valley sequence are marked by interbedded debris flows consisting of tectonic blocks. The Wilbur Springs sedimentary serpentinite in Colusa County is up to 2000 ft thick and consist of ophiolite breccia of olistostrome origin interbedded with basal mudstone and sandstone of the Great Valley sedimentary sequence. Marine invertebrate fossils contained in the serpentinite debris indicate a Hautervian age (125–120 Ma). The breccias contain fragments of serpentinitinized peridotite, greenstone, metachert, gabbro, and basalt in a poorly stratified unsorted mixture within a serpentinite matrix. These rocks are exposed within the southeast plunging Wilbur Springs antiform and are characteristically devoid of grass, forming block-strewn slopes with dense chaparral.

Walker Ridge extends north–south along the boundary between Colusa and Lake counties. Ultramafic rocks are predominant along the entire 18 or 20 km length of Walker Ridge and Red Ridge, which is a northern extension of Walker Ridge. The summit of Walker Ridge is a bit over 1000 m and Red Ridge is about 750 m high. Slopes are gentle to moderately steep near the summits of the ridges and steep on the sides. The mean annual precipitation is 60–75 cm.

The soils on Walker Ridge are mostly shallow Mollisols (Henneke and Okiota series) with argillic horizons (Reed 2003). Soils in both the Henneke and Okiota series are Lithic Argixerolls. Rock land is not extensive on Walker Ridge. Shallow Mollisols lacking argillic horizons occur on ridge summits and shoulder slopes. These are Lithic Haploxerolls in the Montara series. There are only small areas of deeper soils. Moderately deep Alfisols with argillic horizons occur sporadically on north-facing slopes. They are Mollic Haploxeralfs in the Dubakella series. These soils have thermic STRs, except Dubakella, which has a mesic STR. Chamise and mixed chaparral is the dominant cover on these soils, with sparse gray pine trees. There are more trees on the Dubakella soils and grassland on the Montara soils.

Chaparral shrubs on Walker Ridge are mainly leather oak and whiteleaf manzanita, and McNab cypress is common. Shrubs on the warmer slopes are mainly chamise, leather oak, and whiteleaf manzanita.

On the east side of Walker Ridge is a large north–south trending valley, Bear Valley, that lies between Coast Ranges ophiolite on the west and Great Valley sedimentary rocks on the east. Very deep serpentine soils have developed in alluvial fans on the west side of Bear Valley and in basin fill near the center of the Valley. Elevations in Bear Valley are in the range from 400 to 500 m. The mean annual precipitation is 50–60 cm. Most of it falls as rain. The frost-free period is about 225 days.

Soils that are predominant on the serpentine fans are well-drained, loamy Mollisols with sandy subsoils. These soils are in the Leesville series of Pachic Haploxerolls. Soils that are predominant in the serpentine basin fill are poorly drained, black Vertisols. These soils are in the Venado series of Aridic Endoaquerts. The vegetation in uncultivated areas is grassland. Grassland vegetation may take on spectacular wildflower displays in the spring with species such as *Layia gallioides, Lasthenia californica*, and many others. Serpentine creek alluvium debauching from the base of Walker Ridge onto the floor of Bear Valley may have stands of Wright buckwheat (*Eriogonum wrightii* ssp. *subscaposum*) and scattered Brewer willow.

16.17 Harbin Springs-Knoxville, Locality 4-18

South of Wilbur Springs the serpentine belt and subjacent Coast Ranges fault that are continuous to the north through localities 16-12 to 16-15, and 16-17 have been disrupted by Tertiary left lateral strike-slip faulting and by folding. In the Harbin Springs–Knoxville locality, the contact between serpentinite and the base of the Great Valley is marked by a thick sedimentary breccia (Hopson et al. 1981). This breccia consists of large blocks of gabbro, basalt, chert, and serpentinite and is considered to have formed as a huge submarine gravity slide of ophiolitic material deposited on the serpentinite melange. The matrix of the sedimentary breccia consists of varying amounts of detrital serpentinite and mudstone. Mudstone and tuff of Late Jurassic age are deposited on the breccia and grade upward into marine sandstones of the Great Valley sequence. Isolated sheets of units retaining the same stratigraphy as that found near Harbin Springs are found around Mount St. Helena (locality 4-20), about 10 km south of Harbin Springs, and in Sonoma County imbricated within the Franciscan complex (McLaughlin et al. 1988, Robertson 1990).

From ultramafic rocks along the north–south fault zone extending south from the Stony Creek fault and the Coast Ranges thrust fault zone that separates the California Coast Ranges from the Great Valley, several branches extend northwestward along en echelon faults. Harbin Springs is on one of these branches. This whole area from eastern Lake County south into northeastern Napa County is called the Harbin Springs–Knoxville locality. The basal ultramafic unit of the Harbin Springs ophiolite is sheared serpentinite derived from harzburgite and peridotite. Gabbro and breccia, or olistrotomes, occur above the basal ultramafics stratigraphically.

In Napa and Lake counties serpentinite olistostrome flows and breccia containing blocks of greenstone, basalt, chert, schist, and amphibolite extend 20 miles across strike (Phipps 1984). The maximum thickness of this unit is 1 km and has been involved in complex folding and thrusting. These serpentinite olistostromes are deposited directly on the Coast Range Ophiolite serpentinite basal member and grades upward into well bedded sediments of the Great Valley sequence. Invertebrate fossils within this unit are Tithonian (Late Jurassic, 140–130 Ma) and support the idea that this unit developed as a large submarine landslide. It is considered to have formed above the east-directed subduction zone where fragments of the Coast Range ophiolite were uplifted and incorporated into the basal sedimentary unit of the Great Valley sedimentary sequence.

The Harbin Springs–Knoxville area is a low mountainous to hilly area, with many small valleys in which Holocene sediments are accumulating. Elevations range from about 150 m along Putah Creek and its tributaries where they flow into Lake Berryessa up to 600–800 m on some of the higher summits. Mean annual precipitation is in the range of 75–90 cm. Most of it falls as rain.

Serpentine soils in the Harbin Springs-Knoxville locality are mostly shallow Mollisols. They are the Henneke and Okiota series with argillic horizons (Lithic Argixerolls) and the Montara series lacking argillic horizons (Lithic Haploxrolls). There are only small areas of deeper soils. Soils in the Henneke series predominate on steep slopes, and soils in the Okiota and Montara series are more common on moderately steep to gentle slopes (Lambert et al. 1978, Smith and Broderson 1989). These soils have thermic STRs.

The vegetative cover is mostly chaparral on the Henneke and and Okiota and grassland on the Montara soils. Gray pine trees occur sporadically on these soils, but blue oak (*Quercus douglasi*) avoids the serpentine soils.

Chaparral shrubs in the area around Harbin Springs are mainly leather oak and whiteleaf manzanita, with common Jepson ceanothus and toyon (Cuff and Neuns 1952). Chaparral shrubs to the east, toward Knoxville, are mainly leather oak and whiteleaf manzanita, also, with more chamise on the warmer slopes and McNab cypress on the cooler slopes, and with common Jepson ceanothus and silktassel (Colwell et al. 1955). Another notable shrub is a local flannelbush (*Fremontodendron californicum* var. *napense*), and Sargent cypress occurs along streams and above 600 m altitude.

Some of the valleys, most notably Pope Valley in Napa County, contain serpentine alluvium with somewhat poorly drained Vertisols, which are in the Maxwell series of Typic Haploxererts. These were grassland soils that have been cultivated with limited success. The addition of several tons of gypsum per hectare raises the calcium level and exchangeable Ca/Mg ratio enough to produce satisfactory crops of wine grapes. A high-value crop is required to justify the expense of adding so much gypsum. Most vegetation on the serpentine alluvium has been modified, but some patches remain, like those near Middletown. In the spring some of these may be temporarily flooded and or saturated and hold interesting and colorful stands of herbaceous species such as *Lasthenia glabrata* and blue-eyed grass (*Sysrinchium bellum*) along with native perennial grasses and graminoids such as meadow barley (*Hordeum brachyantherum*) and irisleaf rush (*Juncus xiphioides*).

Serpentine wetlands along Bear Creek have stands of Baltic and Mexican rush (*Juncus balticus* and *J. mexicanus*), along with creeping wild rye (*Elymus triticoides*) and saltgrass (*Distichlis spicata*). Some showy wetland species in locality 4-18 are *Delphinium uliginosum, Mimulus guttatus, M. nudatus, Centaurea trichanthum, Aquilegia eximia,* and *Senecio clevelandii*.

A substantial amount of ecological research has been conducted in this locality, much of it centered in the vicinity of the McLaughlin Reserve managed by the University of California, Davis. Several studies have investigated topics including the population biology of rare plants (Harrison et al. 2000, Wolf et al. 2000, Jurjavcic et al. 2002), local and regional species diversity (Harrison 1997, 1999) fire ecology of serpentine chaparral (Safford and Harrison 2004), weeds and revegetation in serpentine (Williamson and Harrison 2002), and grazing effects on serpentine (Safford and Harrison 2001).

16.18 The Cedars, Locality 4-19

Slightly serpentinized peridotite is exposed at the Cedars locality in Sonoma County. It is one of the largest areas of peridotite in the California Coast Ranges (Bailey et al. 1964). The Cedars is named for the thick forests of Sargent cypress that grow over the peridotite. This is a steep mountainous area with some very steep canyons sides that are almost devoid of soils and plants. Elevations range from about 100 m along Big Austin Creek to at least 600 m on the higher ridges. Mean annual precipitation is in the 150–160 cm range.

A fascinating feature of the Cedars locality is the highly alkaline springs laden with calcium hydroxide emanating from peridotite where the olivine and pyroxenes are being altered to produce serpentine. When the very strongly alkaline spring water encounters carbon dioxide in the atmosphere or bicarbonates in surface water, the calcium is precipitated to produce travertine. This travertine has built dams around some of the springs to enclose small pools of spring water. The process of serpentinization that yields calcium and leads to the formation of alkaline spring water is explained more fully in chapters 3 and 4.

Most of the serpentine soils are shallow to moderately deep Inceptisols (DeLapp and Smith 1978, quadrangles 61D-1, 2, 3, 4) that have been mapped as the Huse series of Inceptisols with inclusion of moderately deep Inceptisols (Miller 1972). These are quite reddish soils in The Cedars locality. They have mesic STRs. The cover is mostly mixed chaparral and groves of Sargent cypress. The main chaparral species are leather oak, Eastwood manzanita (*Arctostaphylos glandulosa*), chamise, shrub California bay, and whiteleaf manzanita (DeLapp and Smith 1978, quadrangles 61D-1, 61D-2, 61D-3, 61D-4). There are appreciable amounts of toyon and California coffeeberry also. Some of the deeper soils support Douglas-fir and madrone trees within the cypress groves and chaparral. The rare cedars manzanita (*Arctostaphylos bakeri* ssp. *sublaevis*) is found at this locality. Other serpentine endemics in locality 4-19 are *Calochortus raichei*, a variety of *Streptanthus hoffmanii*, and an unnamed creambush (*Holodiscus* sp.) with small leaves (R. Raiche, personal communication, 2004). Southern populations of pitcher plant (*Darlingtonia californica*) and lady's-slipper (*Cypripedium californicum*) occur in this locality also.

16.19 Mt. Saint Helena, Locality 4-20

One of several places where detrital breccias containing serpentinite occur with outliers of Coast Range ophiolite and Great Valley sediments west of the Harbin Springs-Knoxville locality is around Mount St. Helena (1323 m asl) at the junction of Lake, Napa, and Sonoma counties (McLaughlin et al. 1988, Robertson 1990). The mountain is a volcanic peak with the serpentine rocks on the flanks of the mountain, mostly on the north side, and northwest along the Mayacamas Mountains. There is an exceptionally large area (perhaps a few dozen hectares) of silica carbonate rocks exposed in narrow margins around serpentine bodies southeast of Mt. Saint Helena.

Elevations of the serpentine rocks range from 300 to 900 m. Mean annual precipitation is 100–150 or 200 cm. Much of the precipitation falls as snow at the higher elevations.

The soils are mostly Mollisols–Lithic Haploxerolls of the Montara series and Lithic Argixerolls of the Henneke and Okiota series. They were mapped as a Henneke–Montara–Rock Outcrop complex at lower elevations and an Okiota–Henneke–Dubakella association at higher elevations (Smith and Broderson 1989). These soils have thermic STRs, except Dubakella, which has a mesic STR. The Dubakella series lacks a mollic epipedon and is a Mollic Haploxeralf. The vegetation types are chaparral with sparse gray pine trees on the Henneke and Okiota soils and brush with scattered Douglas-fir, ponderosa pine, madrone, and California bay trees on the Dubakella soil. Dense stands of knobcone pine commonly follow severe fires.

The main chaparral shrubs are leather oak, whiteleaf Manzanita (*A. viscida*), and lesser concentrations of Jepson ceanothus, hoary coffeebery, and California bay (Nelson and Zinke 1952, Zinke 1952). Gray pine is scattered through the chaparral and Sargent cypress trees are present in places.

Vegetation of serpentine chaparral and barrens in the Geyser Peak area, which is in or adjacent to this locality on the northwest, is characterized in a U.S. Bureau of Land Management report (Fritz et al. 1980). Soils in the serpentine chaparral are mostly Mollisols, and the characteristic plants are gray pine, Sargent cypress, chamise, whiteleaf manzanita, leather oak, Jepson ceanothus, minute willowherb (*Epilobium minutum*), serpentine fern (*Apidotis densa*), and bird's-foot fern (*Pellea mucronata*). A rare species of jewelflower (*Streptanthus brachiatus*) occurs on the serpentine barrens (R.S. Osterling n.d., I. Noell, personal communication, 2004).

O'Donnell (2005) found that the silica carbonate in Napa County harbors some serpentine endemic plant species and some that avoid serpentine. In the surface layer, the silica carbonate soils are generally strongly to moderately acid, compared to slightly acid to neutral in the serpentine soils. The exchangeable Ca/Mg ratio is generally higher (molar Ca/Mg >1) in the silica carbonate soils, but the total chromium, cobalt, and nickel are high in many of the silica carbonate soils. Mercury has been taken from mines in the area and some of the silica carbonate soils have high mercury concentrations.

16.20 Healdsburg–Bradford Mountain, Locality 4-21

One of several places where detrital breccias containing serpentinite occur with outliers of Coast Range ophiolite and Great Valley sediments west of the Harbin Springs–Knoxville locality is at Bradford Mountain northwest of Healdsburg in Sonoma County (McLaughlin et al. 1988). This locality is in steep-to-moderately steep hilly terrain around Bradford Mountain. Elevations are >40 m, but not as high as Bradford Mountain (375 m). The mean annual precipitation is in the 100–120 cm range.

Mollisols have been mapped on serpentine substrates in this locality, mainly Lithic Argixerolls of the Henneke Series and some Lithic Haploxerolls of the Montara Series and unclassified moderately deep soils (Miller 1972, DeLapp and Smith 1979, quadrangles

60C-2, 61D1). The dominant vegetation types are chaparral on the Henneke and grassland on the Montara series. The main shrubs in the chaparral are leather oak and, at lower elevations, chamise. Toyon is widespread and common manzanita (*Arctostaphylos manzanita*) and California coffeeberry are common shrubs. Shrubby California bay, Jepson ceanothus, and silktassel are common in some stands of chaparral. Gray pine is sparse, but widely distributed, and coast liveoak (*Quercus agrifolia*) occurs sporadically on the serpentine soils. Two typical serpentine plant communities found north of Bradford mountain are chamise–Jepson ceanothus on south-facing slopes and California bay/leather oak–toyon/California fescue (*F. californica*) with herbs and serpentine fern in the understory on north-facing slopes. The endemic serpentine shrub Siskiyou mat (*Ceanothus pumilus*) reaches its southern known limit at this locality.

16.21 Occidental, Locality 4-22

Discontinuous shear zones within the Franciscan melange near Occidental in Sonoma County contain abundant blueschist tectonic blocks and some eclogite blocks (Coleman et al. 1965). Serpentine soils occur from 1 to 3 km north and east of Occidental. They are on moderately steep slopes at the margins of a low, rolling plateau. Elevations range from 180 to 240 m in this locality. The mean annual precipitation is in the 100–120 cm range.

Shallow Lithic Argixerolls (Henneke series) and Lithic Haploxerolls (Montara series) and moderately deep Mollic Haploxeralfs (Dubakella series) have been mapped on serpentine in the locality (DeLapp and Smith 1979, quadrangle 64B-2). Henneke and Montara have thermic STRs and Dubakella has a mesic STR. The Dubakella series was not recognized in the NRCS soil survey of Sonoma County (Miller 1972). The vegetative cover is shrubs and Sargent cypress on the Henneke soils, grassland on the Montara soils, and forest on the moderately deep Mollic Haploxeralf. The main shrub species are Baker manzanita (*Arctostaphylos bakeri*), toyon, and California coffeeberry (DeLapp and Smith 1978, quadrangle 64B-2). The forest species are California bay, coast liveoak, and Douglas-fir.

A California Department of Fish and Game Ecological Reserve has been established for the protection and study of the local endemic species at Harrison Grade in this locality. Here sizable stands of Sargent cypress are interspersed with chaparral dominated by the rare and local Baker manzanita. Openings in the understory contain many native wildflowers in the spring. Among the serpentine plant communities are *Cupressus sargentii*/*Carex rossii* on summits and concave north-facing slopes, *A. bakeri bakeri*–*Rhamnus tomemtella*/*Melica harfordii* on north-facing slopes, and *A. bakeri*–*Ceanothus jepsonii jepsonii*/*Cordyanthus tenuis tenuis* on south-facing slopes.

16.22 Mount Tamalpais, Locality 4-23

A small area of sheared serpentinite is exposed on the crest of Mount Tamalpais, in the Mount Tamalpais State Park (Rice 1960), in Marin County, California.

Stands of leather oak and mixed leather oak–eastwood manzanita chaparral exist here, as well as several small stands of Sargent cypress (Keeler-Wolf 2003). The endemic Mount Tamalpais manzanita (*Arctostaphylos hookeri* ssp. *montana*) along with several other endemic plants such as serpentine reedgrass (*Calamagrostis ophitidis*) and Tamalpais lessingia (*Lessingia micradenia* var. *micradenia*) form unique stands on the upper slopes of the Mountain.

16.23 Tiburon, Locality 4-24

Serpentinite on Ring Mountain on the Tiburon Peninsula in Marin County contains a variety of inclusions (Page 1968). Huge blocks of blueschists form spectacular tors protruding from the surrounding soft serpentinite (Coleman et al. 1965). The elevation at the summit of Ring Mountain is only 183 m; it is more of a hill than a mountain. Slopes are mostly moderately steep to steep. The mean annual precipitation, practically all rain, is about 75 cm. The frost-free period is about 240 days.

Kashiwagi (1985) mapped the soils as Mollisols in the Henneke series (Lithic Argixerolls). The vegetative cover is grassland and patches of shrubs and small trees. A native grassland on Ring Mountain has been set aside as part of the Nature Conservancy's California Critical Areas Program (Rice and Wagner 1991).

Fiedler and Leidy (1987) described the vegetation of the grasslands on Ring Mountain. There are several very rare endemic plant species known from this site including the Tiberon mariposa lilly (*Calochortus tiboronensis*), Tiburon paintbrush (*Castilleja affinis* ssp. *neglecta*), Tiburon jewelflower (*Streptanthus niger*), and Marin dwarf-flax (*Hesperolinon congestum*). All these species are considered rare (California Natural Diversity Database 2004) and have been treated in a recovery plan developed by the U.S. Fish and Wildlife Service (Elam et al. 1998). These species associate with more common and widespread grassland and rock outcrop inhabitants, including foothill needlegrass (*Nassella lepida*), golden yarrow (*Eriophyllum confertflorum*), hayfield tarweed (*Hemizonia congesta* ssp. *congesta*), purple needlegrass, serpentine reedgrass, sticky calycadenia (*Calycadenia multiglandulosa*), and Tiburon buckwheat (*Eriogonum luteolum* var. *caninum*). Introduced species in the area include Italian ryegrass (*Lolium muitfiorum*), slender wild oat (*Avena barbata*), and soft brome (*Bromus hordeaceus*).

17   Klamath Mountains, Domain 5

Domain 5 has mountains on the west aligned in broad arcs; in the eastern Klamath Mountains this arcuate pattern is not as evident (fig. 16-1). The outer arc curves from the triple junction of the California Coast Ranges, Great Valley, and Klamath Mountains in western Tehama County around through Del Norte County, where it is within a few kilometers of the Pacific Ocean, to Douglas County in Oregon. This western edge of the Klamath Mountains marks a boundary with the California Coast Ranges. The eastern edge of the Klamath Mountains forms a boundary with the Cascade Mountains. The southern, or southeast, boundary is where sediments of the Great Valley lap over the Klamath Mountains in Shasta and western Tehama counties. Serpentine rocks are more extensive in the Klamath Mountains than in any other domain or physiographic province in North America.

Through the middle of the Tertiary, the Klamath Mountains were eroded to a nearly level plain called the Klamath peneplain (Diller 1902). According to Diller, the Klamath peneplain and the submerged coastal area that had been accumulating sediments were uplifted slightly during the Miocene, and erosion reduced the northern California Coast Ranges to a nearly level plain which he called the Bellsprings peneplain, noting that it is practically continuous with the Klamath peneplain. Subsequently discovered sediments of the Weaverville formation were deposited in a depression, or graben, in the Klamath Mountains during the Oligocene, indicating that uplift of the Klamath peneplain must have begun during the Paleogene, before the Miocene. Uplift was intermittent, allowing time for the erosion of broad valleys in less resistant rocks between episodes of uplift.

Concordant summits, or mountains with summits of nearly equal elevation, are the evidence that led Diller (1902) to suspect a former peneplain. Although the mountain summits in any particular area are nearly equal or subequal, there is a general increase in summit elevation from the coast inland to 2.5–2.7 km. The altitude of Mt. Eddy on the eastern edge of the Klamath Mountains is 2751 m and that of Mt. Linn in the South Yolla Bolly Mountains near the eastern edge of the California Coast Ranges is 2466 m, just south of the Klamath Mountains. There are some summits, such as Preston Peak, that rise above adjacent concordant summits either because of original relief on the peneplain or because of faulting and differential uplift more recently. Some areas on the peneplain, such as the Trinity Alps, have been warped upward; Thompson Peak in the Trinity Alps is 2744 m high, even though it is not near the eastern edge of the Klamath Mountains.

The pace of uplift accelerated in the latter part of the Cenozoic Era due to thrusting of the Gorda Plate under the North American Plate along the Cascadia Subduction Zone. Few broad summits remain; most of the peneplain has been reduced to narrow or rounded summits. Nickel-laterite, which takes millions of year to form in a warm climate, is still present on some of the broader ridges. Uplift has left some broad valleys and terraces high above current stream channels and floodplains. The western edge of the Klamath peneplain was covered by the Wimer Formation, a fossiliferous marine sediment of Pliocene age. Elevation of this onlap contact of the Wimer Formation is utilized to estimate an uplift rate of 0.18 mm/yr near the western edge of the Klamath Mountains since Pliocene time. Marine terraces indicate recent uplift about 1 mm/yr, and up to 4 m in 1000 years (4 mm/year) in the vicinity of Cape Mendocino (Kelsey 1987).

Currently, the Klamath Mountains are deeply incised, with predominantly dendritic stream drainage patterns. Glaciers formed at high elevations during the Quaternary and modified the terrain by scouring out cirques, flowing down valleys, and depositing till in moraines and outwash downstream from moraines. During the last major glaciation, the climatic snowline was about 8000 feet altitude, and the orographic snowline was about 6500 feet in the Trinity Alps (Sharp 1960). Cirques formed at somewhat lower elevations nearer the coast. Periglacial climate increased mass movement on slopes around the glaciers and melting glaciers increased fluvial activity downstream from the glaciers. On serpentine rocks, Quaternary glaciation was most extensive in the Mt. Eddy area of extreme northeastern Trinity County, but small glaciers occupied the northern slopes of many other serpentine mountains in the domain.

Most of the Klamath Mountains area drains through the Klamath, Rogue, and a few smaller streams to the coast of the Pacific Ocean. The eastern part of the area, south of the Klamath River, drains through the Sacramento River. Climates range from warm at lower elevations to very cold at higher elevations. Summers are dry, and coastal fog, which reaches the western margin of the Klamath Mountains, has little impact inland from that margin. Mean annual precipitation ranges from about 300 cm near the sea coast to about 50 cm in the driest parts on the east. Summer drought is hardly evident near the coast but pronounced inland (fig. 17-1). Winter snowfall is common, and snow remains on the ground through the winter at the higher elevations. The frost-free period ranges from about 225 days at the lower elevations on the south and near the sea coast to <75 days at the higher elevations.

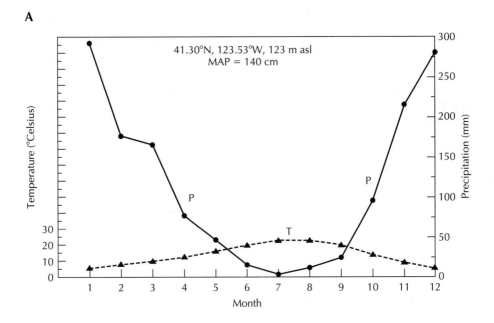
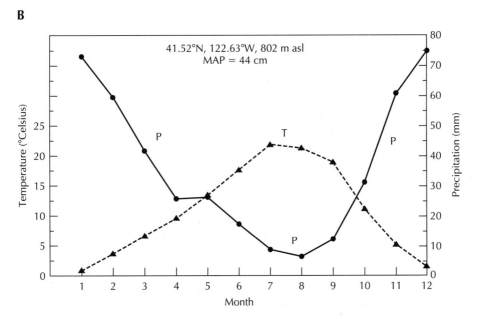

Figure 17-1 Mean monthly temperature and precipitation at two sites in the Klamath Mountains domain. (A) Orleans, California. (B) Yreka, California.

17.1 Geology

The general structure of the Klamath Mountains domain is a series of east-dipping packages of rocks (terranes) decreasing in age westward toward the edge of the continent. Each terrane has undergone more than one period of deformation marked by folding and thrusting that accompanied large-scale rotation of the North American continent (Irwin 1981). These rocks are a complex mixture of igneous and sedimentary rocks that range in age from early Paleozoic to Mesozoic. They formed as part of oceanic island arcs that were formed by volcanic eruptions and deposition of andesitic and basaltic lavas, often interlayered with marine sediments. Extreme erosion of these rocks has exposed deep-seated plutonic rocks of diorite and granodiorite composition. The diverse assemblage of rocks that make up the Klamath Mountain domain has been divided into four separate units based on their ages of formation and rock types. From east to west, these are the Eastern Klamath Belt, the Central Metamorphic Belt, the Paleozoic–Triassic Belt, and the Western Jurassic Belt.

The Eastern Klamath Belt consists of Paleozoic marine sediments that are in thrust contact with the Trinity Ultramafic (locality 5-10) sheet (Lindsley-Griffin 1977a, b). The eastern edge of the Trinity ultramafic sheet is overlain by the Copely greenstone, an altered mafic volcanic of Middle Devonian age.

The Central Metamorphic Belt marks the west boundary of the Eastern Klamath belt where it is thrust over this narrow belt. The Central Metamorphic Belt consists of amphibolite facies metamorphic rocks of Devonian age. Ultramafic rocks are sparse, but at Yreka a thin zone of serpentinite is imbricated within the metamorphic rocks (Hotz 1977).

The Paleozoic–Triassic belt is a huge complex zone of several distinct terranes (Irwin 1979). The North Fork Terrane is a narrow north-trending package containing dismembered ophiolite with narrow zones of sheared serpentinite that was the basement for an island arc assemblage of andesite volcanic flows, chert, tuff, and limestone (Donato 1987). The Eastern and Western Hayfork terranes (locality 5-11) consist of imbricated and accreted island arc andesite rock associated with chert, and argillite (shale) marine sediments. Jurassic calc–alkaline plutons, indicating rapid uplift and deep erosion, intrude all of these associations. There are no significant ultramafic or serpentinite bodies present within these terranes. In the central part of the Paleozoic–Triassic Belt a large irregular circular area of amphibolite facies metamorphism centered around the Condrey Mountain dome has been included in the Rattlesnake Creek terrane (locality 5-12), but some authors prefer to call this area the Marble Mountain terrane. These meta-peridotite bodies are only partially serpentinized. The meta-peridotites are coherent blocks within a tectonic melange that are exposed around the uplifted flanks of the Condrey Mountain dome. Amphbolite, quartz-biotite schist, and marble surround the meta-peridotite blocks in the melange. Many of the meta-peridotites occupy high ridges and have been affected by glaciation. At Preston Peak on the western margin of the Marble Mountain terrane a significant area of tectonite peridotite forms the basement for arc derived from calc–alkaline granite and gabbro. Medium sized bodies of meta-peridotite are found in the Marble Mountain–Kings Castle, Tom Martin, Seiad, and Wrangle Gap–Red Mountain localities. These meta-peridotite bodies are only partially serpentinized.

The Western Jurassic Belt consists of Oxfordian to Early Kimmeridgian (160–153 Ma) marine sediments (Galice Formation) overlying the Josephine ophiolite (locality 5-3) (Harper 1984) that crops out in an arcuate shape paralleling the earlier Klamath trends.

17.2 Soils

Most serpentine soils in the Klamath Mountains are shallow or moderately deep, or, on very steep mountain slopes, shallow to very deep in colluvium. Floodplains and stream terraces are generally narrow and lake plains are not extensive. Salts are leached from well drained soils and carbonates accumulate in very few of them. Clay is leached from surface soils and accumulates in subsoils on relatively stable landforms.

Ochric epipedons predominate in much of this domain, with mollic epipedons predominating in the drier areas, and with some umbric epipedons on nonserpentine soils in the wetter climates. Poorly drained soils are sparse, and few of them are organic or have histic epipedons. Cambic and argillic horizons are the common subsurface diagnostic horizons.

Clay in subsoils is commonly dominated by minerals inherited from soil parent materials. Kaolinite accumulates in some older nonserpentine soils in wetter climates and smectites in many soils in drier areas and in basins where silica and cations are washed in from soils in higher landscape positions. Much serpentine and chlorite are inherited from serpentine parent materials (Istok and Harward 1982). Smectites and iron oxides are more prevalent in serpentine soils than in soils with other kinds of parent materials (Graham et al. 1990). Also, a chlorite/vermiculite interlayer clay mineral forms in some serpentine soils (Lee et al. 2003).

Inceptisols and Alfisols are most extensive on incompletely serpentinized peridotite and Mollisols are common on serpentinite and in drier areas. Ultisols occur on old surfaces in the wetter areas, but are less extensive with serpentine than with nonserpentine parent materials. Small areas of Vertisols occur in basins where the soils are ephemerally wet. Soil temperature regimes are isomesic near the coast and thermic, mesic, frigid, and cryic at increasing elevations inland. Soil moisture regimes are predominantly xeric, with relatively small areas having udic soil moisture regimes (SMRs) near the coast. Also, there are some small areas of poorly drained soils with aquic soil moisture regimes.

17.3 Vegetation

A variety of forest, chaparral, and alpine plant communities occur in this domain, and there are some small serpentine fens. Serpentine plant associations in Oregon have been characterized by Atzet and Wheeler (1984). In California, those in the western Klamath Mountains have been described by Jimerson et al. (1995) and those in the eastern Klamath Mountains by C. (Bud) Adamson (Shasta-Trinity National Forest, Redding, CA, unpublished data).

17.4 Pearsol Peak, Locality 5-1

Locality 5-1 is in steep mountainous terrain, with Pearsol Peak (1828 m) on the western border of Josephine County at the highest point. The mean annual precipitation ranges from about 250 cm at lower elevations on the north to about 400 cm at higher elevations around Pearsol Peak.

Soils of the shallow Pearsol, the moderately deep Dubakella, and the deep Eightlar series are the main serpentine soils that have been mapped in this locality (Borine 1983). These soils are shallow (Lithic) Haploxerepts, moderately deep Mollic Haploxeralfs, and very deep Typic Haploxerepts, all with mesic soil temperature regimens (STRs). The predominant plant communities are open to semi-dense forests with Jeffrey pine, incense-cedar, and Douglas-fir trees in the overstory and Idaho or California fescue in the understory. Whiteleaf manzanita (*Arctostaphylos viscida*) is a common successional shrub.

17.5 Sexton Mountain, Locality 5-2

At Sexton Mountain sheared serpentinite and blocky serpentinized peridotite form the basal part of a dismembered ophiolite that forms the western boundary of the Paleozoic–Triassic belt (Ramp and Peterson 1979). These rocks are thrust westward over the Western Jurassic Belt forming a discontinuous zone of northeastern trending elongate serpentinite bodies as far as Drew, Oregon.

This locality extends from the vicinity of Sexton Mountain in northern Josephine County northeast into southern Douglas County. Most of the elevations are in the 500–1200 m range, but >1400 m on Red Mountain in southern Douglas County. The mean annual precipitation ranges from about 75 cm at lower elevations on the southwest to about 125 cm at higher elevations on the northeast.

Soils of the moderately deep Dubakella and the deep Cornutt series are the main serpentine soils that have been mapped in this locality (Borine 1983). These soils are Mollic and Ultic Haploxeralfs with mesic STRs. They are mostly on steep and moderately steep slopes. Most of the steep slopes face north, according to the soils maps of Josephine County, Oregon (Borine 1983). The predominant plant communities are open to semi-dense forests with Jeffrey pine, incense-cedar, and Douglas-fir trees in the overstory and Idaho or California fescue in the understory.

17.6 Josephine, Locality 5-3

The Josephine locality is in the Western Jurassic Belt on the western margin of the Klamath Mountains, in Oregon and California. It is mostly in Del Norte County, California, and Curry and Josephine counties, Oregon.

The Western Jurassic Belt consists of Oxfordian to Early Kimmeridgian (160–153 Ma) marine sediments (Galice Formation) and the Josephine ophiolite (Harper 1984) that crop out in an arcuate shape paralleling the earlier Klamath trends. They are overthrust westward over the Dothan graywacke trench sediments of comparable age. Partially serpentinized peridotite is the dominant rock type exposed and covers >800 km^2.

Marine sediments of the Galice formation containing *Buchia concentrica* of Early Kimmeridgian age are deposited over the Josephine ophiolite (Harper et al. 1994). The Josephine ultramafic was folded, as it was thrust over the outboard Dothan trench graywackes to the west. During the Early Tertiary the Josephine ophiolite was near sea level and its upper surface was reduced to a relatively flat surface (Klamath peneplane). The ultramafic rocks of the Josephine were exposed to long term weathering on this low relief surface developed "nickel laterite" (Ramp 1978). Small areas of this old surface are preserved on this uplifted surface in the Klamath Mountains and adjacent California Coast Ranges from Nickel Mountain in locality 4-1 to Red Mountain locality 4-8, with the most extense areas occuring in the Josephine locality. The "nickel laterite" deposits attain thicknesses of up to 15 m on this old surface. The Miocene Wimer Formation overlap on these nickel laterites on the west edge of the uplifted Klamath erosion surface and confirm the low stand of the surface which is now elevated up to 1300 m (Coleman and Kruckeberg 1999).

Elevations in the Josephine locality range from 100 m, or less, along the Smith River near Gasquet to about 1000 m over the divide to the Illinois River and down to about 300 m along the Illinois River. Josephine Mountain (1452 m) is one of the higher points in the locality. Mean annual precipitation in the locality ranges from about 200 cm along the Smith River below Gasquet up to 400 cm at the higher elevations and down to 125 cm adjacent to the Illinois Valley. Much of the precipitation is snow, and snow covers the ground continuously through winters at the higher elevations.

Serpentine soils mapped in Del Norte County have been classified as Lithic Haploxerepts, Lithic Haploxeralfs, moderately deep Typic Xerochrepts, and deep Ultic Haploxeralfs and Typic Haploxerults with mesic STRs (Miles et al. 1993). They were assumed to have xeric SMRs, but some of the wetter soils may have udic SMRs. Subsequently, laboratory analyses have indicated that the soils mapped as Haploxerults are actually Kanhaploxeralfs and Kanhaplohumults.

Serpentine soils mapped in western Josephine County are mainly in the Pearsol, Dubakella, and Eightlar series (Borine, 1983). These soils are shallow (Lithic) Haploxerepts, moderately deep Mollic Haploxeralfs, and very deep Typic Haploxerepts, all with mesic STRs and xeric SMRs.

The plant communities in drier areas that drain to the Illinois River are open forests on shallow soils and denser forests on deeper soils with mainly Jeffrey pine and incense cedar in the overstory and fescues (*Festuca idahoensis* and *F. californica*) in the understory. Douglas-fir and sugar pine are common trees in the wetter areas. The white pine in some of the wetter areas is western white (*Pinus monticola*), rather than sugar pine (*P. lambertiana*). The most common broad-leaved trees are tanoak (*Lithocarpus densiflora*), madrone (*Arbutus menziesi*), and California bay (*Umbellularia californica*). Common shrubs in the wetter areas are huckleberry oak (*Quercus vaccinifolia*), California coffeeberry (*Rhamnus californica*), California huckleberry (*Vaccinium ovatum*), and red huckleberry (*V. parvifolium*).

After severe fires, knobcone pine (*P. attenuata*) is common inland and lodgepole pine (*P. contorta*), near the coast.

Steep pitcher-plant (*Darlingtonia californica*) fens are present on some mountain slopes. Although sparse, they are of great botanical interest. The most famous is the gently sloping Howell Fen in colluvium at the foot of Eight Dollar Mountain in Josephine County. About a meter of organic soils has accumulated over alluvium that has collected in a colluvial depression there. The soil is a Histosol (a Terric Haplosaprist). It has a neutral pH (7.0) at the upper margin of the fen, decreasing to pH 6.6 downslope.

17.7 Cave Junction, Locality 5-4

The Cave Junction locality is in moderately steep-to-steep hilly and mountainous terrain on the east side of the Illinois Valley. Elevations range from about 400 m on the edge of the Valley up several hundred meters on the higher mountains. Mean annual precipitation is 100–150 cm.

Serpentinites of the locality are highly sheared and contain numerous tectonic inclusions of rodingite and gabbro. Lizardite and chrysotile are the main serpentine minerals. East of Cave Junction serpentine melange lies at the base of the westward directed, low angle, east-dipping Orleans Thrust fault (Ramp and Peterson 1979), which marks the western boundary of the Rattlesnake Creek terrane (or Marble Mountain terrane) there.

Soils of the moderately deep Dubakella and the deep Cornutt series are the main serpentine soils that have been mapped in this locality (Borine 1983). These soils are Mollic and Ultic Haploxeralfs with mesic STRs. They are mostly on steep and moderately steep slopes. Most of the steep slopes face north, according to the soils maps of Josephine County, Oregon (Borine 1983). The predominant plant communities are open to semi-dense forests with Jeffrey pine, incense-cedar, and Douglasfir trees in the overstory and Idaho and California fescues in the understory.

17.8 Preston Peak, Locality 5-5

Preston Peak is in steep mountainous terrain of western Siskiyou County, California. At Preston Peak on the western margin of the Marble Mountain terrane a large area of tectonite peridotite forms the basement for arc derived from calc–alkaline granite and gabbro (Snoke 1977, Snoke et al. 1981). Preston Peak (2228 m) rises a few hundred meters above surrounding summits of this locality that represent the old Klamath peneplain. Glacial cirques are evident on the north sides of a few of the higher mountains and glaciers flowed short distances down Clear Creek and the West Fork of Indian Creek.

The mean annual precipitation ranges from 150–250 cm. Much of it is snow and snow remains on the ground through winters at the higher elevations.

Soils are mostly shallow and moderately deep-to-very deep Inceptisols (Haploxerepts) with mesic and frigid STRs (Foster and Lang 1982, Soil Survey Staff 1999). Many of these soils have Bt horizons and some of them are thick enough to be argillic horizons, making them Haploxeralfs. Cold, shallow Entisols (Lithic Cyrorthents) and Inceptisols (Lithic Dystrocryepts) are common, but not extensive, at the higher elevations. Plant communities are open to semi-dense forests with Jeffrey pine, incense-cedar, and Douglas-fir trees in the overstory. Sugar pine is common at lower elevations and western white pine at higher elevations. White fir is sparse on serpentine soils. Brewer spruce (*Picea breweriana*) is present on some serpentine soils. Madrone and tanoak trees are sparse on the soils of this locality. Huckleberry oak, pinemat manzanita (*Arctostaphylos nevadensis*), greenleaf manzanita (*A. patula*), Oregon grape (*Berberis nervosa*) and silktassel (*Garrya buxifolia*) are common shrubs at the higher elevations, above 1000 m, and beargrass (*Xerophyllum tenax*) is common, also. Buckbrush (*Ceanothus cuneatus*) and whiteleaf manzanita are common at the lower elevations, below 1200 m. Serviceberry (*Amelanchier* sp.) and California coffeeberry are common shrubs at all elevations. Grasses are not abundant, but besides the more common Idaho and California fescues, purple reedgrass (*Calamagrostis purpurascens*) and oniongrass (*Melica* spp.) are commonly present.

17.9 Seiad, Locality 5-6

The Seiad locality is in Siskiyou County, north from the Klamath River near the mouth of Seiad Creek across the Siskiyou Mountains drainage divide toward the Applegate River. It is steep, mountainous terrane. Medium-sized bodies of partially serpentinized metaperidotite are found in the Seiad locality. Elevations range from 400–2000 m. The summit of Red Butte (2054 m) on the Siskiyou crest is marble.

The mean annual precipitation is 100–150 cm. Much of it is snow, and snow remains on the ground through winters at the higher elevations.

Soils are mostly shallow and moderately deep-to-very deep Inceptisols (Haploxerepts) or Alfisols (Haploxeralfs) with mesic and frigid STRs (Foster and Lang 1982, Soil Survey Staff 1999). The shallow soils were mapped as Lithic Haploxerepts (NRCS pedon S76CA093-6) and Lithic Ruptic-Inceptic Haploxeralfs (Foster and Lang 1982). Plant communities are open to semi-dense forests with Jeffrey pine, incense-cedar, and Douglas-fir trees in the overstory. Sugar pine is common at lower elevations and western white pine at higher elevations. White fir is sparse on serpentine soils. Baker cypress (*Cupressus. bakeri*) is present on some of the shallow serpentine soils. Madrone trees are sparse on the soils of this locality. Huckleberry oak, pinemat manzanita, greenleaf manzanita, Oregon grape, and silktassel are common shrubs at the higher elevations, above 1000 m, and beargrass is common also. Buckbrush and whiteleaf manzanita are common at the lower elevations, below 1200 m. Serviceberry and California coffeeberry are common shrubs at all elevations. Grasses are not abundant, although Idaho and California fescues are more common at lower elevations. Purple reedgrass and oniongrass are sparse.

17.10 Marble Mountain–Kings Castle, Locality 5-7

Locality 5-7 is in the Marble Mountains of Siskiyou County, mostly in a wilderness area. Medium-sized bodies of partially serpentinized meta-peridotite are found in the Marble Mountain-Kings Castle locality (Donato 1987). The locality is mountainous, with many glacial cirques and broad glaciated valleys. Elevations range from about 800 m along the Scott River to 2530 m on Boulder Peak. Boulder Peak is the highest point, other than Mt. Eddy, that is a remnant from the old Klamath peneplain. Serpentine soils were mapped up to 2460 m on Red Mountain, just south of Boulder Peak (Foster and Lang 1982).

The mean annual precipitation is 75–200 cm. Much of it is snow, and snow remains on the ground through winters at the higher elevations.

Soils are mostly shallow and moderately deep-to-very deep Inceptisols (Haploxerepts) and Alfisols (Haploxeralfs) with mesic and frigid STRs and Entisols, Inceptisols, and Mollisols with cryic STRs. The shallow soils are Lithic Ruptic–Inceptic Haploxeralfs, Lithic Cryorthents, Eutrocryepts, and Haplocryolls and the deeper soils are Typic Haploxerepts, Haploxeralfs, and Eutrocryepts (Foster and Lang 1982, Soil Survey Staff 1999).

Plant communities at lower to intermediate elevations are open to semi-dense forests with Jeffrey pine, incense-cedar, and Douglas-fir trees in the overstory, commonly with some sugar pine and madrone (Foster and Lang 1982). White fir is sparse. On cool soils (mesic STRs) at lower elevations, buckbrush and whiteleaf manzanita are common shrubs, and fescues are generally present. On cool soils at intermediate elevations, above ~1000 m, and on cold soils (frigid STRs) below ~1800–2000 m, huckleberry oak, pine-mat manzanita, and greenleaf manzanita are common shrubs, and pipsissawa (*Chimaphila umbellata*) is common on forest floors. Beargrass is common in some places and California coffeeberry, silktassel, and serviceberry are common at all of these elevations. On cold soils at the higher elevations, about 1800–2200 m, Jeffrey pine and western white pine are common trees, silktassel is common along with some of the previously mentioned shrubs, and phlox (*Phlox diffusa*) is common on the ground. The ridge striking eastward from Red Mountain appears to be typical for very cold serpentine soils (cryic STRs) that are above 2100–2200 m. The ridge crest is occupied by an open whitebark pine (*Pinus albicaulis*) stand with pinemat manzanita and forbs in the understory. Foxtail pine (*P. balfouriana*) is sparse. The south-facing slope is nearly barren, with considerable prostrate buckwheat (*Eriogonum umbellatum* and *E. libertini*), phlox, sedge (*Carex* spp.), linanthus (*L. nuttallii*), sedum (*S. lanceolatum*), and daisy (*Erigeron bloomeri*). The north-facing slope has a hemlock (*Tsuga mertensiana*) forest with drifted snow that persists into summer.

17.11 Tom Martin, Locality 5-8

The Tom Martin locality is in Siskiyou County, from about 400 m elevation along the Klamath River south over Tom Martin Peak (2142 m) and down to about 600 m above the Scott River. It is steep, mountainous terrane. Medium-sized bodies of partially serpentinized meta-peridotite are found in the Tom Martin locality.

The mean annual precipitation is 100–150 cm. Much of it is snow, and snow remains on the ground through winters at the higher elevations.

Serpentine soils were mapped up to about 2100 m on the northwest and southeast ends of the short ridge on which Tom Martin Peak is located (Foster and Lang 1982). Soils are mostly shallow and moderately deep-to-very deep Inceptisols (Haploxerepts) or Alfisols (Haploxeralfs) with mesic and frigid STRs. The shallow soils were mapped as Lithic Haploxerepts and Lithic Ruptic-Inceptic Haploxeralfs, and the deeper soils were mapped as Typic Xerochrepts and Haploxeralfs (Foster and Lang 1982, Soil Survey Staff 1999). A small area of very cold soils (cryic STRs), too small to map on the broad survey of the Klamath National Forest, is present just northwest of Tom Martin Peak.

Plant communities are open to semi-dense forests with Jeffrey pine, incense-cedar, and Douglas-fir trees in the overstory. White fir is sparse on serpentine soils. Sugar pine is common at lower elevations and western white pine at higher elevations. Huckleberry oak, pinemat manzanita, greenleaf manzanita, Oregon grape, and silktassel are common shrubs at the higher elevations, above 1000 m, and beargrass is common, also. Buckbrush and whiteleaf manzanita are common at the lower elevations, below 1200 m. Serviceberry and California coffeeberry are common shrubs at all elevations. Fescues are more common at the lower elevations and on south-facing slopes at the higher elevations. An open stand of western white pine and mountain hemlock with greenleaf manzanita and rock-spiraea (*Holodiscus microphyllus*) in the shrub layer and phlox and beargrass on the ground prevails on the very cold soils. Lemmon's swordfern (*Polysticum lemmonii*) is common in this area as well.

17.12 Wrangle Gap–Red Mountain, Locality 5-9

The Wrangle Gap locality extends from about 600 m elevation along the Klamath River in Siskiyou County, California, north over the Siskiyou Mountains drainage divide at Big Red Mountain (2142 m) in Jackson County, Oregon. It is between Condrey Mountain on the west and the Ashland pluton on the northeast. Medium-sized bodies of partially serpentinized meta-peridotite are found in the Wrangle Gap–Red Mountain locality (Kays et al. 1988).

The mean annual precipitation is 75–150 cm. Much of it is snow, and snow remains on the ground through winters at the higher elevations.

Soils are mostly shallow and moderately deep to very deep Inceptisols (Haploxerepts), Mollisols (Argixerolls), and Alfisols (Haploxeralfs) with mesic and frigid STRs. The shallow soils were mapped as Lithic Haploxerepts, Lithic Argixerolls, and Lithic Mollic Haploxeralfs, and the deeper soils were mapped as Typic Xerochrepts, Pachic Argixerolls, Mollic and Typic Haploxeralfs, and Mollic Palexeralfs (Foster and Lang 1982, Soil Survey Staff 1999). Some cold soils (cryic STRs), Lithic Cryorthents and Haplocryolls, were mapped above 2000 m.

Open to semi-dense forests with Jeffrey pine, incense-cedar, and Douglas-fir trees prevail on cool soils at the lower elevations and on cold soils at the intermediate elevations.

Sugar pine is common at lower elevations and western white pine at higher elevations. White fir is sparse on serpentine soils. Buckbrush and whiteleaf manzanita are common shrubs at the lower elevations and huckleberry oak, pinemat manzanita, and greenleaf manzanita are common at the higher elevations. Fescues are common in the understory and beargrass is common at the higher elevations. Serviceberry is a common shrub at all elevations.

Red Mountain (or Big Red Mountain) has a unique combination of plants. Although the ultimate summit of Red Mountain is gabbro, serpentinized peridotite is no more than 25 m away and only 1 or 2 m lower. Colluvium in soils near the summit is commonly mixed, but mostly ultramafic material. Ground cover in the treeless area on the south slope from Big Red Mountain is mostly prostrate buckwheat, and sandwort (*Arenaria congesta*), with much phlox (*P. diffusa*) and daisy, and common coyote mint (*Monardella odoratissima*), hawksbeard (*Crepis pleurocarpa*), lupine (*Lupinus* sp.), and locoweed (*Astragalus* sp.). A few of the less common forbs that are characteristically on the serpentine soils are phacelia (*P. corymbosa*), sandwort (*Minuartia nutallii*), mountain violet (*Viola pinetorum*), Lee's lewisia (*Lewisia leana*), and onion (*Allium* sp.). An open white pine/pinemat manzanita/beargrass forest with sparse white fir (*Abies concolor*) is characteristic of the north slope from Red Mountain, with sparse ferns (*Aspidota densa* and *Polystichum lemonii*).

17.13 Trinity, Locality 5-10

The Trinity locality is a large ultramafic body that is spread over 2000 km^2 from southern Siskiyou County south into Shasta and Trinity counties. The area of serpentine rocks is nearer 1200 km^2 because of large silicic and mafic plutons in the locality.

The Trinity ultramafic sheet is a subhorizontal slab made up of peridotite, gabbro, and diabase and considered to have formed in the Devonian period 437 Ma. The ultramafic unit consists of partially serpentinized harzburgite, lherzolite, and dunite. The gabbro bodies are intrusive into the ultramafic rocks and have nearly the same age. Numerous petrologic studies of the Trinity ultramafic have referred to this complex as an ophiolite, which implies that it originated as oceanic crust.

The Trinity ultramafic body is in a mountainous area (fig. 17-2). Elevations range from about 750 m along the Sacramento River and 900 m along the Trinity River up to 2751 m on Mt. Eddy. The highest serpentine terrain in western North America and most distinctly alpine south of the Northern Cascade Mountains is on Mt. Eddy. Much of the northern part of the area has been glaciated, and glaciers from the Trinity Alps have crossed eastward over ultramafic rocks in Trinity County. The mean annual precipitation is about 50 cm on the northern edge of the area to 175 cm in the mountains south of Mt. Eddy that are between the Sacramento and Trinity rivers. Much of the precipitation falls as snow, and it accumulates and persists though winters at the higher elevations. The frost-free period is from <25 days at higher elevations to 100 days or more at lower elevations.

Figure 17-2 A view across High Camp basin to Mt. Eddy, with Mt. Shasta on the left. Ultramafic rocks and glacial drift from them dominate the landscape, which is in locality 5-10 (Trinity), except for the Mt. Shasta volcano in the Southern Cascade Mountains. The timber stand in the foreground is foxtail pine (*Pinus balfouriana*).

The serpentine soils in this locality are mainly shallow and moderately deep Alfisols, Mollisols, and Inceptisols, although many in colluvium and glacial drift are deeper. Most have mesic, some have frigid, and few have cryic STRs. Bud Adamson (unpublished data) has described the serpentine soils and plant communities at many sites on the Shasta-Trinity National Forest, allowing some generalizations to be made about relationships among soils and other environmental parameters.

Rock outcrop and very cold Inceptisols (Cryepts) prevail above 2000 m (Lanspa 1993). Trees are mostly in open stands or sparse, although somewhat dense stands of mountain hemlock (*Tsuga mertensiana*) and Shasta red fir (*Abies magnifica* var. *shastensis*) occur in sheltered locations, commonly on gabbro. Jeffrey pine, western white pine, and white fir are common on serpentine soils, and in places foxtail pine. Huckleberry oak, pinemat manzanita, and bush chinquapin (*Chrysolepis sempreviens*) are common shrubs, and there is some Klamath manzanita (*A. klamathensis*) that is endemic in the area. Leatherleaf mountain mahogany (*C. ledifolius*) occurs on many narrow ridges. Coyote mint is abundant in the understory of some forest stands. Some unique and highly diverse conifer stands are codominated by as many as six species, including *Pinus albicaulis*, *P. balfouriana*, *P. contorta* ssp. *murryana*, *P. monticola*, *Tsuga mertensiana*, and *Abies magnifica* var. *shastensis* (Keeler-Wolf 1987). Shaded glacial cirques hold herbaceous plant communities dominated by *Penstemon rupicola* and *Juncus parryi*, with many localized and endemic species and ericaceous subshrubs such as pink mountain-heather (*Phyllodoce empetriformis*) and white mountain-heather (*Cassiope mertensiana*) and many localized or endemic herbaceous species such as *Carex gigas, Lewisia leana, Minuartia nutalli, Saxafraga mertensiana*, and *Romanzoffia sitchensis*.

Whipple and Cole (1979) described two phases of alpine vegetation on Mt. Eddy. One, characterized by bladder-pod (*Lesquerella occidentallis*) with *Poa pringlei, Erigeron compositus, Hulsea nana, Potentilla fruticosa*, and *Ivesia gordoni*, occurs on the highest exposed ridges. The other, which occurs on southerly facing slopes, is characterized by the endemic *Eriogonum siskiyouensis* with *Crepis pleurocarpa, Phlox diffusa, Sedum lanceolatum*, and *Arenaria congesta*. Many locally or regionally rare taxa occur in this setting, including *Draba aureola, Eriogonum alpinum, Campanula scabrella*, and *Polemonium chartaceum*.

Between 2000 m and 1600 m, much higher on south-facing and lower on north-facing slopes, the soils have frigid STRs. They are mainly Inceptisols and Alfisol, with Mollisols in the drier areas. These Inceptisols (Haploxerepts) and Alfisols (Haploxeralfs) have open to semi-dense forests with common Jeffrey pine, sugar pine, and white fir and less commonly Douglas-fir trees in the overstory. Huckleberry oak is a common shrub, and pipsissewa is common to abundant in the understories of many open forests. Open to sparse Jeffrey pine forests with leatherleaf mountain mahogany and bluebunch wheatgrass (*Pseudoroegneria spicata*) are common on the Mollisols (Lithic and Typic Argixerolls).

Most extensive are soils at lower elevations, with mesic STRs. Alfisols (Mollic and Ultic Haploxeralfs) predominate with higher precipitation (mean annual precipitation [MAP] >125–150 cm) and Mollisols (Lithic and Typic Argixerolls) predominate with lower precipitation (MAP <100–125 cm). Open to sparse stands of Jeffrey pine and incense-cedar, with buckbrush, fescues, and sparse but ubiquitous oniongrass are the most common plant communities on the Mollisols. Douglas-fir is common, along with Jeffrey pine and incense-cedar on the Alfisols, and there is more California coffeeberry than on the Mollisols. The yellow pine is ponderosa (*Pinus ponderosa*) rather than Jeffrey pine around Castle Crags, because the ultramafic rocks there have more calcium and the precipitation is high enough to leach much of the excess magnesium from the soils.

Serpentine fens and wet meadows are small, but fairly common on steep to gentle slopes in this locality. They are rich with localized and endemic species, including pitcher-plant, *Cypripedium californicum, Tofieldia glutinosa, Narthecium californicum, Veronica copelandi, Kalmia polifolia* var. *microphylla, Vaccinium uliginosum, Drosera, rotundifolia, Carex buxbaumii, Menyanthes trifoliata*, and *Dulichium arundinaceum*. Surrounding many of the meadows and along many streams are the most continental and diverse stands of Port Orford cedar (*Chamaecyparis lawsoniana*). Jimerson et al. (1995) have described six plant association in the Port Orford-cedar alliance, from very cold stands containing mountain hemlock to warm stands with western spicebush (*Calycanthus occidentalis*) in the understory.

A few small alkali seeps in this locality have a dominant cover of saltgrass (*Distichlis spicata*) with many local mineral or halophitic species, including *Triglochin maritima, Cirsium breweri, Sisyrinchium idahoensis*, and *Hastingia alba*. The alkali seeps are presumed to be downslope from alkaline springs flowing from masses of peridotite that are being serpentinized presently, as described by Barnes et al. (1967). Water sampled from these kinds of springs in the Tuolumne locality (2-7), Canyon Mountain locality (6-1), and the

Cedars locality (4-19) contained mostly calcium, sodium, and chlorine with mean concentrations of 2.3(+), 1.8(+), and 1.0(−) mmol/L. Presumably the ion balance was maintained with hydroxyl ions and a lesser amount of silicon oxyanions (for example, $H_3SiO_4^-$). Calcium is readily precipitated as a carbonate in the atmospheric environment (chapters 3 and 4), but sodium remains in solution and is carried downslope from the springs to create halophytic habitats.

17.14 Hayfork, Locality 5-11

The Eastern and Western Hayfork terranes (locality 67) consist of imbricated and accreted island arc andesite rock associated with chert, and argillite (shale) marine sediments. Jurassic calc-alkaline plutons, with indications of their deep erosion and rapid uplift, intrude all of these associations. There are no significant ultramafic or serpentinite bodies present within these terranes.

17.15 Rattlesnake Creek, Locality 5-12a

The Rattlesnake Creek locality is the southern exposure of the Rattlesnake Creek terrane (RCT), from western Tehama County through Trinity County to eastern Humboldt County (Wright and Wyld 1994). Terrane identified as Rattlesnake Creek terrane farther north in the Klamath Mountains may be different—some investigators prefer to call it Marble Mountain terrane (see section 17.1).

The southern exposure of the Rattlesnake Creek terrane is a narrow strip, 10–12 km wide and 120 km long, of serpentine melange and overlying early Mesozoic deposits with intrusions of gabbro and more silicic Mesozoic plutons (Wright and Wyld 1994). It is between the Salt Creek and the Bear Wallow faults, which are parallel to the South Fork fault along the southwestern margin of the Klamath Mountains. Large slabs of somewhat serpentinized peridotite occur adjacent to the Salt Creek fault from Tedoc Mountain in Tehama County up through central Trinity County. With few exceptions, such as Red Mountain (one of two Red Mountains in Trinity County), all large bodies of peridotite are near the Salt Creek fault. Ultramafic rocks in the serpentine melange are mostly massive-to-sheared serpentinite. The peridotite is mostly in large blocks. Pyroxenite is sparse.

Most of the soils are cool (mesic STR, fig.17-3). Areas of cold (frigid STR) and warm (thermic STR) soils are relatively small and there are no very cold (cryic STR) soils in the RCT of the southwestern Klamath Mountains. Elevations range from 250 m along the South Fork of the Trinity River and 700 m along Beegum Creek to 1793 m on Dubakella Mountain, although there are no ultramafic rocks near the summit of the mountain, and up to about 1790 m, where there is serpentine on Rat Trap Ridge. Mean annual precipitation ranges from 75 cm or less along Beegum Creek on the southeast to 150 or 200 cm in the higher mountains and on the northwest. The entire RCT is mountainous, with few broad valleys, and landslides are common.

Figure 17-3 A typical serpentine soil and plant community at 1160 m elevation in the Rattlesnake Creek terrane, Trinity County, California. The soil is a moderately deep Ultic Haploxeralf (tentative Hyampom series) and the plant community is Jeffrey pine–incense-fir–Douglas-fir/buckbrush (*Ceanothus cuneatus*)/Idaho fescue (*Festuca idahoensis*).

Serpentine soils of the locality have been mapped in detail (Alexander 2003), about 12,000 ha (30,000 acres). It is a unique survey in that soils from partial serpentinized peridotite were differentiated from those with completely serpentinized peridotite (serpentinite) parent materials. Slopes in the area are about 24% very steep (slope >60%), 61% steep, and 15% moderately to gently sloping (slope <30%). In general, the peridotite terrain is steeper (fig. 17-4) because slopes in it are more stable than those in the serpentinite terrain. The soils are predominantly shallow to moderately deep: 10% rock outcrop, 50% shallow soils, 30% moderately deep soils (0.5–1 m deep), and 6% deep to very deep soils on very steep slopes and 4% in landslides. Barren areas are common in 12%–15% of the area, mostly on both shallow and very deep soils in very steep peridotite terrain and on shallow soils in moderately steep to gently sloping serpentinite terrain. Carbonates have been leached from the soils, but plenty of basic cations remain to maintain neutral to slightly acid reaction. A few serpentine soils on the older land surfaces have moderately acid surfaces. Mollic epipedons are common, more in soils on serpentinite than in soils on peridotite. There are no diagnostic subsoil horizons other than cambic and argillic horizons. Soils on peridotite commonly have redder hues than those on serpentinite.

Soils on peridotite are predominantly Lithic Mollic Haploxeralfs (tentative Wildmad Series) and moderately deep Mollic Haploxeralfs (Dubakella Series), with much rock outcrop, Lithic Xerorthents (Hiltabidel Series), and Typic Haploxerepts on very steep slopes (table 17-1). There are some Haplic Palexeralfs on colluvial footslopes and landslides. Soils on serpentinite are predominantly Lithic Ultic Argixerolls (tentative Bramlet Series) and moderately deep Ultic Haploxeralfs (tentative Hyampom series) and Typic or Ultic Argixerolls, with some Lithic Xerorthents, Lithic Haploxerolls, Typic Haploxerepts, and

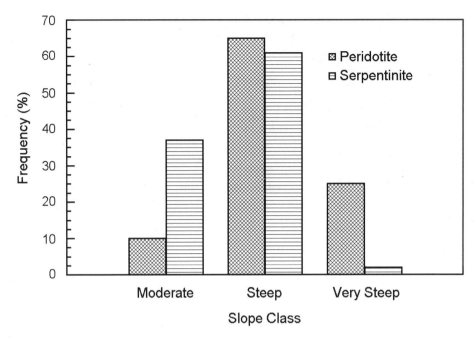

Figure 17-4 Slope class frequencies for peridotite and serpentinite soils in the Rattlesnake Creek terrane. The slope classes are moderately sloping, slopes <30%; steep, 25%–65%; and very steep, >60%.

Typic Haploxerolls on eroded summits and very steep slopes. Soils in serpentine barrens are predominantly Lithic Xerorthents and Lithic Haploxerolls. Warm soils on steep slopes are similar to cool soils, but they commonly have higher clay concentrations. Warm soils on moderately steep slopes are predominantly Lithic Argixerolls (Henneke Series) and moderately deep Typic Palexerolls. There are some closed depressions in landslides with wet soils. Many of the wet soils are Entisols and Mollisols, one is a Histosol, and some that are only seasonally wet are Vertisols.

Trees on the serpentine soils are predominantly Jeffrey pine and incense-cedar. Douglas-fir is common on extremely stony soils. Ponderosa pine and black oak are absent from the serpentine soils, unless the soils are intensively leached, and white fir does not occur on the serpentine soils with mesic soil temperature regimes. Sugar pine and madrone are sparse here—less common than on serpentine soils farther north in the Klamath Mountains. Canyon live oak appears to avoid serpentine soils of the RCT, with a few exceptions. White oak or scrub white oak grow on some soils where the serpentine syndrome is not too intense. Gray pine is the only common conifer on warm soils. Although it is present on most warm serpentine soils, it is sparsely distributed.

Shrubs are generally more common on the more shallow soils, although stand densities may be reduced on very shallow soils. Leather oak and whiteleaf manzanita are

Table 17-1 Soil distributions in steep and very steep serpentinite and peridotite terrains with soils in mesic soil temperature regimes.

Slope gradient	Peridotite		Serpentinite	
	N-facing slopes	S-facing slopes	N-facing slopes	S-facing slopes
>60%	Rock Outcrop	Rock Outcrop	T. Haploxerepts	T. Haploxerolls
	L. Xerorthents	L. Xerorthents	L. Xerorthents	L. Ultic Argixerolls
	T. Haploxerepts	Haplic Palexeralfs		
30%–60%	M. Haploxeralfs	L.M. Haploxeralfs	U. Haploxeralfs	L Ultic Argixerolls
	L.M. Haploxeralfs	M. Haploxeralfs	L. Ultic Argixerolls	U. Haploxeralfs
				Ultic Argixerolls
Summit <15%	Ultic Haploxeralfs		Lithic Haploxerolls	
	Mollic Palexeralfs		L. Ultic Argixerolls	

Very steep slopes (slopes > 60%) are common on peridotite but sparse on serpentinite. Gently sloping mountain summits are not extensive. Soil subgroup abbreviations: L., Lithic; M., Mollic; T., Typic; and U., Ultic. Typic Palexerolls are common on gentle slopes at low elevations (thermic soil temperature regimes).

common and abundant on cool serpentine soils of the RCT and huckleberry oak is common at the higher elevations. Leather oak, however, occurs only in the eastern half of the RCT. Buckbrush is common and widely distributed at all elevations. Whiteleaf manzanita is common on disturbed soils, and there is some Stanford manzanita (*Arctostaphylos stanfordiana*) on the east and much hoary manzanita (*A. canescens*) on the west in the RCT. Greenleaf and common manzanitas occur on serpentine soils only at the higher and lower elevations and on pyroxenite and leached soils. Coffeeberry is widely distributed, but sparse, on serpentine soils. It is mostly hoary coffeeberry (*Rhamnus tomentella crassifolia*) in the eastern part and California coffeeberry in the western part of the RCT. Silktassel is less widely distributed, but it is abundant on some soils—valley silktassel (*Garrya congdoni*) on warm soils and mountain silktassel (*G. buxifolia*) on cool soils. Serviceberry is widely distributed, but sparse. White oak (*Quercus garryana*) occurs on serpentine soils that have been leached moderately to intensively. Azalea (*Rhododendron occidentale*) and willow (*Salix* sp.) occur around wet areas, and spicebush is common where groundwater seeps to the surface. Common species of ceanothus other than buckbrush—that is, deerbrush (*C. integerrimus*), mountain whitethorn (*C. cordulatus*), tobaccobrush (*C. velutinus*), and prostrate ceanothus (*C. prostratus*)—avoid the serpentine soils. Serpentine goldenbush (*Ericameria ophitidis*) and yerba santa (*Eriodictyon californicum*) are common on many disturbed serpentine soils. Bastard toad-flax (*Comandra umbellata californica*) is fairly common in warmer areas, on soils with thermic STRs.

Many rare and sensitive forbs occur on serpentine soils of the RCT. Susan Erwin (U.S. Forest Service) is making an inventory of the plants by soil map unit. She anticipates discovering many plant–habitat relationships that were not documented in the survey of serpentine soils in the RCT. Most forbs are sparse, with some exceptions on disturbed

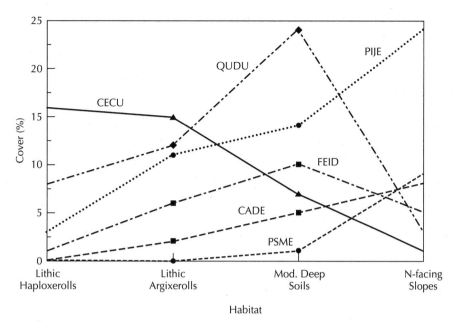

Figure 17-5 Visually estimated plant cover for the most abundant species on serpentinite in the eastern half of the Rattlesnake Creek terrane, means for 33 sites. Plant distributions are similar on peridotite. Lithic Haploxerolls are generally very shallow (depth <25 cm) and most Lithic Argixerolls are deeper (18–50 cm deep). Soils on north-facing slopes are most commonly moderately deep (50–100 cm), with fewer shallow soils.

sites, but many of them are botanically important species. Yarrow and wooly sunflower, two of the most common species of composite plants (Asteraceae), have no habitat indicator value. Deltoid balsamroot and yampa are composites that are conspicuous where they occur on serpentine soils, but neither is widely distributed on serpentine soils in the RCT. Lemon fawn lily, scarlet fritillary, spring beauty, mountain violet, western dog violet, Tolmie's pussy ears, lomatium, death camas, maiden blue-eyed Mary, showy phlox, blue-headed gilia, bowl-tubed iris, flame butterweed, California Indian pink, bedstraw (including Yolla Bolly bedstraw, among others), wintergreen, serpentine Phacelia, sticky calycadenia, and peanut sandwort are widely distributed on serpentine soils, and on disturbed serpentine soils naked buckwheat, vinegar weed, bird's-beak, and dogbane are common. Some plants common or abundant in the spring, such as lemon fawn lily, are not visible aboveground during the summer. Starflower is common on the moderately leached hummocky Palexeralfs. Wavy-leaf soap plant is widely distributed on warm serpentine soils.

Idaho and California fescues are the most abundant grasses on the serpentine soils, and squirreltail (*Elymus elymoides*) is the most widely distributed one. Oniongrass is widely, but sparsely, distributed also. Annual fescue (*Vulpia* spp.) is common on disturbed soils, and soft chess (*Bromus hordaceous*), red brome (*B. rubens*), and silver hairgrass (*Aira caryophyllea*) occur at lower altitudes. Tufted hairgrass (*Deschampsia cespitosa*) is common on wet soils, and annual hairgrass (*D. danthoniodes*) is common on Vertisols. California oatgrass (*Danthona californica*) is common on ephemerally wet soils with mixed serpentine–nonserpentine parent materials at Round Bottom. Kentucky bluegrass (*Poa pratensis*) is found on ephemerally wet soils and pine bluegrass (*P. secunda*) is common at lower elevations. Upland sedges, *Carex multicaulis* and *Carex rossi*, are widely distributed, but not abundant in any of the serpentine habitats. Indian's dream (*Aspidotis densa*), the most widely distributed fern, is common, although not abundant, on stony serpentine soils. Imbricate sword fern (*Polystichum imbricans* ssp. *imbricans*) is also common, but less widely distributed than Indian's dream. Nonvascular plants were not identified in the serpentine soil survey.

One characteristic of serpentine soils that is relevant to plant species distributions is the occurrence of barren areas (Kruckeberg 1999). A barren area has no more than a sparse cover of vascular plants and plant detritus, or "litter." Serpentine barrens are most common on very shallow soils and soils in active colluvium, or talus. Prostrate buckwheat (*Eriogonum libertini* and other species) and jewelflower (mostly *Streptanthus barbatus*) are common on the serpentine barrens, and serpentine montia (*Claytonia saxosa*) is common in a barren on Tedoc Mountain. Prostrate milkweed (*Asclepias solaniana*) is sparse and appears to be confined to south-facing slopes on serpentinite soils.

17.16 Yolla Bolly Junction, Sublocality 5-12b

Sublocality 5-12b is at the southern extreme of the Klamath Mountains, where the South Fork and Bear Wallow faults end at the north end of the Coast Range fault zone. The Coast Range fault separates the Coast Ranges on the west from the Great Valley on the east (Jennings 1994, Blake et al. 1999). Jayko and Blake (1986) have called the juncture of the Coast Range fault to the south, the South Fork fault to the northwest, and the Cold Fork fault to the east a triple junction and named it the Yolla Bolly junction. Serpentinite of the Yolla Bolly junction sublocality, which extends a few kilometers west from the Yolla Bolly junction along the north side of the Bear Wallow fault, may be in the Smith River terrane of the Western Jurassic Belt, rather than in the Rattlesnake Creek terrane (Blake et al. 1999).

Serpentine rocks in the Yolla Bolly junction sublocality are serpentinite that is intricately associated with gabbro, diabase, greenstone, metagraywacke, chert, and minor limestone. There is little peridotite in the mix, or melange. They are separated from RCT serpentine rocks in the Wells Creek area to the northwest by nearly 10 km of melange lacking serpentine rocks. The Yolla Bolly junction sublocality is warmer and drier than the RCT to the northwest, and these differences are reflected in the plant community compositions.

Elevations in this sublocality are 700–950 m. The area might be characterized as hilly, rather than mountainous, with mostly moderately steep-to-steep slopes and a few gently sloping flats. The mean annual precipitation is 50–75 cm. Much of it falls as snow that melts between storms.

The topography at the Yolla Bolly junction is hilly and the altitude ranges from 700 to 920 m. Most of the slopes are steep, with predominantly Lithic Argixerolls (Henneke series) and, on north-facing slopes, some moderately deep Pachic Haploxerolls. Some moderately deep Mollic Palexeralfs are present on the more-gentle slopes of Poker Flat. All of these soils have thermic soil temperature regimes. The soils on Red Flat, which is a short distance west of the Yolla Bolly junction and is occupied by an abandoned landing strip, are not serpentine soils.

Plant communities on south-facing slopes and flats are chamise (*Adenostema fasciculatum*) chaparral with appreciable amounts of valley silktassel, leatheroak, buckbrush, and whiteleaf manzanita. Those on north-facing slopes are mixed chaparral with leatheroak, buckbrush, valley silktassel, and whiteleaf ceanothus and a substantial herbaceous understory in which red brome, oniongrass, valley sandwort (*Minuartia douglasii*), and cryptantha (*Cryptantha* sp.) are the most abundant species. Other herbaceous plants common in chamise chaparral on flats and disturbed slopes are soft chess, squirreltail, annual fescue, threadstem flax (*Hesperolinon micranthum*), another species in the flax family (Polemoniaceae), serpentine lotus (*Lotus wrangelianus*), woolly sunflower (*Eriophyllum lanatum*), bedstraw (*Galium* sp.), wavy-leaf soap plant (*Chorogalum pomeridianum*), lomatium (*Lomatium* sp.), and traces of many other species.

18   Blue Mountains, Domain 6

The Blue Mountains domain is mostly in northeastern Oregon (fig 18-1). It is the name that we and others (Orr and Orr 1996) have adopted for the Central Highlands subprovince of the Columbia Intermountain province (Freeman et al. 1945). Small areas of Blue Mountains ultramafic rocks are exposed in an arcuate trend from central Oregon through northeastern Oregon into western Idaho (fig. 18-1). They are in the Baker and Wallowa terranes (Vallier and Brooks 1995). These terranes with the ultramafic rocks are covered or surrounded by Tertiary volcanic flows, largely Columbia River basalt. The ultramafic rocks are exposed in the Canyon Mountain and Sparta complexes and in smaller areas from the edge of the Idaho Batholith near Riggins in Idaho south–southwest across northeastern Oregon to the Aldrich Mountains south of Dayville. The Snake River has cut a deep gorge through the Blue Mountains domain. At Hells Canyon it is >2000 m deep. Strawberry Mountain southeast of John Day rises to 2755 m. Ultramafic rocks are exposed from about 975 m at the foot of the Strawberry Range, near Canyon City, to 2243 m on Baldy Mountain in the Strawberry Range and a bit higher on Vinegar Hill, which is about 45 km northeast of the Strawberry Range, although the summit of Vinegar Hill (2478 m above sea level) is not composed of ultramafic rocks.

Summers are hot and dry and winters are cold (fig. 18-2), with snow that persists through winters at the higher elevations. Mean annual temperatures are mostly in the 3°C–9°C range, and mean annual precipitation ranges from 25 to 100 cm. The frost-free period is about 150 days at lower elevations and <60 days at higher elevations.

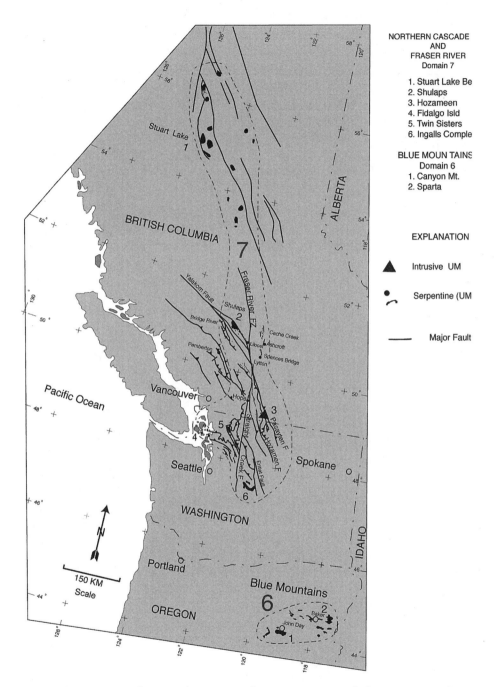

Figure 18-1 A map showing domain 6, Blue Mountains, and domain 7, Northern Cascades–Fraser River.

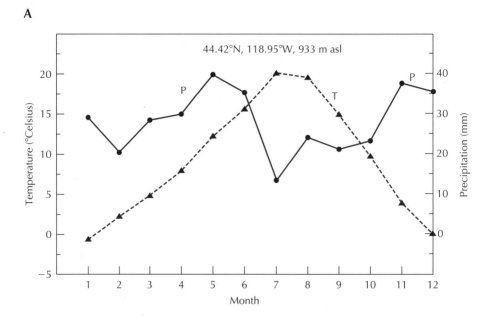

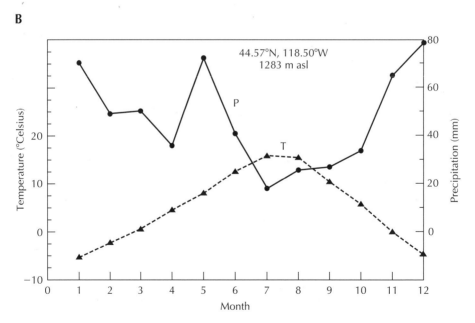

Figure 18-2 Mean monthly temperature and precipitation at two sites in the Blue Mountains. (A) John Day, Oregon (mean annual precipitation [MAP] = 33.9 cm). (B) Austin, Oregon (MAP = 53.2 cm).

18.1 Geology

The ultramafic rocks were exposed by late Tertiary uplift and erosion of the overlying volcanic sequence. The older rocks are composed of a volcanic island arc complex that contains marine sediments interlayered with mafic volcanic flows. Deep erosion of this area has exposed the roots of the volcanic arc. The roots contain gabbro and peridotite–serpentine at their lowest levels. Seven-thousand-year-old volcanic ash from Mt. Mazama has covered serpentine soils on the western edge of the Canyon Mountain locality, in the Aldrich Mountains, and there are subtantial amounts of the ash in some serpentine soils at least as far east as Vinegar Hill.

18.2 Soils

The major soil-forming processes in the Blue Mountains domain are the accumulation of organic matter in surface soils and leaching of clay from surface to subsoils. Although salts and carbonates are leached from most of the soils, carbonates may accumulate in some of the drier soils at lower elevations.

The serpentine soils generally have mollic epipedons, except for Cryepts and Cryalfs with ochric epipedons. Generally, they have either cambic or argillic subsurface horizons.

Most of the soils are cold or very cold, with frigid or cryic soil temperature regimes (STRs)—serpentine soils with mesic STRs are not extensive in the Blue Mountains domain. They are predominantly Mollisols, but with Inceptisols and Alfisols in the higher, colder forests. The established serpentine soils are generally in families with loamy–skeletal (Grell, Overholt, and Stussi series) or clayey–skeletal (Marksbury) particle-size classes.

18.3 Vegetation

Forests and woodland prevail in serpentine habitats of the Blue Mountains domain, except for high mountain summit areas that are dominated by grasses, shrubs, and forbs. Although sagebrush (*Artemesia tridentata*) alliances are extensive at lower elevations, they are not common on the serpentine soils of this domain. The alliances on serpentine soils range from juniper (*Juniperus scopulorum*) woodland at lower elevations through ponderosa pine (*Pinus ponderosa*), Douglas-fir (*Pseudotsuga menziesii*), and grand fir (*Abies grandis*) forests to white pine (*Pinus monticola*), subalpine fir (*Abies lasiocarpa*), and whitebark pine (*P. albicaulis*) at higher elevations. The higher serpentine elevations of Baldy Mountain and Vinegar Hill are nearly devoid of trees. Kruckeberg (1969) has suggested that the serpentine flora in higher parts of the Blue Mountains has some similarities with that of the Wenatchee Mountains (Ingalls Complex, locality 7-6), even though the Canyon Mountain complex in Blue Mountains domain is about 400 km southeast of the Ingalls complex, with no serpentine between the two localities. There is a "disjunction" between the two serpentine areas.

18.4 Canyon Mountain, Locality 6-1

The Canyon Mountain complex is exposed in a thick section of island arc crust that was elevated along the John Day Basin and Range fault of late Tertiary age (Thayer and Himmelberg 1968). The lowest part of the Canyon Mountain complex consists of tectonite harzburgite overlain by cumulate peridotite and gabbro. Extensive tracts of serpentine are exposed on the north flanks of Strawberry Mountain above the John Day River, and other exposures of sheared serpentine are present along Canyon Creek south of the town of John Day.

Peridotite of the Canyon Mountain complex consists of olivine (80%) and enstatite (10%–30%), with variable amounts of clinopyroxene and chromite. Numerous chromite pods are randomly present in the tectonite peridotite. Olivine-rich layers are present in the lower part of the cumulate gabbro. The serpentine consists mainly of chrysotile–lizardite and in some zones is highly sheared, obscuring all original structures of the peridotite. The Canyon Mountain complex is considered to be part of a larger island arc complex that was accreted to North America during the Triassic to Early Jurassic time.

On the western flanks of the Aldrich Mountains, just south of Dayville, the South Fork of the John Day River has cut an exposure of serpentine melange.

Serpentine soils at the foot of the Strawberry Range and Aldrich Mountains are Lithic Haploxerolls with A-R horizon sequences (Overholt Series) and Typic Argixerolls with A-Bt-R sequences. Open stands of juniper woodland and grass predominate on these soils. Wheatgrass (*Pseudoroegneria spicata*) and Idaho fescue (*Festuca idahoensis*) are the dominant grasses, with the concentration of fescue in the shade of juniper trees (C. Brooks, personal communication, 2004). Thurber's needlegrass (*Achnatherum thurberianum*) is present in at least some of the juniper stands (fig. 18-3).

The juniper woodlands grade upward into ponderosa pine–Douglas-fir forests with grasses and elk sedge (*Carex geyeri*) dominating the understory. Ponderosa pine is more dominant at lower elevations (fig. 18-4) and Douglas-fir at higher elevations. Lodgepole pine is a common successional tree, at least at the higher elevations. Mountain mahogany (*Cercocarpus ledifolius*) is scattered sparsely through these forests. The main grasses are Idaho fescue, wheatgrass, and bluegrass (*Poa secunda*). Stonecrop (*Sedum lanceolatum*) is common. The soils are predominantly Typic Argixerolls with frigid soil temperature regimes.

As ponderosa pine diminishes in abundance upward, the Douglas-fir-dominated forests grade into grand fir forests. These forest have some Pacific silver fir (*Abies amabilis*), at least on the south slope from Vinegar Hill (rock specific gravity [SG] = 2.9). Lodgepole pine is a common successional tree in these forests. The understory is dominated by grouseberry (*Vaccinium scoparium*) in many places and by pinemat manzanita (*Arctostaphylos nevadensis*) in others. Elk sedge is common in the understory. The serpentine soils are mainly Eutrocryepts and Haplocryalfs.

Serpentinized peridotite (SG = 2.58) is exposed up to about 2320 m on a ridge that trends northwest from the north end of Vinegar Hill. The north slope from this ridge has been glaciated, but the upper part of that slope is free of glacial drift. Upper convex slopes

Figure 18-3 A Lithic Haploxeroll–Typic Argixeroll complex on serpentinized peridotite at 1450 m altitude in the Aldrich Mountains ~9 km south of Mt. Vernon. The plant community on the shallow soil is juniper (*J. occidentalis*)/mountain mahogany (*Cercocarpus ledifolius*)/Idaho fescue (*Festuca idahoensis*)–bluegrass (*Poa secunda*) and the one on the deeper soil is ponderosa pine (*P. ponderosa*)–juniper (*J. occidentalis*)/Idaho fescue (*Festuca idahoensis*)–elk sedge (*Carex geyeri*). Spike-moss (or lesser club-moss, *Selaginalla wallacei*) grows over rocks on the shallow soil.

of the ridge are rocky and have very shallow to shallow (Lithic subgroups) Mollisols (Haplocryolls) and Inceptisols (Eutrocryepts). Southward from the ridge the characteristic soils are moderately deep Alfisols (Mollic Haplocryalfs) on linear backslopes and deep Mollisols on lower concave slopes. Although rocks collected from the upper convex slope contain almost exclusively serpentine, fine sand from a surface soil sample contains appreciable glass and other nonserpentine minerals, in addition to serpentine and chlorite, indicating the influence of volcanic ash. The reaction of the sample is pH 6.8. The upper, convex slopes are dominated by sagebrush, cushion phlox, grasses, sedges, and a variety of forbs. Sparse whitebark pine and low juniper are present along the rocky ridge, and patches of stunted subalpine fir forest occur on the upper north-facing slope. The south-facing concave slopes are dominated by sagebrush and mountain sorrel (*Rumex* sp.), with patches of buckwheat (*Eriogonum flavum* and *E. heracleoides*). C. Johnson (unpublished document, Wallowa-Whitman National Forest, Baker City, OR) has identified some of the more common forbs in this area as showy polemonium (*Polemonium pulcherrimum*), sandwort (*Arenaria aculeata*), wandering daisy (*Erigeron peregrinus*), lupine (*Lupinus leucophyllus leucophyllus*), cymopterus (*C. terebinthinus foeniculaceus*), and lovage (*Ligusticum grayi*).

The serpentine vegetative cover on the broad treeless summit of Baldy Mountain (2243 m) in the Strawberry Range is grasses, subshrubs (*Eriogonum* spp.), and forbs. The grasses are mainly Idaho fescue and wheatgrass. Some of the more common forbs are cymopterus (*C. terebinthinus*), chickweed (*Cerastium arvense*), cushion draba (*D. densifolia*), frasera (*F. albicaulis*), and fringed-onion (*Allium fibrillum*). The soils are Lithic Haplocryolls and Typic Argicryolls. Rocks on the summit are adorned with a yellow-to-bright orange lichen (*Caloplaca* sp.) and a pale green one (*Lecanora* sp.). On the upper north slopes of Baldy Mountain there are sparse woodlands of white pine and Douglas-fir, but most of the area is open. Kruckeberg (1969) remarked that at similar elevation on the adjacent nonserpentine areas of Strawberry Mountain the major vegetation included many woody species such as snowberry (*Symphorocarpos* sp.), some species of *Ribes*, *Pachistema myrsinites*, and a number of herbs not locally found on serpentine. On the more sheltered north-facing slopes lacking conifers, the peridotite crevices contain dense aggregations of the faithful serpentine ferns *Polystichum lemmonii* and *Aspidotis densa*. Other species that are local serpentine indicators also found in the Wenatchee Mountains include *Thalaspi alpestre*, *Cerastrium arvense*, and *Arenaria obtusiloba*. Other wandering ("bodenvag") species, those that occur both on and off of serpentine, mentioned by Kruckeberg (1969) in a subalpine releve at 7400 feet (2556 m altitude) include *Elymus elymoides*, *Melica* sp., *Eriophyllum lanatum*, *Achillea millefolium*, *Heuchera cylindrica*, *Arenaria obtusiloba*, *Viola* sp., *Horkelia* sp., *Potentilla flabellifolia*, *Penstemon procerus*, *Eriogonum umbellatum*, *Draba* sp., *Polygonum bistortoides*, and *Silene douglasii*.

Kruckeberg (1969) has described vegetation in the Fields Creek drainage (~30 km southeast of Dayville). On the lowest serpentine exposures at Fields Creek there is a strong reduction of plant cover relative to adjacent nonserpentine areas. These most xeric exposures are so dry that they do not contain the nearly ubiquitous serpentine faithful *Aspidotis densa*. These are true barrens and have no woody species and relatively sparse cover of herbaceous species, including the non-native *Bromus tectorum*, as well as unidentified species of *Lepidium* and *Grimmia*. A large outcrop of serpentine on Buck Cabin Creek (a tributary of Fields Creek) has isolated individuals of *Cercocarpus ledifolius*, *Pseudotsuga menziesii*, and *Pinus ponderosa* (insufficient cover to constitute a woodland) with an herbaceous vegetation dominated and characterized by *Aspidotis densa*, *Pseudoroegneria spicatum*, *Selaginella densa* var. *scopulorum*, *Sedum divergens*, *Eriogonum umbellatum*, *Silene douglasii*, and *Cirsium utahense*. The *Cirsium* appears to be ubiquitous on serpentine substrates in all of the sites visited, but was absent from the nonserpentine sites locally. Farther up slope, presumably in more mesic settings, *Cercarpus ledifolius* composes a woodland with a relatively dense understory of grasses and forbs. Here *Aspidotis densa* is common along with *Arenaria capillaris* and a falcate-leaved *Allium*.

18.5 Sparta, Locality 6-2

At Sparta, east of Baker, is a dismembered base of an island arc complex consisting of serpentine and layered clinopyroxenite (Vallier and Brooks 1995). This complex is coeval with the Canyon Mountain complex and was also accreted to North America in Permian–Triassic times. Smaller mafic gabbro and serpentine exposures have been mapped in the Blue Mountain area and are also considered to be in island arc complexes, similar to the Canyon Mountain complex, that were accreted to North America in the same time period.

19 Northern Cascade–Fraser River, Domain 7

The Northern Cascade–Fraser River domain conforms to the Northern Cascade Mountains physiographic province in northwestern Washington and southern British Columbia, the San Juan Islands between the southern tip of Vancouver Island and the Northern Cascade Mountains, and much of the Interior Plateau province of British Columbia. The thread that connects these areas is the north–south Straight Creek–Fraser River fault system that runs through the Northern Cascade Mountains and northward along the Fraser River. The localities of domain 7 are along faults that branch off from this major fault system.

The Northern Cascade Mountains are indeed mountainous, and the Interior Plateau of British Columbia is an area of dissected plateaus and scattered mountains. The Fraser River flows northwest in the Rocky Mountain Trench, which separates the North American craton on the northeast from accreted terranes on the southwest; then it turns around the northwest end of the Cariboo Mountains to the Interior Plateau. In the Interior Plateau, the Fraser River flows from Prince George south about 500 km to the Northern Cascade Mountains before turning westward toward the Pacific Coast. The northern part of domain 7 is in that part of the Fraser River basin, including tributaries northwest of Prince George, which is in the Interior Plateau province. Low, hilly terrain dominates the San Juan Islands. All of these areas in domain 7, except the Ingalls complex on southeast margin of the Northern Cascade Mountains, were covered by the Cordilleran ice sheet during the last stage of the Pleistocene glaciation, leaving <15 ka years for soil development on the current ground surfaces. Although alpine glaciers formed in the southeastern margin of the Northern Cascade Mountains, they did not cover all of the soils, allowing some of them longer time for development.

Elevations in domain 7 range from sea level on San Juan Islands to mostly in the 600–1500 m range on the Interior Plateau of British Columbia, and up to 4392 m on

Mt. Rainier in the Northern Cascade Mountains. The highest exposures of ultramafic rocks in the Northern Cascade Mountains, however, are just >2000 m on Twin Sisters Mountain, 2335 m on Ingalls Peak in the Ingalls complex, 1798 m on the summit of Olivine Mountain in the Tulameen complex. The altitude on Shulaps Peak, which is near the margin of the coastal mountains of British Columbia, is 2880 m.

The climate ranges from cool to very cold and from dry to wet. Summers are cool or warm west of the Northern Cascade Mountains to hot inland. The mean annual precipitation is up to about 250 cm on Twin Sisters Mountain and about 100 cm on the San Juan Islands to the west and on the Ingalls complex in the Wenatchee Mountains on the east. The Interior Plateau of British Columbia is in a rain shadow with mean annual precipitation 50–75 cm around Prince George, and northwest from there, to <30 cm farther south along the Fraser River where the coastal mountains are higher and create a greater rain shadow effect. Precipitation is greater in winter than in summer, and snow accumulates and persists through winters, except on the San Juan Islands, where snow does not persist throughout winters. The frost-free period is about 220 days on the San Juan Islands, <100 to 110 days on Twin Sisters Mountain, 100–175 days in the Wenatchee Mountains, and 100 to <60 days on the Interior Plateau of British Columbia.

19.1 Geology

All the ultramafic rocks in the northern part of domain 7 are present as thrust slices imbricated with the Cache Creek and Bridge River rock assemblages. These rocks are a mixture of chert, shale, limestone, and basalts that are considered to have formed in an oceanic basin marginal to the North American continent. These rocks range in age from Paleozoic to Early Mesozoic and were accreted onto North America in Late Triassic to Late Jurassic times (Monger 1977). Dextral strike–slip motion along the Yalakom fault and Fraser River fault has displaced the earlier thrust slices formed during continental accretion to produce a highly disjunct distribution of ultramafic rocks.

The Straight Creek fault splits the Northern Cascade Mountains into an eastern half dominated by terranes that developed along the margin of the North American continent and an eastern half with terranes that developed in oceanic settings (Orr and Orr 1996).

19.2 Soils

The major soil-forming process in the Northern Cascade Mountains and most of the interior plateau of British Columbia is podzolization. This is a process in which organic matter, iron, and aluminum, rather than clay, are leached from the surface and deposited in subsoils. Podzolization is not active in serpentine soils, however, except in soils where the ultramafic materials are overlain by surface layers of nonultramafic materials, as explained in chapter 6. Leaching of clay and its deposition in subsoils is an important soil-forming process only at lower elevations on the east side of the Ingalls Complex. Some of the soils in this area, where it has not been glaciated, have argillic horizons.

Surface soil diagnostic horizons are predominantly ochric epipedons and most serpentine soils have cambic horizons. Many serpentine soils, however, have nonultramafic surface layers with eluvial horizons and spodic horizons in underlying ultramafic materials.

Most of the soils are very cold, with cryic STRs, except on Fidalgo Island and on the lower east side of the Ingalls complex where the soils are just cold, with frigid STRs. The very cold serpentine soils are mostly Inceptisols, and some are Spodosols.

19.3 Vegetation

Forests prevail in most of this domain, except where there is barren rock or talus or very shallow soils. The alpine timberline is about 2300 m elevation in the Shulaps (locality 7-2) area (Bulmer and Lavkulich 1994). Kruckeberg (1969) has described the vegetation at many sites in localities 7-2, 7-3, 7-4, 7-5, and 7-6. In this part of northwestern North America, as in others, the vegetation on serpentine is distinguishable primarily through its contrasting physical structure from surrounding nonserpentine vegetation. Open or stunted woodlands on serpentine substrates contrast with dense, tall coniferous forests off of it. In the southern part of this domain, serpentine endemics appear, as the climate is less cold than in domains and localities that are farther north in western North America. Several local serpentine endemic plants occur in the Wenatchee Mountains (Ingalls Complex, locality 7-6), including *Poa curtifolia* and *Lomatium cuspidatum*. Two widespread ferns that are mostly on serpentine are *Aspidotis densa* and *Polystichum scopulinum*. The serpentine-faithful relative of *P. scopulinum*, *P. kruckebergii* occurs in the northern portion of this domain.

19.4 Stuart Lake Belt, Locality 7-1

The Stuart Lake belt of ultramafic rocks lies northwest of Fort St. James. The largest concentrations of ultramafic rocks are southwest of the Middle River, which runs from Takla Lake to Trembleur Lake, and in Rubyrock Lake Provincial Park. These rocks are predominately serpentinized peridotites and dunites (Monger 1977). Remnant tectonite fabrics typical of oceanic peridotites are found on some glaciated surfaces. Large areas of talc–carbonate rock are present south of Mt. Sidney Williams, indicating strong hydrothermal alteration by Cretaceous granite intrusions. These bodies were eroded prior to Late Triassic as evidenced by layers of ultramafic conglomerates in the surrounding Cache Creek Triassic sediments. These rocks were accreted during the Triassic and imbricated with the Cache Creek assemblage at that time.

Elevations of the ultramafic rocks range from about 750 m to nearly 2000 m. The highest are above tree limit, but below the 1986 m summit of Mt. Sidney Williams.

Two small serpentine soil sites were observed at about 1000 and 1100 m altitude on the southeast side of Mt. Sidney Williams. The forests adjacent to these sites are dominated by subalpine fir and white spruce, with understories of alder (*Alnus crispa sinuata*) and black huckleberry (*Vaccinium membranaceum*). Both serpentine soil sites had been logged in the past year or two. The soils are shallow to moderately deep Inceptisols (Eutrocryepts). Both sites had been dominated by subalpine fir forest. Although not present here, unless some tree stumps were overlooked, white spruce was found to be common on serpentine soils near the lower (north) end of Dease Lake, which is also in northwestern British Columbia. The understory in the serpentine subalpine fir forest was mostly Alaska spirea (*S. beauverdiana*), buffaloberry (*Shepherdia canadensis*), roses (*Rosea* spp.), saskatoon (*Amelanchier alnifolia*), and dwarf blueberry (*Vacciniun caespitosum*). The main forb was northern starflower (*Trientalis borealis*). Mosses were common. A very shallow soil had a cover of low juniper (*J. communis*) and lichen (*Cladina* sp.), with sparse wheatgrass (*Elymus trachycaulus* ssp. *trachycaulus*). Aspen (*Populus tremuloides*) and lodgepole pine (*Pinus contorta*) are the main successional tree species in this area, and both grow on the serpentine soils. The serpentine faithful *Polystichum kruckebergii* is known from the Cassiar Mountains and other more northerly serpentine habitats in British Columbia (Kruckeberg 1982, Cody and Britton 1989, Douglas et al. 1998).

19.5 Shulaps, Locality 7-2

Ultramafic bodies of the Bridge River ophiolite occur along faults that branch off from the Fraser River fault and extend into the coastal mountains of British Columbia. Prominent among these bodies are the large Shulaps body along the Yalakom fault and the relatively small Pioneer peridotite near Bralorne (Wright et al. 1982).

Along the eastern edge of the Coast Mountain granitic plutons, the Shulaps ultramafic body is thrust westward over the Bridge River Group. The body was emplaced teconically in the Early Cretaceous. This elongate body is 30 km by 15 km and one of the largest in British Columbia (Wright et al. 1982). The unserpentinized rock is mainly tectonite harzburgite containing olivine and orthopyroxene and interlayered with conformable layers of dunite. Along the boundaries the sheared peridotite is pervasively serpentinized and contains tectonic blocks of gabbro and greenstone forming a major melange zone. A smaller ultramafic called the Pioneer peridotite is found southwest of the Shulaps body.

Inceptisols (Cryepts) are the predominant serpentine soils (Bulmer et al. 1992). A blanket of fine pumice about 20 cm thick was laid over the soils of the Shulaps body about 2400 years ago. Podzolization has advanced where there are pumice deposits over the serpentine materials, but the soils do not have the qualifications of Spodosols, at least not yet. Soils described by Bulmer et al. (1992) had a Douglas-fir (*Pseudotsuga menziesii*)/juniper (*Juniperus* sp.)/kinnikinnick (*Arctostaphylos uva-ursi*) plant community at 1101 m, a lodgepole pine/white-flowered rhododendron (*Rhododendron albiflorum*)/low blueberry (*Vaccinium myrtillus*) plant community at 1550 m, and a krumholtz subalpine fir (*Abies lasiocarpa*)-whitebark pine (*Pinus albicaulis*)/willow (*Salix* spp.)–white mountain heather (*Cassiope mertensiana*)/arctic lupine (*Lupinus arcticus*) plant community at 2150 m elevation.

A more general assessment of the vegetation on the Shulaps body might be open Douglas-fir forests on south-facing slopes at low elevations and dense forest from there up to the subalpine area that is dominated by sparse whitebark pine and low shrubs. The understory in the open Douglas-fir forest is commonly dominated by wheatgrass, low juniper, kinnikinnick, buffaloberry, stonecrop (*Sedum lanceolata*), and moss. Dense forests are predominantly subalpine fir and lodgepole pine grading to whitebark pine near the upper elevation limit of forest. The main forest shrubs are white rhododendron and black huckleberry on north-facing slopes at lower elevations and otherwise low juniper with common kinnikinnick and buffaloberry. Low juniper, dwarf willow, and kinnikinnick are the main subalpine shrubs, and buckwheat is a common subshrub. Some of the main subalpine forbs are stonecrop, lupine (*Lupinus* sp.), spotted saxifrage (*Saxifraga brachialis*), paintbrush (*Castilleja* sp.), and sandwort (*Minuartia* sp.).

Kruckeberg (1969) described vegetation of the Pioneer peridotite, which he referred to as the Bralorne and Choate area. He noted dense forests of *Tsuga mertensiana-Chamacyparis nootkatensis* surrounding the sites off serpentine, but he described openings with *Carex stichensis* in seeps on pyroxenite, and on mesic pyroxenite talus he described a *Rubus leucodermis/Adiantum pedatum aleuticum-Aspidotis densa* stand. He also defined a stunted woodland which could be called *Pinus contorta latifolia/Juniperus communis–Shepherdia canadensis/Aspidotis densa–Polystichum kruckebergii*. Although farther south in this domain *Paxistema myrsinites* is considered a stringent avoider of serpentine, here it does occur on pyroxenite talus.

19.6 Hozameen, Locality 7-3

19.6.1 Locality 7-3a (Coquihalla)

The ultramafic rocks near Hope, British Columbia, are present within a north-trending fault zone that extends southward into Washington (Ray 1986). These narrow bodies are almost completely serpentinized, with a maximum width of 1.5 km, and characterized by tectonic blocks of rodingtized gabbro. The serpentine is mainly lizardite and antigorite, with only rare amounts of olivine. The Hozameen Fault marks a major boundary between the oceanic crust assemblages and the North American continental basement. To the north the Hozameen Fault merges with the Straight Creek and Fraser River Faults.

Serpentinite is exposed in small slivers between the East and West Hozameen faults (Ray 1986). It is covered by patches of glacial drift. A serpentine Inceptisol (Eutrocryept or Dystric Eutrudept, depending on the soil temperature regime [STR]) observed a few hundred meters northwest of the Carolin Mine, where there is minimal glacial drift, has a dense forest cover of amabilis fir (*Abies amabilis*), western hemlock (*Tsuga heterophylla*), and yellow cedar (*Chamaecyparis nootkatensis*) with an understory of black huckleberry, mosses, and star flower (*Trientalis latifolia*). The trees in this forest are the same as those recognized by Bulmer and Lavkulich (1994) on serpentine soils of this area. The upper 30 cm of the soil sampled by them in the Coquihalla area has characteristics of a spodic horizon, but the soil lacks a very acid E horizon.

19.6.2 Locality 7-3b (Tulameen Complex)

Between Hope and Princeton, British Columbia, along the Tulameen River an Alaskan-type intrusion has a dunite core about 4 by 1 km (Findlay 1969). This is the most southerly occurrence of this type in North America.

The Tulameen complex is exposed mostly on Olivine Mountain, which is south of the Tulameen River and has predominantly north-facing slopes, and on Grasshopper Mountain, which is north of the river and has south-facing slopes. Glacial drift of mixed lithologies covers the lower mountain slopes, adjacent to the Tulameen River, and is present in patches all of the way up the mountains. It has been completely removed from very steep and most steep slopes. Inceptisols (Haploxerepts and Eutrocryepts) are common on moderate to steep slopes and very shallow to very deep, extremely stony Entisols are common on very steep slopes. There are some Mollisols, which appear to be associated with pyroxenite, rather than the dunite core of the Tulameen complex, but observations were insufficient to verify soil parent material related differences. The vegetation on very steep slopes of Grasshopper Mountain is dominated by sparse Douglas-fir, low juniper, and bluebunch wheatgrass (*Pseudoroegneria spicatum*). These same species occur on Inceptisols, with denser Douglas-fir, sparse ponderosa pine, common kinnikinnick, and some saskatoon (*Amelanchier alnifolia*), roses (*Rosea* sp.), bittercherry (*Prunus* sp.), prostrate buckwheats (*Eriogonum* spp.), sandwort (*Arenaria capillaris*), pod fern (*Aspidotis densa*), and sparse yew (*Taxus brevifolia*). Subalpine fir is common on upper north-facing slopes, and there is some white pine.

Bulmer and Lavkulich (1994) sampled serpentine soils at 1260 m altitude on Grasshopper and 1730 m altitude on Olivine Mountain. Only small amounts of clay have accumulated in these soils. Mineralogical analyses showed the very fine sand fraction of the soil from Olivine Mountain to be olivine, and the clay fraction was dominated by serpentine.

Kruckeberg (1969) reported an impressive array of serpentine effects on the vegetation in the Tulameen area. He noted the xeric nature of the serpentine vegetation relative to the surrounding vegetation. The widespread serpentine-faithful ferns *Aspidotis densa*, *Polystichum kruckebergii*, and *Polystichum lemmonii* are common. He sampled several releves from the vicinity and described the vegetation in open woodlands as *Pseudotsuga menziesii/Juniperus communis–Prunus virginiana demissa/Aspidotis densa–Polystichum kruckebergii*.

A detailed floristic and ecological analysis of the Grasshopper Mountain area was completed by Lewis and Bradfield (2003). They sampled 70 circular plots of 10-m radius (314 m^2), with 34 on serpentine and 26 off serpentine in adjacent terrain of similar topography. Species diversity was not significantly different (Shannon or Simpson diversity indices) between serpentine and nonserpentine sites. This conclusion differs from general statements made by Kruckeberg (1969), Brooks (1987), and others on the overall reduction of species diversity on serpentine substrates. Of all species recorded, 28% (49 taxa) were found only on serpentine. The majority of these species are not true endemics but are species that are generally out of their normal distribution locally, as a result of the unique substrate. They are local serpentine indicators. Plants in several families including Apiaceae, Caryophyllaceae, Poaceae, and fern families (Pteridophytes) are more common on serpentine than off, while those in the Liliaceae, Rosaceae, Ranunculaceae, Betulaceae, Caprifoliaceae, Grossulariaceae, and Salicaceae are more common off serpentine, locally. Thirty-five species were identified as local indicators of serpentine by Lewis and Bradfield (2003). Several species were thought to reach their westward extension of range in this area, including xerophytes such as *Pseudoroegneria spicata* and *Eriogonum ovalifolium* var. *nivale*. Two conifers, *Pinus albicaulis* and *P. ponderosa*, reach the lower and upper limits of their elevation ranges, respectively, on serpentine soils in this locality.

Eight rare (in British Columbia) plant taxa occur at Grasshopper Mountain only on serpentine. As appears to be the general pattern in northwestern North America, many of these are ferns and include *Aspidotis densa*, *Polystichum kruckebergii*, *P. scopulinum*, and *Adiantum aleuticum* (*A. pedatum calderi*). Other species (nonpteridophytes) considered rare are *Arabis holboellii* var. *pinetorum*, *Crepis atrabarba* ssp. *atrabarba*, *Lupinus arbustus* ssp. *pseudoparviflorus*, and *Melica bulbosa*.

Lewis et al. (2004) identified 43 species of bryophytes (mosses and liverworts) in the serpentine area of the Tulameen complex and an additional 21 species in the surrounding area. They suggested that species distributions are more closely related to physical habitat than to rock or soil chemistry.

Lewis and Bradfield (2004) collected subsurface samples from 26 serpentine and 45 nonserpentine soils on Grasshopper Mountain. They performed multidimensional scaling and principal component analyses on the soil chemical data. The fist axis, which was assumed to represent fertility, was positively correlated ($\alpha < 0.05$) with soil calcium, potassium, and phosphorus and negatively correlated with magnesium, nickel, and cobalt.

19.7 Fidalgo Island, Locality 7-4

The San Juan Islands (fig. 19-1) are composed of a diverse assemblage of rocks from five terranes that were thrust over the Wrangelia terrane as it collided with the North American continent during the Late Cretaceous (Brandon et al. 1988). The topography has been modified by regional glaciation during the Quaternary. Low, rounded hills that are generally elongated from north to south (or more precisely north, northeast–south, southwest, with local exceptions), which was the dominant direction of ice flow, rise above a glacial plain that has been largely inundated as sea level has risen during the Holocene.

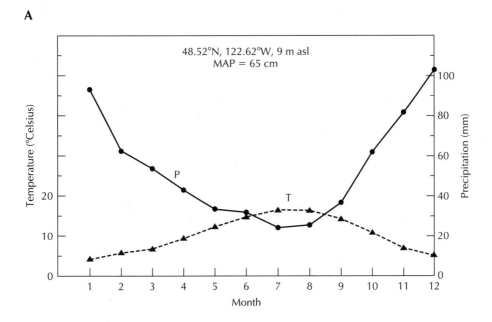

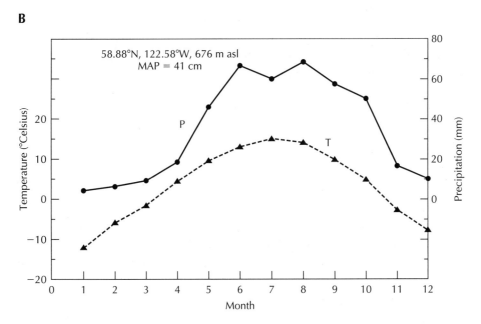

Figure 19-1 Mean monthly temperature and precipitation at two sites in the Northern Cascades–Fraser River domain. (A) Anacortes on Fidalgo Island, Washington. (B) Prince George in British Columbia.

Ultramafic rocks are exposed in a linear trend from Cypress Island on the north to Fidalgo Head on Fidalgo Island and to Burrows and Allan Islands south of Fidalgo Head. At Fidalgo Island in Skagit County, Brown (1977) described a complete ophiolite section with base plagiogranites that have been dated as Jurassic. The ultramafic rocks are mainly mildly serpentinized dunite and harzburgite that have a tectonite fabric. The lithic associations suggest that this is an island arc complex accreted to the North American continental margin.

Typic Haploxeralfs of the Guemes Series are the main soils in the ultramafic terrains of the San Juan Islands (Klungland and McArthur 1989). On Fidalgo Island, the ultramafic rocks are exposed on and around Fidalgo Head, which is on the northwest corner of the island. They are limited to a narrow fringe from the shoreline inland no more than a few hundred meters. The soils are mostly Lithic Haploxerolls with much bedrock showing through the soil cover. The vegetation on the shallow serpentine soils of Fidalgo Island is juniper (*J. scopulorum*) grassland, with much tufted phlox (*Phlox diffusa*) and pod fern. Denser clumps of trees occupy deeper but still relatively shallow soils in and around the juniper grassland. Trees in the clumps are Douglas-fir, juniper, and madrone (*Arbutus menziesii*). Wild blue rye (*Elymus glaucus*) is common under the clumped trees. The most abundant shrub in denser Douglas-fir forest in Washington Park, which includes Fidalgo Head, is salal (*Gaultheria shallon*), which seems to avoid the serpentine soils. Kruckeberg (1969) contains two excellent photographs of serpentine meadows and juniper, with Douglas-fir in the background, and a similar photo on the lower slopes of Olivine Hill on Cypress Island that shows beach pine (*Pinus contorta*) in addition to the other species.

Kruckeberg (1969) sampled in two releves in grassy balds, one representing a deeper soil and the other a steeper stony soil. Chappell (2006) recently sampled and described these as a *Festuca rubra—Festuca roemeri—Aspidotis densa* association known only from this area. The deeper soil sites have a relatively high cover of the non-native grasses *Bromus hordaceous* and *Aira caryophyllea*, as well as native *Festuca* and *Koeleria macrantha*. Kruckeberg did not indicate actual soil depths, so by "deeper soil" he was likely referring to deeper shallow (depth <50 cm) soil.

19.8 Twin Sisters, Locality 7-5

Along the Cascade Mountains east of the San Juan Islands ultramafic rock occurs within an accretional belt of Paleozoic rocks (fig 19-1). This belt consists of imbricate thrusts and on the upper side of the Shuksan thrust the Twin Sisters dunite body is a mass about 260 km^2 (Thompson and Robinson 1975). The rocks are mainly dunite interlayered with small amounts of harzburgite, with little or no serpentinization except along the faulted boundaries. The dominant mineral is olivine with minor amounts of enstatite and clinopyroxene with chromite.

The higher reaches (up to 2065 m) of the Twin Sisters locality are mostly barren rock and talus lacking fine earth (soil particles <2 mm) at the ground surface. There are sparse patches of Cryepts and very shallow Cryents with sparse stands of mountain hemlock

(*Tsuga mertensiana*) and subalpine fir and semidense understory of pink and white mountain heather (*Phyllodoce empetriformis* and *Cassiope mertensiana*) and black huckleberry and sparse patches of very deep Cryepts with sparse stands of subalpine fir and dense understory of common juniper. Colluvial footslopes commonly have Cryepts with semidense stands of mountain hemlock and subalpine fir and an understory of pink mountain heather and oval-leaved blueberry (*Vaccinium ovalifolium*, or *V. alaskaense*), patches of white-flowered rhododendron, and sparse rusty Menziesia (*Menziesia ferruginea*).

At the lower elevations (~300–1000 m) around Twin Sisters, the serpentine soil parent materials are generally covered by layers of glacial till and volcanic ash, which are commonly mixed with serpentine materials in colluvium (Goldin 1992). The till and ash are commonly thin enough, or serpentine materials are sufficiently concentrated (but not too concentrated) in the colluvium, that serpentine Spodosols (Cryorthods) have formed and dominate the landscape around the western margin of Twin Sisters. The natural vegetative cover on the serpentine Spodosols is conifers and shrubs and the timber site index for growth of western hemlock is about 17 or 18 m in 50 years (Goldin 1992).

Kruckeberg (1969) found no local endemics, but the widespread regional serpentine indicators *Aspidotis densa* and *Polystichum lemmonii* are common. He described the following vegetation types from several releves taken in the area near Orsano Creek, between 1500 and 1610 m elevation. Krummholz (timberline) woodland: *Pinus contorta latifolia-Abies lasiocarpa/Phyllodoce glandulifera/Silene acaulis-Sibbaldia procumbens* (snow melt basins). Xeric moraine upland: *Pinus contorta latifolia–Abies lasiocarpa/Phyllodoce glandulifera/Silene acaulis–Polystichum lemmonii*. Mesic heath-sedge woodlands: *Tsuga mertensiana–Abies lasiocarpa–Chamaecyparis nootkatensis/Cassiope mertensiana–Luetkea pectinata*. Xeric open woodlands: *Abies lasiocarpa–Chamaecyparis nootkatensis/Juniperus communis/ Aspidotis densa–Polystichum lemmonii*. Other than the higher than normal elevation representation of *Pinus contorta latifolia* at timberline in the Twin Sisters locality, Kruckeberg described the floristic contrasts with nonserpentine areas as less pronounced than in the Ingalls Complex area (locality 7-6). However, at about 1800 ft elevation on the South Fork of the Nooksak River, a more pronounced vegetation matrix exists where a dwarf sparse woodland/scrub is present. This formation has occurred in part as a result of a massive dunite slip. It is actively slipping and eroding and contains a large boulder-field scrub. The vegetation could be defined as a *Pinus contorta latifolia* (dwarfed)/*Juniperus communis-Arctostaphylos nevadensis/Rhacomitrium canescens ericoides* (a moss) association and is surrounded by a dense and typical regional forest of *Psedotsuga menziesii* and *Tsuga heterophylla*.

19.9 Ingalls Complex, Locality 7-6

The Ingalls complex is located in the southeastern part of northern Cascade Mountains approximately 300 km east of Seattle. It is thrust over the Cascade metamorphic core and is intruded by the Mount Stuart granitic batholith. This intrusion has strongly deformed and metamorphosed the Ingalls complex (Miller 1985). Earlier-formed east–west shear zones are related to preaccretion oceanic transform faults. The ultramafics exposed are serpentinized tectonite peridotite. The mineral assemblage is olivine (75%) orthopyroxene (10%–25%) with minor clinopyroxene and chromite. Within the shear zones the peridotite is completely serpentinized and contains tectonic blocks of greenstone and marine sediments that have well-developed rodingite borders in contact with the serpentine. The Ingalls complex appears to have formed in a marginal oceanic basin along the North American continent. After its accretion onto the North American continent, low-angle thrusting and faulting has modified its original oceanic sequence.

On the west side of the complex, west of Ingalls Peak, the soils are predominantly Inceptisols. Spodosols develop where sufficient volcanic ash has been deposited over the ultramafic materials. Three soils were described in what appear to be the more common serpentine plant communities southwest of Ingalls Peak. A Typic Eutrocryept was described at 1800 m elevation in an open Douglas-fir–subalpine fir–whitebark pine/grouseberry plant community (fig. 19-2); a Lithic Eutrochrept at 1650 m elevation in a low juniper/bluebunch wheatgrass plant community, and a Lithic Cryorthent at about 1800 m elevation in a sparse forb plant community. Some of the plant communities on Eutrocryepts have much pinemat manzanita (*Arctostaphylos nevadensis*) and pink mountain heather (*Phyllodoce empetriformis*) in the understories. Volcanic ash has been incorporated into the surface of the Typic Eutrocryept resulting in a moderately acid (pH 5.8) surface soil, which is unusually low for serpentine Inceptisols in western North America. Serpentine colluvium on lower footslopes west of Ingalls Peak, in the Fortune Creek drainage, generally have dense, mixed conifer forest plant communities with much black huckleberry in the understory. Conifers in the mix are Douglas-fir, hemlock (*Tsuga* sp.), Engelmann spruce (*Picea engelmanni*), and fir (*Abies* sp.). It is difficult to ascertain which

Figure 19-2 A treeless serpentine hill on the east side of Mt. Sidney Williams, locality 7-1, from a recently logged area that has only patches of serpentine.

footslope soils have enough ultramafic material to be considered serpentine where non-serpentine colluvium and glacial drift has been mixed in with the serpentine materials. A description of a shallow serpentine soil on a very steep (72% gradient) slope follows:

Pedon IC2: 47.457°N, 120.974°W, 1650 m above sea level

O   0 cm, absent

A   0–5, dark brown (10YR 3/3, 6/3 dry) gravelly loam, common very fine and fine roots, pH 6.8

Bw 5–22, brown (10YR 4/3, 6/4 dry) very gravelly loam, few fine and medium roots, pH 6.9

C   22–42, brown (10Y 5/3, dry) very gravelly sandy loam, pH 7.0

R   42+, moderately fractured serpentinite bedrock

Trees: sparse Douglas-fir and whitebark pine. Shrubs: common low juniper.

Forbs: sparse yarrow (*Achillea millefolium*), spreading phlox (*Phlox diffusa*), and other unidentified forbs

Graminoids: common bluebunch wheatgrass

Ferns: sparse pod fern (*Aspidota densa*) and Shasta fern (*Polystichum lemmomii*).

Figure 19-3 An open forest–grassland on serpentinized peridotite at ~1050 m altitude on the south side of the ridge that slopes eastward from Iron Mountain toward Peshastin Creek in the Ingalls complex, Washington. The soils are Argixerolls and the plant community, which has been logged, is ponderosa pine (*Pinus ponderosa*)/bluebunch wheatgrass (*Pseudoregnaria spicata*).

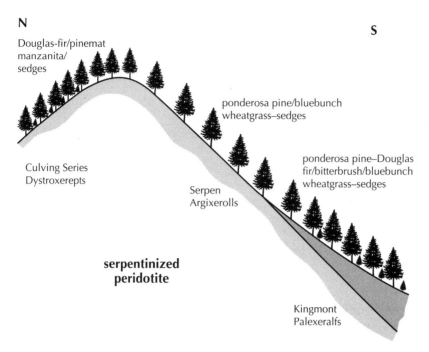

Figure 19-4 A conceptual north–south landscape profile across a mountain in the eastern part of the Ingalls complex, near Peshastin Creek.

At lower elevations on the east side of the Ingalls complex, serpentine soils have been mapped by T. Aho and others of the National Resources Conservation Service. Moderately deep Typic Dystroxerepts (Culving series), Typic Argixerolls (Serpen series), and deep Typic Palexeralfs (Kingmont series) have been mapped on serpentine below about 1500 m in the Ingalls complex (R. Myhrum, personal communication, 2003). Slopes facing north to northeast are dominated by Typic Dystropepts (Culving Series) with Douglas-fir/pinemat manzanita /sedge plant communities containing some ponderosa pine on drier slopes and some white pine on cooler or more moist slopes. Lodgepole pine (*Pinus contorta*) is a common successional species following severe fires. Occurrences of Oregon grape (*Berberis aquilinum*) and spirea (*S. betulifolia*) are spotty or sparse. All slope aspects other than north to northeast are dominated by Argixerolls with open-to-sparse conifer forest. The conifer forests are either ponderosa pine and Douglas-fir with bitterbrush (*Purshia tridentata*) and bluebunch wheatgrass or ponderosa pine with bluebunch wheatgrass. Sulfur buckwheat (*Eriogonum umbellatum*) and other buckwheat species (possibly *E. compositum lancifolium*), balsamroot (*Balsamorhiza* sp.), and lupine (*Lupinus* sp.) are common in the open conifer forests on Argixerolls. These Argixerolls are generally moderately deep to shallow on convex slopes and deep to very deep on concave slopes. Where the argillic horizons are buried deeply by colluvium, the soils are Typic Palexeralfs (Kingmont Series).

Kruckeberg (1969) was particularly attracted to the serpentine terrain in the western part of the Ingalls complex in the Wenatchee Mountains, which he considered to be the most spectacular and unique serpentine area in northwestern North America. He described it in greater detail floristically and vegetationally than other serpentine areas. The following are vegetation types sampled by Kruckeberg:

Southwest serpentine exposures: *Pinus albicaulis/Juniperus communis/Poa curtifolia* woodland.

Mesic slopes: *Pseudotsuga menziesii/Juniperus communis-Ledum glandulosum*.

Moderately dry woodland on slopes: *Abies lasiocarpa-Pseudotsuga menziesii/Pseudoroegneria spicatum*.

Moderately dry woodland on alluvium: *Pinus contorta latifolia–Pseudotsuga menziesii–Pinus ponderosa/Arctostaphylos nevadensis*.

Serpentine barrens: *Eriogonum pyrolaefolium–Arenaria obtusiloba–Poa curtifolia* herbland.

Kruckeberg (1969) noted the upward elevation shift of several regional species in the Ingalls complex, such as *Pinus contorta latifolia, P. ponderosa, P. monticola*, and *Taxus brevifolia*. All of the regional conifers including *Pinus contorta latifolia, P. monticola, P. albicaulis, P. ponderosa, Pseudotsuga menziesii, Abies lasiocarpa, Tsuga mertensiana, Juniperus communis*, and *Picea engelmannii* occur on serpentine, especially on the more mesic aspects. *Arctostaphylos nevadensis* and *Juniperus communis* are the two shrub species most representative of the local serpentine landscapes. Herbaceous species vary depending on moisture conditions. Many more bodenvag species exist on the more mesic sites than on xeric sites. Some of the most common generalists appear to be the grasses such as *Pseudoroegneria spicata, Elymus elymoides*, and *Festuca viridula* and herbs such as *Achillea millefolium, Fragaria virginiana, Eriophyllum lanatum, Senecio pauperculus*, and *Erisimum torulosum*. Kruckeberg listed 12 species as strong indicators of serpentine in this area. Two additional plant species, *Poa curtifolia* and *Lomatium cuspidatum*, are considered serpentine endemics to this locality, whereas *Polygonum newberryi* and *Eriogonum pyrolaefolium* are local indicators of serpentine, though they may be found on Cascadean volcanics farther south. Other serpentine indicators include *Aspidotis densa, Polystichum lemmonnii* (endemic to serpentine throughout northwestern North America), *Thlaspi alpestre, Douglasia nivalis dentata, Castilleja elmeri, Claytonia nivalis, Ivesia tweedei, Anemone drumnmondii, Claytonia nivalis*, and *Arenaria obtusiloba*. A large number of serpentine avoiders occur adjacent to the Ingalls complex. Kruckeberg lists 32 with tendencies to avoid serpentine. Some of the most characteristic of these are *Pachistema mysinites, Ceanothus velutinus, Prunus emarginata, Cheilanthes gracillima, Phacelia leptosepala, Penstemon fruticosus, Penstemon davidsonii*, and *Balsamorrhiza saggitata*. Species such as *Pseudoroegneria spicatum*, characteristic of more arid Great Basin and steppe environments, are given by Kruckeberg (1969) as examples of species that tend to exemplify the shift of more xeric species to serpentine substrates in a generally moist climatic zone.

20 Gulf of Alaska, Domain 8

The Gulf of Alaska domain extends eastward from Kodiak Island across the Kenai Peninsula, around the Chugach Mountains, and beyond Tonsina, curves southward across the glacier-covered St. Elias Mountains. The southeastern segment of the domain is west of the Canadian Coastal Mountains and includes the coastline and islands in southeastern Alaska (fig. 20-1). Ultramafic rocks occur sporadically in this region from British Columbia northwestward through southeastern Alaska and around the Gulf of Alaska to the Kenai Peninsula.

Most of this domain is mountainous, especially in areas where there are ultramafic rocks. Elevations range from sea level to >5000 m, although ultramafic rocks are not found at the highest elevations. This area was glaciated during the Pleistocene. Many glaciers persist at the higher elevations, and some descend to sea level.

The present climate ranges from cold to very cold and from humid to very humid, or perhumid (fig. 20-2). Mean annual precipitation ranges from 30 or 40 cm on the north side and west end of the Chugach Mountains, to >400 cm along the coast around Prince William Sound and in southeastern Alaska. Southeastern Alaska is a humid-to-perhumid area with dense forests at lower elevations that grade upward into alpine areas. The north side of the Chugach Mountains is drier, but still humid. Precipitation exceeds evapotranspiration in all months of every year. The frost-free period ranges from no more than a few days or weeks at the higher elevations to about 220 days in sheltered areas near sea level in southeasern Alaska. Some of the highest snowfall in North America is in this domain, and ice caps persist in some of the higher mountains with many glaciers flowing into the sea.

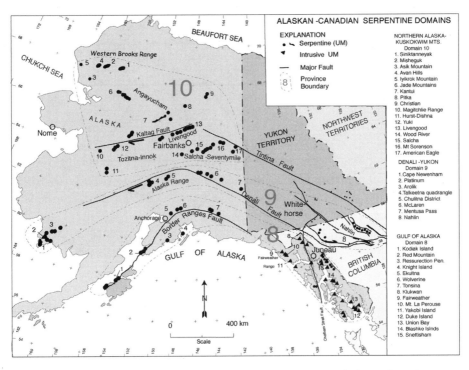

Figure 20-1 A map showing serpentine domains 8, 9, and 10 in Alaska and northwestern Canada.

20.1 Geology

Serpentine rocks of the Gulf of Alaska domain occur in both ophiolites and in concentric bodies, and some are from the roots of volcanic arcs complexes.

West of the Chatham Straight fault (fig. 20-1), a discontinuous belt of ultramafic bodies extends for >1600 km from Kodiak Island across the Kenai Peninsula, and around the Chugach Range, arching southward as far as Baranof Island (Burns 1985). This belt is nearly coincident with the Border Ranges fault, a major structure in this belt. The mafic–ultramafic complexes in this belt are now considered to be the exposed roots of island arc complexes formed in the ocean offshore of Alaska prior to their accretion. Ophiolites are present in the Prince William Sound area.

In southeastern Alaska a belt of concentrically zoned ultramafic complexes extends 560 km along the length of the Coast Range batholith, from Klukwan near Haines to Duke Island, south of Ketchikan. Localities 8-8 and 8-12 through 8-15 represent these concentrically zoned bodies. The greatest concentrations of serpentine rocks in the Gulf of Alaska domain are in these concentric intrusions of southeastern Alaska. They generally have areas <18 km^2. These concentric bodies have cores of dunite that grade outward through peridotite and clinopyroxenite to hornblendite and are surrounded by

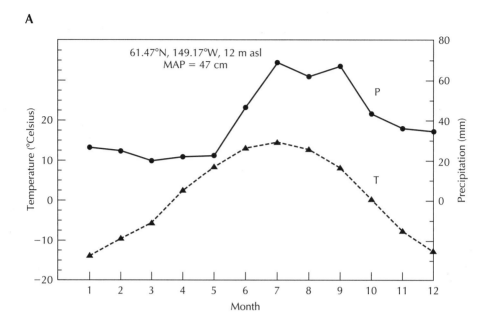
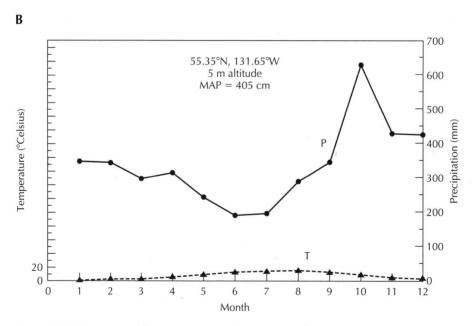

Figure 20-2 Mean monthly temperature and precipitation at two stations in the Gulf of Alaska domain. (A) Eklutna at the western end of the Chugach Mountains. (B) Ketchikan in southeast Alaska.

gabbro (Irvine 1974). These complexes are often referred to as Alaskan-type intrusive mafic–ultramafic bodies that are distinct from most of the other ultramafic rocks in Alaska, which are pieces of displaced oceanic crust or island arc complexes accreted to the continent. Most of the dozens of Alaskan-type intrusions in southeastern Alaska are in the Klukwan–Duke Island mafic–ultramafic belt (Himmelberg and Loney 1995).

20.2 Soils

Spodosols generally develop throughout the domain, with the exception of Andisols in volcanic deposits, but Entisols, Mollisols, and Inceptisols prevail in serpentine materials. Cambic horizons and ochric, mollic, and histic epipedons are the common diagnostic horizons in the serpentine soils. There is insufficient aluminum in serpentine material for the development of Andisols, but some of the Inceptisols may be in andic subgroups. Spodosols occur where there are enough nonserpentine materials over the serpentine materials for E horizons to develop. Albic (bleached) E horizons do not develop in serpentine materials, but enough organic matter and iron can accumulate in serpentine materials to form spodic horizons. Soil temperature regimes (STRs) are frigid at low elevations in southeastern Alaska and cryic elsewhere in this domain. There is no permafrost, except for isolated patches in the Chugach Mountains (Ferrians 1965).

20.3 Vegetation

The main vegetation types are conifer forest at lower elevations and alpine plant communities above a few hundred meters. Also, there are many, and some extensive, wet meadows with herbaceous vegetation. Kruckeberg holly fern has been reported to occur on serpentine in southeastern Alaska, but it is not a serpentine endemic (A. Kruckeberg, personal communication, 2002). Ungulate browsers, especially moose, can have profound effects on the plant communities.

20.4 Kodiak Island, Locality 8-1

This locality contains several mafic–ultramafic complexes located along the Border Ranges fault, along the west side of Kodiak Island. These are layered gabbro and ultramafic bodies. The ultramafic bodies are kilometer-size and consist of layered peridotite, dunite, clinopyroxenite and orthopyroxenite (Hill and Brandon 1976; Connelly 1978; Patton et al. 1989, 1994).

The locality is in a steep mountainous area. Elevations range from near sea level up to 1343 m on Kodiak Island, but the serpentine bodies are much lower. The mean annual precipitation is 60–150 cm on the west side of the island.

20.5 Red Mountain, Locality 8-2

This locality is an oval shaped body about 7 km long near the southern tip of the Kenai Peninsula. It consists of layered dunite and clinopyroxenite. The exposed area is approximately 49 km^2 (Guild 1942; Magoon et al. 1976; Toth 1981; Burns 1983, 1985). Elevations range from about 300 m up to 1074 m on the summit of Red Mountain. The mean annual precipitation is about 150 cm.

20.6 Resurrection Peninsula, Locality 8-3

Locality 8-3 is on the east side of the Kenai Peninsula, near Seward. It consists of serpentinized ultramafic rocks and associated mafic rocks. The area is approximately 8 km by 27 km, or 216 km^2 (Miller 1984). Elevations range from sea level up to 885 m. The mean annual precipitation is 200–250 cm.

20.7 Knight Island, Locality 8-4

This island is just east of the Kenai Peninsula. The locality contains complete ophiolites, from top down: pillow lavas, sheeted dikes, layered gabbros, serpentinized dunite, and harzburgite (Miller 1984, Tysdal et al. 1977). Elevations range from sea level up to 672 m on Iron Mountain, but they are somewhat lower on serpentine. The mean annual precipitation is about 200 cm.

20.8 Eklutna, Locality 8-5

This locality is at the west end of the Chugach Mountains, about 40 km northeast of Anchorage. Serpentine rocks in this locality consist of weakly serpentinized, layered wherlite, dunite, and pyroxenite in a fault-bounded body about 3 × 15 km (Rose 1966; Patton et al. 1989, 1994). Elevations range from near sea level adjacent to the mouth of the Knik River to 1253 m on Mount Eklutna. The mean annual precipitation is about 40 cm.

North of the Eklutna River, soils and vegetation were observed across a ridge that extends northwest from West Twin Peak. The typical bedrock is clinopyoxenite with a high specific gravity (3.2) that confirms the visual impression that there has been little alteration of it. Slopes are steep on the north and very steep to steep on the southwest side of the ridge. The soils are Inceptisols (Eutrocryepts) and Mollisols (Haplocryolls). A forest of predominantly white spruce (*Picea glauca*) and paper birch (*Betula papyrifera*) prevails on the north-facing slope, with much bunchberry (*Cornus canadensis*) and mosses in the understory. The southwest-facing slopes have shallow soils on very steep inclines and deeper soils on steep inclines. The vegetation on the very steep slopes is grasses, largely alpine fescue (*Festuca altaica*), kinnikinnick (*Arctostaphylos uva-ursi*), prickly saxifrage

(*Saxifraga tricuspinata*), low juniper (*J. communis*), buffaloberry (*Shepherdia canadensis*), and saskatoon (*Amelanchier alnifolia*), with sparse lichen (*Cladina* sp.). Semidense to open forest prevails on the steep southwest-facing slopes. It is predominantly aspen (*Populus tremuloides*) and paper birch, with much buffaloberry, rose (*Rosa acicularis*), fireweed (*Epilobium angustifolium*), and other forbs in the understory, and with sparse white spruce and mosses. Balsam poplar (*Populus balsamifera*) is present in forests of the lower slopes on both northern and southwestern sides of the ridge.

20.9 Wolverine Complex, Locality 8-6

The Wolverine complex locality is near the northwest edge of the Chugach Mountains, about 80 km northeast of Anchorage. Serpentine rocks in this locality consist of layered gabbro, dunite, clinopyroxenite, and peridotite. They are exposed for approximately 8 km parallel to the Border Ranges fault (Clark 1972; Burns 1985; Burns et al. 1991; Patton et al. 1989, 1994). Glaciers lap over the serpentine rock outcrops. Elevations range from 900–1900 m. The mean annual precipitation is about 50 cm, or greater.

20.10 Tonsina, Locality 8-7

The Tonsina locality is near the north edge of the Chugach Mountains, between the Richardson Highway and the Copper River near Chiltina. The Tonsina island arc complex is a layered gabbro and ultramafic body, approximately 60 km^2. Dunite, harzburgite, wherlite, and websterite are the most abundant ultramafic units (Hoffman 1974; Winkler et al. 1981; Burns 1985; Coleman and Burns 1985; DeBari and Coleman 1989; Patton et al. 1989, 1994). Elevations range from 900–1738 m. The mean annual precipitation is about 30 cm.

20.11 Klukwan, Locality 8-8

This locality is on a steep southwest-facing slope of the Takshanuk Mountains facing Klukwan, which is on the Chilkat River, northwest of Haines. The ultramafic body is hornblende pyroxenite, which is surrounded by gabbro and diorite (MacKevett et al. 1974). Elevations on the hornblende pyroxenite range from 300 to 1900 m. The mean annual precipitation is 75–80 cm.

20.12 Fairweather, Mt. La Perouse, and Yakobi Island, Localities 8-9, -10, and -11

The Fairweather and Mt. La Perouse localities are in the Fairweather Range in Glacier Bay National Park, and the Yakobi Island locality is a few kilometers southeast of the

Fairweather Range. Intrusions of layered mafic rocks occur in these localities. A small body of layered gabbro is present in the Fairweather locality (Plafker and MacKevett 1970; Brew et al. 1978). The La Perouse locality has 6000 m of layered mafic rocks that consist of olivine gabbro, norite, gabbronorite, and troctolite. They formed as an intrusion 40 Ma (Rossman 1963, Brew et al. 1978, Himmelberg and Loney 1981, Loney and Himmelberg 1983). The Yakobi Island locality contains a Tertiary, layered gabbro–norite intrusion similar to the La Perouse intrusion, but much smaller in extent (Loney et al. 1975).

20.13 Duke Island, Locality 8-12

Locality 8-12 is on a small island about 50 km south of Ketchikan. The serpentine rocks in this locality are in a concentrically zoned body (Taylor 1967, Irvine 1973). The land on Duke Island is low, mostly below 150 m elevation and wet, but with elevations up to nearly 500 m on Mt. Lazaro. The rocks on Mt. Lazaro are mafic, rather than ultramafic.

20.14 Union Bay, Locality 8-13

The Union Bay locality is north of Ketchikan, on the northwestern side of the Cleveland Peninsula. The serpentine rocks in this locality are in a concentrically zoned body (Ruckmick and Noble 1959). The dunite core of the body is centered on Golden Mountain, which has a maximum elevation of 773 m, and the outer zones are at lower elevations

Figure 20-3 A view from the Cleveland Peninsula north–northwestward across Golden Mountain in locality 8-13 toward Vixen Inlet of Ernest Sound. Open meadow occupies the dunite core of the Alaskan-type Union Bay complex on the summit of Golden Mountain. Glacial drift around the mountain is densely forested, with some fens and bogs. Photo by P. Krosse.

down practically to sea level. Golden Mountain is 5 or 6 km east of Mt. Burnett, at the opposite end of a ridge that joins them (topographic map in Alexander et al. 1989, 1994a). The older literature places Mt. Burnett at the current location of Golden Mountain. Glaciers from the east covered the area during the last ice age, but any glacial drift that might have cover the summit of Golden Mountain has been eroded away. Currently the mean annual precipitation is about 200 cm.

Alexander et al. (1989, 1994a) described an Inceptisol on dunite at 740 m elevation on Golden Mountain and at 420 m a Spodosol in serpentine glacial till with a thin cover of nonserpentine material. The Inceptisol has an Oe-A-Bw-C-R horizon sequence with a friable, brown (7.5YR 4/5, moist), slightly acid Bw horizon. It is an Andic Eutrocryept. It is in an Andic subgroup, even though there is little aluminum in the soil, because there is plenty of iron in the acid oxalate extract (Alexander et al. 1994a). Bulk densities are low in surface soil horizons, corresponding to high organic matter concentrations. The organic carbon concentration in the Bw horizon of the Inceptisol is 102 g/kg (10.2%), yet the color value is four and the chroma is five.

The serpentine Spodosol described by Alexander et al. (1989, 1994a) has an A-E-2Bs-2C-2Cm soil horizon sequence, with the A and E horizons in nonserpentine material. It is an Oxyaquic Duricryod. Although there are a few strong brown (7.5YR 4/6) mottles in the albic E horizon, these sparse redoximorphic features were not considered to be adequate for an Aquod. A nonserpentine Spodosol on glacial till near the serpentine Spodosol is appreciably wetter and is definitely an Aquod, a Typic Cryaquod. The nonserpentine Spodosol has an O-E-Bhs-Bsg-C-Cd soil horizon sequence. There is no Bhs horizon in the serpentine Spodosol, in which the Bs horizon occupies a comparable profile position. Acid oxalate extractable iron is 82 g/kg from the Bs horizon of the serpentine Spodosol, but it is only 43 g/kg from the Bhs horizon of the nonserpentine Spodosol. Silica cementation in glacial till beneath the serpentine Spodosol was verified by laboratory investigations (Alexander et al. 1994b).

The nonserpentine Spodosol on Golden Mountain supports a Sitka spruce–hemlock–western red cedar–yellow cedar forest, with red alder and many species of shrubs in the understory. Douglas-fir does not reach this far north in the Alaska Coast Ranges. The serpentine soils have more open plant communities, with sparse shore pine (*Pinus contorta*) trees. Also, there are some yellow cedar (*Chamaecyparis nootkatensis*) and mountain hemlock (*Tsuga mertensiana*) trees on the serpentine Spodosol, but no western hemlock (*T. heterophylla*), which occurs on the nonserpentine Spodosol. The serpentine Spodosol was sampled on a ridge and the nonserpentine just below the ridge–the slightly more protected aspect of the nonserpentine Spodosol may or may not account for some of the differences in plant communities, but the major differences are judged to be related to substrate differences. Shrubs on the serpentine Inceptisol were white mountain heather (*Cassiope mertinsiana*), crowberry (*Empetrum nigrum*), yellow mountain heather (*Phyllodoce glanduliflora*), and bog blueberry (*Vaccinum uliginosum*), and those on the serpentine Spodosol were crowberry, common juniper, and Labrador tea (*Ledum groenlandicum*). Several species of shrubs that occurred on the nonserpentine Spodosol and are common in the area were not found on the serpentine soils. None of forbs or grasses is endemic.

Figure 20-4 A very steep south-facing slope in locality 20-5, Eklutna.

The forb species are listed by Alexander et al. (1989). Bryophytes were not identified. Maidenhair fern (*Adiantum aleuticum*) on the Spodosol was the only fern found on the serpentine soils, but lady fern (*Athyrium filix-femina*) and deer fern (*Blechnum spicant*) were found on the nonserpentine soil.

20.15 Blashke Islands, Locality 8-14

The Blaske Islands are all part of one concentrically zoned body (Himmelberg et al. 1986). They are between Ketchikan and Petersburg, just east of the northern end on Prince of Wales Island. Elevations range from sea level to 147 m, with the highest elevation on gabbro, rather than on ultramafic rocks. The mean annual precipitation is about 150 cm.

20.16 Snettisham, Locality 8-15

The Snettisham locality is on a steep northwest-facing mountain slope above Port Snettisham, about 50 km southeast of Juneau. Elevations range from sea level to about 300 m. The mean annual precipitation is about 120 cm. The main ultramafic body is pyroxenite, or hornblende–pyroxenite (Thorne and Wells 1956). This part of the Snettisham Peninsula was recognized on early navigational charts of the coast and geodetic survey for its magnetic anomaly, which can be attributed to large amounts of magnetite in the ultramafic body.

21 Denali–Yukon, Domain 9

The Denali-Yukon domain occupies a broad arc that, in general, follows the path of the Denali Fault along the Alaska Range and southwestward into the Yukon Territory. An ophiolite in the northwestern corner of British Columbia that is northeast of the projected Denali fault is included in this locality (fig. 20-1). A projection of the Denali fault system southwestward from the Alaska Range passes through the southwestern part of the Ahklun Mountains physiographic province, as the province was defined by Wahrhaftig (1965), to Kuskokwim Bay between the mouth of the Kuskowim River and Cape Newenham. Three mafic–ultramafic complexes on the southwestern edge of the Ahklun Mountains province are included in this domain. Glaciers covered this entire domain during the Pleistocene, and mountain glaciers and ice caps are still present at the higher elevations. Permafrost is currently discontinuous.

The highest mountain in North America (Mt. McKinley, 6194 m) is in the Alaska Range, but the ultramafic rocks are all at much lower elevations. The climate is very cold throughout the domain, with severe winters and short summers. The mean annual precipitation ranges from 45 to 150 cm in the Ahklun Mountains, from 30 to 60 cm in the Alaska Range, and from 30 to 75 cm, or more, in the Atlin area of northwestern British Columbia, which is in the rain shadow of the Coast Mountains. The greatest precipitation is during summers, from June or July to September or October (fig. 21-1). The frost-free period is on the order of 60–90 days, or shorter, but it may be longer in some of the Atlin area of British Columbia.

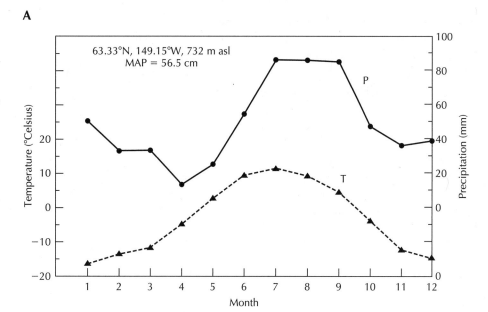

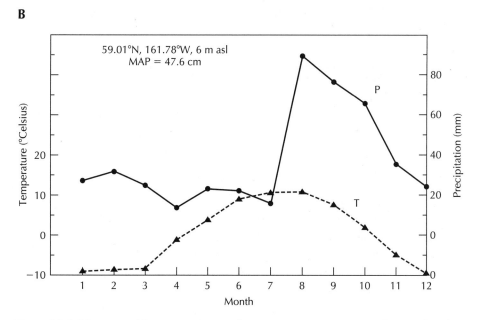

Figure 21-1 Mean monthly temperatures and precipitation at two sites in the Denali-Yukon domain. (A) Summit, Alaska. (B) Platinum, Alaska.

21.1 Geology

Localities 9-1 through 9-3 are from Cape Newenham northeastward in the Ahklun Mountains. The ultramafic rocks in the Cape Newenham area were accreted to North America by north directed thrust faults during the Late Triassic and Middle Jurassic time. Localities 9-4 through 9-7 are in the Alaska Range. Locality 9-8 is along a projection of the Denali fault to the eastern edge of the Coast Ranges in British Columbia.

21.2 Soils

Pergelic subgroups of Cryepts, Aquepts, and Cryods have been mapped in this domain (Rieger et al. 1979), but there was no mention of serpentine soils. Many of the higher and steeper mountains are barren, where they are not covered by ice. The serpentine soils observed in this domain are Mollisols (Haplocryolls).

21.3 Vegetation

The serpentine vegetation in much of this domain is alpine tundra. Vegetation of the alpine tundra was described in locality 9-8 (Nahlin) and vegetation in a transition between forest and alpine tundra was described in locality 9-6 (Maclaren–Rainbow).

21.4 Cape Newenham, Locality 9-1

The mafic-ultramafic complexes in this locality are about 10% serpentinized harzburgite and dunite and 90% nonlayered gabbro. They are exposed over an area of aproximately $90\,km^2$ (Box 1985, Patton et al. 1989, 1994). The mean annual precipitation is about 100 cm.

21.5 Platinum, Locality 9-2

The mafic–ultramafic complexes in this locality are zoned Alaskan-type intrusive bodies with cores of serpentinized dunite grading outward to gabbro–diorite with rims of hornblendite (187–176 Ma intrusive age). Their surface exposure is approximately $25\,km^2$ (Mertie 1940, Bird and Clark 1976). The ultramafic rocks are exposed on Red Mountain (575 m) and the northwest side of Suzie Mountain, which are 8–10 km south of Goodnews Bay. The mean annual precipitation is about 75 cm, with the greatest amounts in late summer and autumn (fig. 21-1).

21.6 Arolik, Locality 9-3

The mafic–ultramafic complexes in this locality are about 25% harzburgite and minor dunite and 75% pyroxene gabbro. Their surface exposure is approximately 70 km^2 (Patton et al. 1989, 1994). The complexes are interpreted as thrust slices overlying greenschist–blueschist metamorphic rocks. Small ultramafic bodies are distributed just north of the fault as narrow slivers in the fault zone.

21.7 Talkeetna, Locality 9-4

Sublocality 9-4a (Mt. Russell)

The ultramafic rocks in the Mt. Russell locality are dunite and serpentinite, with small amounts of talc schist, exposed discontinuously for 25 km along minor northeast–southwest-trending faults between the Lacuna and Yentna Glaciers and Mt. Russell (Reed and Nelson 1980). The Mt. Russell sublocality is 90–100 km northwest of Talkeetna, and southeast of the Denali fault. Also, there is a small, fault-bounded slice of serpentinite and associated talc schist about 6 km northwest of the Denali fault (~15 km northwest of Mt. Russell) and parallel to the Denali fault system. Elevations of the ultramafic rock exposures are mostly in the 900–1800 m range.

Sublocality 9-4b (Willow Creek)

Several small bodies of serpentinite occur discordantly in pelitic mica schist on the south side of Willow Creek, 10–20 km east of Willow, Alaska (Csejtey and Evarts 1979). This serpentinite is in the southwestern part of the Talkeetna Mountains, about 80 km south–southeast of Talkeetna. Elevations range up to 975 m on the higher hills. Much of the serpentinite and associated gabbro is covered by glacial drift and the serpentinite not covered by drift is in alpine tundra. Cespitose willow *(Salix* sp.) and alpine manzanita (*Arctostaphylos alpina*) are abundant in the alpine tundra.

21.8 Chulitna, Locality 9-5

The Chulitna locality is in the Upper Chulitna district on the south side of the Alaska Range, northwest of the Chulitna River, just southwest of Summit in Broad Pass. It is in steep-to-moderately steep mountainous terrain.

A tectonically dismembered Upper Devonian ophiolite assemblage in the Upper Chulitna district contains serpentine, gabbro, basalt, and chert (Clark et al. 1972). It is associated with the southwest-trending Upper Chulitna fault, an off-shoot from the Cantwell fault, which is an off-shoot from the Denali fault. The serpentine is highly sheared clinochrysotile and lizardite (Hawley and Clark 1974). It is in narrow bodies along the Upper Chulitna fault zone, with the largest masses between the Eldridge Glacier to Long Creek. Elevations range from 600–1500 m, and the mean annual precipitation is about 50 cm.

21.9 Maclaren–Rainbow, Locality 9-6

There are two belts of serpentinized dunite-peridotite and gabbro (Triassic in age) in this locality, south of the Denali fault zone. One belt is at the north end of Rainbow Mountain, near the Denali fault zone, and the other is from the Richardson Highway west to the Maclaren Glacier, some distance south of the Denali fault zone (Hanson 1964, Rose 1965, Nokleberg et al. 1982, Stout 1976). Elevations range from 1000–1750 m, and the mean annual precipitation is about 40 cm.

Much of this locality is in alpine tundra, but serpentine soils and vegetation were observed only in the transition from conifer forest along the Delta River to alpine tundra on Rainbow Mountain. They were observed on steep to very steep southeast-facing slopes at the lower (850–950 m altitude) northwest end of Rainbow Mountain. The soils are shallow to very deep Mollisols (Haplocryolls). An alder (*Alnus crispa*)–paper birch (*Betula papyrifera*)–willow (*Salix* sp.)–scrub birch (*Betula nana*) scrub thicket covers the slopes. The understory is mostly Labrador tea (*Ledum palustre*), roses (*Rosa* spp.), and a lichen (*Cladonia decorticata*). Wintergreen (*Pyrola secunda*) and cloudberry (*Rubus chamaemorus*) are common. Low juniper and grasses, predominantly fescue, prevail on shallow soils.

21.10 Mentasta Pass, Locality 9-7

A small exposure of a dunite body, about 3 km long and <1 km wide, is present in the Denali fault zone near the headwaters of the Slana River, just west of Gillet Pass (Richter 1967). Its elevation is 1200–1500 m, in steep terrain, and the mean annual precipitation is about 50 cm. From Gillet Pass to Canada there are several small ultramafic bodies containing serpentinized peridotite and dunite, north of the Denali fault zone. These may be Devonian in age (Richter 1967; Patton et al. 1989, 1994).

21.11 Nahlin, Locality 9-8

The Nahlin locality is represented by an ophiolite body in northwest British Columbia, east of Atlin. It is 100 km long and up to 8 km wide. It consists of foliated tectonite peridotites, ranging from harzburgite, with up to 40% pyroxene, to dunites. The ultramafic rocks are associated with a well-preserved ophiolite sequence consisting of deformed and metamorphosed gabbro and diabase. The ultramafic has been serpentinized along its faulted boundaries. The age of the associated gabbro is 325 Ma (Mississippian) (Aitken 1953, Souther 1971, Monger 1977).

Figure 21-2 Alpine tundra on the 1850-m-high summit of a mountain in locality 9-8, Nahlin.

This locality appears to have more affinity to the Intermountain Plateau of British Columbia that is between the Rocky Mountains and the coastal mountains than to the Alaska Range. Elevations are up to 2087 m on Mt. Barnham, 15–18 km northeast of Atlin, but they are mostly much lower in this locality.

Ultramafic rocks are relatively extensive northwest of Surprise Lake. Soils and vegetation were observed on southwest-facing slopes from Ruby Creek up to the 1850-m high summit of the first mountain south of Mt. Barnham. The soils are very shallow to very deep Mollisols (Haplocryolls) with some rock outcrop and much barren talus, although the talus is far from barren when considering the abundant lichens on the rocks. Most of the vegetation on the mountain is alpine tundra (fig. 21-2). The upper slopes are covered by a carpet of caespitose willows (*Salix* spp.) and yellow mountain-avens (*Dryas drummondii*) with sedges and grasses, including alpine fescue, and patches of mountain-heather (*Cassiope tetragona*), a cushion plant (*Silene acaulis*), and stonecrop (*Sedum lanceolatum*). The alpine tundra grades downward to alpine fescue meadow with crowberry and cespitose willows, with patches of erect willow shrubs (*Salix* spp.) and scrub birch, and scattered low juniper and subalpine fir (*Abies lasiocarpa*). Among the more common serpentine forbs in this alpine meadow are lupine (*Lupinus* sp.), burnet (*Sanguisorba stipulata*), fireweed (*Epilobium angustifolium*), ragwort (*Senicio lugens?*), and monkshood (*Aconitum delfinifolium*). Farther downward, the patches of fir coalesce to a discontinuous forest around the fringe of Ruby Valley. There is no forest below about 1300 m altitude in Ruby Valley. The vegetation in Ruby Valley, which has a filling of glacial drift with mixed lithologies, is willow–birch scrub with a soft ground cover of mosses.

22 Northern Alaska–Kuskokwim Mountains, Domain 10

The ultramafic rocks in this domain are in the western part of the Brooks Range, the interior Alaska lowlands of the Koyukuk–Yukon Basin, the interior Alaska highlands of the Tanana–Yukon Upland, and the Kuskokwim Mountains (fig. 20-1). This domain extends east to the Seventymile River, a tributary of the Yukon River that is near the Canadian border, and presumably to the Clinton Creek area in the Yukon Territory.

Although the highest elevations in the Brook Range are near 2700 m, those in the western mountains of the range are mostly <1400 m. Flatlands, hills, and low mountains dominate the Koyukuk–Yukon Basin and Tanana–Yukon Uplands, Elevations in the Kuskokwim Mountains are mostly <1000 m. Some of the mountains in uplands of the Tanana–Yukon Uplands are higher than 1600 m.

Although the Brooks Range was glaciated during the Pleistocene, there was no glaciation in the Koyukuk–Yukon Basin, and only the higher elevations in the Tanana-Yukon Upland were glaciated during the Quaternary. Today, permafrost prevails throughout the Brooks Range, but it is discontinuous in the Koyukuk–Yukon Basin and Tanana–Yukon Upland and in the Kuskokwim Mountains (Ferrians 1965). Loess is extensive in the basins of interior Alaska and at lower elevations in the Kuskokwim Mountains, with some deposits >60 m thick (Péwé 1975).

The climate is very cold throughout the domain, with severe winters and relatively short summers, although mean maximum summer (July) temperatures are >20°C (up to 24°C or 25°C) in the interior basins. With latitudes from 61°N to 68°N, days are very long during summers and very short during winters. The mean annual precipitation is 15–45 cm, with the greatest precipitation during summers (fig. 22-1). Even though the precipitation is low, the climate is not arid because evapotranspiration is limited by short and relatively cool summers. The freeze-free period is on the order of 60–90 days.

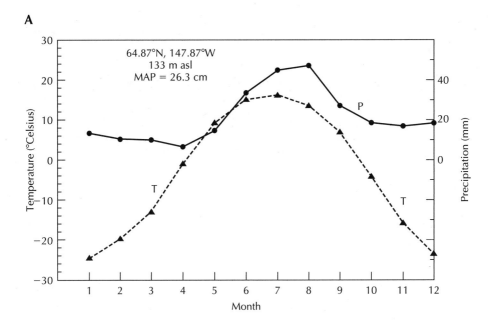

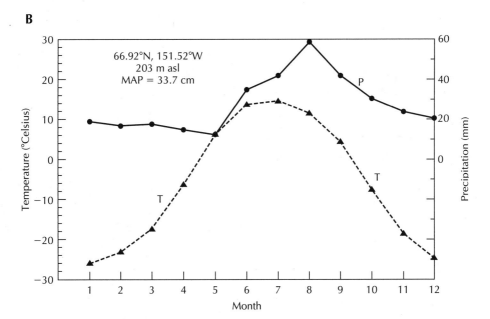

Figure 22-1 Mean monthly temperature and precipitation at two sites in the Northern Alaska–Kuskokwim Mountains domain. (A) Fairbanks in central Alaska. (B) Bettles, near the southern edge of the Brooks Range.

22.1 Geology

The northern and interior Alaska ultramafics (serpentine) consist of Paleozoic and Mesozoic thrust slices emplaced onto Precambrian and Paleozoic marine sediments. They all belong to well-defined belts and are related to low-angle thrust faults or to later high strike–slip faults. They have been tectonically transported from an original oceanic setting and have been accreted during the Late Jurassic to Late Cretaceous.

The serpentine localities of Northern Alaska–Kuskokwim Mountains domain are grouped into five belts, or areas (fig. 20-1). Localities 10-1 through 10-5 are in the western Brooks Range. The western Brooks Range contains some of the largest ultramafic bodies in Alaska. Localities 10-6, 10-7, and 10-10 are in a nearly continuous belt of oceanic rocks with limbs along the northern and southeastern edges of the Koyukuk-Yukon Basin. Localities 10-8, 10-9, 10-11, and 10-12 are in the Tozitna-Innoko belt of ocean crust and mantle rocks that were thrust onto the Ruby geanticline. The Tozitna–Innoko belt occupies a physiographically diverse area from the Dishna and Hurst complexes (locality 10-11) in the Kuskokwim Mountains northeastward across the Yukon River to the Christian complex (locality 10-9) north of the Arctic Circle. Locality 10-13 represents a series of east–northeast to west–southwest-trending ultramafic complexes in hilly terrain just north of Fairbanks. Localities 10-14 through 10-17 are in the discontinuous Salcha River–Seventymile belt of ultramafic rocks in the Tanana-Yukon Upland.

22.2 Soils

Upland soils of the western Brooks Range have been mapped mostly as Lithic Cryorthents and Pergelic Cryaquolls (Rieger et al. 1979). Soils mapped in interior Alaska were mainly Typic Cryochrepts, Typic Cryorthods, and Histic Pergelic Cryaquepts. Most of these names have been changed, and frozen serpentine soils, such as those the Brooks Range and parts of the interior Alaska, are now called Gelisols (Soil Survey Staff 1999).

Weathering to produce cambic horizons and accumulation of organic matter are the serpentine soil-forming processes. Cryepts and Cryolls are generally the results of these processes in serpentine materials. Podzolization, leaching of organic matter and iron and aluminum, is an active process where serpentine materials are covered by loess. Soils with serpentine parent materials are Gelisols, Cryepts, and Cryolls. The only known investigation of serpentine soils in interior Alaska, up to the Brooks Range, is that of Alexander (2004).

22.3 Vegetation

The plant cover is of low stature in the Brooks Range and at higher elevations south of the Brooks Range, with much moss (and other bryophytes) and shrubs and few forbs and stunted trees. Reindeer lichens are abundant. Open spruce–birch forest is prevalent on serpentine soils at lower elevations south of the Brooks Range. Bryophytes and leafy or feathery lichens are abundant. Grasses are common on some serpentine soils.

22.4 Siniktanneyak Mountain, Locality 10-1

The Siniktanneyak Mountain locality consists of 20% dunite, cumulate layers of wherlite, harzburgite, olivine pyroxenite and 80% layered gabbro. Surface exposure is about 300 km^2 (Nelson and Nelson 1982; Patton et al. 1989, 1994).

22.5 Misheguk Mountain, Locality 10-2

Locality 10-2 consists of 10% serpentinized ultramafic rocks and 90% gabbro with minor ultramafics. Surface exposure is about 460 km^2 (Curtis et al. 1984; Patton et al. 1989, 1994).

22.6 Asik Mountain, Locality 10-3

Locality 10-3 consists of 5% ultramafic and 95% gabbro. Surface exposure is about 100 km^2, with 300 km^2 covered (Zimmerman and Soustek 1979; Zimmerman et al. 1981; Frank and Zimmerman 1982; Patton et al. 1989, 1994).

22.7 Avan Hills, Locality 10-4

The Avan Hills locality consists of 10% ultramafic tectonic dunite–harzburgite and 90% layered gabbro. Exposure is about 200 km^2 with 100 km^2 covered (Patton et al. 1989, 1994; Mayfield et al. 1990).

22.8 Iyikrok Mountain, Locality 10-5

The mafic–ultramafic complex of Iyikrok Mountain spans approximately 30 km^2, with 20 km^2 covered (Mertie 1940; Barnes and Tailleur 1970; Mayfield et al. 1988; Patton et al. 1989, 1994).

22.9 Jade Mountains, Locality 10-6

The Jade Mountain ultramafic body is completely serpentinized and derived from harzburgite tectonite. Surface exposure is 40–50 km^2. Cosmos Hills upper thrust panel is completely serpentinized, 130 m thick, and 8 km long (Loney and Himmelberg 1985).

22.10 Kanuti, Locality 10-7

This body is made up of 35% mantle tectonite harzburgite-dunite and 65% cumulate gabbro. Surface exposure is about 300 km^2 (Loney and Himmelberg 1985).

Soils and vegetation were observed from 400 to 700 m altitude on the northeast end of Caribou Mountain (969 m), which is practically all ultramafic rock. A stone from this area had a specific gravity of 3.0, indicating no more than slight serpentinization of the peridotite. Stepped slope profiles, nivation or solifluction hollows, and stone stripes are common. The soils are mostly Inceptisols–Eutrocryepts, with Typic and Histic Cryaquepts in the hollows. No permafrost was found close enough to the surface to call any of the serpentine soils Gelisols. Permafrost was found well within a meter depth in grus (weathered granitic rock) at the foot of the mountain. The vegetation is semidense to open white spruce forest on the lower slopes and alpine tundra on the upper mountain slopes (fig. 22-2).

The vegetation in the spruce forests is mostly white spruce (*Picea glauca*), alder (*Alnus crispa*), crowberry (*Empetrum nigrum*), alpine blueberry (*Vaccinium uliginosum*), Labrador tea (*Ledum palustre*), mosses, and lichens (mostly *Cladina* sp. and *Flavocetraria* sp.). White mountain heather (*Cassiope tetragona*), shrub birch (*Betula nana*), and fescue (*Festuca altaica*) and other grasses are common. Shrub birch is more common in the transition to alpine tundra.

Figure 22-2 A summit (~700 m high) at the northeast end of Caribou Mountain. It is all partially serpentinized peridotite.

The vegetation in the alpine tundra is most commonly crowberry, alpine blueberry, dwarf willow (*Salix* sp.), Labrador tea, mountain-avens (*Dryas octapetala*), white mountain heather, mosses, and lichens (especially *Cladina* sp.). On mountain or ridge summits, mountain-avens, cushion silene (*Silene acaulis*), and lichens are dominant plants, and arctic sandwort (*Minuartia arctica*) and arctic lupine (*Lupinus arcticus*) are common. Polar grass (*Arctagrostis* sp.) is the most common graminoid on the high ridges, and it occurs with sedges (especially *Carex bigelowii*) in wet hollows. Wet meadows are present in some of the hollows, but graminoids are sparse on most of the remainder of the upper mountain slopes.

22.11 Pitka, Locality 10-8

The Pitka locality consists of a metmorphosed mafic–ultramfic sequence with basal garnet amphibolite. Surface exposure is $25\,km^2$.

22.12 Christian, Locality 10-9

Locality 10-9 consists of layered ultramafics from leuco gabbros to peridotite with a well-developed garnet amphibolite metamorphic basal zone. Exposure is about $170\,km^2$ (Brosgé and Reiser 1962; Patton et al. 1989, 1994).

22.13 Magitchlie Range, Locality 10-10

The ultramafic body of the Magitchlie Range consists of a serpentinized cumulate ultramafic with approximately $10\,km^2$ of poor exposures with lots of rubble (Patton et al. 1989, 1994).

22.14 Hurst-Dishna, Locality 10-11

Locality 10-11 is serpentinized harzburgite 33% and 66% cumulate ultramafics. Surface exposure is approximately $30\,km^2$ (Chapman et al. 1985; Miller and Angeloni 1985; Patton et al. 1989, 1994).

22.15 Yuki, Locality 12

The Yuki locality is partly serpentinized harzburgite tectonite and ultramafic cumulates including dunite, wehrlite, and olivine clinopyroxenite. Surface exposure is approximately $140\,km^2$ (Patton et al. 1989, 1994).

22.16 Livengood, Locality 13

The Livengood locality contains completely serpentinized harzburgite and dunite. Surface exposure is approximately 150 km². The age is Pre-Ordovician (Loney and Himmelberg 1988).

The Livengood ultramafic body is elongated in a west–southwest to east–northeast direction in low mountainous terrain with topographic orientation similar to that of the ultramafic body. Most of the serpentine soils are Eutrocryepts. They were sampled on a steep south-facing slope, on a broad ridge at about 550 m altitude, and on a very steep north-facing slope (Alexander 2004). The corresponding soils on these landscape positions were a deep, loamy–skeletal over fragmental, magnesic Typic Eutrocryept with an open white spruce/kinnikinnick (*Arctostaphylos uva-ursi*)–juniper (*Juniperus communis*)/arctic sandwort plant community; a moderately deep, loamy–skeletal, magnesic Typic Eutrocryept with a paper birch (*Betula papifera*)–black spruce (*Picea mariana*)/labrador tea–lingonberry *(Vaccinium vitis-idaea)*–alpine blueberry/bluejoint reedgrass (*Calamagrostis canadensis*)–fescue grass plant community; and a very deep, loamy–skeletal, magnesic, Oxyaquic Eutrocryept with a black spruce–paper birch/labrador tea–alpine blueberry/sphagnum plant community. The soils on south-facing and ridge slopes are on frost-shattered bedrock and have about 0.2 m of loess incorporated into their surfaces. The soil on the north-facing slope is in colluvium with a 0.25 m cover of nonserpentine colluvium. Only the serpentine soil on the south-facing slope has a plant community that is distinctly different from those on adjacent nonserpentine soils. The common plant community on steep nonserpentine south-facing slopes is with a relatively dense white fir (*P. glauca*)–paper birch/lingonberry/northern comandra (*Geocaulon lividum*) woodland. There is no northern comandra on the serpentine soils, nor any juniper on adjacent nonserpentine soils. The common lichens on the serpentine soils are crinkled snow lichen (*Flavocentraria nivalis*) and antlered powderhorn (*Cladonia sublata*) on the south-facing slope, and antlered powderhorn, curled snow lichen (*Flavocetraria cucullata*), thorn lichen (*Cladonia unicaulis*), and sieve lichen (*Cladonia multiformis*) and lesser amounts of studded leather-lichen (*Peltigera aphthosa*) on the broad ridge. No permafrost was found in the Livengood area, at least not within 1 m depth. The loess on the Livengood soils was probably deposited thousands of years ago. It is a slightly plastic silt loam, easily molded into 3 mm-diameter threads when wetted to the plastic limit. Recent loess in the Yukon basin is nearly pure silt (Dennis Mulligan, personal communication, 2003), which is nonplastic.

Plants have been inventoried in the Serpentine Slide Research Natural Area, which is in the Livengood locality. The rock types in the research natural area are clastic sedimentary rocks, dolomite, chert, greenstone, and serpentinite. Much of the serpentinite has been exposed and is maintained in barren condition by erosion. The serpentinite is in the 500–900 m altitude range. Juday (1992) found that the plant species with greatest tolerance of the serpentine conditions are thoroughwax (*Bupleurum tiradiatum*) and arctic sandwort, and reported that sorrel (*Rumex acetosa*) grows on the serpentine barrens. Only a crustose lichen inhabits the most severe barrens.

Figure 22-3 A serpentine (serpentinized peridotite) landscape on the west side of American Creek a few kilometers south of Eagle, Alaska. The lateral ridges sloping down toward the east have open forest on south-facing sides and semidense forest on north-facing sides.

22.17 Wood River, Locality 10-14, Salcha, Locality 10-15, Mt. Sorenson, Locality 10-16, and American Eagle, Locality 10-17

Localities 10-14 through 10-17 are serpentinized harzburgite, with minor dunite and clinopyroxenite. These ultramafics are closely associated with greenstones forming a 200-km-long belt (Patton et al. 1989, 1994).

Soils and vegetation were observed along the Taylor Highway where American Creek cuts through ultramafic rocks of the American Eagle locality just south of the village of Eagle, which is on the Yukon River (fig. 22-3). Because of very steep slopes, soils in the canyon are mostly either shallow to bedrock or very deep in colluvium. West of American Creek south-facing slopes have open forest, and north-facing slopes have semi-dense forest cover. Two soils were described on west-facing slopes on the east side of American Creek: a Lithic Haplocryoll with an open white spruce–paper birch/mountain alder (*Alnus crispa crispa*)–alpine blueberry/moss plant community and a very deep Pachic Haplocryoll with a semi-dense paper birch–white spruce/mountain alder/purple reedgrass (*Calamagrostis purpurascens*)/moss plant community. Forbs and sedges are sparse on both soils. Shrubby cinquefoil (*Potentilla fruticosa*) is widely distributed on adjacent landslides and serpentine roadcuts, but it does not occur on moss-covered soil. Silt loam fine-earth textures in Lithic Haplocryoll surface soil indicate that some loess from the floodplain of the Yukon River, about 12 km to the north, has reached the site. The loess influx is more obvious in the Pachic Haplocryoll, which is only 8 km from the Yukon River. An E-horizon has formed in the extremely stony Pachic Haplocryoll surface soil.

Part IV References

Aitken, J.D. 1953. Greenstones and associated ultramafic rocks of the Atlin map area, B.C. Ph.D. thesis, University of California, Los Angeles.

Alexander, E.B. 1988. Morphology, fertility, and classification of productive soils on serpentinized peridotite in California (U.S.A.). Geoderma 41: 337–351.

Alexander, E.B. 1995. Silica cemented serpentine soils in the humid Klamath Mountains, California. Soil Survey Horizons 36: 154–159.

Alexander, E.B. 2003. Trinity Serpentine Soil Survey. Shasta-Trinity National Forest, Redding, CA, unpublished manuscript, maps in a GIS.

Alexander, E.B. 2004. Serpentine soils of northern Alaska. Soil Survey Horizons 45: 120–129.

Alexander, E.B., P. Cullen, and P.J. Zinke. 1989. Soils and plant communities of ultramafic terrain on Golden Mountain, Cleveland Peninsula. Pages 47–56 in E.B. Alexander (ed.), Proceedings of Watershed '89. USDA Forest Service, Alaska Region, Juneau, AK.

Alexander, E.B., C.L. Ping, and P. Krosse. 1994a. Podzolization in ultramafic materials in southeast Alaska. Soil Science 157: 46–52.

Alexander, E.B., R.C. Graham, and C.L. Ping. 1994b. Cemented ultramafic till beneath a Podzol in southeast Alaska. Soil Science 157: 53–58.

Arroues, K.D. 2005. Soil Survey of Western Fresno County, California. U.S. Government Printing Office, Washington, D.C.

Arroues, K.D. and C.H. Anderson Jr. 1986. Soil Survey of Kings County, California. U.S. Government Printing Office, Washington, DC.

Atzet, T. and D.L. Wheeler. 1984. Preliminary Plant Associations of the Siskiyou Mountain Province. USDA Forest Service, Pacific Northwest Region, Portland, OR.

Bailey, E.H. and D.L. Everhart. 1964. Geology and ore deposits of the New Almaden district, Santa Clara County, California. Professional Paper 360, U.S. Geological Survey.

Bailey, E.H., W.P. Irwin, and D.L. Jones. 1964. Franciscan and related rocks, their significance in the geology of western California. California Division of Mines and Geology, Bulletin 183: 1–177.

Barnes, D.F. and I.L. Tailleur. 1970. Preliminary interpretation of geophysical data from the lower Noatak River basin, Alaska. Open-File Report 70–18, U.S. Geological Survey.

Barnes, I., V.C. LaMarche, and G. Himmelberg. 1967. Geochemical evidence of present-day serpentinization. Science 156: 830–832.

Beauchamp, R.M. 1986. A flora of San Diego County, California. Sweetwater Press, National City, CA.

Begg, G.L. 1968. Soil Survey of Glenn County, California. U.S. Government Printing Office, Washington, DC.

Bird, M.L. and A.L. Clark. 1976. Microprobe study of olivine chromitites of the Goodnews Bay ultramafic complex, Alaska, and the occurrence of platinum. U.S. Geological Survey, Journal of Research 4: 717–725.

Blake, M.C., A.S. Jayko, and T.E. Moore. 1984. Tectonostratigraphic terranes of Magdalena Island, Baja California Sur. Pages 183–191 in V.A. Frizzell (ed.), Geology of the Baja California Peninsula. Pacific Section, Society of Economic Paleontologists and Mineralogists. Los Angeles, CA.

Blake, M.C., D.S. Harwood, E.J. Helley, W.P. Irwin, A.S. Jayko, and D.L. Jones. 1999. Geological Map of the Red Bluff 30' X 60' Quadrangle, California. Map I-2542, U.S. Geological Survey. Washington, DC.

Bonilla, M.G. 1971. Preliminary geologic map of the San Francisco South quadrangle and part of Hunters Point quadrangle. Miscellaneous Field Studies Map MF-311, U.S. Geological Survey.

Borine, R. 1983. Soil Survey of Josephine County. USDA Soil Conservation Service, Washington, DC.

Bowman, R.H. 1973. Soil Survey the San Diego Area, California. USDA Soil Conservation Service and Forest Service, Washington, DC.

Brabb, E.E. and E.H. Pampeyan. 1983. Geologic map of San Mateo County, California. Miscellaneous Field Investigations, Map I-1257A, U.S. Geological Survey.

Brandon, M.T., D.S. Cowan, and J.A. Vance. 1988. The Late Cretaceous San Juan thrust system, San Juan Islands, Washington. Geological Society of America, Special Paper 221.

Brew, D.A., B.R. Johnson, D. Grybeck, A. Griscom, and D.F. Barnes. 1978. Mineral Resources of the Glacier Bay National Monument wilderness study area, Alaska. Open-File Report 62-15, U.S. Geological Survey.

Britton, L.A. 1975. Soil Survey of Nevada County Area, California. U.S. Government Printing Office, Washington, DC.

Brosgé, W.P. and H.N. Reiser. 1962. Preliminary geological map of the Christian Quadrangle, Alaska. Open-File Report 62-15, U.S. Geological Survey.

Brown, R.D.J. 1964. Geological map of the Stonyford quadrangle, Glen, Colusa, and Lake Counties, California. Mineral Investigations Field Study Map, MF-279, U.S. Geological Survey.

Brown, E.H. 1977. Ophiolite on Fidalgo Island. Oregon Department of Geology and Mineral Industries, Bulletin 95: 67–74.

Bulmer, C.E. and L.M. Lavkulich. 1994. Pedogenic and geochemical processes of ultramafic soils along a climatic gradient in southwestern British Columbia. Canadian Journal of Soil Science 74: 165–177.

Bulmer, C.E., L.M. Lavkulich, and H.E. Schreier. 1992. Morphology, chemistry, and mineralogy of soils derived from serpentinite and tephra in southwestern British Columbia. Soil Science 154: 72–82.

Burch, S.H. 1968. Tectonic emplacement of the Burro Mountain ultramafic body, Santa Lucia Range, California. Geological Society of America, Bulletin 60: 527–544.

Burke, M.T. 1985. An ecological survey of Organ Valley candidate research natural area, Cleveland National Forest. Pacific Southwest Research Station, Albany, CA.

Burns, L.E. 1983. The Border Ranges ultramafic and mafic complex: plutonic core of an intraoceanic island arc. Ph.D. thesis, Stanford University, Stanford, California.

Burns, L.E. 1985. The Border Ranges ultramafic and mafic complex, south-central Alaska: cumulate fractionates of island-arc volcanics. Canadian Journal of Earth Sciences 22: 1020–1038.

Burns, L.E., G.H. Pessel, T.A. Little, T. L. Pavlis, R.J. Newberry, G.R. Winkler, and J. Decker. 1991. Geology of the northern Chugach Mountains, southcentral. Alaska Division of Geological and Geophysical Surveys, Professional Report 94: 1-63, maps.

Burt, R., M. Fillmore, M.A. Wilson, E.R. Gross, R.W. Langridge, and D.A. Lammers. 2001. Soil properties of selected pedons on ultramafic rocks in the Klamath Mountains, Oregon. Communications in Soil Science and Plant Analysis 32: 2145–2175.

Butler, C.S. and J.S. Griffith. 1974. Soil survey of Mariposa County, California. U.S. Government Printing Office, Washington, DC.

California Natural Diversity Data Base. 2004. Wildlife and Habitat Data Analysis Branch, California Department of Fish and Game, Sacramento.

Chapman, R.M., W.W. Patton, and E.J. Moll. 1985a. Reconnaissance geologic map of the Ophir quadrangle. Miscellaneous Field Studies, MF 85-203,U.S. Geological Survey.

Chappell, C. 2006. Plant associations of balds and bluffs of Western Washington. Natural Heritage Report 2006–02. Washington State Department of National Resources.

Cheng, S. (ed.). 2004. Research Natural Areas in California. General Technical Report PSW-GTR-188. USDA Forest Service.

Chesterman, C.W. 1963. Intrusive ultrabasic rocks and their metamorphic relationships at Leech Lake Mountain, Mendocino County, California. California Division of Mines and Geology, Special Report 82: 5–10.

Clark, A.L., S.H.B. Clark, and C.C. Hawley. 1972. Significance of Upper Paleozoic crust in the Upper Chilutna district, west-central Alaska Range. Professional Paper 800-C, pages 95–101, U.S. Geological Survey.

Clark, S.H.B. 1972. The Wolverine complex, a newly discovered layered ultramafic body in the western Chugach Mountains, Alaska. U.S Geological Survey, Open-File Report 72-70.

Cody, W.J. and D.M. Britton. 1989. Ferns and Fern Allies of Canada. Publication 1829/E. Agriculture Canada, Ottawa.

Coleman, R.G. 1972. The Colebrooke schist of southwestern Oregon and its relation to the tectonic evolution of the region. U.S. Geological Survey Bulletin 1339: 1–61.

Coleman, R.G. 1991. Serpentines of the Santa Clara Valley, California. Geological Society of America, Abstracts with Programs 23: 15.

Coleman, R.G. 1996. New Idria serpentinite: a management dilemma. Environmental and Engineering Geoscience 2: 9–22.

Coleman, R.G. 2004. Geologic nature of the Jasper Ridge Biological Preserve, San Francisco peninsula, California. International Geology Review 46: 629–637.

Coleman, R.G. and L.E. Burns. 1985. The Tonsina high-pressure mafic-ultramafic cumulate sequence, Chugach Mountains, Alaska. Geological Society of America, Abstracts with Programs 17: 348.

Coleman, R.G. and T.E. Keith. 1971. A chemical study of serpentinization—Burro Mountain, California. Journal of Petrology 12: 311–328.

Coleman, R.G. and A.R. Kruckeberg. 1999. Geology and plant life of the Klamath-Siskiyou mountain system. Natural Areas Journal 19: 320–340.

Coleman, R.G. and M.A. Lanphere. 1971. Distribution and age of high-grade blueschists associated eclogites and amphibolites from Oregon and California. Geological Society of America, Bulletin 82: 2397–2412.

Coleman, R.G., D.E. Lee, L.B. Beatty, and W.E. Brannock. 1965. Ecologites and eclogites—their differences and similarities. Geological Society of America Bulletin 76: 483–508.

Colwell, W., W. McGee, and G. McClellan. 1955. Vegetation-Soil, quadrangles 59B-1,2,3,4, Lake County. California Department of Forestry, Sacramento.

Connelly, W. 1978, Uyak complex, Kodiak Islands, Alaska: an early Mesozoic subduction zone complex. Geological Society of America, Abstracts with Programs 8: 364.

Cook, T.D. 1978. Soil Survey of Monterey County, California. U.S. Government Printing Office, Washington, DC.

Csejtey, B. and R.C. Evarts. 1979. Serpentinite bodies in the Willow Creek district, southwestern Talkeetna Mountains, Alaska. U.S. Geological Survey Circular 804-B: 92–93.

Cuff, K., W. Hoffman, and P. Zinke. 1951. Vegetation-Soil, quadrangle 43D-2, Mendocino County. California Department of Forestry, Sacramento.

Cuff, K. and R. Neuns. 1952. Vegetation-Soil, quadrangle 60D-2, Lake County. California Department of Forestry, Sacramento.

Curtis, S.M., I. Ellerseick, C.F. Mayfield, and I.L. Tailleur. 1984. Reconnaissance geologic map of southwestern Misheguk Mountain Quadrangle. Miscellaneous Field Investigation Series, Report I-1502, U.S. Geological Survey.

Day, H.W. and E.M. Moores. 1985. Structure and tectonics of the northern Sierra Nevada. Geological Society of America Bulletin 96: 436–450.

Day, H.W., P. Schiffman, and E.M. Moores. 1988. Metamorphism and tectonics of the northern Sierra Nevada. Pages 737–763 in W.G. Ernst (ed.), Metamorphism and Crustal Evolution of the Western United States. Prentice-Hall, Englewood Cliffs, NJ.

DeBari, S.M. and R.G. Coleman. 1989. Examination of the deep levels of an island arc: Evidence from the ultramafic assemblage, Tonsina, Alaska. Journal of Geophysical Research 94: 4373–4391.

DeLapp, J.A. and B.F. Smith. 1978, 1979. Soil-Vegetation Map(s), Sonoma County. California Department of Forestry, Sacramento.

Dickinson, W.R. 1966. Table Mountain serpentinite extrusion in the California Coast Ranges. Geological Society of America Bulletin 77: 451–472.

Dickinson, W.R., C.A. Hopson, and J.B. Saleeby. 1996. Alternate origins of the Coast Range ophiolite (California): introduction and implications. GSA Today 6(2): 1–10.

Dilek, Y. 1989. Structure and tectonics of an early Mesozoic oceanic basement in the northern Sierra Nevada metamorphic belt, California: evidence for transform faulting and ensimatic arc evolution. Tectonics 8: 999–1014.

Diller, J.S. 1902. Topographic development of the Klamath Mountains. U.S. Geological Survey Bulletin 196.

Donato, M.M. 1987. Evolution of an ophiolite tectonic melange, Marble Mountains, northern California Klamath Mountains. Geological Society of America Bulletin 98: 448–464.

Douglas, G.W., G.B. Straley, and D. Meidinger. 1998. Rare native vascular plants of British Columbia. British Columbia Ministry of Environment, Lands, and Parks, Victoria.

Duffield, W.A. and R.A. Sharp. 1975. Geology of the Sierra Foothills melange and adjacent areas, Amador County, California. Professional Paper 827, U.S. Geological Survey, Washington, DC.

Dukes, J.S. 2001. Productivity and complementarity in grassland microcosms of varying diversity. Oikos 94: 468–480.

Dukes, J.S. 2002. Comparison of the effect of elevated CO_2 on an invasive species (*Centaurea solistialis*) in monoculture and community settings. Plant Ecology 160: 225–234.

Durrell, C. 1966. Tertiary and Quaternary geology of the northern Sierra Nevada. California Division of Mines and Geology, Bulletin 190: 185–197.

EA Engineering, Science, and Technology. 1995. Ecological Unit Inventory, Trabuco Ranger District, Cleveland National Forest. EA Engineering, Science, and Technology, Lafayette, California.

Eckel, E.B. and W.B. Myers. 1946. Quicksilver deposits of the New Idria district, San Benito County, California. California Division of Mines and Geology, Bulletin 42: 81–124.

Ehrenberg, S.M. 1975. Feather River ultramafic body, northern Sierra Nevada, California. Geological Society of America, Bulletin 86: 1235–1243.

Ehrlich, P.R. and I. Hanski. 2004. On the Wings of the Checkerspots: A Model System for Population Biology. Oxford University Press, New York.

Elam, D.R., B. Goettle, and D.H. Wright. 1998. Draft Recovery Plan for Serpentine Soil Species of the San Francisco Bay Area. U.S. Fish and Wildlife Service.

Eric, J.H., E.J.H. Stromquist, and C.M. Swinney. 1955. Geology and mineral deposits of the Angels Camp and Sonora quadrangles, Calaveras and Tuolumne Counties, California. Special Report 41. California Division of Mines and Geology, Sacramento.

Ernstrom, D.J. 1984. Soil Survey of San Luis Obispo County, California, Central Part. U.S. Government Printing Office, Washington, DC.

Ertter, B. and M.L. Bowerman. 2002. The Flowering Plants of Mount Diablo. California Native Plant Society, Sacramento.

Evans, J.R. 1966. Nephrite in Mariposa County, California. California Division of Mines and Geology, Mineral Information Service 19: 135–148.

Evarts, R. 1977. The geology and petrology of the Del Puerto ophiolite, Diablo Range, central California. Oregon Department of Geology and Mineral Industries Bulletin 95: 161–139.

Evens, J. and S. San. 2004. Vegetation associations of a serpentine area: Coyote Ridge, Santa Clara County, California. California Native Plant Society, Sacramento.

Evens, J.M., S. San, and J. Taylor. 2004. Vegetation classification and mapping of Peoria Wildlife Area, South of New Melones Lake, Tuolumne County, CA. California Native Plant Society, Sacramento.

Ferrari, C.A. and M.A. McElhiney. 2002. Soil Survey of Western Stanislaus County. U.S. Government Printing Office, Washington, DC.

Ferrians, J.C. 1965. Permafrost Map of Alaska. Miscellaneous Geological Investigations, Map I-445, U.S. Geological Survey.

Fiedler, P.L. and R.A. Leidy. 1987. Plant communities of the Ring Mountain Preserve, Marin County, California. Madroño 34: 173–196.

Findlay, D.C. 1969. Origin of the Tulameen ultramafic gabbro complex, southern British Columbia. Canadian Journal of Earth Science 6: 399–425.

Foster, C.M. and G.K. Lang. 1982. Soil Survey of Klamath National Forest Area, California. USDA Forest Service, Klamath National Forest, Yreka, CA.

Frank, C.O. and J. Zimmerman. 1982. Petrography of nonultramafic rocks from the Avan Hills complex, DeLong Mountains, Alaska. U.S. Geological Survey Circular 844: 22–26.

Freeman, O.W., J.D. Forrester, and R.L. Lupher. 1945. Physiographic divisions of the Columbia Intermontane Province. Annals of the Association of American Geographers 35: 53–75.

FWS (U.S. Fish and Wildlife Service). 1998. Endangered or threatened status for three plants from the chaparral and scrub of Southwestern California. Federal Register 63: 54956–54971.

Giaramita, M., G.L. MacPherson, and S.P. Phipps. 1998. Petrologically diverse basalts from a fossil ocean forearc in California: the Llanada and Black Mountain remnants of the Coast Range ophiolite. Geological Society of America Bulletin 110: 553–571.

Goldin, A. 1992. Soil Survey of Whatcom County Area, Washington. U.S. Government Printing Office, Washington, DC.

Gordon, H. and T. White. 1994. Ecological guide to southern California chaparral plant series: Transverse and Peninsular Ranges: Angeles, Cleveland, and San Bernardino National Forests. Technical Report R5-ECOL-TP-005, USDA Forest Service, Pacific Southwest Region.

Gowans, K.D. 1967. Soil Survey of Tehama County, California. U.S. Government Printing Office, Washington, DC.

Gowans, K.D. and H.S. Hinkley. 1964. Reconnaissance Soil Survey of Tuolumne County, California. California Agricultural Extension Service, Davis.

Graham, R.C., M.M. Diallo, and L.J. Lund. 1990. Soils and mineral weathering on phyllite colluvium and serpentine in northwestern California. Journal of the Soil Science Society of America 54: 1682–1690.

Gram, W.K., E.T. Borer, K.L. Cottingham, E.W. Seabloom, V.L. Boucher, L. Goldwasser, F. Micheli, B.E. Kendall, and R.S. Burton. 2004. Distribution of plants in a California serpentine grassland: are rocky hummocks spatial refuges for native species? Plant Ecology 172: 159–171.

Graymer, R.W., D.L. Jones, and E.E. Brabb. 1995. Geologic map of the Hayward fault zone. Open-file Report 95-597, 1:50,000 map, U.S. Geological Survey.

Guild, P.W. 1942. Chromite deposits of Kenai Peninsula, Alaska. U.S. Geological Survey Bulletin 931-G: 139–175.

Hanes, R.O. 1994. Soil Survey of Tahoe National Forest Area, California. USDA Forest Service, Pacific Southwest Region, Vallejo, California.

Hanson, L.G. 1964. Bedrock geology of the Rainbow Mountain area, Alaska Range, Alaska. Report 2. Alaska Division of Geological Survey, Anchorage, AK.

Harrison, S. 1997. How natural habitat patchiness affects the distribution of diversity in Californian serpentine chaparral. Ecology 78: 1898–1906.

Harrison, S. 1999. Local and regional diversity in a patchy landscape: native, alien and endemic herbs on serpentine soils. Ecology 80: 70–80.

Harrison, S., D.D. Murphy, and P.R. Ehrlich. 1988. Distribution of the Bay checkerspot butterfly, *Euphydryas editha bayensis*: evidence for a metapopulation model. American Naturalist 132: 360–382.

Harrison, S., J.L. Viers, and J.F. Quinn. 2000. Climatic and spatial patterns of diversity in the serpentine plants of California. Diversity and Distributions 6: 153–161.

Harper, G.D. 1984. The Josephine ophiolite, northwestern California. Geological Society of America Bulletin 95: 1009–1026.

Harper, G.D., J.B. Saleeby, and M. Heizler. 1994. Formation and emplacement of the Josephine ophiolite and the Nevadan orogeny in the Klamath Mountains, California-Oregon: U/Pb and $^{40}Ar/^{39}Ar$ geochronology. Journal of Geophysical Research 99(B3): 4293–4321.

Hastings, J.R. and R.R. Humphrey. 1969. Climatological data and statistics for Baja California. Technical Reports for Meteorology and Climatology, No. 18. Institute of Atmospheric Physics, University of Arizona, Tucson.

Hastings, J.R. and R.M. Turner. 1965. Seasonal precipitation regimes in Baja California. Geografiska Annaler 47A: 204–223.

Hawley, C.C. and A.L. Clark. 1974. Geology and mineral resources of the Upper Chulitna district, Alaska. Professional Paper 758-B, U.S. Geological Survey.

Hietanen, A. 1973. Geology of the Pulga and Bucks Lake quadrangles, Butte and Plumas Counties, California. Professional Paper 731, U.S. Geological Survey.

Hill, B.B. and J. Brandon. 1976. Layered basic and ultrabasic rock, Kodiak Island, Alaska: the lower portion of a dismembered ophiolite? EOS, American Geophysical Union Transactions 57:1027.

Himmelberg, G.R. and R.A Loney. 1981. Petrology of the ultramafic and gabbroic rocks of the Grady glacier nickel deposit, Fairweather Range, southeastern Alaska. Professional Paper 1195, U.S. Geological Survey.

Himmelberg, G.R. and R.A. Loney. 1995. Characteristics and petrogenesis of Alaskan-type ultramafic-mafic intrusions, southeastern Alaska. Professional Paper 1564, U.S. Geological Survey.

Himmelberg, G.R., R.A. Loney, and J.T. Craig. 1986. Petrogenesis of the ultramafic complex at the Blaske Islands, southeastern Alaska. U.S. Geological Survey Bulletin 1662.

Hobbs, R.J. and H.A. Mooney. 1995. Spatial and temporal variability in California annual grassland: results from a long-term study. Journal of Vegetation Science 6: 43–56.

Hoffman, B.L. 1974. Geology of Bernard Mountain area, Tonsina, Alaska. M.S. thesis, University of Alaska, Fairbanks.

Holland, R.F. 1986. Preliminary Descriptions of the Terrestrial Natural Communities of California. California Department of Fish and Game, Sacramento.

Hopson, C.A., J.M. Mattinson, and E.A. Pessagna. 1981. Coast Range ophiolite, western California. Pages 418–510 *in* W.G. Ernst (ed.), Metamorphism and Crustal Evolution of the Western United States. Prentice-Hall, Englewood Cliffs, NJ.

Hotz, P.E. 1964. Nickeliferous laterites in southwestern Oregon and northwestern Callifornia. Economic Geology 59: 355–396.

Hotz, P.E. 1977. Geology of the Yreka quadrangle, Siskiyou County, California. U.S. Geological Survey Bulletin 1736: 1–72.

Howard, R.F. and R.H. Bowman. 1991. Soil Survey of Mendocino County, Eastern Part, and Trinity County, Southwestern Part, California. USDA Soil Conservation Service, Washington, DC.

Hunter, J.C. and J.E. Horenstein. 1992. The vegetation of the Pine Hill area (California) and its relation to substratum. Pages 197–206 *in* A.J. M. Baker, J. Proctor, and R.D. Reeves (eds.), The Vegetation of Ultramafic (Serpentine) Soils. Intercept, Andover, Hampshire, UK.

Huntington, G.L. 1971. Soil Survey of the Eastern Fresno County Area, California. U.S. Government Printing Office, Washington, DC.

Irvine, T.N. 1973. Bridget Cove volcanics, Juneau area, Alaska: possible parent magma of Alaskan type ultramafic complexes. Carnegie Institution of Washington Yearbook 72: 478–491.

Irvine, T.N. 1974. Petrology of the Duke Island ultramafic complex, southeastern Alaska. Geological Society of America Memoir 138.

Irwin, W.P. 1979. Ophiolitic terranes of part of the western United States. Geological Society of America, Map and Chart Series MC-33: 2–4.

Irwin, W.P. 1981. Tectonic accretion of part of the Klamath Mountains. Pages 29–49 *in* W.G. Ernst (ed.), Metamorphism and Crustal Evolution of the Western United States. Prentice–Hall, Englewood Cliffs, NJ.

Isgrig, D. 1969. Soil Survey of San Benito County, California. U.S. Government Printing Office, Washington, DC.

Istok, J.D. and M.E. Harward. 1982. Influence of soil moisture on smectite formation in soils derived from serpentine. Soil Science Society of America, Journal 46: 1106–1108.

James, O.B. 1971. Origin and emplacement of the ultramafic rocks of the Emigrant Gap area, California. Journal of Petrology 12: 523–560.

Jayko, A.S. and M.C. Blake. 1986. Significance of Klamath rocks between the Franciscan complex and Coast Range ophiolite, northern California. Tectonics 5: 1055–1071.

Jennings, C.W. 1994. Fault Activity Map of California and Adjacent Areas. Geologic Data Map No. 6. California Department of Conservation, Division of Mines and Geology, Sacramento.

Jimerson, T.M., L.D. Hoover, E.A. McGee, G. DeNitto, and R.M. Creasy. 1995. A Field Guide to Serpentine Plant Associations and Sensitive Plants in Northwestern California. Technical Report R5-ECOL-TP-006. USDA Forest Service, Pacific Southwest Region.

Juday, G.P. 1992. Alaska Research Natural Areas. 3: Serpentine Slide. General Technical Report PNW-GTR-271, USDA Forest Service.

Jurjavcic, N.L., S. Harrison, and A.T. Wolf. 2002. Abiotic stress, competition, and the distrubution of the native grass *Vulpia microstachys* in a mosaic environment. Oecologia 130: 555–562.

Kashiwagi, J.H. 1985. Soil Survey of Marin County. U.S. Government Printing Office, Washington, DC.

Kays, M.A., M. Fern, and L. Beskow. 1988. Complimentary meta-gabbros and peridotites in the northern Klamath Mountains. Oregon Department of Geology and Mineral Industries Bulletin 96: 91–107.

Keeler-Wolf, T. 1983. An Ecological Survey of the Frenzel Creek Research Natural Area, Mendocino National Forest, Colusa County, California. USDA Forest Service, Pacific Southwest Region, Vallejo, CA.

Keeler-Wolf, T. 1987. An ecological survey of the proposed Crater Creek Research Natural Area, Klamath National Forest, Siskiyou County, California. USDA Forest Service, Pacific Southwest Research Station, Berkeley.

Keeler-Wolf, T. 1990a. An ecological survey of the King Creek Research Natural Area, Cleveland National Forest, San Diego County, California. Unpublished report prepared for The Nature Conservancy, on file, Pacific Southwest Research Station, Albany, California.

Keeler-Wolf, T. 1990b. Ecological surveys of Forest Service Research Natural Areas in California. General Technical Report PSW-GTR-125.USDA Forest Service.

Keeler-Wolf, T. 2003. Plant community descriptions: Point Reyes National Seashore, Golden Gate National Recreation Area, San Francisco Water Department Watershed Lands, Mount Tamalpais, Tomales Bay, and Samuel P. Taylor State Parks. Unpublished report on file at Wildlife and Habitat Data Analysis Branch, California Department of Fish and Game, Sacramento.

Kelsey, H.M. 1987. Geomorphic processes in the recently uplifted Coast Ranges of northern California. Geological Society of America, Centennial Special Volume 2: 550–560.

Klungland, M.W. and M. McArthur. 1989. Soil Survey of Skagit County Area, Washington. U.S. Government Printing Office, Washington, DC.

Kruckeberg, A.R. 1969. Plant life on serpentinite and other feromagnesian rocks in northwestern North America. Syesis 2: 15–114.

Kruckeberg, A.R. 1982. Noteworthy collections: British Columbia. Madroño 29: 271.

Kruckeberg, A.R. 1984. California Serpentines: Flora, Vegetation, Geology, Soils, and Management Problems. University of California Press, Berkeley.

Kruckeberg, A.R. 1999. Serpentine barrens of western North America. Pages 309–321 in R.C. Anderson, J.S. Fralish, and J.M. Baskin (eds.), Savannas, Barrens, and Rock Outcrop Plant Communities of North America. Cambridge University Press, Cambridge, UK.

Lambert, G., J. Kashiwagi, B. Hansen, P. Gale, and A. Endo. 1978. Soil Survey of Napa County. U.S. Government Printing Office, Washington, DC.

Lanspa, K. 1993. Soil Survey of Shasta-Trinity National Forest Area, California. USDA Forest Service, Redding, CA.

Latimer, H. 1984. Serpentine flora:notes on prominent sites in California, eastern Fresno County. Fremontia 11(5): 29–30.

Lee, B.D., S.K. Sears, R.C. Graham, C. Amrhein, and H. Vali. 2003. Secondary mineral genesis from chlorite and serpentinite in an ultramafic soil toposequence.Journal of the Soil Science Society of America 67: 1309–1317.

Leney, G.W. and E.E. Loeb. 1971. The geology and mining operations at Pacific Asbestos Corporation, Copperopolis, California. Mining and Engineering 23: 78.

Lewis, G.J. and G.E. Bradfield. 2003. A floristic and ecological analysis at the Tulameen ultramafic (serpentine) complex, southern British Columbia, Canada. Davidsonia 14: 121–128, 131–134, 137–144.

Lewis, G.J. and G.E. Bradfield. 2004. Plant community-soil relationships at an ultramafic site in southern British Columbia, Canada. Pages 191–197 in R.S. Boyd, A.J.M. Baker, and J. Proctor (eds.), Ultramafic Rocks: Their Soils, Vegetation, and Fauna. Science Reviews, St. Albans, Herts, UK.

Lewis, G.J., J.M. Ingram, and G.E. Bradfield. 2004. Diversity and habitat relationships of bryophytes at an ultramafic site in southern British Columbia, Canada. Pages 199–204 in R.S. Boyd, A.J.M. Baker, and J. Proctor (eds.), Ultramafic Rocks: Their Soils, Vegetation, and Fauna. Science Reviews, St. Albans, Herts, UK.

Lindsay, W.L. 1974. Soil Survey of Eastern Santa Clara County Area, California. U.S. Government Printing Office, Washington, DC.

Lindsley-Griffin, N. 1977a. Paleogeographic implications of ophiolites: the Ordovician Trinity complex, Klamath Mountains, California. Pages 409–420 in J.H. Stewart et al. (eds.), Paleozoic Paleogeography of the Western United States. Pacific Section, Society of Economic Paleontologists and Mineralogists. Los Angeles, CA.

Lindsley-Griffin, N. 1977b. The Trinity ophiolite, Klamath Mountains, California. Oregon Department of Geology and Mineral Industries, Bulletin 95: 107–120.

Loney, R.A., D.A. Brew, L.J.P. Muffler, and J.S. Pomeroy. 1975. Reconnaissance Geology of Chichagof, Baranof, and Kurzof Islands, southeastern Alaska. Professional Paper 792, U.S. Geological Survey, Washington, DC.

Loney, R.A. and G.R. Himmelberg. 1983. Structure and petrology of the La Perouse gabbro intrusion, Fairweather Range, southeastern Alaska. Journal of Petrology 24: 377–423.

Loney, R.A. and G.R. Himmelberg. 1985. Ophiolitic ultramafic rocks of the Jade Mountain-Cosmos Hills area, southwestern Brooks Range, Alaska. U.S. Geological Survey Circular 967: 13–15.

Loney, R.A. and G.R. Himmelberg. 1988. Ultramafic rocks of the Livengood terrane. U.S. Geological Survey, Circular 1016: 68–70.

MacKevett, E.M., E.C. Robertson, and G.R. Winkler. 1974. Geology of the Skagway B-3 and B-4 quadrangles, southeastern Alaska. Professional Paper 832, U.S. Geological Survey.

Magoon, L.B., Adkison, W.L., and Egbert, R.M. 1976. Map showing, geology, wildcat wells, Tertiary plant fossil localities, K-Ar age dates and petroleum, operations Cook Inlet area, Alaska. Miscellaneous Investigations Series, Map I-1019, U.S. Geological Survey.

Mayfield, C.F., S.M. Curtis, I. Ellersieck, and I.L. Tailleur. 1990. Reconnaissance geologic map of the DeLong Mountains A3, B3, and parts of A4, B4 quadrangles, Alaska. Miscellaneous Investigation Series, Map I-1929, U.S. Geological Survey.

Mayfield, C.F., I.L. Tailleur, and I. Ellersieck. 1988. Stratigraphy, structure, and palinspastic synthesis of the western Brooks Range, northwestern Alaska. Professional Paper 1399: 143–186, U.S. Geological Survey.

Mayfield, J.D. and H.W. Day. 2002. Ultramafic rocks in the Feather River Belt, northern Sierra Nevada, California. California Division of Mines and Geology, Special Publication 122: 1–15.

McCarten, N. 1992. Community structure and habitat relations in a serpentine grassland in California. Pages 207–211 in A.J.M. Baker, J.Proctor, and R.D. Reeves (eds.), The Vegetation of Ultramafic (Serpentine) Soils. Intercept, Andover, NH.

McColl, J.G., L.K. Monette, and D.S. Bush. 1982. Chemical characteristics of wet and dry atmospheric fallout in northern California. Journal of Environmental Quality 11: 585–590.

McLaughlin, R.J., M.C. Blake, A. Griscom, C.D. Blome, and B. Murchey. 1988. Tectonics of formation, translation, and dispersal of the Coast Range ophiolite of California, Tectonics 7: 1033–1039.

McLaughlin, R.J., H.N. Ohlin, D.J. Thormahlen, D.J. Jones, J.W. Miller, and C.D. Blome. 1985. Geologic map and structure sections of the Little Indian Creek Valley-Wilbur Springs geothermal area, northern Coast Ranges, California. Open-File Report 85-285, U.S. Geological Survey.

Medeiros, J. 1984. Serpentine flora: notes on prominent sites in California, the Red Hills. Fremontia 11(5): 28–29.

Mertie Jr., J.B. 1940. The Goodnews platinum deposits. U.S. Geological Survey Bulletin 918.

Miles, S.R., B.A. Roath, and A.M. Parsons. 1993. Soil Survey of Six Rivers National Forest, California. USDA Forest Service, Pacific Southwest Region, Vallejo, CA.

Miller, M.L. 1984. Geology of the Resurrection Peninsula. Pages 25–40 in G.R. Winkler, M.L. Miller, R.B. Hoekzema, and J.A. Dumoulin (eds.), A guide to the bedrock geology of a traverse of the Chugach Mountains from Anchorage to Cape Resurrection, Alaska. Alaska Geological Society, Anchorage.

Miller, M.L. and L.M. Angeloni. 1985. Ophiolitic rocks of the Iditarod quadrangle, west central Alaska. American Association of Petroleum Geologists, Bulletin 69: 669–670.

Miller, DC. R.B. 1985. The ophiolitic Ingalls complex, north-central Cascade Mountains, Washington. Geological Society of America Bulletin 96: 27–48.

Miller, V.C. 1972. Soil Survey of Sonoma County, California. U.S. Government Printing Office, Washington, DC.

Monger, J.W.H. 1977. Ophiolitic assemblages in the Canadian cordillera. Oregon Department of Geology and Mineral Industries, Bulletin 95: 74–92.

Moore, T.E. 1986. Petrology and tectonic implications of the blue-schist bearing Puerto Nuevo melange complex, Vizcaíno Peninsula, Baja California Sur, Mexico. Geological Society of America Memoir 164: 43–58.

Morgan, B.A. 1976. Geologic map of the Chinese Camp and Moccasin quadrangles, Tuolumne County, California. U.S. Geological Survey, Washington, DC.

Nelson, R. and P. Zinke. 1952. Soil-Vegetation Map, quadrangle 60D-2, Lake County. California Department of Forestry, Sacramento.

Nelson, S.W. and W.H. Nelson. 1982. Geology of the Siniktanneyak Mountain ophiolite, Howard Pass quadrangle, Alaska. Miscellaneous Field Studies, MF-1441, U.S. Geological Survey.

Neuns, R. 1950. Vegetation-Soil, quadrangle 61A-1, Mendocino County. California Department of Forestry, Sacramento.

Nokleberg, W.J., N.R.D. Albert, G.C. Bond, R.L. Herzon, R.T. Miyoka, S.W. Nelson, D.H. Richter, T.E. Smith, L.H. Stout, W. Yeend, et al. 1982. Geologic map of the southern Mount Hayes quadrangle, Alaska. Open-File Report 82-52, U.S. Geological Survey.

Oberbauer, T.A. 1993. Soils and plants of limited distribution in the Peninsular Ranges. Fremontia 21(4): 3–7.

O'Donnell, R. 2005. Edaphic infidelity along the Oat Mine trail. The Four Seasons 12(3): 2–50.

O'Hare, J.P. and B.G. Hallock. 1980. Soils Survey of Los padres National Forest Area, California. USDA Forest Service, Santa Barbara, CA.

Orr, E.L. and W.N. Orr. 1996. Geology of the Pacific Northwest. McGraw-Hill, New York.

Page, N.J. 1968. Serpentinization in a sheared serpentinite lense, Tiburon Peninsula, California. Professional Paper 600-B, pages 21–28, U.S. Geological Survey.

Page, B. 1972. Oceanic crust and mantle fragment in subduction complex near San Luis Obispo, California. Geological Society of America Bulletin 83: 957–972.

Page, B.M. 1992. Tectonic setting of the San Francisco Bay region, California California Division of Mines and Geology, Special Publication 113: 1–7.

Page, B.M., L.A. DeVito, and R.G. Coleman. 1999. Tectonic emplacement of serpentine southwest of San Jose, California. International Geology Review 41: 494–505.

Page, B.M. and L.L. Tabor. 1967. Chaotic structure and decollement in Cenozoic rocks near Stanford University, California. Geological Society of America Bulletin 78: 1–12.

Page, B.M., G.A. Thompson, and R.G. Coleman. 1997. Tectonics of the central and southern Coast Ranges, California. Geological Society of America Bulletin 110: 846–876.

Pampeyan, E.H. 1963. Geology and mineral deposits of Mount Diablo, Contra Costa County, California. California Division of Mines and Geology, Special Report 80.

Pampeyan, E.H. 1993. Geologic map of the Palo Alto and part of the Redwood Point 1/2 minute quadrangle. Miscellaneous Investigations Series, Map I-2371, U.S. Geological Survey.

Patton Jr., W.W., S.E. Box, and D.J. Grybeck. 1989. Ophiolites and other mafic-ultramafic complexes in Alaska. Open-File Report 89-648, U.S. Geological Survey.

Patton Jr., W.W., S.E. Box, and D.J. Grybeck. 1994. Ophiolites and other mafic-ultramafic complexes in Alaska. Pages 671–686 *in* G. Plafker and H.C. Berg (eds.), The Geology of North America, vol. G-1, The Geology of Alaska. Geological Society of America, Boulder, CO.

Péwé, T.L. 1975. Quaternary geology of Alaska. Professional Paper 835, U.S. Geological Survey.

Phipps, S.P. 1984. Ophiolitic olistostromes in the basal Great Valley sequence, Napa County, northern California Coast Ranges. Geological Society of America, Special Paper 198: 103–125.

Plafker, G. and E.M. MacKevett. 1970. Mafic and ultramafic rocks from a layered pluton at Mt. Fairweather, Alaska. Professional Paper 700-B, pages 21–26, U.S. Geological Survey.

Radelli, L. 1989. The ophiolites of Calmalli and the Olivdada nappe of northern Baja California and west-central Sonora, Mexico. Pages 79–85 in P.L. Abbot (ed.), Geological Studies in Baja California. Pacific Section, Society of Economic Paleontologists and Mineralogists, Los Angeles, CA.

Ramp, L. 1978. Investigations of nickel in Oregon. Oregon Department of Geology and Mineral Industries, Miscellaneous Paper 20: 1–60.

Ramp, L. and N.V. Peterson. 1979. Geology and mineral resources of Josephine County, Oregon: Oregon Department of Geology and Mineral Industries Bulletin 100: 1–45.

Ray, G.E. 1986. The Hozameen fault system and related Coquihalla serpentine belt of southwestern British Columbia. Canadian Journal of Earth Sciences 23: 1022–1041.

Raymond, L.A. 1984. Classification of melanges. Geological Society of America, Special Paper 198: 7–20.

Reed, B.L. and S.W. Nelson. 1980. Geologic map of the Talkeetna quadrangle, Alaska. Miscellaneous Investigations MI-1174, U.S. Geological Survey.

Reed, W.R. 2003. Soil Survey of Colusa County, California. U.S. Government Printing Office, Washington, DC.

Reiser, C.H. 1994. Rare Plants of San Diego County. Aquafir Press, San Diego, CA.

Rice, S.J. 1957. Nickel. California Division of Mines and Geology Bulletin 176: 391–399.

Rice, S.J. 1960. Tourmalinizated Franciscan sediments at Mount Tamalpais. Geological Society of America Bulletin 71: 2073.

Rice, S.J. and G.B. Cleveland. 1955. Laterite silification of serpentinite in the Sierra Nevada, California. Geological Society of America Bulletin 66: 1660.

Rice, S.J. and D. Wagner. 1991. Geology and mineralogy of Ring Mountain: a popular nature reserve. California Geology 44: 99–106.

Richter, D.H. 1967. Geology of the upper Slana-Mentasta Pass area, southcentral Alaska. Report 30. Alaska Division of Geological and Geophysical Surveys, Juneau.

Rieger, S., D.B. Schoephorster, and C.E. Furbush. 1979. Exploratory Soil Survey of Alaska. USDA Soil Conservation Service, Washington, DC.

Robertson, A.H.F. 1990. Sedimentology and tectonic implications of ophiolite-derived clastics overlying Jurrasic Coast Range ophiolite, northern California. American Journal of Science 290: 109–163.

Rogers, J.H. 1974. Soil Survey of El Dorado Area. U.S. Government Printing Office, Washington, DC.

Rose, A.W. 1965. Geology and mineral deposits of the Rainy Creek area, Mt. Hayes quadrangle Alaska. Geologic Report 14. Alaska Division of Mines and Minerals, Juneau.

Rose, A.W. 1966. Geology of the chromite-bearing rocks near Eklutna, Anchorage quadrangle, Alaska. Geology Report 18. Alaska Division of Mines and Minerals, Juneau.

Rossman, D.L., 1963, Geology and petrology of two stocks of layered gabbro in the Fairweather Range, Alaska: U.S. Geological Survey Bulletin 1121-F: 1–50.

Ruckmick, J.C. and J.A. Noble. 1959. Origin of the ultramafic complex at Union Bay southeastern Alaska. Geologic Society of America Bulletin 70: 981–1018.

Safford, H. and S. Harrison, 2001. Ungrazed road verges in a grazed landscape: interactive effects of grazing, invasion, and substrate on grassland diversity. Ecological Applications 11: 1112–1122.

Safford, H. and S. Harrison. 2004. Fire effects on plant diversity in serpentine versus sandstone chaparral. Ecology 85: 539–548.

Saleeby, J.B. 1978. Kings River ophiolite, southwest Sierra Nevada foothills, California. Geological Society of America Bulletin 89: 617–636.

Saleeby, J.B. 1979. Kaweah serpentinite melange, southwest Sierra Nevada foothills, California. Geological Society of America Bulletin 90: 129–146.

Saleeby, J.B. 1982. Polygenetic ophiolite belt of the California Sierra Nevada: geochronological and tectonostratigraphic development. Journal of Geophysical Research 87: 1803–1824.

Schlising, R.A. 1984. Serpentine flora: notes on prominent sites in California, Magalia in Butte County. Fremontia 11(5): 25–26.

Schlocker, J. 1974. Geology of the San Francisco North quadrangle, California. Professional Paper 782, U.S. Geological Survey.

Schweickert, R.A. and D.S. Cowan. 1975. Early Mesozoic tectonic evolution of the western Sierra Nevada, California. Geological Society of America Bulletin 86: 1329–1336.

Schweickert, R.A., C. Merguerian, and N.L. Bogen. 1988. Deformational and metamorphic history of Paleozoic and Mesozoic basement terranes in the western Sierra Nevada metamorphic belt. Pages 789–822 in W.G. Ernst (ed.), Metamorphism and Crustal Evolution of the Western United States. Prentice-Hall, Englewood Cliffs, NJ.

Seabloom, E.W., E.T. Borer, V.L. Boucher, R.S. Burton, K.L. Cottingham, L. Goldwasser, W.K. Gram, B.E. Kendall, and F. Micheli. 2003. Competition, seed limitation, disturbance, and reestablishment of California native annual forbs. Ecological Applications 13: 575–592.

Sedlock, R.L. 1993. Mesozoic geology and tectonics of blueschist and associated oceanic terranes in the Cedros-Vizcaino-San Benito and Magdalena-Santa Margarita regions, Baja California, Mexico. Pages 113–126 in G. Dunne and K. McDougall (eds.), Mesozoic Paleogeography of the Western United States-II, Pacific Section, Society of Economic Paleontologists and Mineralogists. Los Angeles, CA. Book 71.

Sedlock, R.L. 2003. Geology and tectonics of the Baja California peninsula and adjacent areas. Geological Society of America, Special Paper 374: 1–42.

Sedlock, R.L., F. Ortega-Gutierrez, and R.C. Speed. 1993. Tectonostratigraphic Terranes and Tectonic Evolution of Mexico. Special Paper 278, U.S. Geological Survey.

Sharp, R.P. 1960. Pleistocene glaciation in the Trinity Alps of northern California. American Journal of Science 258: 305–340.

Shreve, F. 1951. Vegetation of the Sonoran Desert. Publication 591. Carnegie Institute of Washington, Washington, DC.

Smith, C. 1998. A Flora of the Santa Barbara Region, California. Santa Barbara Botanic Garden and Capra Press, Santa Barbara, CA.

Smith, D.W. and W.D. Broderson. 1989. Soil Survey of Lake County, California. U.S. Government Printing Office, Washington, DC.

Snoke, A.W. 1977. A thrust plate of ophiolite rocks in the Preston Peak area, Klamath Mountains, California. Geological Society of America Bulletin 88: 1641–1659.

Snoke, A.W., J.E. Quick, and H.R. Bowman. 1981. Bear Mountain igneous complex, Klamath Mountains, California: an ultrabasic to silicic calc-alkaline suit. Journal of Petrology 22: 501–552.

Soil Survey Staff. 1999. Soil Taxonomy—a Basic System for Making and Intrepreting Soil Surveys. USDA, Agriculture Handbook No. 436. U.S. Government Printing Office, Washington, DC.

Souther, J.G. 1971. Geology and mineral deposits of Tulsequah map-area, British Columbia. Geological Survey of Canada Memoir 362.

Springer, R.K. 1980. Geology of the Spring Hill intrusive complex, a layered gabbroic body in the western Sierra Nevada foothills, California. Geological Society of America Bulletin 91: 381–385.

Stebbins, G.L. 1984. Serpentine flora: notes on prominent sites in California, the northern Sierra Nevada. Fremontia 11(5): 26–28.

Stout, J.H. 1976. Geology of Eureka Creek area, east-central Alaska Range, Alaska. Geologic Report 46. Alaska Division of Geological and Geophysical Surveys, Juneau.

Swensen, M. 1950. Vegetation-Soil, quadrangle 61A-2, Mendocino County. California Department of Forestry, Sacramento.

Taylor, H.P. 1967. The zoned ultramafic complexes of southeastern Alaska. Pages 97–121 *in* P.J. Wylie (ed.), Ultramafic and Related Rocks. Wiley, New York.

Thayer, T.P. and G.R. Himmelberg. 1968. Rock succession in the alpine-type mafic complex at Canyon Mountain, Oregon. Pages 175–186 *in* Earth's Crust and Upper Mantle (Geological Processes), Proceedings of section 1, 23rd International Geological Congress, Prague.

Thompson, G.A. and R. Robinson. 1975. Gravity and magnetic investigation of the Twin Sisters dunite, northern Washington. Geological Society of America Bulletin 86: 1413–1422.

Thorne, R.F. 1977. Montane and subalpine forests of the Transverse and Peninsular Ranges. Pages 537–557 *in* M.G. Barbour and J. Major (eds.), Terrestrial vegetation of California. Wiley-Interscience, New York [reprinted by the California Native Plant Society, 1988, Sacramento].

Thorne, R.L. and R.R. Wells. 1956. Studies of the Snettisham magnetite deposit, southeastern Alaska. U.S. Bureau of Mines, Report of Investigations 5195.

Toth, M.I. 1981. Petrology, geochemistry, and origin of the Red Mountain ultramafic body near Seldovia, Alaska. Open-File Report 81-514, U.S. Geological Survey.

Tysdal, R.G., J.A. Case, G.R. Winkler, and S.H.B. Clark. 1977. Sheeted dikes, gabbro, and pillow basalt in flysch of coastal southern Alaska. Geology 5: 377–383.

Vallier, T.L. and H.C. Brooks (eds.). 1995. Geology of the Blue Mountains Region of Oregon, Idaho, and Washington: Petrology and Tectonic Evolution of the Pre-Tertiary Rocks of the Blue Mountains Region. Professional Paper 1438, U.S. Geological Survey.

Vogl, R.J. 1973. Ecology of knobcone pine in the Santa Ana Mountains, California. Ecological Monographs 43: 125–143.

Wachtell, J.K. 1978. Soil Survey of Orange and Western Part of Riverside County, California. USDA Soil Conservation Service and Forest Service, Washington, DC.

Wahrhaftig, C. 1965. Physiographic Divisions of Alaska. Professional Paper 482, U.S. Geological Survey.

Wakabayashi, J. and T.L. Sawyer. 2001. Stream incision, tectonics, and evolution of topography of the Sierra Nevada, California. Journal of Geology 109: 539–562.

Walawender, M.J. 1976. Petrology and emplacement of the Los Pinos pluton, southern California. Canadian Journal of Earth Science 13: 1288–1300.

Weiss, S.B. 1999. Cars, cows, and checkerspot butterflies: nitrogen deposition and management of nutrient-poor grasslands for a threatened species. Conservation Biology 13: 1476–1486.

Welch, L.E. 1981. Soil Survey of Alameda County, California, Western Part. U.S. Government Printing Office, Washington, DC.

Whipple, J. and E. Cole. 1979. An ecological survey of the proposed Mount Eddy Research Natural Area. USDA Forest Service, Pacific Southwest Research Station, Berkeley, CA.

Wiggins, I.L. 1980. Flora of Baja California. Stanford University Press, Stanford, CA.

Williamson, J.N. and S. Harrison. 2002. Biotic and abiotic limits to the spread of exotic revegetation species in oak woodland and serpentine habitats. Ecological Applications 12: 40–51.

Winkler, G.R., R.J. Miller, M.L. Siberman, A. Grantz, J.E. Case, and W.J. Pickthorn. 1981. Layered gabbroic belt of regional extent in the Valdez quadrangle. U.S. Geological Survey Circular 823-B: 74-B76.

Wolf, A.T., S.P. Harrison, and J.L. Hamrick. 2000. Influence of habitat patchiness on genetic diversity and spatial structure of a serpentine endemic plant. Conservation Biology 14: 454–463.

Wright, J.E. and S.J. Wyld. 1994. The Rattlesnake Creek terrane, Klamath Mountains, California: an early Mesozoic volcanic arc and its basement of tectonically disrupted ocean crust. Geological Society of America Bulletin 106: 1033–1056.

Wright, R.L., J. Nagel., and K.C. Taggart. 1982. Alpine ultramafic rocks of southwestern British Columbia. Canadian Journal of Earth Sciences 19: 1156–1173.

Zimmerman, J., C.O. Frank, and S. Bryn. 1981. Mafic rocks in the Avan Hills ultramafic complex, De Long Mountains. U.S. Geological Survey Circular 823-B: 14–15.

Zimmerman, J. and P.G. Soustek. 1979. The Avan Hills ultramafic complex, De Long Mountains, Alaska. U.S. Geological Survey Circular 804-B: 8–11.

Zinke, P. 1952. Soil-Vegetation Map, quadrangle 60D-1, Lake County. California Department of Forestry, Sacramento.

Part V Social Issues and Epilogue

23 Serpentine Land Use and Health Concerns

23.1 Environmental Aspects

Soils developed from serpentine (ultramafic) substrates are noted for their meager and strange biomass. The chemical infertility is the main controlling factor in the development of plants in serpentine soils (Proctor and Woodell 1975, Kruckeberg 1984, Brooks 1987). Botanists have recognized the unusual nature of the endemic plants and this has led to preserving serpentine tracts that contain rare plant species. The evolution of plant species that are restricted to serpentine has produced remarkable adaptations to survival on serpentine substrates. Kruckeberg (1984) pointed out that the long-term habitat attrition on these rare natural serpentine ecosystems requires conservation initiatives to insure their preservation. In California, private and public land managers are required to develop environmental impact studies before disturbing tracts containing serpentine bedrock and its overlying soils (Clinkenbeard et al. 2003). The U.S. Fish and Wildlife Service (USFWS 1998) carried out a recovery plan for 28 species of plants and animals that occur exclusively or primarily on serpentine soils and grasslands in the San Francisco Bay area. The strategy was to provide detailed actions needed to achieve self-sustaining populations of endangered species so they will no longer require protection under the Endangered Species Act.

Serpentine land tracts within metropolitan areas have come under closer regulation, as there is concern of releasing naturally occurring asbestos during construction disturbances. Typical examples of disturbance would be construction sites, new road construction, and quarry excavation. Of particular concern are the large amounts of dust produced in quarry operations or unpaved gravel roads consisting of crushed serpentine rock. The dust from such sites may contain airborne asbestos fibers released from the serpentine. This asbestos-bearing dust may pose a toxic threat to the construction workers and to later occupants of homes, schools, and office buildings occupying serpentine tracts.

Asbestos is the blanket term for a group of naturally occurring silicate minerals that can be separated into fibers. The fibers are strong, durable, and resistant to extreme heat. Because of these qualities, asbestos has been used in industrial, maritime, automotive, scientific, and building products. During the twentieth century, some "100 million tons" of asbestos have been used in industrial sites, homes, schools, shipyards, and commercial buildings in the United States. There are several types of asbestos fibers used for commercial applications: (1) Chrysotile, or white asbestos, which comes mainly from Canadian mines. It is white-gray in color and found in serpentinized peridotites. (2) Amosite, or brown asbestos, from southern Africa. (3) Crocidolite, or blue asbestos, comes from southern Africa and Australia. Amosite and crocidolite are amphiboles (Ross 1981, Skinner and Ross 1988). (4) Tremolite-actinoilite, gray to white, is often mixed with chrysotile asbestos. Serpentines of western North America often contain thin veins of chrysotile asbestos which rarely exceed more than 5% of the serpentine host rock.

Why is asbestos still a problem? Asbestos is still a problem because a great deal of it has been used in buildings and manufactured products in the United States and elsewhere. Many asbestos-containing products remain in buildings, ships, industrial facilities, and other environments where the fibers can become airborne. The suspected human health hazard of inhaling asbestos fibers in the workplace is now strongly regulated by EPA, requiring frequent surveys of fiber content in public buildings (Ross and Skinner 1994).

It is clear that asbestos is a health hazard, but it is unclear whether the six main fibrous minerals composing the generic asbestos grouping each exert the same harmful effects. The fibers are released by erosion and carried by the wind; thus, depending on where you live, it is estimated that you are most likely inhaling between 10,000 and 15,000 fibers each day (Abelson 1990). Water also contains asbestos. In the regions of Québec where the world's largest asbestos mines are located, the drinking water contains up to 170 million fibers/L (Toft et al. 1981). The main source of drinking water from the San Francisco Bay is from the Hetch Hetchy viaduct that brings water draining from the Sierra Nevada granite bedrock. The water from this source is stored in the Crystal Springs Reservoir where the bedrock includes extensive areas of sheared chrysotile-bearing serpentine. An epidemiological study of this drinking water in the San Francisco Bay estimates that this water contains 1,300,00–5,800,00 asbestos fibers/L (Conforti et al. 1981). This is nothing to be alarmed about, however, because asbestos is harmless in water; the problem is not ingesting the fibers, but inhaling them (Hayward 1984, Commins 1989, Ross and Skinner 1994).

Most asbestos mines in serpentine require more than 15 volume percentage asbestos fiber to be profitably mined. Not all serpentine lands (ultramafic rocks) contain asbestos fibers. It is, however, important to establish the presence or absence of asbestos by a registered geologist who has had experience identifying natural occurring asbestos. Serpentine landowners can use commercial testing laboratories to establish the approximate amount of asbestos fiber within natural serpentine land tracts. During active excavation and disturbance of serpentine lands, air monitors should be arrayed to determine the concentration of asbestos fibers in the excavation dust. Regulations concerning the disturbance of serpentine may be different in city, county, state, or federal lands. There is no scientific consensus regarding health risks of exposure to asbestos at low levels found in the air around natural serpentine rock. In western North America, there are significant tracts of naturally occurring asbestos-bearing serpentine. These tracts are present in residential areas, state and national parks, underlay schoolyards and reservoirs, and occupy significant tracts of land under control of local, state, and federal governments. It is unknown if there is a real health risk from exposure to asbestos fibers at the low levels encountered in these serpentine tracts. There has been no systematic evaluation of the health risks related to exposure of asbestos fibers released from these naturally occurring serpentine tracts.

Furthermore, these same serpentine tracts support endemic plants species of great rarity. We now recognize the importance of preserving these natural serpentine tracts, but at the same time, urban development threatens pollution by spreading dust laden with naturally occurring asbestos fibers. The California Geological Survey has published guidelines for evaluating serpentine tracts for land use decisions, land acquisitions, and property development (Clinkenbeard et al. 2003). The rising concern over potential exposure to naturally occurring asbestos in developing serpentine tracts initiated the publication of these guidelines by the California Department of Conservation.

23.2 Toxic Chemical Elements in Well Water

The chemical elements that are more concentrated in ultramafic rocks than in others and might be toxic in drinking water are the first transition elements from vanadium through nickel. Those that are sufficiently abundant and mobile to be concentrated enough in drinking water to become major concerns are chromium(VI), manganese(II), and iron(II). Manganese and iron are more commonly nuisances than health risks, but high concentrations of manganese become neurotoxic with prolonged exposures. Although chromium is more toxic, it is considerably less toxic than some elements such as mercury, cadmium, arsenic, and lead, which are not concentrated in ultramafic rocks.

Chromium in drinking water is hazardous. Long-term exposures at levels above the MCL (maximum contaminant level) of 0.1 mg/L established by the U.S. Environmental Protection Agency may damage liver, kidney, circulatory, and nerve tissues, and cause dermatitis (Scharfenaker 2001, Makeig and Nielsen 2006). The California Department of Health Services has an action level for chromium >50 µg (0.05 mg)/L. Hundreds of wells in California have been monitored and tested for chromium, but only a few have chromium >50 µg/L, and all of those with very high chromium are in Los Angeles and San Bernardino counties, southern California. There are no ultramafic rocks in these counties; therefore, the sources of chromium are presumed to be industrial. Records of chromium up to 44 µg/L have been reported for wells at Davis and up to 35 µg/L for wells at Woodland in Yolo County, California. Davis is near Putah Creek, and Woodlands is near Cache Creek; both creeks drain areas in the northern California Coast Ranges with much serpentine. The main source of chromium in these wells in Yolo County might be ultramafic rocks. Most of the chromium in ground-water is Cr(VI); it ranged from 5 to 39 µg/L and averaged 84% of the total chromium in water samples from 10 sites in the Aromas Red Sands aquifer in Santa Cruz County, CA (Gonzales et al. 2005). Ingested Cr(VI) is reduced to Cr(III) in saliva and the acidic environment of the stomach and may not be toxic to people in concentrations well above 0.1 mg/L (100 ppb), possibly even as high as the 1 or 2 mg, which causes water to appear yellow (Paustenbach et al. 2003). Enrichment of chromium in soils, some of which is Cr(VI), has been documented (Oze et al. 2003, 2004). This Cr(VI) can be leached from soils to the groundwater from which it can enter drinking water systems.

Manganese is not an immanent hazard in drinking water, but it can be toxic if contaminated water is ingested over long periods (Moore 1991). The California Department of Health Services has an action level for manganese >0.5 mg/L, and a secondary level of 0.05 mg/L for aesthetic reasons, because manganese may cause discoloration of the water. Manganese is a neurotoxic risk when levels >0.5 mg/L are present in drinking water. Levels of manganese >0.5 mg/L have been reported from 231 well systems in 41 of the 58 counties in California. Some of these are in southern California and east of the Sierra Nevada, where there are practically no ultramafic rocks. Therefore, it is uncertain that serpentine is the source of manganese in any of the wells with high values.

24   Synthesis and Future Directions

Ultramafic rocks come from deep within the earth. Most rocks on the surface of the earth are quite different from them. Unique rocks make unique soils and support special plants. Exploring the links and interactions among these unique rocks, soils, and vegetation is an interdisciplinary endeavor that has been accomplished by experts in three areas. It has helped elucidate serpentine rock–soil–plant relationships and provide a rationale for the unusual soil properties and vegetation associated with ultramafic rocks. Examples from arctic tundra to temperate rainforest and hot desert in western North America provide a framework for the investigation of serpentine geoecosystems around the world.

24.1 The Role of Geology in Serpentine Geoecology

The unusual character of most serpentine vegetation is readily apparent even to an untrained eye. Although a vast number of rock and soil types make up the earth's surface, few have as dramatic and visible effects on ecosystems as do ultramafic, or serpentine materials. Most ultramafic rocks in western North America have been derived from the mantle of earth via ocean crust. Magnesium is highly concentrated in the mantle and calcium, potassium, and phosphorous are relatively low. Calcium and potassium are further depleted from peridotite in the partial melting of ultramafic rock at the base of the ocean crust. As oceanic plates drift from spreading centers, most of the ocean crust is subducted and returns to the mantle (chapter 2). Only relatively small fragments of ocean crust are added to the continents. Because eukaryotic organisms, from protozoa to plants and animals, have evolved on continental crust, they are adapted to soils with higher concentrations of calcium, potassium, and phosphorus (elements with higher concentrations in continental crust than in ultramafic rocks from the base of the ocean crust) and much lower concentrations of magnesium. Having evolved on continents, plants depend on relatively high ratios of calcium and potassium to magnesium, elements that they use for a wide range of physiological functions. Although there has been a long history of evolutionary adaptation to the chemistry of the continental crust, special adaptations have allowed some plants to colonize the atypical conditions of serpentine.

There is quite a variety of ultramafic rocks, based on their mineral compositions (chapter 3). Most of those in western North America are either a variety of peridotite, called harzburgite, or serpentinite. Harzburgite is composed of olivine and pyroxenes. Most of the pyroxenes in harzburgite are orthopyrozenes—pyroxenes that lack calcium. Dunite, which is nearly all olivine, contains even less calcium. Pyroxenites, rocks that are nearly all pyroxene, commonly have clinopyroxenes that contain considerable calcium (chapter 3 and appendix A). The vegetation on dunite and harzburgite soils is distinctly different from that on pyroxenite soils. Differences in vegetation among different classes of peridotite have not been documented. This is an area that awaits detailed investigation. Also, we lack information about vegetation differences from peridotite to serpentinite soils.

Peridotite and serpentinite are chemically similar, but about 12%–15% water is added in the conversion of peridotite to serpentinite. This leads to great differences in their mineralogical and physical properties. Serpentinite is weaker than the much stronger peridotite, and it is commonly deformed and sheared compared to the more massive peridotite. The weaker serpentinite is more susceptible to massive failure and sliding, making very steep slopes highly unstable. Very steep slopes are more common where the bedrock is peridotite (chapter 6). Peridotite falling down the very steep slopes yields talus that is commonly barren. In serpentinite landscapes, barrens are more commonly on finer colluvial deposits, rather than on coarse talus, or on soils that have been eroded practically to bedrock.

Most of the iron in peridotite is in highly weatherable minerals, whereas in serpentinite most of it is in minerals that are more resistant to weathering (chapter 5). Large proportions of the iron released by weathering form iron oxides that add red color to the soils. Peridotite soils generally have more of the iron oxides and redder colors than serpentine soils (chapter 6). These iron oxides promote soil aggregation that reduces the erodibility of the soils. Although it appears that serpentinite soils are more erodible than peridotite soils, there have been no definitive investigations to document this observation.

Highly deformed and sheared serpentinite acts as an aquifer, although not as great a water collector and dispenser as basalt and limestone (chapter 4). Runoff and stream flow is delayed in serpentinite watersheds, and there is more stream flow later into the dry season, which is summer in the Sierra Nevada and Cascade (and Blue) Mountains and in coastal areas west of these mountains. Springs and water seeps seem to be more plentiful from serpentinite landscapes than from most others, but these observations have not been documented quantitatively.

Some springs in peridotite landscapes are highly alkaline, with much calcium and sodium, rather than magnesium (chapters 3 and 4). Most of the calcium is precipitated as a carbonate (travertine) near the sources of the springs, but sodium, and chlorine, can remain in solution and can be carried farther from the sources. The alkaline springs support calciphilic and halophitic plant species among other serpentine wetland flora.

Gabbro is a geoecological enigma. It is between peridotite and granitic rocks in mineralogical and chemical composition. Gabbro is formed in the ocean crust, above peridotite, and in intrusions of magma into continental crust. It is higher in iron and magnesium than granitic rocks, but not as high as peridotite in magnesium. It is called a mafic rather than an ultramafic rock. There are different varieties of gabbro. Some gabbro rocks and soils of the Sierra Motherlode and the Peninsular Ranges support highly distinctive and endemic-rich plant communities (chapters 13 and 14), but other gabbro rocks and soils may have vegetation that is little different from that on many other kinds of rocks and soils. A better understanding of the geoecology of gabbro is hindered by the fact that few geologic maps distinguish among the different varieties of gabbro, and few botanists are aware of how to identify gabbro, let alone to distinguish among the different varieties of gabbro.

24.2 The Role of Soils in Serpentine Geoecosystems

A rich variety of serpentine soil taxa is found from Baja to Alaska, and there are also local differences related to topographic relief and climatic differences from the Pacific Coast inland (chapter 6). Serpentine soils have different effects on vegetation in different climatic and topographic settings. These effects do not appear to be related to any differences in capacities of serpentine and nonserpentine soils to hold water available to plants (chapter 8). To understand the influences of different kinds of soils on the distribution of vegetation, it is necessary to observe vegetation patterns at a local scale. Locally soil differences may be related to parent material differences and to topographic differences. Vegetation differences on soils with similar topography, but different parent material, can generally be attributed to parent material differences. Local topographic differences that control soil development and influence vegetation distribution are slope gradient and aspect and position on a landscape. For example, abrupt boundaries between serpentine chaparral and serpentine grassland are common at the transition from hillsides to valley bottoms in the North California Coast Ranges where serpentine alluvium in valley bottoms commonly forms clay-rich soils that exclude trees and shrubs. In other cases, more subtle shifts between different types of shrub vegetation may be the result of position on a slope and soil depth differences across a serpentine landscape (fig. 24-1). In some cases, fire history and other disturbances may be overriding influences; for example, in the distribution of serpentine stands of scrub and grassland on Coyote Ridge in the Southern California Coast Ranges (Evens and San 2004).

Figure 24-1 Serpentine vegetation in the Northern California Coast Ranges showing a shift from a stand of leather oak (*Quercus durata*) and Jepson ceanothus (*Ceanothus jepsonii*) on stony soils of the upper slope to a stand of McNab cypressus (*Cupressus macnabiana*) on deeper nonstony soils at the base of the slope. Line shows approximate boundary between soils. The off-highway vehicle track on the right shows that the soil is much stonier above than below the line. Napa County, May 2003.

24.3 Serpentine Endemism: An Enigma that Geoecology Can't Explain?

One of the most striking features of western North American serpentine plant life remains difficult to explain in geoecological terms. From south to north, the distinctiveness of the serpentine flora and vegetation is low in Baja California, increases to one of the highest levels in the world throughout the California Floristic Province, and then drops again north of the Cascade Mountains and remains low to northern Alaska. This does not seem to be explained by variation in geology or soils. Instead, it may involve an interactive effect of climate and the serpentine substrate on plant productivity. When water or temperature is strongly limiting to plant growth and survival, the serpentine substrate appears to have much less effect on the vegetation. The evolutionary component of these effects is not well understood, but it follows the same pattern. Whatever factors have caused the proliferation of plant species in the California Floristic Province seem to have dictated that the serpentine endemics of western North America are almost completely confined to the province. Past climate changes and glaciation may also have played important roles.

Another mysterious pattern that does not seem to have a geologic or soil explanation is the drop in diversity of unique serpentine plant species in the Sierra Motherlode compared with the Coast Ranges of California. In the Sierra, although vegetation structure is generally distinctive on serpentine, endemism is relatively low. Plant diversity as a whole is lower in the Sierra Motherlode than the Coast Ranges, so there may be an explanation that is relatively independent of serpentine geology or soils.

24.4 Future Research Directions

One of the most rapidly developing fields in ecology today is the study of below-ground interactions, especially those among plants, mycorrhizae, and root-associated bacterial communities. Molecular and biochemical techniques are allowing better characterization of fungal and microbial individuals and species, and soil inoculation experiments are elucidating the effects of these biotas on plants. We should soon know much more about the role that belowground interactions play in mediating the ability of individual plant species to grow on serpentine.

New molecular and analytic tools are also reopening the study of plant adaptations to serpentine soil and providing new insights into specific biochemical traits that are responsible. At the same time, phylogenetic studies are being used to determine the taxonomic distribution of such traits, which will greatly improve our understanding of the evolutionary history of serpentine adaptation. The long-standing observation that some taxa appear preadapted to evolve serpentine tolerance may soon have a mechanistic explanation. Finally, serpentine endemics are proving to be valuable study organisms in the rapidly developing study of mechanisms of speciation.

Biological invasions are a growing field of study, and serpentine environments are of particular interest because they are highly resistant to invasion yet not completely immune. One of the most interesting and important questions in serpentine plant ecology is to what extent the barrier to invasion posed by serpentine environments will be breached through the rapid evolution of serpentine tolerance in invasive species. Additional threats include the arrival of invaders that are already serpentine-tolerant, and the increased invasibility in urban serpentine areas due to atmospheric nitrogen pollution.

Among the interesting future directions for exploration across suites of plant species (e.g., stands of vegetation), one involves quantitative analyses of the shifts in vegetation stand similarity between adjacent serpentine and nonserpentine exposures. In many parts of western North America there appear to be abrupt shifts between serpentine and nonserpentine stands (fig. 24-2). By attempting to hold many environmental variables such as soil depth, slope gradient and aspect, elevation, and microtopography constant, it may be possible to investigate the restrictive effects of soil chemistry versus soil depth and other possible confounding effects on species composition on adjacent serpentine and nonserpentine sites. This type of investigation conducted on many replicate pairwise comparisons on and off serpentine could yield interesting generalizations about local and regional tolerance of suites of species to serpentine.

Basic research on developing detailed vegetation classifications for all serpentine parts of western North America is needed in Baja California, in the Northern California Coast Ranges, and in the Sierra Motherlode.

One of the most interesting aspects of natural community ecology involves a philosophical broadening from what traditionally has been strictly an animal/plant dichotomy. Expanding from the philosophy of this book that an interaction between the biotic and the abiotic explains the whole, ecologists should also consider quantifying and more deeply investigating the relationships among both plants and animals in serpentine environments. Certain relationships, such as those between endemic butterflies and their host plants on serpentine, have been pointed out already. However, interdisciplinary teams of plant and animal ecologists quantifying and comparing entire suites of interacting species could have much to tell us about serpentine ecosystems. For example, are patterns of endemism mirrored more closely in certain types of organisms than others? What is the diversity of serpentine and nonserpentine herbivorous invertebrates in

Figure 24-2 The open conifer woodland on serpentine soils (S) is distinctly different from the dense forest on nonserpentine soils (N) with similar slopes and microclimate. North Fork of Begum Creek, Shasta County, California, October 2004.

different areas? Does that diversity vary depending upon soil chemistry and other abiotic factors?

24.5 Conclusions

This book posits that there is more to be gleaned from a subject if one explores the broad relationships in an interdisciplinary way. Although we are left with many questions about serpentine geoecology of western North America, let alone the rest of the world, it has been inspiring to collaborate on such an effort, and we hope the inspiration is contagious. With the volume of literature expanding rapidly on many aspects of the serpentine syndrome, we expect there will be cause to develop a second edition, in which we will be able to clarify many of the issues that we were only able to touch upon here.

Part V References

Abelson, P.H. 1990. The asbestos removal fiasco. Science 247:1017.
Brooks, R.R. 1987. Serpentine and its Vegetation: A Multidisciplinary Approach. Dioscorides Press, Portland, OR.

Clinkenbeard, J.P., R.K. Churchill, and L. Kiyoung. 2003. Guidelines for geologic investigations of naturally occurring asbestos in California. California Geological Survey, Special Publication 124: 1–70.

Commins, B.T. 1989. Estimations of risk from environmental asbestos in perspective. IARC Science Publications 90: 476–485.

Conforti, P.M., M.S. Kanarek, A. Jackson, R.C. Cooper, and J.C. Murchio. 1981. Asbestos in drinking water and cancer in the San Francisco Bay area: 1969–1974 incidence. Journal of Chronic Disease 34: 211–224.

Evens, J. and S. San. 2004. Vegetation associations of a serpentine area: Coyote Ridge, Santa Clara County, California. California Native Plant Society, Sacramento.

Gonzales, A., K. Ndurg'u, and A.R. Flegal. 2005. Hexavalent chromium in the Aromas Red Sands aquifer. Environmental Science and Technology 39: 5505–5511.

Hayward, S.B. 1984. Field monoriting of chrysotile asbestos in California waters. Journal of the American Water Works Association 76: 66–73.

Kruckeberg, A.R. 1984. California Serpentines: Flora, Vegetation, Geology, Soils, and Management Problems. University of California Press, Berkeley.

Makeig, K.S., and D.M. Nielsen. 2006. Regulatory mandates for ground-water monitoring.Pages 1–34 *in* D.M. Nielsen (ed.), Practical Handbook of Environmental Site Characterization and Ground-Water Monitoring. Taylor and Francis, Boca Raton, FL.

Moore, T.E. 1991. Inorganic Contaminants of Surface Water. Springer-Verlag, Berlin.

Oze, C., S. Fendorf, D.K. Bird, and R.G. Coleman. 2004. Chromium geochemistry in serpentinized ultramafic rocks and serpentine soils from the Franciscan Complex of California. American Journal of Science 305: 67–101.

Oze, C.J., M.J. LaForce, C.M. Wentworth, R.T. Hanson, D.K. Bird, and R.G. Coleman. 2003. Chromium geochemistry of serpentinous sediment in the Willow core, Santa Clara County, CA. Open-File Report OF 03-0251, U.S. Geological Survey.

Paustenbach, J.A., B.L. Finley, F.S. Mowat, and B.D. Kerger. 2003. Human health risk and exposure assessment of chromium in tap water. Journal of Toxicology and Environmental Health, Part A 66: 1295–1339.

Proctor, J. and S.R.J. Woodell. 1975. The plant ecology of serpentine soil. Advances in Ecological Research 9: 375–395.

Ross, M. 1981. The geologic occurrences and health hazards of amphibole and serpentine asbestos. Pages 279–232 *in* D. Veblen (ed.), Amphiboles; Petrology and Experimental Phase Relations. Mineralogical Society of America, Washington, DC.

Ross, M. and H.C.W. Skinner. 1994. Geology and health. Geotimes 39: 10–12.

Scharfenaker, M.A. 2001. Chromium VI: a review of recent developments. American Water Works Association, Journal 93(11): 20–26.

Skinner, H.C.W. and M. Ross. 1988. Fibrous minerals, mining, and disease. Report GPP 012. Geological Society of America, Denver, CO.

Toft, P., D. Wigle, J.C. Meramger, and Y. Mao. 1981. Asbestos and drinking water in Canada. Science of the Total Environment 18: 77–89.

USFWS. 1998. Draft recovery plan for serpentine soil species of the San Francisco Bay area. U.S. Fish and Wildlife Service, Pacific Region, Portland, OR.

Appendices

APPENDIX A: NATURE OF MINERALS IN SERPENTINE ROCKS AND SOILS

The common minerals in rocks and soils are composed of more or less closely stacked oxygen atoms, or anions, with enough cations among them to balance the electronic charges. Although oxygen atoms are the most common and abundant anions in minerals in rocks and soils, notable exceptions are sulfides (e.g., pyrite, FeS_2) and halides (e.g., halite, NaCl, and fluorite, CaF_2). None of these sulfides or halides is abundant minerals in rocks and soils, but sulfides may be common in anaerobic environments and halides in saline environments. Oxygen atoms are larger than most of the cations that are common in rocks and soils (silicon, aluminum, iron, magnesium, titanium, phosphorus, and manganese)—although calcium, sodium, and potassium are nearly as large or larger than oxygen (table A-1). Arrangements of the oxygen atoms depend largely on the nature of the cations that are distributed among the oxygen atoms, although the temperature and pressure at the time of crystallization are also important factors.

Two of the most basic arrangements of oxygen atoms are tetrahedral and octahedral. A tetrahedron has four sides and an octahedron has eight sides. There are oxygen atoms at each of the four apices of a tetrahedron and at each of the six apices of an octahedron (fig. A-1). A cation at the center of a tetrahedron is coordinated with four oxygen atoms, and a cation at the center of an octahedron is coordinated with six oxygen atoms. Silicon has a small ion (Si^{4+}, radius = 0.34 Å) that fits nicely into the space between four oxygen atoms; therefore, its coordination number is four. Magnesium has a larger ion (Mg^{2+}, radius = 0.80 Å) that fits well into the space between six oxygen atoms; its coordination number is six. Aluminum has an ion (Al^{3+}) of intermediate size that will fit in either tetrahedral (fourfold coordination) or octahedral (sixfold coordination) positions. The aluminum atom is allotted more space in the octahedral position (0.61 Å) than in the tetrahedral position (0.47 Å). Cations the size of calcium (Ca^{2+}, radius = 1.20 Å) and sodium (Na^+, radius = 1.24 Å) are coordinated with eight or more oxygen atoms. Potassium (K^+, radius = 1.68 Å) is larger than the oxygen ion and fits into a space between 12 oxygen atoms.

Table A-1 Radii (Å) of common ions in mafic and ultramafic igneous rocks and serpentinite and in soils derived from them.

Ion	IV	VI	VIII	XII
Si^{4+}	0.34			
Cr^{6+}	0.38			
V^{5+}	0.44	0.62		
Al^{3+}	0.47	0.61		
Co^{3+}		0.61, 0.69		
Mn^{4+}		0.62		
Fe^{3+}	0.57	0.63, 0.73		
Mn^{3+}		0.66, 0.73		
V^{4+}		0.67		
Ti^{4+}		0.69		
Fe^{2+}		0.69, 0.86		
Cr^{3+}		0.70		
V^{3+}		0.72		
Co^{2+}		0.73, 0.83		
Mn^{2+}		0.75, 0.91		
Ni^{2+}		0.77		
Mg^{2+}		0.80		
Cu^{2+}		0.81		
Zn^{2+}		0.83		
Ca^{2+}			1.20	1.43
Na^{+}			1.24	
O^{-2}	1.30	1.32	1.34	
K^{+}				1.68

Data from Whittaker and Muntus (1970).
[a]Two numbers separated by a comma represent radii for ions with unpaired electrons of low, high spin.

Silicon is the most abundant cation in most rocks; therefore the silicon–oxygen tetrahedron is a common structural unit in the most abundant minerals in common rocks. Silicon tetrahedra (SiO_4) can exist independently of each other, they can be joined in pairs, they can be linked in single or double chains, they can form rings, or they can be joined in layers. They can also form rings or they can be in framework structures. Independent tetrahedra, chains, and layer silicate minerals are the most common in serpentine rocks and soils (fig. A-2).

Silicon tetrahedra are independent of each other in olivine, $(Mg,Fe)_2SiO_4$. The Mg^{2+} and Fe^{2+} ions are between the tetrahedra. Silicon tetrahedra in pyroxenes are linked in single

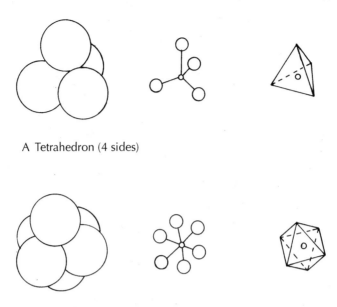

A Tetrahedron (4 sides)

B Octahedron (8 sides)

Figure A-1 Closed and expanded views of a silicon–oxygen tetrahedron (A) and a magnesium–oxygen octahedron (B). Cations are visible only in the expanded views.

chains. Negative charges on the chains are balanced by cations between them. With only Mg^{2+} and Fe^{2+} between chains, a pyroxene is in the orthorhombic crystal system (three orthogonal axes). Larger cations such as Ca^{2+} between chains distort chain arrangements to place pyroxenes containing them in the monoclinic crystal system (two orthononal axes and a third axis inclined to the plane of the orthogonal axes). Therefore, no orthopyroxenes (pyroxenes in the orthorhombic system) contain more than traces of calcium. All pyroxenes containing calcium are clinopyroxenes (pyroxenes in the monoclinic system). Silicon tetrahedra in amphiboles are arranged in double chains. Like pyroxenes, amphiboles with only Mg^{2+} and Fe^{2+} between chains (for example, anthophyllite) are in the orthorhombic crystal system. Amphiboles with calcium, sodium, or potassium between chains are in the monoclinic crystal system (for example, hornblende).

Oxides lacking silicon are common in serpentine rocks and soils. Common examples are chromite ($FeCr_2O_4$) in peridotite, magnetite ($FeFe_2O_4$) in serpentinite, and goethite ($FeOOH$) and hematite (Fe_2O_3) in soils. Chromite and magnetite, along with spinel ($MgAl_2O_4$), are in the spinel group of minerals. Hexagonally close-packed oxygen ions in spinel have cations between planes of oxygen ions. Cations in alternate sheets have divalent cations in sixfold coordination in one and trivalent cations in fourfold coordination in the other. In hematite, the oxygen ions are in slightly deformed hexagonal packing, forming slightly distorted octahedra that share all faces with other octahedra. Two-thirds of the octahedral positions are occupied by Fe^{3+} and the rest are vacant. Some trivalent ions such as aluminum, manganese, chromium, vanadium, and cobalt can substitute for Fe(III) without modifying the oxide structure.

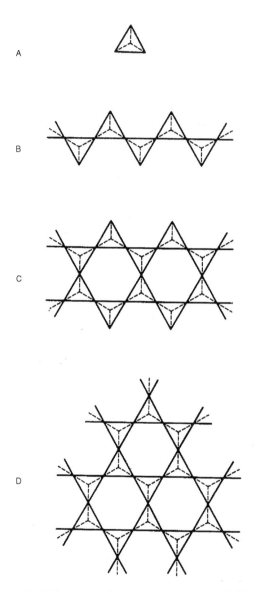

Figure A-2 Plan views showing arrangements of silicon tetrahedra in silicate minerals that are common in serpentine rocks and soils. (A) An individual tetrahedron, as in olivine. (B) A chain of tetrahedra, as in pyroxenes. (C) A double chain of tetrahedra, as in amphiboles. (D) A sheet of tetrahedra, as in serpentine. (From table 1-3 of Schulze, 1989.)

Goethite is actually an oxyhydroxide, containing both oxygen and hydroxyl (OH⁻) ions. The size of the hydroxyl ion is essentially the same as that of the oxygen ion, because space occupied by the hydrogen ion is negligible. The oxygen atoms (O^{2-} and OH⁻ ions) are arranged in hexagonal close packing. Goethite and some of the other mineral structures are difficult to describe, but they are easy to comprehend in diagrams such as those of Deer et al. (1966) and Schwertmann and Taylor (1989). Substitution of other cations for iron is common in goethite.

Having hexagonal close packing, hematite and goethite are α-phases (α-Fe_2O_3 and α-FeOOH). Corresponding minerals with approximately cubic close packing are maghemite and lepidocrocite, referred to as having γ-phases (γ-Fe_2O_3 and γ-FeOOH). Lepidocrocite and maghemite are much less common than hematite and goethite in soils, but are not rare.

Hydroxides of magnesium and aluminum are present as minerals in some rocks and soils. Brucite, $Mg(OH)_2$, is a product of serpentinization of dunite and other peridotites, but gibbsite, $Al(OH)_3$, and its polymorphs are not present in serpentine rocks and are seldom found in serpentine soils. Brucite and gibbsite have layered (single sheet) structures. In brucite, the OH⁻ ions are in hexagonal close packing in two planes with a plane containing Mg^{2+} ions between them. The Mg^{2+} ions are in sixfold coordination and occupy all of the octahedral positions among the OH⁻ ions (fig. A-3). In gibbsite, only two-thirds of the octahedral positions are occupied because it takes less Al^{3+} to balance the negative charges of the OH⁻ ions. Each sheet consisting of two planes with OH⁻ ions and one with Mg^{2+} or Al^{3+} ions is affixed to adjacent sheets by weak forces that allow easy separation, or cleavage, of the layers in brucite and gibbsite.

Layered silicate minerals—those in which three of the oxygen atoms in each SiO_4 tetrahedron are shared with adjacent tetrahedra to form a sheet of silica tetrahedra—always have oc-

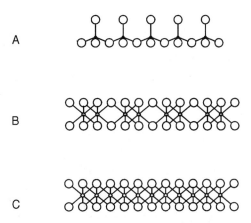

Figure A-3 Lateral views of tetrahedral and octahedral sheets, which are the structural components of layer silicate minerals. (A) A sheet of tetrahedra. (B) A sheet of octahedra in which two-thirds of the octahedral positions are occupied as in gibbsite, $Al(OH)_3$. (C) A sheet of octahedra in which all of the octahedral positions are occupied as in brucite, $Mg(OH)_2$ (From figs. 1-4 and 1-5 of Schulze, 1989.)

tahedral sheets, which lack silicon, along with the SiO_4 tetrahedral sheets. Micas, serpentine, talc, kaolinite, and smectites are prominent layered silicate minerals in rocks and soils. Serpentine, $Mg_3Si_2O_5(OH)_4$, consists of a tetrahedral sheet and a parallel octahedral sheet that share O^{2-} or OH^- ions (fig. A-4). Serpentine is sometimes called a trioctahedral mineral because all octahedral sites are occupied—in this case by Mg^{2+} ions. Kaolinite, $Al_2Si_2O_5(OH)_4$, is similar to serpentine, except that the octahedral sites contain Al^{3+} rather than Mg^{2+} ions. It is a dioctahedral mineral because only two-thirds of the octahedral sites are occupied by Al^{3+} ions. Because the O^{2-} atoms on the surface of a silicon-tetrahedral sheet are not a precise match for those on the surface of a magnesium-octahedral sheet, the layers are slightly distorted when they join together in serpentine minerals.

Three different modes of accomodation produce three different serpentine minerals—chrysoltile, lizardite, and antigorite. In chrysotile, which is one variety of serpentine, the sheets are not planar but curved to allow them to roll up and form tubes. Whereas serpentine is a one tetrahedral-to-one octahedral (or 1:1) layer mineral, talc, $Mg_3Si_4O_{10}(OH)_2$, is a two tetrahedal-to-one octahedral (2:1) layer mineral containing the same elements but with a reduced proportion of octahedral sites and Mg^{2+} ions. Forces between these 2:1 layers are weak, allowing them to slip past each other and produce the slippery feel of talc, which is sometimes called soapstone. Micas are also 2:1 layer minerals, but with Al^{3+} ions substituted for some of the Si^{4+} ions in tetrahedral positions, reducing the positive charges and giving the layers net negative charges. These negative charges are balanced by large cations, generally K^+, which are accommodated in large spaces between 2:1 layers. The oxygen atom coordination of K^+ ions in spaces between tetrahedral layers of mica layer complexes is 12. The electrostatic bonds between mica layer complexes are much stronger than those between layer complexes in serpentine and talc. Nevertheless,

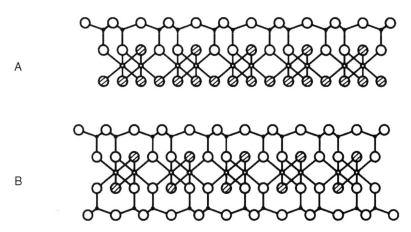

Figure A-4 Lateral views of a 1:1 layer mineral with juxtaposed tetrahedral and octahedral sheets sharing oxygen atoms in the mutual plane between them and a 2:1 layer mineral consisting of an octahedral sheet sandwiched between two parallel tetrahedral sheets. (A) Kaolinite, a trioctahedral 1:1 layer silicate mineral. (B) Pyrophyllite, a 2:1 layer silicate mineral. Serpentine (1:1) and talc (2:1) are similar to kaolinite and pyrophyllite, except that they have trioctahedral sheet containing Mg, rather than dioctahedral sheets containing Al. (From fig. 1-6 of Schulze, 1989.)

structural ruptures between layer complexes give micas essentially perfect cleavage and platy appearance. Muscovite is a dioctahedral mica with Al^{3+} ions in two-thirds of the octahedral sites, and biotite is a trioctahedral mica with Mg^{2+} or Fe^{2+} occupying all of the octahedral sites. Smectites, which are common clay minerals in serpentine soils, have 2:1 layer structures much like those of micas. The main differences are that weathering and leaching have replaced enough K^+ ions with Ca^{2+}, Mg^{2+}, and Na^+ ions that forces between layer complexes no longer prevent entry of water and expansion when the minerals are saturated with water. Thus, smectites are swelling clay minerals. Generally, they are dioctahedral smectites with much Mg^{2+} and Fe^{3+} in octahedral positions. Although most of the negative charge responsible for the relatively high CEC of smectites is derived from the substitution of Al^{3+} for Si^{4+} in tetrahedal positions, aluminum is sparse in serpentine soils. Substitution of Fe^{3+} in tetrahedral positions is limited to small proportions (Stucki 1988). Much of the negative charge on smectites in serpentine soils may be derived from the substitution of Mg^{2+} for Fe^{3+} in octahedral positions.

Chlorites are layer silicate minerals (or phyllosilicates) that are produced both in rocks and in soils. They consist of layers of two tetrahedral sheets to one octahedral sheet alternating with octahedral sheets, or layers, that do not share oxygen atoms with 2:1 layers (fig. A-5). The octahedral cations in the 2:1 layers are generally divalent magnesium or iron, while those in octahedral positions of the sheets between 2:1 layers are commonly Al^{3+}, Mg^{2+}, Fe^{2+}, or Fe^{3+}. In soils the octahedral sheets are commonly incomplete, allowing water to enter the

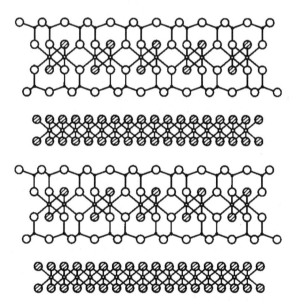

Figure A-5 Chlorite–a mineral with 2:1 layers alternating with hydroxyl sheets. The 2:1 layers are formed by combinations of two tetrahedral sheets and one octahedral sheet. Smectites are similar, except the space occupied by the hydroxyl sheet in chlorite is occupied by exchangeable cations and water. The open circles represent oxygen ions and the cross-hatched ones represent hydroxyl ions. (From figure 1-8 of Schulze, 1989.)

space between 2:1 layers. These imperfect chlorites are commonly referred to as hydroxy-interlayer vermiculite, or, if they swell, hydroxy-interlayer smectite.

Not All Serpentine Is the Same

Serpentinite rock is made up of mixtures of three main green serpentine minerals: (1) chrysotile, (2) lizardite, and (3) antigorite. Serpentine minerals are trioctahedral hydrous phyllosilicates based on 1:1 layered structure consisting of alternating silica tetrahedral and magnesia octahedral sheets (see above). Chrysotile is characteristically fibrous and under the electron microscope shows a tubelike structure. Lizardite and antigorite crystallize as plates and do not assume the fibrous habit (asbestos) of chrysotile. Structurally all three serpentine minerals consist of two sheets of tetrahedral silica (tridymite, SiO_4) joined to a trioctahedral sheet of oxygen and hydroxyl ions (Coleman and Jove 1993). The differences in dimensions between the brucite and tridymite layers is compensated by substitution in either or both layers or by warping, and this is considered to be the main reason for the diverse character of the serpentine minerals (Wicks and O'Hanley 1988).

Page (1968) showed that the three serpentine species have distinct chemical compositions or could be part of a complex solid solution series. There are variations of the ideal serpentine formula $Mg_3Si_2O_5(OH)_4$ by substitution and change in valence state as follows: (1) antigorite has a much higher silica content than chrysotile or lizardite; (2) chrysotile has only traces of aluminum; (3) lizardite has a much higher ferrous to ferric iron (Fe^{2+} to Fe^{3+}) ratio; (4) antigorite contains less magnesium and water; and (5) antigorite has large amounts of trivalent ions in octahedral coordination. These disparities in composition point to the fact that chrysotile, lizardite, and antigorite are not true polymorphs, and the presence or absence of one species over the other cannot be used to specify a particular pressure–temperature (P-T) condition (O'Hanley 1996). There are some cases, however, where mineral assemblages associated with these various serpentine minerals may provide clues to the P-T environment (Chernosky et al. 1988).

References

Borchardt, G. 1989. Smectites. Pages 675–727 in J.B. Dixon and S.B. Weed (eds.), Minerals in Soil Environments. Soil Science Society of America, Madison, WI.

Chernosky Jr., J.V., R.G. Berman, and L.T. Bryndzia. 1988. Stability, phase relations, and thermodynamic properties of chlorite and serpentine group minerals. Mineralogical Society of America, Reviews in Mineralogy 19: 295–346.

Coleman, R.G. and C. Jove. 1993. Geological origin of serpentinites. Pages 1–17 in J. Procter, A.J.M. Baker, and R.D. Reeves (eds.), The Vegetation of Ultramafic (Serpentine) Soils. Intercept, Andover, NH.

Deer, W.A., R.A. Howie, and J. Zusman. 1966. An Introduction to the Rock-Forming Minerals. Wiley, New York.

O'Hanley, D.S. 1996. Serpentinites: Records of Tectonic and Petrological History. Oxford University Press, New York.

Page, N.J. 1968. Serpentinization in a sheared serpentinite lens, Tiburon Penninsula, California. Professional Paper P 600-B, pages B21–B28, U.S. Geological Survey.

Schulze, D.G. 1989. An introduction to soil mineralogy. Pages 1–34 *in* J.B. Dixon and S.B. Weed (eds.), Minerals in Soil Environments. Soil Science Society of America, Madison, WI.

Schwertmann, U. and R.M. Taylor. 1989. Iron oxides. Pages 379–438 *in* J.B. Dixon and S.B. Weed (eds.), Minerals in Soil Environments. Soil Science Society of America, Madison, WI.

Stucki, J.W. 1988. Structural iron in smectites. Pages 625–675 *in* J.W. Stucki, B.A. Goodman, and U. Schwertmann (eds.), Iron in Soils and Clay Minerals. Reidel, Dordrecht.

Whittaker, E.J.W. and R. Muntus. 1970. Ionic radii for use in geochemistry. Geochimica et Cosmochimica Acta 34: 945–956.

Wicks, F.J. and D.S. O'Hanley. 1988. Serpentine minerals; structures and petrology. Mineralogical Society of America, Reviews in Mineralogy 19: 91–167.

APPENDIX B: CHARACTERISTICS OF THE CHEMICAL ELEMENTS— IONIC PROPERTIES AND TOXICITIES

The atoms of chemical elements are composed of neutrons, protons (positive charge), and electrons (negative charge). These components were once considered to be fundamental particles. Although they are now known to be composed of even smaller particles, it is the neutrons, protons, and electrons that determine the character of chemical elements. The behavior of the elements has been successfully predicted using knowledge of the numbers of the neutrons, protons, and electrons in the atoms of the element. Therefore it is appropriate to review the structures of atoms and their distributions in a periodic table of the elements before utilizing their properties to explain their behaviors. Much of this appendix is a review and may not be appropriate for all readers.

Nature of the Chemical Elements

The chemical elements are distinguished by the numbers of protons in each atom. They are numbered from 1–92, or more, by the numbers (Z) of these protons. Atomic weights are proportional to the numbers of neutrons and protons in the atoms. Electron mass is relatively small, only 0.0005 times that of neutrons or protons. The numbers of neutrons are equal, or greater than, the numbers of protons—except in hydrogen. Hydrogen generally has no neutrons; if it has one neutron it is called deuterium and with two neutrons it is called tritium. Deuterium and tritium are isotopes of hydrogen.

Neutral atoms have equal numbers of protons and electrons. The neutrons and protons are in a central nucleus and the electrons revolve around it in shells of different energy levels. The electrons occur alone or in pairs in four kinds of orbitals that are labeled s, p, d, and f. The shell with the lowest energy level contains an s orbital, the second and third each contain s and p orbitals, the fourth and fifth each contain s, p, and d orbitals, the sixth contains s, f, d, and p orbitals, and the seventh contains s and partially filled f orbitals. This structure is reflected in

I	II											III	IV	V	VI	VII	VIII
1 H																	2 He
3 Li	4 Be											5 B	6 C	7 N	8 O	9 F	10 Ne
11 Na	12 Mg											13 Al	14 Si	15 P	16 S	17 Cl	18 Ar
19 K	20 Ca	21 Sc	22 Ti	23 V	24 Cr	25 Mn	26 Fe	27 Co	28 Ni	29 Cu	30 Zn	31 Ga	32 Ge	33 As	34 Se	35 Br	36 Kr
37 Rb	38 Cs	39 Y	40 Zr	41 Nb	42 Mo	43 Tc	44 Ru	45 Rh	46 Pd	47 Ag	48 Cd	49 In	50 Sn	51 Sb	52 Te	53 I	54 Xe
55 Cs	56 Ba	57 La →	70														
		71 Lu	72 Hf	73 Ta	74 W	75 Re	76 Os	77 Ir	78 Pt	79 Au	80 Hg	81 Tl	82 Pb	83 Bi	84 Po	85 At	86 Rn
87 Fr	88 Ra	89 Ac	90 Th	91 Pa	92 U												

Figure B-1 Periodic table of the chemical elements. Elements 57–70 are Lanthanides, elements in which f-orbitals are filled before filling the third set of transition elements with d-orbitals (elements 71–80).

the grouping of chemical elements in a periodic table (fig. B-1). The table is periodic because the elements in each column have similar characteristics. Elements in columns I and II of the table, and helium, are created by adding electrons in s-orbitals. Those in the six columns III through VIII are created by adding one to six electrons in p-orbitals. Helium and the elements in column VIII are all inert gasses, which is a good example of periodicity that is reflected in the table. Elements 21–30, 39–48, and 71–80 are created by adding 1–10 electrons in d-orbitals. Elements 57–70 (lanthanides) and 89–102 (actinides) are created by adding one to 14 electrons in f-orbitals. Actinides are unstable and decay by losing alpha particles (two protons+two neutrons) and beta particles (electrons) to produce elements with lower atomic numbers. Elements with atomic numbers > 92 decay so rapidly that they are too ephemeral to be considered naturally occurring elements.

The numbers of electrons are equal to the numbers of protons for electronic balance. Electrons may be gained or lost from atoms to form positively or negatively charged ions. The charge on the atoms of an element is called its valence. Valence is a major factor in governing reactions with other atoms. Valence can be predicted from the periodic table. Elements in column I of the table each have one unpaired electron in an s-orbital that can be removed to form positively charged ions with valences of +1. The paired s-orbital electrons in the elements of column II can be removed to form positively charged ions with valences of +2. Columns III through VIII represent the filling of p-orbitals. The elements in these six columns commonly have valences of +3, +4, +5, −2, −1, and 0. Electrons from a set of three filled p-orbitals (3 orbitals with 2 electrons in each orbital) are not easily removed, limiting their reactions with other elements. Thus the elements in column VIII are inert. Elements with filled s- and p-orbitals and incompletely filled d-orbitals are called transition elements, except that Cu^{+1}, Ag^{+1}, Au^{+1}, Zn^{+2}, Cd^{+2}, and Hg^{+2} have filled d-orbitals and have lost electrons from an s-orbital. The first two transition elements, scandium and titanium and other d-block elements in the same columns, have valences of +3 and +4. All other transition elements are generally represented by more than one valence in geoecosystems. The f-block elements are not important in natural geoecosystems, but uranium is a hazardous contaminant in some industrial waste.

Magnesium and transition elements from vanadium through nickel are more abundant and important in ultramafic rocks than in most other kinds of rocks.

Ionic Potentials for Cations in Serpentine Rocks and Soils

When considering the behavior of scores of chemical elements in geoecological systems, it is convenient to group the elements into classes. Prime examples are the chalcophile, siderophile, lithophile, and atmophile classes of Goldschmidt (1954) and the hard and soft acid scheme popularized by Pearson (1963). A simple algorithm commonly used in the characterization of cation dispositions in aqueous systems is the ionic potential. Ionic potential is so simple and so helpful in categorizing chemical elements to predict their ionic bonding behaviors that it deserves special mention. Other considerations, such as hard or soft Lewis-acid character, are necessary for predicting covalent bonding behavior.

The ionic potential (IP) is simply the charge of a ion divided by its ionic radius (fig. B-2). Cations do not have definite perimeters, but distances between ions can be ascertained in crystals. Some assumptions must be made—mainly the radius of oxygen or other anions in the crystals. With different assumptions, different investigators have obtained slightly different results (Whittaker and Muntus 1970, Shannon 1976). The cation radius ratios of Whittaker and Muntus used in figure B-2 are slightly larger than those of Shannon because Whittaker and Muntus chose smaller values for oxygen (and for fluorine) ions. Either set of values is good for comparing ionic radii of cations, if values from only one set are used in each application.

Small cations fit into the space between 4 anions of oxygen, but successively larger cations are surrounded by 6, 8, and 12 anions. The number of neighboring anions is called the coordination number. The same cations with different coordination numbers have different radii; for example, the radius of Al(III) is 0.47 with a coordination number of 4 and 0.61 with a number of 6 (fig. B-2).

The most consequential segregation of cations that can be made by IP involves their reactions with water and likely fate in aqueous environments. Cations with IP <32 charge (+)/nm (approximate) are those that are hydrated and readily soluble in water as cations; cations with IP >75 charge (+)/nm are those that form oxyanions that are commonly soluble in water; and cations with intermediate IP generally hydrolyze to form insoluble oxides and hydroxides. Thus, cations on the lower right in figure B-2 are relatively mobile as cations, those on upper left are generally somewhat mobile as oxyanions, and those on in the middle are immobile. Notice that, according to this classification, iron is mobile in the reduced state, Fe(II), and immobile in the oxidized state, Fe(III), whereas chromium is immobile in the reduced state, Cr(III), and mobile in the oxidized state, Cr(VI). Likewise, predictions are good for some other cations with multiple oxidation states, such as molybdenum and vanadium.

The common cations, other than silicon, in ultramafic rocks are grouped together in figure B-2. They have coordination numbers of 4 or 6 and oxidation states of 2(+) as magnesium and iron in olivine and pyroxene and some iron in spinel group minerals or 3(+) as aluminum, chromium, and some iron in spinel group minerals. These minerals are common in rocks de-

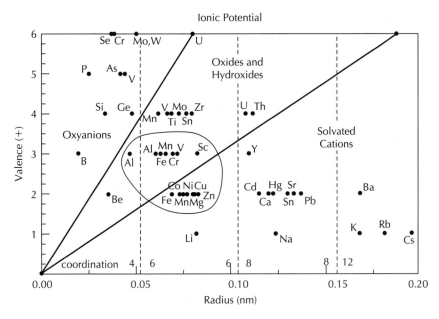

Figure B-2 Ionic potentials of geoecologically important cations, other than C, N, and S. Ionic potentials are ratios of ionic charge to radius, thus increasing from lower right to upper left. The diagonal lines represent somewhat arbitrary ionic potentials of 32 and 75 charge (+) per nanometer. A line with an IP value slightly lower than 75 would place U(VI) with the oxyanions, which is a more appropriate category for it. The vertical dashed lines are approximate divisions between cations that have 4-, 6-, 8-, and 12-fold coordination with oxygen atoms.

rived directly from the mantle. Cations with larger radii or higher charges are not compatible in olivine and orthopyroxenes (Appendix A). The only elements that are conspicuously out of place in the grouping with the cations of ultramafic rocks are copper and zinc, and they are on the margin of this grouping. These elements, plus cadmium and mercury, have full compliments of d-orbital electrons and are "soft" acids that associate with sulfur, rather than with oxygen. Aluminum and calcium are less depleted from primary ultramafic rocks (mainly peridotite) than most elements, other than the compatible ones, because aluminum is a major constituent of many spinel group minerals and calcium is a major constituent of clinopyroxene minerals. Calcium, however, is not compatible in serpentine minerals.

Chemical Elements Toxic to Living Organisms

Toxicities of cations, or metals, in geoecosystems can be related to their IPs (ionic charge/ionic radius) and to their tendencies to form covalent bonds. Electronegativity has been used in an index for covalent bond tendency (Nieboer and Richardson 1980), and alternatively, ionization potential has been used in a Lewis-acid softness index (Misono et al. 1967). A classification of metals similar to that of Nieboer and Richardson can be shown on a rectangular plot of IP versus a Lewis-acid softness index (Sposito 1989; fig. B-3). Cations

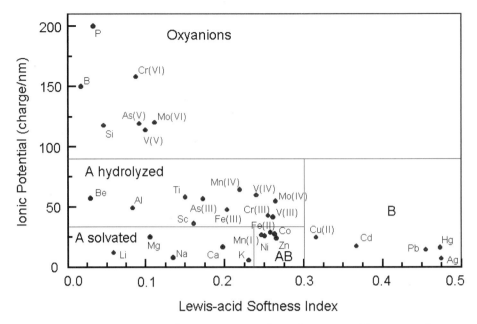

Figure B-3 A partitioning of cations among those that react with water to form oxyanions, those than hydrolyze to form insoluble oxides and hydroxides, those soluble cations that are hard Lewis-acids (A), those that are soft Lewis-acids (B), and those relatively soluble cations that are intermediate ("borderline") between hard and soft Lewis-acids. The Lewis-acid softness scale X is $X = [r(I_Z/I_{Z+1})]Z^{-0.5}$, where r is the ionic radius of a cation in Å units (Å = 10 nm), Z is the cationic charge, I_Z is the ionization potential for that charge, and I_{Z+1} is the ionization potential for the next higher charge of the cation (Misono et al. 1967).

with the greatest covalent tendencies are soft Lewis-acids (softness index >0.3) in class B and those with low Lewis softness indices are hard Lewis-acids in class A of Pearson (1963). Cations with intermediate indices were called borderline (AB). The class A and borderline cations can be separated into a solvated group of soluble cations with low IP, a hydrolyzed group of insoluble cations, and a group with high IPs that form oxyanions, as shown above. Class A cations are more reactive with hard Lewis-bases, and class B cations are more reactive with soft Lewis-bases. Organic ligands with oxygen and aliphatic nitrogen donors are considered to be hard, those with sulfur donors are deemed soft, and those with aromatic nitrogen donors are intermediate, or borderline. Class B (fig. B-3) metals, which are not among the more abundant ones in serpentine soils, are the most toxic, largely as a result of reactions with enzymes.

Some oxyanions, such as As(V) and Cr(VI), and some borderline and hydrolyzed metals are moderately toxic. Besides those that react with the functional groups of enzymes, some of these metals exhibit toxic effects by replacing metals that are essential, some promote the formation of reactive oxygen species (free radicals), and some affect plasma membranes, altering membrane permeability. Class A, solvated cations, are not toxic, or are toxic only in very high concentrations. Although the reactions of some elements are contradictory to predictions from this classification, it provides a helpful grouping of the chemical elements.

References

Goldschmidt, V.M. 1954. Geochemistry. Clarendon Press, Oxford.

Misono, M., E. Ochiai, Y. Saito, and Y. Yoneda. 1967. A dual parameter scale for the strength of Lewis acids and bases with the evaluation of their softness. Journal of Inorganic and Nuclear Chemistry 29: 2635–2691.

Nieboer, E. and D.H.S. Richardson. 1980. The replacement of the nondescript term "heavy metals" by a biologically and chemically significant classification of metal ions. Environmental Pollution B1: 3–26.

Pearson, R.G. 1963. Hard and soft acids and bases. American Chemical Society Journal 85: 3533–3539.

Shannon, R.D. 1976. Revised effective radii and systematic studies of interatomic distances in halides and chalcogenides. Acta Crystallographica A 32: 751–767.

Sposito, G. 1989. The Chemistry of Soils. Oxford University Press, New York.

Whittaker, E.J.W. and R. Muntus. 1970. Ionic radii for use in geochemistry. Geochimica et Cosmochimica Acta 34: 945–956.

APPENDIX C: SOIL CLASSIFICATION

Classification is necessary for efficient discourse about any set of populations in which each population has different kinds of individuals. A population of individual trees, for example, that can interbreed with each other, but not with individuals in other populations, is called a species. Similar species are grouped into genera, and similar genera are grouped into families, and so on. A classification system with many levels (or categories) is a hierarchical system.

Most soils do not have discrete "individuals," so the recognition of individual soils is somewhat arbitrary. In lieu of discrete individuals, 1–10 m^2 units of soil called *pedons* are described and classified, with an area >1 m^2 described only where necessary to include lateral variations in a soil (Soil Survey Staff 1999). It is important to concentrate on pedons in soil classification, because larger landscape units, such as landforms, generally contain more than one kind of soil, just as plant communities generally contain more than one species. The complexities of soil distribution in landscapes are dealt with in mapping rather than in classification.

Many countries have developed soil classification systems. Two systems are widely used around the world: the World Reference Base for Soil Resources (FAO/ISRIC/ISSS 1998) and the Soil Taxonomy of the U.S. Department of Agriculture (Soil Survey Staff 1999). This section introduces these systems to persons unfamiliar with the concepts and nomenclature of soil taxonomy. The Soil Taxonomy of the USDA is the more comprehensive system and is the primary one used in this book.

USDA Soil Taxonomy

Soil Taxonomy is a hierarchical system containing six levels, or categories—order, suborder, great group (or group), subgroup, family, and series. Phases of soil series are used in mapping, but they are not a part of Soil Taxonomy.

Table C-1 Diagnostic horizons of the USDA Soil Taxonomy.

Horizon	Explanatory phrase[a]
Surface layers, or Epipedons	
Anthropic	Evidence of human disturbance or high acid-soluble P_2O_5, otherwise similar to a mollic epipedon
Folic	Thick layer of well-drained organic material, generally 20 to 40 cm thick
Histic	Thick layer of poorly drained organic material, generally 20 to 40 cm thick
Melanic	Thick (thickness > 30 cm) black layer with much organic matter (carbon > 6%) and andic soil properties
Mollic	Thick dark-colored layer with base saturation > 50%, thickness > 25 cm, with thinner exceptions
Ochric	Thin or lighter-colored layer that lacks the diagnostic features of any other epipedons
Umbric	Thick dark colored layer with base stauration < 50%, thickness > 25 cm, with thinner exceptions
Subsurface horizons, or layers	
Albic	Light-colored eluvial horizon
Argillic	Illuvial accumulation of clay, generally with clay skins in tubular pores or on ped faces
Calcic	Illuvial accumulation of lime, or calcium carbonate
Cambic	Evidence of incipient soil development, such as higher chroma or soil structure
Duripan	Layer cemented by illuvial accumulation of silica
Gypsic	An illuvial layer of secondary gypsum accumulation
Kandic	Similar to an argillic horizon, but with low CEC
Natric	Similar to an argillic horizon, but with high exchangeable sodium content
Ortstein	A layer of cemented iron compounds or spodic materials
Oxic	Horizon of low weatherable mineral content and low CEC, lacking the illuvial clay content of a kandic horizon
Petrocalcic	Horizon cemented by illuvial accumulation of calcium carbonate
Petrogypsic	Horizon cemented by illuvial accumulation of gypsum
Placic	Thin hardpan cemented by iron compounds, and commonly manganese compounds, containing organic matter > 1%
Salic	Accumulation of salts more soluble than gypsum

(*continued*)

Table C-1 (*continued*)

Horizon	Explanatory phrase[a]
Sulfuric	Mineral or organic material, pH < 3.5 from oxidation of sulfides, horizon contains sulfates
Special materials, conditions, and soil properties	
Andic soil	Very high amorphous to poorly crystallized aluminum-silicate content, or high content and 5% or more volcanic glass in coarse silt and sand, bulk density < 0.9 Mg/m^3
Anhydrous conditions	Dry soil, dry more than one-half time, with mean annual soil Temperature < 0°C
Aquic conditions	Continuous or periodic saturation and reduction, generally recognizable by redoximorphic features such as mottles in a bleached matrix or nodules of iron and manganese
Organic soil materials	Organic carbon from 12% with no clay to 18% with clay > 60% in very poorly drained soils, or carbon > 20% in well drained soils

[a]Complete definitions are commonly complex (Soil Survey Staff 1999).

Currently there are 12 orders, dozens of suborders, hundreds of groups, thousands of subgroups, and probably tens of thousands of families and hundreds of thousands of soil series. There is no definite of count how many soil series have been mapped on earth.

Diagnostic Horizons

Diagnostic horizons are fundamental in Soil Taxonomy. They are precisely defined layers, whereas the horizons of pedon description, other than O-horizons, are not defined quantitatively. A diagnostic horizon incorporates one or more horizons recognized in a pedon description; for example, both Bt and Btk horizons in a particular pedon may be included in one argillic horizon. Diagnostic horizons occurring in surface soils are called epipedons. Epipedons and diagnostic subsurface horizons are listed in table C-1. Complete definitions of some of the diagnostic horizons are complex, and the details need not concern us here. More important is an appreciation of the pedological processes that have influenced the development of the diagnostic horizons (discussed below, and in chapter 7).

Epipedons

Root concentrations and animal activity are greatest in surface soils. Animals facilitate the decomposition of organic matter, and some of them carry organic matter from above-ground plant detritus down into soils. Root decay and incorporation of organic matter into soils by animals generally adds enough organic matter to surfaces soils to make them dark brown or dark grayish brown to black.

Different kinds of surface horizons, or epipedons, are recognized based largely on their color, concentration of organic matter, and thickness (table C-1). An organic surface horizon

20—40 cm thick resting on mineral (inorganic) soil is called a *folic* epipedon if it is well drained or a *histic* epipedon if it is saturated with water during the growing season. It can be thicker, up to 60 cm, if the organic matter is no more than slightly decomposed. A thick (thickness >30 cm) black horizon with concentrated organic matter (organic carbon >6%) and andic soil properties is called a *melanic* epipedon. Other inorganic, or mineral, surface layers are *mollic* or *umbric* epipedons if the dark-colored surface is thicker than 25 cm. Thinner mollic and umbric epipedons occur in shallow soils, but the thickness requirements are too complex to explain here (Soil Survey Staff 1999). The base saturation, based on cation-exchange capacity (CEC) at pH 7, is >50% in mollic epipedons and <50% in umbric epipedons—no other distinctions are made between them. Mineral surface layers that do not meet the requirements of mollic or umbric epipedons are called *ochric* epipedons.

Most serpentine soils in western North America have either ochric or mollic epipedons. No serpentine soils are expected to have andic soil properties and melanic epipedons, unless much volcanic ash has been deposited on them.

Subsurface Diagnostic Horizons, or Layers

There are many kinds of subsurface diagnostic horizons, resulting from many soil processes acting on different soil parent materials. These soil processes involve weathering, leaching, local transport of materials, accumulation of materials, and shrink–swell.

Weak expression of any of the soil processes is evidence for a *cambic* horizon. This evidence may be increased chroma (intensity of color) resulting from weathering that releases iron which then concentrates on surfaces or in spots and is oxidized, loss of carbonates by leaching, or development of blocky or prismatic structure by repeated shrinkage and swelling of soil. The expression, or grade, of blocky or prismatic structure may be increased by accumulation of illuvial (transported) clay on ped faces. When the increase of clay is at least 3% and at least 20% of the amount initially present in the parent material, and some of the clay is illuvial, the layer is called an *argillic* horizon. Two variations from the argillic horizon are recognized as diagnostic horizons: an argillic horizon in which sodium has accumulated is called a *natric* horizon, and one in which weathering has reduced the CEC (pH 7) of clay to <16 me/100 g is called a *kandic* horizon.

An extremely weathered layer with sparse evidence of illuvial clay is called an *oxic* horizon. The clay minerals in oxic horizons are predominantly kaolinite, gibbsite, goethite, and hematite. These are all low-activity (low CEC) clay minerals and iron oxides lacking the basic cations present in magnesium-chlorites, hydrous micas, and smectites that are common in less weathered soils.

The majority of subsurface diagnostic horizons are recognized by accumulations of materials, mostly illuvial accumulations. These horizons and their definitive materials include, along with nonilluvial cambic horizons, argillic horizons with clay, calcic horizons with calcium carbonate; gypsic horizons with calcium sulfate; natric horizons with clay and sodium; salic horizons with salts more soluble than gypsum; and spodic horizons with aluminum, iron, and organic matter accumulations. Few of these horizons are known to be present in serpentine soils. Where these horizons are cemented they are petrocalcic horizons with calcium carbonate, or calcite; duripans with silica in opal or chalcedony (cryptocrystalline quartz); ortstein with iron or iron–organic complexes; and placic horizons with iron–organic and aluminum–organic complexes. Placic horizons are thinner (thickness <25 mm) than ortstein layers, and the surfaces of placic horizons are generally wavy or convoluted. A subsurface root-restricting layer that is hard or very hard when dry, brittle when moist, and slakes in water is called a *fragipan*. Placic horizons, fragipans, and petrocalcic horizons are not known to be present in serpentine soils, at least not in western North America. Duripans are present below cambic and argillic horizons in some soils of the Klamath Mountains (Alexander 1995).

Soil Temperature and Moisture Regimes

Climate is a major factor in soil development, plant community dynamics, and soil management. Because soils are classified by soil properties rather than by external factors such as above ground climate, soil climate is used in Soil Taxonomy. Soil temperature and moisture regimes have been established to represent soil climate. Reliably ascertaining these regimes requires data collected over several years because differences from one year to another are often great.

Soil temperature regimes (STRs).

Differentiation of soil temperature regimes is based primarily on mean annual temperature at 50 cm depth, or at the soil–bedrock contact in shallow soils (table C-2). The regimes, from equator to poles, or to high mountains, are hyperthermic or hot tropical, thermic or warm temperate, mesic or cool temperate, frigid or cold temperate, cryic or very cold boreal (or very cold austral), and pergelic or frozen polar. The cryic regime is differentiated from the frigid regime by being colder during summer. Regimes where the means of the coldest and warmest months differ by < 5°C are given iso- prefixes: isohyperthermic, isothermic, isomesic, and isofrigid. All STRs from pergelic to hyperthermic are believed to be represented by serpentine soils of western North America, although none with a pergelic regime has been described.

Soil moisture regimes

Soil moisture regimes (SMRs) are primary differentiating features at the order level in one order and the suborder level in seven orders, or in nine orders when aquic conditions are considered along with SMRs. This reflects the great importance of soil moisture in soil development, plant distribution and productivity, and soil management.

Table C-2 Soil temperature regimes.

Subjective aspect	Soil temperature at 50 cm depth annual range[a]		
	Mean annual (°C)	Range >5°C	Range <5°C
Hot	T > 22	Hyperthermic	Isohyperthermic
Warm	15–22	Thermic	Isothermic
Cool	8–15	Mesic	Isomesic
Cold	0–8	Frigid	Isofrigid
Very cold	0–8	Cryic[b]	
Frozen	T < 0	Pergelic	

[a]Iso-regimes occur in soils with <5°C between means for the coldest and warmest months.
[b]Soils in cryic regimes are never warm, even in summer.

Soil moisture regimes are based on the moisture status of the moisture control section during growing seasons, or when the soil temperature is 5°C or greater at 50 cm depth. The soil moisture control section is the depth of wetting by 2.5–7.5 cm of water, which is between 10 and 30 cm deep in silty soils to about 50 to 150 cm or deeper in stony sandy soils. The moisture regimes are very dry, arid; dry, ustic; summer dry, xeric; moist, udic; always moist, perudic; and wet, aquic regime (table C-3). Aquic conditions, as interpreted by redoximorphic features, are used in defining soil classes, rather than the aquic soil moisture regime. The perudic regime is not used in differentiating soil classes.

Between the aridic regime of deserts and the udic regime of humid regions, are the ustic and xeric regimes of semi-arid to subhumid regions. Ustic soils are moist long enough during a growing season to produce many kinds of crops without irrigation. Xeric soils, which occur in Mediterranean climates, are too dry following the summer solstice to grow more than a few kinds of crops, mostly small grains, without irrigation. Soils with cryic STRs are generally not considered to have either ustic or xeric soil moisture regimes, because the growing season is too short (<90 days). And xeric soils, which are moist for at least 1.5 mo when soils are cold following the winter solstice, are not recognized in hyperthermic STRs where soils are practically never cold.

Most serpentine soils in western North America that do not have cryic STRs have xeric SMRs. Some serpentine soils along the Pacific coast in northwestern California and southwestern Oregon have udic SMRs. Serpentine soils in Baja California and Baja California Sur have aridic SMRs. Aquic conditions occur in some serpentine soils of western North America, but they are not extensive.

Order and Lower Levels of Soil Taxonomy

A person can become familiar with all taxa at the order level, because there are only 12 orders. The number of taxa increases so markedly with each step down the hierarchy, that it is not practical to attempt familiarity with every suborder, and it is virtually impossible to

Table C-3 Soil moisture regimes.

Moisture regime	Definitive characteristics in moisture control section[a]
Aquic	Saturated with water in growing season long enough to form redoximorphic features
Aridic	Dry throughout for more than one-half of growing season and no consecutive 90 days in growing season when soils are moist in some part
Udic	Not dry in any part for any 90 days in growing season
Ustic	Dry in some part for any 90 days in growing season
Xeric	Dry throughout for at least 45 consecutive days in 4 m following summer solistice

[a]Depth of the soil moisture control section is about 10–30 cm in fine–silty soils and ranges to 50–100 cm, or more, in sandy soils that hold only 5%, or less, plant-available water.

become familiar with every subgroup. However, many pedologists are familiar with, or cognizant of, most of the diagnostic soil horizons and other features that are used to differentiate and characterize the classes. These features are so closely linked to the taxonomic names that an experienced pedologist can readily form a conceptual construction of a soil and infer many of its properties and characteristics based simply on the name of the subgroup or family. There is no need to remember every class of soil to know something about each class of soil.

Orders of Soil Taxonomy

The orders of Soil Taxonomy are designed to reflect processes of soil development. Only the Histosols are restricted to a single soil parent material, organic material, although most Andisols develop in volcanic parent materials. Climate, or soil climate, appears at the order level only in definitions of Aridisols and Gelisols. Soil temperature and moisture regimes are important at the suborder and lower levels of Soil Taxonomy.

Soils in initial phases of soil development, lacking diagnostic horizons other than an ochric epipedon, are Entisols. Vertisols may lack diagnostic epipedons, too, but because they are mixed by shrinkage and swelling, rather than because they are in initial phases of soil development. Soil development in Entisols may lead directly to Aridisols, Mollisols, Andisols, Spodosols, or Inceptisols with mollic, umbric, melanic, or histic epipedons or cambic, calcic, gypsic, salic, or placic horizons or fragipans. Inceptisols are generally intermediate in a succession from Entisols to Alfisols or Ultisols with argillic horizons. Aridisols and Mollisols may have argillic horizons, too, and many other kinds of diagnostic horizons. Both Aridisols and Gelisols may have many different kinds of diagnostic horizons because they are defined by dry or freezing climates, rather than by soil development, although Aridisols must have some diagnostic horizons that are indicative of pedological development. There are many alternative paths in the long (on the order of 10^5 or 10^6 years) succession from Entisols lacking soil development to Ultisols with kandic horizons and Oxisols with oxic horizons.

Suborders, Groups, and Subgroups

Soils in each subgroup have the properties required of the great group, suborder, and order above it in the hierarchy of Soil Taxonomy. Naming a taxon consists of beginning with an order ending (table C-4), adding on the left formative elements for a suborder and a great group (table C-5), and attaching an adjective for a subgroup (table C-6).

As an example, consider a warm, summer dry soil with an argillic horizon, a mollic epipedon, and a shallow hard bedrock contact. We find in the key to orders (Soil Survey Staff 1999) that this soil is an Mollisol with a formative ending "oll." In the key to Mollisols, we find that the soil is a Xeroll (formative element "xer" + "oll") because it has a xeric SMR; in the Argixeroll group because it has an argillic horizon; and in a Lithic Argixeroll subgroup because it is shallow and rests on hard bedrock. If we had only the name, we could infer everything about the soil that we needed to know to classify it. It must have a mollic epipedon because it is a Mollisol; it must be dry during summers because it is a Xeroll, and so on. Thus, there is considerable information about a soil in its taxonomic name, and we do not have to have previous knowledge about a particular soil to infer what it is like. We can infer, without guessing, that the Lithic Argixeroll has a subsoil with illuvial clay accumulation (argillic horizon) and is shallow to hard bedrock.

Table C-4 Orders of Soil Taxonomy and properties or characteristics that define them.

Order	Formative ending	Special characteristics
Alfisols	-alf	Argillic (or natric) horizon with base saturation (pH 8.2) > 35%
Andisols	-and	Andic properties and no spodic horizon
Aridisols	-id	Dry soil climate (aridic soil moisture regime) and some soil development
Entisols	-ent	Lack of soil development
Gelisols	-gel	Permafrost within 1 or 2 m depth
Histosols	-ist	Organic material > 40 or 60 cm thickness
Inceptisols	-ept	Incipient soil development
Mollisols	-oll	Mollic epipedon and high base saturation (>50% at pH 7) in subsoil as well as in epipedon
Oxisols	-ox	Highly weathered soil without argillic (or kandic) horizon
Spodosols	-od	Spodic horizon
Ultisols	-ult	Argillic (or kandic) horizon with base saturation (pH 8.2) <35%
Vertisols	-ert	Cracking clay soils mixed by shrink-swell

Soil Families and Series

The family level is the most complex in Soil Taxonomy. A family name consists of the subgroup name plus terms for a textural class, a mineralogical class, and an STR. A clay activity class is added for soils with mixed mineralogy classes. For example, a Mollic Haploxeralf might be in a fine-loamy, mixed, superactive, frigid family of that subgroup. Mixed is a mineralogy class in which the silt and sand fractions are not dominated by any one kind of mineral; and superactive is an activity class in which clay is very active (CEC >60 me/100 g of clay). Most serpentine soils are in the "magnesic" mineralogy class and lack a clay activity class designation in the family name. Some soils have soil reaction (pH) classes in addition to other kinds of classes. All of the detail about families is in Soil Taxonomy if one wants to learn more about the designation of soil families.

All families contain at least one soil series, and some families have several. Families are most commonly differentiated into soil series by soil depth class, but there are many other soil features that are used to differentiate series within a family. Any soil property or characteristic that might be important in soil use and management can be used to separate soil series. Soil series are given local names from locations where they are found; for example, the San Joaquin Series is mapped in the San Joaquin Valley. Fortunately, series names stand alone, rather than being added to the commonly very long family names. Examples of some serpentine soils in the California Region are given in table C-7.

In mapping, slope phases and commonly other phases, such as erosion, are added to a series. These phases are not considered to be part of Soil Taxonomy. Many phases indicate landscape features that are not strictly soil properties or characteristics. They are mapped for use in land use planning and management.

Table C-5 Some formative elements used in naming suborders and great groups.

Element	Origin	Brief explanation of meaning in Soil Taxonomy
Acr	Gr. *akros*, at the end	Weathered completely
Agr	L. *ager*, field	Having an agric horizon
Al	*Aluminium* or *aluminum*	Aluminum present, without iron
Alb	L. *albus*, white	Bleached by leaching of iron
Anhy	Gr. *anydros*, waterless	Anhydrous conditions
Anthr	Gr. *anthropos*, man	Having an anthropic epipedon
Aqu	L. *aqua*, water	Wet long enough to form redoximorphic features
Arg	L. *argilla*, clay	Having an argillic horizon
Bor	Gr. *boreas*, northern	A cold soil
Calc	L. *calcis*, lime	Having a calcic horizon
Camb	L.L. cambiare, to exchange	Having a cambic horizon
Cry	Gr. *kryos*, icy cold	Cryic soil temperature regime
Dur	L. *durus*, hard	A duripan present
Dys, dystr	Gr. *dys*, ill or bad	Low basic cation content
Endo	Gr. *endon*, within	Saturation from below
Epi	Gr. *epi*, at, on, over	Saturation from above, as in a perched water table
Eu	Gr. *eu*, good or well	High basic cation status
Ferr	L. *ferrum*, iron	Iron-cemented nodules present
Fibr	L. *fibra*, fiber	Organic material with a high fiber content
Fluv	L. *fluvus*, river	Fluvial stratification that affects vertical sequence of soil properties
Fol	L. *folia*, leaf	Well-drained organic soil formed by accumulation of leaves
Frag	L. *fragilis*, brittle	A fragipan present
Gloss	Gr. *glossa*, tongue	A glossic horizon present
Gyps	L. *gypsum*, gypsum	A gypsic horizon present
Hal	Gr. *hals*, salt	High exchangeable sodium concentration
Hapl	Gr. *haplous*, simple	No special features
Hem	Gr. *hemi*, half	Organic material with moderate fiber content
Hum	L. *humus*, earth	High humus, or organic matter, concentration
Hydr	Gr. *hydor*, water	Continuously under water
Kandi	*Kandite*, kaolinite group	A kandic horizon present
Kanhapl	Kan(di) + hapl	Lacking depth requirements of a kandic horizon
Luv	Gr. *louo*, to wash	Contains humilluvic material, illuviated organic matter

Table C-5 (*continued*)

Element	Origin	Brief explanation of meaning in Soil Taxonomy
Med	L. *media*, middle	Moderate temperature, neither hot tropical nor very cold
Melan	Gr. *melas*, black	A melanic epipedon present
Natr	Gr. *nitron*, niter	A natric horizon present
Ochr	Gr. *ochros*, pale	An ochric epipedon present
Orth	Gr. *orthos*, true	No special features
Pale	Gr. *paleos*, old	Advanced soil development
Per	L. *per*, thorougly	A perudic soil moisture regime (SMR)
Petr	Gr. *petra*, rock	A hardpan present
Plac	Gr. *plax*, flat stone	A placic horizon present
Plinth	Gr. *plinthos*, brick	Plinthite present
Psamm	Gr. *psammos*, sand	Sandy
Quartz	Ger. *quarz*, quartz	Very high quartz content in sand fractions
Rend	Rus. *rendzina*, lime-rich soil	Calcium carbonate in a mollic epipedon or just below it
Rhod	Gr. *rhodon*, rose	Reddish colors of high chroma and a dry value nearly equal to the moist value
Sal	L. *sal*, salt	A salic horizon present
Sapr	Gr. *sapros*, rotten	Decomposed organic material with a low fiber content
Sphag	Gr. *sphagnos*, bog	Fibric organic material containing mostly *Sphagnum* spp.
Sulfi	*Sulfide*	Sulfidic materials present
Sulfo	L. *sulfur*, sulfur	A sulfuric or a sufidic horizon present
Torr	L. *torridus*, hot and dry	Very dry soil
Trop	Gr. *tropikos*, of the solstice	Equable soil temperatures, lacking annual extremes
Ud	L. *udus*, humid	Soil moist through most of growing season, udic SMR
Umbr	L. *umbra*, shade	Presence of umbric epipedon
Ust	L. *ustus*, burnt	Soil dry during some of growing season, ustic SMR
Verm	L. *vermes*, worm	Abundant worm holes or worm casts or filled animal burrows
Xer	Gr. *xeros*, dry	Soil dry during summer, xeric SMR

Table C-6 Subgroup adjectives preceding Great Group names.

Subgroup name	Origin	Meaning
Abruptic	L. *abruptum*, turn off	Abrupt texture change, vertically
Aeric	Gr. *aer*, air	Aerated ephemerally saturated soil
Albic	L. *albus*, white	With an albic horizon
Anthropic	Gr. *anthropos*, man	With an anthropic epopedon
Aquic	L. *aqua*, water	Ephemerally saturated soil
Arenic	L. *arena*, sand	Sandy epipedon
Argic	L. *argilla*, clay	An argillic horizon
Aridic	L. *aridus*, dry	An aridic soil moisture regime (SMR)
Calci	L. *calcis*, lime	A calcic horizon
Chromic	Gr. *chroma*, color	High color value
Cumulic	L. *cumulus*, heap	Thickened epipedon
Duric	L. *durus*, hard	Silica cementation
Dystric	Gr. *dys*, bad; *trophia*, nutrition	
Entic	Recent	Minimal development
Eutric	Gr. *eu*, good; *trophia*, nutrition	
Fibric	L. *fibra*, fiber	
Glossic	Gr. *glossa*, tongue	With a glossic horizon
Gypsic	L. *gypsum*, gypsum	
Halic	Gr. *hals*, salt	
Haplic	Gr. *haplous*, simple	
Hemic	Gr. *hemi*, half	
Histic	Gr. *histos*, tissue	With a histic epipedon
Humic	L. *humus*, earth	
Hydric	Gr. *hydor*, water	Continually saturated with water
Leptic	Gr. *leptos*, thin	Thin soil, moderately deep
Lithic	Gr. *lithos*, stone	A shallow lithic contact
Mollic	L. *mollis*, soft	Some characteristics of a mollic epipedon
Natric	Sp. *natron*, Na-carbonate	With a natric horizon, containing sodium
Oxic	*Oxygen*	
Oxyaquic	Oxy + aquic	Saturation without appreciable reduction
Pachic	Gr. *pachys*, thick	Thickened epipedon on slopes
Petrocalcic	Gr. *petra*, rock; L. *calcis*, lime	A petrocalcic horizon

Table C-6 (continued)

Subgroup name	Origin	Meaning
Petroferric	Gr. *petra*, rock; L. *ferrum*, iron	Indurated by cementation with iron compounds
Petronodic	Gr. *petra*, rock; L. *nodus*, a knot	Indurated nodules (L. *nodulus*, a little knot)
Plinthic	Gr. *plinthos*, brick	Plinthite, iron concentrations that harden on drying
Psammentic	Gr. *psammos*, sand	Sandy soil
Rhodic	Gr. *rhodon*, rose	
Ruptic-	L. *ruptum*, broken	Irregular thickness or discontinuous
Salidic	L. *sal*, salt	Salty soil, with a salic horizon
Sapric	Gr. *sapros*, rotten	Well-decomposed organic material
Sodic	sodium	High sodium concentration
Spodic	Gr. *spodos*, wood ash	
Sulfic	*Sulfide*	
Sulfuric	*Sulfuric* (acid)	
Terric	L. *terra*, earth	A mineral substratum in an organic soil
Typic	L. *typicus*, Gr. *typikos*, typical	
Udic	L. *udus*, humid	
Umbric	L. *umbra*, shade	
Vermic	L. *vermes*, worm	
Vertic	L. *verto*, turn	Cracking clay, homogenized vertically
Xanthic	Gr. *xanthos*, yellow	Yellowish color
Xeric	Gr. *xeros*, dry	Nearly dry enough for a xeric SMR

World Reference Base for Soil Resources

The World Reference Base for Soil Resources is a development from the Legend of the Soil Map of the World (FAO-UNESCO-ISRIC 1988). It is a cooperative effort among the International Society of Soil Science (ISSS), the International Soil Reference and Information Centre (ISRIC) in Wageningen, and the Food and Agriculture Organization of the United Nations (FAO) and is cited accordingly (FAO/ISRIC/ISSS 1998). The ISSS has been superceded by the International Union of Soil Science (IUSS).

The World Reference Base (WRB) has only two levels, or categories. The first level has 30 major soil groups. At this level, classes are differentiated by soil properties that are indicative of the primary pedogenic processes. Some of the major soil groups correspond closely to orders of Soil Taxonomy. A major difference is that the WRB has no soil temperature or moisture regimes; therefore there is no equivalent of Aridisols. The WRB has major soil groups for

Table C-7 Some extensively mapped serpentine soils and counties in California and southwestern Oregon that have been mapped as soil series (S) or families (F).[a]

	Soil series (S) and families (F)												
State/county	Beau	Clim	Corn	Duba	Eigh	Grav	Henn	Mont	Pear	Serp	Toad	Waln	Weitc
California													
Alameda							S	S					
Amador							S						
Butte													
Calaveras[b]													
Colusa							S						
Del Norte												F	F
El Dorado[b]													
Fresno[c]													
Glenn				S			S	S					F
Humboldt							S					F	F
Kings[c]							S						
Lake				S			S	S					F
Kern								S					
Marin							S	S					
Mariposa							S						
Mendocino				S			S	S					F
Monterey	S	S					S	S					

454

Napa					S		S
Nevada	S				S		S
Placer	S				S		S
Plumas							
San Benito		S			S		S
S. Luis Obispo				S	S		S
San Mateo		S			S		
Sta. Barbara		S			S		
Santa Clara					S		S
Shasta	F				S	F	
Sierra					S		
Siskiyou	F		S		S	F	
Sonoma					S		S
Tehama	F				S		S
Trinity	F					F	
Tulare							
Tuolumne[b]	S				S		
Yuba					S		
Oregon							
Coos							S
Curry			S		S		S
Douglas			S		S		S

(*continued*)

Table C-7 (continued)

State/county	Soil series (S) and families (F)											
	Beau	Clim	Corn	Duba	Eigh	Grav	Henn	Mont	Pear	Serp	Toad	Waln	Weitc
Jackson				S		S	S		S				
Josephine			S		S				S				

^aSoil series are mapped in areas of more intensive inventory and families in areas of less intensive inventory. Soil subgroup classifications: Alfisols—Cornutt (Corn) and Walnett (Waln) (Ultic Haploxeralfs), Dubakella (Duba) (Mollic Haploxeralfs), and Toadlake (Toad) (Typic Haploxeralfs); Inceptisols—Eightlar (Eigh) and Weitchpec (Weitc) (Typic Haploxerepts), Gravecreek (Grav) (Dystric Haploxerepts), Pearsoll (Pear) (Lithic Haploxerepts), and Serpentano (Serp) (Dystric Eutrudepts); Mollisols—Beaughton (Beau) and Henmeke (Henn) (Lithic Argixerolls), and Montara (Mont) (Lithic Haploxerolls); and Vertisols—Climara (Clim) (Aridic Haploxererts).

^bSoils of the Delpiedra Series, shallow Mollic Haploxeralfs, were the main serpentine soils mapped in the foothills of the Sierra Nevada in several California counties.

^cSerpentine soils are present only on the western edge of Kings county, in the California Coast Ranges, and serpentine soils present on the western edges of San Joaquin and Stanislaus counties may be similar.

Table C-8 Major soil groups of the World Reference Base for Soil Resources.

Soil group	Major features	Corresponding classes in Soil Taxonomy
Histosols	Organic soils	Histosols
Cryosols	Frozen soils	Gelisols
Anthrosols	Anthropogenically modified soils	Anthrepts, Anthracambids, etc.
Leptosols	Shallow soils (depth < 25 cm) or fragmental soils	Some Entisols, Aridisols, Alfisols, and Mollisols
Vertisols	Cracking-clay soils	Vertisols
Fluvisols	Soils in stratified sediments, lacking cambic, argic (argillic), and natric horizons	Some Entisols, Aridisols, Inceptisols, Mollisols, and Spodosols
Solonchaks	Salty soils	Salids (suborder of Aridisols)
Gleysols	Wet soils with high groundwater, lacking argic (argillic) and natric horizons	Some Entisols, Inceptisols, and Mollisols
Andosols	Black soils with amorphous aluminosilicates (andic horizons)	Andosols
Podzols	Soils with spodic horizons	Spodosols
Plinthosols	With soft reddish Fe-rich concentrations that can hardened to plinthite or petro-plinthite upon repeated wetting and drying	Many Ultisols and some poorly drained (epiaquic conditions) Alfisols and Oxisols
Ferralsols	Containing a ferralic (or oxic) horizon	Oxisols
Solonetz	Soils with natric horizons	Natralbolls, Natrargids, Natricryolls, Natrixeralfs, Natrixerolls, Natrudolls, etc.
Planosols	An ephemerally wet eluvic horizon with an abrupt clay increase to an argic horizon	Albaqualfs, Albaquults, Argialbolls
Chernozems	Mollic epipedon, thickness > 50 cm, and lime concentrations above 125 cm depth	Many Mollisols, especially Ustolls and Xerolls
Kastanozems	Mollic epipedon, thickness < 50 cm, and lime concentrations above 125 cm depth	Many Mollisols, especially Calciustolls and Calcixerolls

(continued)

Table C-8 (*continued*)

Soil group	Major features	Corresponding classes in Soil Taxonomy
Phaeozems	Mollic epipedon and no free lime	Many Mollisols, especially Udolls
Gypsisols	Soils with gypsic or petrogypsic horisons	Gypsids, Gypsiargids, Gypsiusterts, etc.
Durisols	Soils with duric or petroduric horizons	Durids, Durixeralfs, Durixerolls, etc.
Calcisols	Soils with calcic or petrocalcic horizons	Calcids, Calcixerolls, Calixerepts, etc.
Albeluvisols	Soils with albic horizons and albeluvic tonguing (glossic horizon)	Glossocryalfs, Fraglossudalfs, etc.
Alisols	Argic horizons with somewhat active clay (cation–exchange capacity [CEC] > 24 mol/kg) highly Al saturated	Some Ultisols
Nitisols	Soils with argic (argillic) and nitic (blocky structured) horizons	Some Kandiudox and some Kandiudults
Acrisols	Soils with subactive clay (low CEC) in argic horizons and low base saturation	Some Udults, Ustults, Kandihumults, etc.
Luvisols	Soils with more active clay (CEC > 0.24 mol/kg) in argic horizons	Haplocryalfs, Haploxeralfs, etc.
Lixisols	Soils with subactive clay (low CEC) in argic horizons and high base saturation	Some Palexeralfs, etc.
Umbrisols	Soils with an umbric horizon (epipedon) lacking other diagnostic horizons except possibly a cambic horizon	Humic Haploxerepts, Humic Eutrocryepts, Humic Dystrocryepts, etc.
Cambisols	Soils lacking diagnostic horizons other than a cambic horizon	Typic Haploxerepts, Typic Eutrocryepets, Typic Dystrocryepts, etc.
Arenosols	Sandy soils lacking diagnostic horizons	Psamments
Regosols	Nonsandy soils lacking diagnostic horizons	Orthents

Soils are in the order of a key. A soil is classified as the first from the top that fits.

shallow or fragmental soils (Leptosols, depth <25 cm or coarse rock fragments >90%) and wet soils, which have no corresponding orders in Soil Taxonomy. Soils wetted by high groundwater are Gleysols and those with perched water tables that are ephemerally wet are Stagnosols (discontinued soil group, FAO/ISRIC/ISSS 1998). These soils may be represented in several different orders of Soil Taxonomy.

Soil units consist of the name of a major soil group plus one or more adjectives; for example Mollic Leptosol for a shallow soil with a mollic horizon. The mollic horizon of the WRB is comparable to a mollic epipedon of Soil Taxonomy.

Most major soil groups are characterized in table C-8. Some groups, such as Anthrosols and Nitisols, that are not known to have serpentine soils are omitted from the table.

References

FAO/ISRIC/ISSS. 1998. World Reference Base for Soil Resources. World Soil Resources Report no. 84. Food and Agricultural Organization of the United Nations, Rome.

FAO-UNESCO-ISRIC. 1988. Revised Legend for Soil Map of the World. World Soil Resources Report no. 60. Food and Agricultural Organization of the United Nations, Rome.

Soil Survey Staff. 1999. Soil Taxonomy—a Basic System for Making and Interpreting Soil Surveys. USDA, Agriculture Handbook No. 436. U.S. Government Printing Office, Washington, DC.

APPENDIX D: KINGDOMS OF LIFE

When biological classification was first developed, there were only two kingdoms of life—plants and animals. Now up to eight or more kingdoms are recognized by different authorities. The five-kingdom scheme of Margulis and Schwartz (1998) is followed in this book (table D-1). It is practically a six-kingdom scheme because two subkingdoms are recognized in the Prokaryotae Kingdom: Archaea (or Archaebacteria) and the Eubacteria. The most controversial kingdom is the Protoctista Kingdom, which is a catch-all for phyla that do not fit into other kingdoms. This controversy is not important for our purposes.

The main level of classification below the kingdom is the phylum. Within the plant kingdom, there is a major distinction between the kingdom and phyla levels that is very important in determining the functions of plants in geoecosystems. Vascular plants with lignified conducting systems and roots can have long stems, and their roots can extract water and nutrients from deep into soils. Nonvascular plants, which lack these features, are much more dependent on dust and precipitation for sustenance. They are all very small compared to trees and shrubs. They dry up when the atmosphere and ground surface dry out, while vascular plants remain active from stored water or from water that their roots extract from soils. Margulis and Schwartz (1998) call the nonvascular plants Bryata and vascular plant Tracheata (table D-1). The Bryata include mosses and liverworts, and the Tracheata include club mosses, horsetails, ferns, conifers, and flowering plants.

Microbiologists commonly call the prokaryote phyla of Margulis and Schwartz (1998) "kingdoms" (Madigan et al. 2000) based on genetic differences, whereas Margulis and Schwartz focus more on morphological differences. Madigan et al. (2000) recognize three kingdoms in the Archaebacteria, which they call Archaea, and 14 kingdoms in the Eubacteria, which they call Bacteria. Divisions of the Protoctists are somewhat arbitrary.

Table D-1 Some geoecologically important phyla in the kingdoms of life.

Kingdom or Subkingdom	Division	Phylum
Archaebacteria		Euryarchaeota (methanogens and halophils)
		Crenarchaeota (thermoacidophils)
Eubacteria	Gracilicutes, gram (−)	Proteobacteria (*Azotobacter, Desulfovirio, Nitrobacter, Nitrosomonas, Rhizobium, Thiobacillus*, etc.)
		Cyanobacteria (*Anabaena, Nostoc*, etc.)
	Tenericutes, wall-less	Aphragmabacteria (*Mycoplasma*, etc.)
	Firmicutes, gram (+)	Endospora (*Bacillus*, etc.)
		Actinobacteria (*Actinomyces, Frankia, Streptomyces*, etc.) including Coryneforms (Arthrobacter, Corynebacteria)
Protoctista	Protozoa	Euglenozoa (Euglena and many other flagellates)
		Ciliophora (Paramecium, Vorticella, and many other ciliates)
		Rhizopoda (*Amoeba* and other amastigote amebas)
	Slime molds	Acrasiomycotes (cellular slime molds)
		Myxomycota (plasmodial slime molds)
	Water molds	Oomycota (Phytophora and many other molds)
	Algae	Diatoms, or Bacillariophyta (*Navicula, Pinnularia*, etc.)
		Chlorophyta (green algae)
		Chrysophyta (golden-brown algae)
		Xanthophyta (yellow algae)
Animalia		Nematoda (thread worms or roundworms)
		Annelida (earthworms, potworms, etc.)
		Mollusca (snails, slugs, etc.)
		Arthropoda
		Chelicerata (spiders, etc.)
		Mandibulata (insects, centipedes, millipedes, etc.)
		Crustacea (crayfish, woodlice, etc.)
		Urochordata (ascidian tunicates, or sea squirts, etc.)
		Craniata (reptiles, amphibians, mammals, etc.)

(continued)

Table D-1 (*continued*)

Kingdom or Subkingdom	Division	Phylum
Fungi		Zygomycota (includes arbuscular mycorrhiza)
		Basidiomycota (includes arbutoid mycorrhiza and ectomycorrhiza)
		Ascomycota (includes ericoid mycorrhiza and most fungi of lichens)
Plantae	(Bryata)	Bryophyta (mosses)
		Hepatophyta (liverworts)
		Anthocerophyta (hornworts)
	(Tracheata)	Lycophyta (*Lycopodium*, *Selaginella*, etc.)
		Sphenophyta, horsetails (*Equisetum*)
		Filicinophyta, ferns (*Aspidotis*, *Pteridium*, etc.)
		Coniferophyta, conifers (*Calocedrus*, *Pinus*, *Pseudotsuga*, etc.)
		Gnetophyta (*Ephedra*, etc.)
		Anthophyta, flowering plants (*Ceanothus*, *Festuca*, *Zigadenus*, etc.)

References

Madigan, M.T., J.M. Martinko, and J. Parker. 2000. Brock Biology of Microoganisms. Prentice Hall, Upper Saddle River, NJ.

Margulis, L. and K.V. Schwartz. 1998. Five Kingdoms. Freeman, New York.

APPENDIX E: WATER BALANCE BY THE THORNTHWAITE METHOD

The Thornthwaite method for computing water budgets has been widely used for decades. Only simple mean monthly temperature and precipitation diagrams are used to represent climates in the serpentine domains of western North America. How well do these diagrams represent water balances? A temperature—precipitation diagram is compared to a water budget computed by the Thornthwaite method in order to examine this question.

The water balance at Auburn (394 m above sea level [asl]) in the foothills of the Sierra Nevada (domain 2) is illustrated in figure E-1. It is a site with cool, relatively wet winters and hot, dry summers. The water balance was calculated, following Thornthwaite (1948), from mean monthly temperature and precipitation for 24 years, from 1951–1974 (National Climatic Center, Asheville, NC, 1977). The actual evapotranspiration (AET) depends on the amount of water available, as well as on the climate. Figure E-1A, B shows the results for soils holding 100 and 300 mm of water that is available for evapotranspiration. Major (1977) assumed 100 mm of storage in his water balance diagrams, and Thornthwaite (1948) assumed 300 mm. A storage capacity, or available-water capacity (AWC), of about 100–150 mm is reasonable for moderately deep, stony soils, but only some very deep soils store 300 mm of water that is available to plants for evapotranspiration. Auburn is typical of the California Region in that most of the precipitation falls during winter. There is an excess, or surplus, of water at the end of winter, which is indicated by precipitation (P) greater than potential evaporatranspiration (PET) through April. Water used, or evaporated, is represented by the space below the actual evapotranspiration (AET) lines in the figures. As long as precipitation exceeds PET, AET equals PET. Once PET exceeds precipitation, which is from May until October at Auburn, evapotranspiration begins to deplete the store of water in soils. Thornthwaite (1948) has assumed, based partly on principle and partly on evidence, that the rate of depletion will be proportional to the amount of water available for extraction from soil. Therefore, as soil becomes drier during summer, when there is little precipitation, evapotranspiration decreases gradually, rather than proceeding at the potential rate to deplete the stored water quickly. When the AWC is

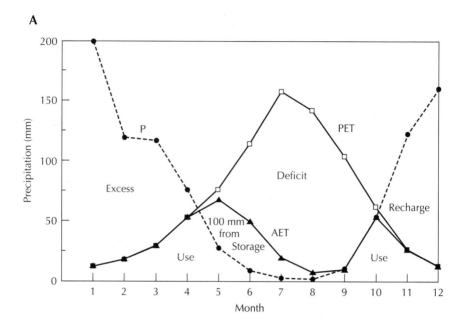

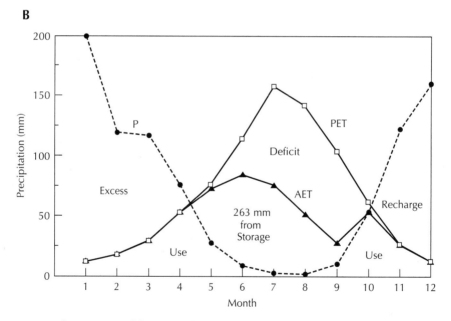

Figure E-1 The mean monthly water balance for a year at 394 m altitude and 38.9°N latitude in Auburn on the lower west slope of the Sierra Nevada. Potential evapotranspiration (PET) and actual evapotranspiration (AET) were calculated by Thornthwaite's method. (A) Calculated AET assuming 100 mm of plant available water storage capacity (AWC). (B) Calculated AET assuming 300 mm of plant AWC. (C) The relationship of PET (mm) to temperature (°C) on a scale of 10°C = 20 mm.

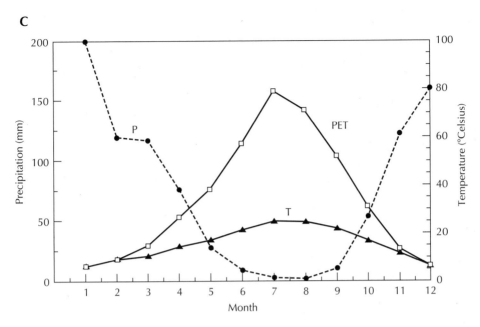

Figure E-1 (*continued*)

100 mm, the soil at Auburn becomes very dry in August and is essentially devoid of available water in September. When the AWC is 300 mm, the soil still retains 37 mm of available water at the end of August, but that water will be below the level of most of the roots. Only deeply rooted plants will be benefit from any water that is available in a soil at Auburn late in summer. When 100 mm of stored water is lost from the soil during summer, it takes until December to replenish it, or recharge the soil. With greater storage capacity and nearly 300 mm water loss, it takes until January to replenish it at Auburn. The difference between PET and AET, which is the amount of water that would be used if plenty were available, is commonly called the deficit. It is a crude indicator of the amount of drought stress on a plant community because the amount of energy required to extract water increases as the concentration of soil water decreases.

Areas below the AET lines in figure E-1 represent the amounts of water used by plants. At Auburn, it is 359 mm per year from a soil with an AWC of 100 mm and 509 mm per year from a soil with an AWC of 300 mm. These numbers are good indicators of the relative productivities of natural plant communities on fertile soils because plant production increases in proportion to water use by evapoptranspiration. Water use, or actual evapotranspiration is not such a good index of productivity on serpentine soils because nutritional imbalances commonly limit production.

Mean monthly temperatures and precipitation are sometimes plotted on the same diagram, with a scale of $10°C = 20$ mm (Walter 1974), as a crude indication of when plants are likely to be stressed by drought. At Auburn, the temperature line is above the precipitation line over about the same span of time that a water deficit is indicated by Thornthwaite's method, but the temperature line is far below a PET line during the summer (fig. E-1C). Thus temperature and precipitation plots, as represented by figure E-1C, are conservative indicators of seasonal drought. They greatly underestimate potential evapotranspiration in hot, dry weather, but not in cool, wet weather (fig. E-1C).

References

Major, J. 1977. California climate in relation to vegetation. Pages 11–75 *in* M.G. Barbour and J. Major (eds.), Terrestrial Vegetation of California. Wiley, New York.

Thornthwaite, C.W. 1948. An approach toward a rational classification of climate. Geographical Review 38: 55–94.

Walter, H. 1974. Vegetation of the Earth in Relation to Climate and the Ecophysiological Conditions. Springer-Verlag, New York.

APPENDIX F: WESTERN NORTH AMERICAN PROTECTED AREAS WITH SERPENTINE

Many serpentine areas in western North America are entirely or partly on publicly managed land. Table F-1 represents a selection of publicly accessible lands throughout western North America with serpentine substrate and associated serpentine biota. These areas have some level of protected status and are not subject to multiple uses such as mining and logging. The list is arranged by the domains and the localities within them—domains and localities described in part IV. Some general characteristics or distinctive features are listed for each area. Contact information is limited to the agency or organization managing the area. For further information on exact location and any appropriate steps that may be necessary to obtain access permission, the organization listed in the table may be contacted. Some wilderness areas with extensive areas of serpentine, such as the Kalmiopsis Wilderness on the Siskiyou National Forest, are not listed in table F-1.

Table F-1 Protected areas in the ten serpentine domains of western North America that have some serpentine within their boundaries.

Domain	Locality	Name of reserve or protected area	Contact	Features
1 Baja California	1-2 Puerto Nuevo, 1-3 Calmalli, 1-4 Magdalena–Margarita	El Vizcaino Biosphere Reserve	Direccion General de Aprovechamiento Ecologico de los Recursos Naturales	Serpentine desert
	1-5 Los Pinos	Guatay Mountain RNA	USFS, Cleveland NF	Gabbro endemics
	1-5 Los Pinos	King Creek RNA	USFS, Cleveland NF	Gabbro endemics
	1-5 Los Pinos	McGinty Mountain ER	USFS, Cleveland NF	Gabbro endemics
	1-5 Los Pinos	Organ Valley RNA	USFS, Cleveland NF	Gabbro endemics
2 Sierra Mother Lode	2-1 Feather River	Butterfly Valley Botanical SIA	USFS, Plumas NF	Serpentine fen, pitcher plants
	2-2 Jarbo Gap	Green Island Lake (recommended RNA)	USFS, Plumas NF	Mountain serpentine vegetation
	2-5 Pine Hill	Pine Hill Preserve	CDFG and BLM, Folsom FO	Rare plants and gabbro plant communities
	2-7 Tuolumne	Peoria Wildlife Area	US Bureau of Reclamation	Jepson onion, *Allium tuolumnensis*
	2-7 Tuolumne	Red Hills ACEC	BLM, Folsom FO	*Astragalus rattanii* var. *jepsonianus*
3 Southern California Coast Ranges	3-1 Presidio	Golden Gate National Recreation Area	NPS	Serpentine endemics
	3-3 Burri Burri–Edgewood–Jasper Ridge	Jasper Ridge Biological Reserve	Stanford University	Serpentine grassland, bay checkerspot butterfly

	3-3 Burri Burri–Edgewood–Jasper Ridge	Edgewood County Park	Santa Clara County	Serpentine grassland and chaparral
	3-4 Oakland–Hayward	Redwood Regional Park	East Bay Regional Parks	Serpentine endemics
	3-7 Mt. Diablo	Mt. Diablo State Park	California State Parks	Serpentine vegetation
	3-12 New Idria	Clear Creek ACEC	BLM, Holister FO	Serpentine barrens, endemics
	3-12 New Idria	San Benito Mountain RNA	BLM, Holister FO	Serpentine mountain vegetation
	3-19 Figueroa–Stanley Mountains	Sedgwick Ranch Reserve	UC Natural Reserve System	Serpentine endemics
4 Northern California Coast Ranges	4-7 Lassic Mountain	Lassics Botanical Area SIA	USFS, Six Rivers NF	Serpentine barrens
	4-8 Red Mountain–Cedar Creek	Red Mountain ACEC	BLM, Arcata FO	Rare serpentine plants
	4-8 Red Mountain–Cedar Creek	Little Red Mountain ER	CDFG	Serpentine vegetation
	4-9 Red Mountain–Red Mountain Creek	Yolla Bolly–Middle Eel Wilderness	USFS, Mendocino NF	Serpentine vegetation
	4-15 Stonyford	Frenzel Creek RNA	USFS, Mendocino NF	Serpentine endemics
	4-17 Wilbur Springs–Walker Ridge	Bear Creek Botanical Management Area	California Department of Transportation	Serpentine grassland with wildflowers
	4-17 Wilbur Springs–Walker Ridge	Indian Valley Wildlife Area	CDFG	Serpentine endemics
	4-18 Harbin Springs–Knoxville	Knoxville Wildlife Area and Ceader Roughs RNA	CDFG and BLM, Ukiah FO	Serpentine vegetation, Sargent cypress
	4-18 Harbin Springs–Knoxville	McLaughlin Reserve	UC Natural Reserve System	Serpentine vegetation, investigations

(continued)

Table F-1 (continued)

Domain	Locality	Name of reserve or protected area	Contact	Features
	4-19 The Cedars	Cedars ACEC	BLM, Ukiah FO	Serpentine endemics
	4-22 Occidental	Harrison Grade ER	CDFG	Rare serpentine plants
	4-23 Mount Tamalpais	Mt. Tamalpais State Park	California State Parks	Serpentine chaparral, endemics
	4-24 Tiburon	Ring Mountain Preserve	Marin County Open Space	Serpentine grassland, endemics
5 Klamath Mountains	5-3 Josephine	Rough & Ready Creek Botanical Wayside	Oregon State Parks	Serpentine vegetation, stream terraces
	5-3 Josephine	Bear Basin Butte Botanical SIA	USFS, Six Rivers NF	Diversity of conifers
	5-3 Josephine	Beatty Creek RNA	BLM, Medford District	Jeffrey pine savanna
	5-3 Josephine	Craig's Creek RNA	USFS, Siskiyou NF	Gabbro and serpentine endemics
	5-3 Josephine	Horse Mountain Botanical SIA	USFS, Six Rivers NF	Serpentine vegetation
	5-3 Josephine	L.E. Horton RNA	USFS, Six Rivers NF	Serpentine vegetation
	5-3 Josephine	North Fork Smith River Botanical SIA	USFS, Six Rivers NF	"Crown jewel" of Klamath serpentine endemism
	5-3 Josephine	Myrtle Creek Botanical SIA	USFS, Six Rivers NF	Pitcher plants
	5-3 Josephine	Eight Dollar Mountain Preserve and ACEC	TNC and BLM, Medford District	Serpentine fen, many rare plants
	5-3 Josephine	Rough and Ready Creek Preserve and ACEC	TNC and BLM, Medford District	Serpentine endemics
	5-3 Josephine	Whetstone Savanna Preserve	TNC	Serpentine vegetation

5-3 Josephine	Woodcock Bog RNA	BLM, Medford District	Serpentine wetland
5-4 Cave Junction	Cedar Log Flat RNA	USFS, Siskiyou NF	Port Orford cedar
5-4 Cave Junction	North Fork Silver Creek RNA	BLM, Medford District	Serpentine vegetation
5-4 Cave Junction	Pipe Fork RNA	BLM, Medford District	Serpentine endemics
5-7 Marble Mountain–Kings Castle	Marble Mountains Wilderness	USFS, Klamath NF	Serpentine vegetation, broad elevation range
5-9 Wrangle Gap–Red Mountain	Gray Back Glade RNA	BLM, Medford District	Serpentine endemics
5-10 Trinity	Cedar Basin RNA	USFS, Shasta–Trinity NF	Serpentine endemics, Port Orford cedar
5-10 Trinity	Crater Creek RNA	USFS, Klamath NF	Dry interior Klamath Mountains
5-10 Trinity	Mount Eddy RNA	USFS, Shasta–Trinity NF	High elevation serpentine vegetation
5-10 Trinity	Preacher Meadow	USFS, Shasta–Trinity NF	Serpentine wetland
5-12a Rattlesnake Creek	Smokey Creek RNA	USFS, Shasta–Trinity NF	Serpentine endemics
6 Blue Mountains			
6-1 Canyon Mountain	Strawberry Wilderness	USFS, Malheur NF	Serpentine Vegetation
7 Northern Cascade–Fraser River			
7-1 Stuart Lake Belt	Rubyrock Lake	BCPP	Serpentine habitat
7-4 Fidalgo Island	Washington Park	City of Anacortes	Coastal serpentine
7-5 Twin Sisters	Olivine Bridge Natural Area Preserve	Washington Department of Natural Resources	Serpentine Vegetation

(continued)

471

Table F-1 (continued)

Domain	Locality	Name of reserve or protected area	Contact	Features
8 Gulf of Alaska	7-6 Ingalls Complex	Alpine Lakes Wilderness	USFS, Wenatchee NF	Serpentine endemics
	8-9 Fairweather	Glacier Bay National Park	NPS	Glaciers
	8-10 Mt. La Perouse	Glacier Bay National Park	NPS	
9 Denali–Yukon	9-4 Talkeetna (9-4a Mt. Russell)	Denali National Park	NPS	
10 Northern Alaska–Kuskokwim Mountains	10-13 Livengood	Serpentine Slide RNA	BLM, White Mountain National Recreation Area	Serpentine barrens

Abbreviations: ACEC, Area of Critical Environmental Concern; BCPP, British Columbia Provincial Parks; BLM, US Bureau of Land Management; CDFG, California Department of Fish and Game; FO, Field Office; USFS, US Forest Service; NPS, US National Park Service; RNA, Research Natural Area; SIA, Special Interest Area; TNC, The Nature Conservancy; UC, University of California.

APPENDIX G: CALIFORNIA PLANT TAXA ENDEMIC TO SERPENTINE

Serpentine endemics (species and subspecies) were identified by reviewing all available sources, including Kruckeberg (1984), Hickman (1993), county and regional floras, online databases, and herbarium records, and scoring each taxon for the degree of serpentine affiliation reported by each source. Scores were then averaged for each taxon, and a cutoff score was determined that corresponded with Kruckeberg's (1984) definition of serpentine endemics (i.e., those species having roughly 85% or more of their occurrences on serpentine; table G-1). See Safford et al. 2005 for further details.

References

Hickman, J.C. (ed.). 1993. The Jepson Manual: Higher Plants of California. University of California Press, Berkeley.

Kruckeberg, A.R. 1984. California Serpentines: Flora, Vegetation, Geology, Soils, and Management Problems. University of California Press, Berkeley.

Safford, H.D, J.H. Viers, and S. Harrison. 2005. Serpentine endemism in the Calfornia flora: a database of serpentine affinity. Madroño 52: 222–257.

Table G-1 Plants in California that are endemic to serpentine.

Taxon	Family	KL[a]	NC	BA	SC	SN
Lomatium ciliolatum	Apiaceae		1	1	1	
Lomatium congdonii	Apiaceae					1
Lomatium engelmannii	Apiaceae	1	1			
Lomatium hooveri	Apiaceae		1			
Lomatium howellii	Apiaceae	1				
Lomatium marginatum	Apiaceae	1	1			1
Lomatium tracyi	Apiaceae	1	1			
Perideridia bacigalupii	Apiaceae					1
Perideridia leptocarpa	Apiaceae	1				
Sanicula peckiana	Apiaceae	1				
Asclepias solanoana	Asclepiadaceae		1			
Antennaria suffrutescens	Asteraceae	1	1			
Arnica cernua	Asteraceae	1				
Arnica spathulata	Asteraceae	1				
Balsamorhiza sericea	Asteraceae	1				
Calycadenia pauciflora	Asteraceae		1			
Chaenactis suffrutescens	Asteraceae	1				
Cirsium fontinale var. *campylon*	Asteraceae			1	1	
Cirsium fontinale var. *fontinale*	Asteraceae			1	1	
Cirsium fontinale var. *obispoense*	Asteraceae				1	
Cirsium hydrophilum var. *vaseyi*	Asteraceae		1	1		
Ericameria ophitidis	Asteraceae	1	1			
Erigeron angustatus	Asteraceae		1			
Erigeron bloomeri var. *nudatus*	Asteraceae	1				
Erigeron petrophilus var. *sierrensis*	Asteraceae					1
Erigeron serpentinus	Asteraceae		1			
Eriophyllum latilobum	Asteraceae			1	1	
Harmonia guggolziorum	Asteraceae		1			
Helianthus exilis	Asteraceae	1	1		1	1
Lagophylla minor	Asteraceae		1			1
Layia discoidea	Asteraceae				1	
Lessingia arachnoidea	Asteraceae			1		
Lessingia micradenia var. *glabrata*	Asteraceae			1		
Lessingia micradenia var. *micradenia*	Asteraceae			1		

Table G-1 (continued)

Taxon	Family	KL[a]	NC	BA	SC	SN
Lessingia ramulosa	Asteraceae		1			1
Madia doris-nilesiae (= *Harmonia*)	Asteraceae	1				
Madia hallii (= *Harmonia*)	Asteraceae		1			
Madia stebbinsii (= *Harmonia*)	Asteraceae		1			
Pyrrocoma racemosa var. *congesta*	Asteraceae	1				
Raillardella pringlei	Asteraceae	1				
Rudbeckia californica var. *glauca*	Asteraceae	1	1			
Senecio clevelandii var. *clevelandii* (= *Packera*)	Asteraceae		1			
Senecio clevelandii var. *heterophyllus* (= *Packera*)	Asteraceae					1
Senecio eurycephalus var. *lewisrosei* (= *Packera*)	Asteraceae					1
Senecio greenei (= *Packera*)	Asteraceae	1	1			
Senecio layneae (= *Packera*)	Asteraceae					1
Senecio macounii (= *Packera*)	Asteraceae	1				
Solidago guiradonis	Asteraceae				1	
Vancouveria chrysantha	Berberidaceae	1	1			
Cryptantha hispidula	Boraginaceae	1	1			
Cryptantha mariposae	Boraginaceae					1
Arabis aculeolata	Brassicaceae	1				
Arabis constancei	Brassicaceae					1
Arabis koehleri var. *stipitata*	Brassicaceae	1				
Arabis macdonaldiana	Brassicaceae	1	1			
Cardamine nuttallii var. *gemmata*	Brassicaceae	1				
Cardamine pachystigma var. *dissectifolia*	Brassicaceae		1			1
Caulanthus amplexicaulis var. *barbarae*	Brassicaceae				1	
Draba carnosula	Brassicaceae	1				
Streptanthus albidus ssp. *albidus*	Brassicaceae			1	1	
Streptanthus barbatus	Brassicaceae	1				
Streptanthus barbiger	Brassicaceae		1			
Streptanthus batrachopus	Brassicaceae		1	1		
Streptanthus brachiatus var. *brachiatus*	Brassicaceae		1			
Streptanthus brachiatus var. *hoffmanii*	Brassicaceae		1			
Streptanthus breweri var. *breweri*	Brassicaceae	1	1	1	1	

(continued)

Table G-1 (continued)

Taxon	Family	KL[a]	NC	BA	SC	SN
Streptanthus breweri var. *hesperidus*	Brassicaceae		1			
Streptanthus drepanoides	Brassicaceae	1	1			1
Streptanthus glandulosus ssp. *pulchellus*	Brassicaceae		1	1		
Streptanthus howellii	Brassicaceae	1				
Streptanthus morrisonii ssp. *elatus*	Brassicaceae		1			
Streptanthus morrisonii ssp. *hirtiflorus*	Brassicaceae		1			
Streptanthus morrisonii ssp. *kruckebergii*	Brassicaceae		1			
Streptanthus morrisonii ssp. *morrisonii*	Brassicaceae		1			
Streptanthus niger	Brassicaceae		1	1		
Streptanthus polygaloides	Brassicaceae					1
Thlaspi californicum	Brassicaceae		1			
Campanula griffinii	Campanulaceae		1	1	1	
Campanula rotundifolia	Campanulaceae	1				
Campanula sharsmithiae	Campanulaceae			1	1	
Githopsis pulchella ssp. *serpentinicola*	Campanulaceae					1
Nemacladus montanus	Campanulaceae		1	1	1	
Minuartia decumbens	Caryophyllaceae	1	1			
Minuartia howellii	Caryophyllaceae	1				
Minuartia rosei	Caryophyllaceae	1	1			
Minuartia stolonifera	Caryophyllaceae	1				
Silene campanulata ssp. *campanulata*	Caryophyllaceae	1	1			
Silene hookeri ssp. *bolanderi*	Caryophyllaceae	1	1			
Silene serpentinicola	Caryophyllaceae	1				
Calystegia collina ssp. *collina*	Convolvulaceae		1	1		
Calystegia collina ssp. *oxyphylla*	Convolvulaceae		1			
Calystegia collina ssp. *tridactylosa*	Convolvulaceae		1			
Calystegia collina ssp. *venusta*	Convolvulaceae				1	
Dudleya abramsii ssp. *bettinae*	Crassulaceae				1	
Dudleya abramsii ssp. *murina*	Crassulaceae				1	
Dudleya setchellii	Crassulaceae			1	1	
Sedum albomarginatum	Crassulaceae					1
Sedum eastwoodiae	Crassulaceae		1			
Cupressus macnabiana	Cupressaceae		1			1
Cupressus sargentii	Cupressaceae		1	1	1	
Carex gigas	Cyperaceae	1				1

Table G-1 (continued)

Taxon	Family	KL[a]	NC	BA	SC	SN
Carex obispoensis	Cyperaceae				1	
Carex serpentinicola (new taxon)	Cyperaceae	1				
Carex serratodens	Cyperaceae		1	1	1	1
Polystichum lemmonii	Dryopteridaceae	1	1			1
Arctostaphylos bakeri ssp. *bakeri*	Ericaceae		1			
Arctostaphylos bakeri ssp. *sublaevis*	Ericaceae		1			
Arctostaphylos hispidula	Ericaceae	1	1			
Arctostaphylos hookeri ssp. *franciscana*	Ericaceae			1	1	
Arctostaphylos hookeri ssp. *montana*	Ericaceae			1	1	
Arctostaphylos hookeri ssp. *ravenii*	Ericaceae			1	1	
Arctostaphylos obispoensis	Ericaceae				1	
Arctostaphylos viscida ssp. *pulchella*	Ericaceae	1	1			
Astragalus clevelandii	Fabaceae		1		1	
Astragalus whitneyi var. *siskiyouensis*	Fabaceae	1	1			
Lathyrus biflorus	Fabaceae		1			
Lathyrus delnorticus	Fabaceae	1				
Lupinus constancei	Fabaceae		1			
Lupinus spectabilis	Fabaceae					1
Trifolium longipes var. *elmeri*	Fabaceae	1	1			
Quercus durata var. *durata*	Fagaceae		1	1	1	1
Garrya buxifolia	Garryaceae	1	1			
Garrya congdonii	Garryaceae	1	1	1		1
Centaurium tricanthum	Gentianaceae		1	1		
Gentiana setigera	Gentianaceae	1	1			
Phacelia breweri	Hydrophyllaceae			1	1	
Phacelia corymbosa	Hydrophyllaceae	1	1			
Phacelia dalesiana	Hydrophyllaceae	1				
Phacelia greenei	Hydrophyllaceae	1				
Iris bracteata	Iridaceae	1				
Iris innominata	Iridaceae	1	1			
Acanthomintha duttonii	Lamiaceae			1	1	
Monardella antonina ssp. *benitensis*	Lamiaceae				1	
Monardella follettii	Lamiaceae					1
Monardella palmeri	Lamiaceae				1	

(continued)

Table G-1 (continued)

Taxon	Family	KL[a]	NC	BA	SC	SN
Monardella stebbinsii	Lamiaceae				1	
Trichostema rubisepalum	Lamiaceae		1		1	1
Pinguicula vulgaris ssp. *macroceras*	Lentibulariaceae	1				
Allium diabloense	Liliaceae				1	
Allium fimbriatum var. *purdyi*	Liliaceae		1			
Allium hoffmanii	Liliaceae	1	1			
Allium jepsonii	Liliaceae					1
Allium sanbornii var. *congdonii*	Liliaceae					1
Allium sharsmithiae	Liliaceae			1	1	
Allium tuolumnense	Liliaceae					1
Brodiaea coronaria ssp. *rosea*	Liliaceae		1			
Brodiaea pallida	Liliaceae					1
Brodiaea stellaris	Liliaceae		1			
Calochortus clavatus var. *clavatus*	Liliaceae				1	
Calochortus greenei	Liliaceae	1				
Calochortus obispoensis	Liliaceae				1	
Calochortus raichei	Liliaceae		1			
Calochortus tiburonensis	Liliaceae		1	1		
Chlorogalum grandiflorum	Liliaceae					1
Chlorogalum pomeridianum var. *minus*	Liliaceae		1	1	1	
Chlorogalum purpureum var. *reductum*	Liliaceae				1	
Erythronium citrinum var. *roderickii*	Liliaceae	1				
Erythronium helenae	Liliaceae		1			
Fritillaria biflora var. *ineziana*	Liliaceae			1		
Fritillaria falcata	Liliaceae			1	1	
Fritillaria purdyi	Liliaceae	1	1			
Fritillaria viridea	Liliaceae				1	
Hastingsia serpentinicola	Liliaceae	1	1			
Lilium bolanderi	Liliaceae	1				
Triteleia crocea var. *modesta*	Liliaceae	1				
Triteleia ixioides ssp. *cookii*	Liliaceae				1	
Hesperolinon adenophyllum	Linaceae		1			
Hesperolinon bicarpellatum	Linaceae		1			
Hesperolinon congestum	Linaceae		1	1		
Hesperolinon didymocarpum	Linaceae		1			

Table G-1 (continued)

Taxon	Family	KL[a]	NC	BA	SC	SN
Hesperolinon disjunctum	Linaceae		1	1	1	
Hesperolinon drymarioides	Linaceae		1			
Hesperolinon serpentinum	Linaceae		1			
Hesperolinon spergulinum	Linaceae		1	1		
Hesperolinon tehamense	Linaceae		1			
Sidalcea hickmanii ssp. *anomala*	Malvaceae				1	
Sidalcea hickmanii ssp. *viridis*	Malvaceae		1	1		
Camissonia benitensis	Onagraceae				1	
Clarkia franciscana	Onagraceae			1	1	
Clarkia gracilis ssp. *tracyi*	Onagraceae		1			
Epilobium rigidum	Onagraceae	1				
Epilobium siskiyouense	Onagraceae	1				
Cypripedium californicum	Orchidaceae	1	1	1		1
Dicentra formosa ssp. *oregana*	Papaveraceae	1				
Achnatherum lemmonii var. *pubescens*	Poaceae	1	1			
Calamagrostis ophitidis	Poaceae		1	1		
Poa piperi	Poaceae	1				
Collomia diversifolia	Polemoniaceae		1	1		
Linanthus ambiguus	Polemoniaceae			1	1	
Linanthus nuttallii ssp. *howellii* (= *Leptosiphon*)	Polemoniaceae		1			
Navarretia jaredii	Polemoniaceae				1	
Navarretia jepsonii	Polemoniaceae		1			
Navarretia rosulata	Polemoniaceae		1	1		
Phlox hirsuta	Polemoniaceae	1				
Chorizanthe breweri	Polygonaceae				1	
Chorizanthe palmeri	Polygonaceae				1	
Chorizanthe ventricosa	Polygonaceae				1	
Eriogonum alpinum	Polygonaceae	1				
Eriogonum congdonii	Polygonaceae	1				
Eriogonum hirtellum	Polygonaceae	1				
Eriogonum kelloggii	Polygonaceae		1			
Eriogonum libertini	Polygonaceae	1	1			
Eriogonum luteolum var. *caninum*	Polygonaceae		1	1		

(continued)

Table G-1 (continued)

Taxon	Family	KL[a]	NC	BA	SC	SN
Eriogonum nervulosum	Polygonaceae		1			
Eriogonum pendulum	Polygonaceae	1				
Eriogonum siskiyouense	Polygonaceae	1				
Eriogonum strictum var. *greenei*	Polygonaceae	1	1			
Eriogonum strictum var. *proliferum*	Polygonaceae	1	1			1
Eriogonum ternatum	Polygonaceae	1	1			
Eriogonum tripodum	Polygonaceae		1			1
Eriogonum umbellatum var. *humistratum*	Polygonaceae	1	1			
Calyptridium quadripetalum	Portulacaceae		1			
Lewisia oppositifolia	Portulacaceae	1				
Lewisia stebbinsii	Portulacaceae		1			
Aspidotis carlotta-halliae	Pteridaceae			1	1	
Delphinium uliginosum	Ranunculaceae		1			
Ceanothus ferrisae	Rhamnaceae			1		
Ceanothus jepsonii	Rhamnaceae		1	1		
Ceanothus pumilus	Rhamnaceae	1	1			
Rhamnus californica ssp. *occidentalis*	Rhamnaceae	1	1			1
Rhamnus tomentella ssp. *crassifolia*	Rhamnaceae	1	1			
Horkelia sericata	Rosaceae	1				
Ivesia pickeringii	Rosaceae	1				
Galium ambiguum var. *siskiyouense*	Rubiaceae	1	1			
Galium andrewsii ssp. *gatense*	Rubiaceae			1	1	
Galium hardhamiae	Rubiaceae				1	
Galium serpenticum ssp. *scotticum*	Rubiaceae	1				
Salix breweri	Salicaceae		1	1	1	
Salix delnortensis	Salicaceae	1				
Castilleja affinis ssp. *neglecta*	Scrophulariaceae		1	1		
Castilleja hispida ssp. *brevilobata*	Scrophulariaceae	1				
Castilleja miniata ssp. *elata*	Scrophulariaceae	1				
Castilleja rubicundula ssp. *rubicundula*	Scrophulariaceae		1			
Collinsia greenei	Scrophulariaceae	1	1			
Cordylanthus nidularius	Scrophulariaceae			1	1	
Cordylanthus pringlei	Scrophulariaceae		1			
Cordylanthus tenuis ssp. *brunneus*	Scrophulariaceae		1			

Table G-1 (continued)

Taxon	Family	KL[a]	NC	BA	SC	SN
Cordylanthus tenuis ssp. *capillaris*	Scrophulariaceae		1			
Cordylanthus tenuis ssp. *viscidus*	Scrophulariaceae	1	1			1
Mimulus nudatus	Scrophulariaceae		1			
Orthocarpus pachystachyus	Scrophulariaceae	1				
Penstemon filiformis	Scrophulariaceae	1				
Veronica copelandii	Scrophulariaceae	1				
Verbena californica	Verbenaceae					1
Viola cuneata	Violaceae	1	1			
Viola primulifolia ssp. *occidentalis*	Violaceae	1				

From: Safford, H.D, J.H. Viers and S. Harrison. 2005. Serpentine endemism in the California flora: a database of serpentine affinity. Madroño 52: 222–257.

[a]KL, Klamaths; NC, North Coast Range; BA, San Francisco Bay Area; SC, South Coast Range; SN, Sierra Nevada.

GLOSSARY

accreted terrane Allocthonous terrane that has been added to a continent within the last one or two billion years.
algae Immobile green aquatic organisms that fix carbon by photosynthesis. Some are eubacteria, some are protoctists, and some are plants.
allochthonous Something that has been formed in one place and is found in another; something that is foreign, or non-native.
alluvial, alluvium Sediment deposited from suspension in moving water.
amphibole A group of ferromagnesian silicate minerals composed of double chains of silicon tetrahedra. Aluminum can substitute for silicon in the tetrahedra. Iron, magnesium, calcium, and sodium are the most common cations that occur between the bands of double chains.
anion A negatively charged ion.
anorthosite A white to gray monomineralic plutonic rock composed almost entirely of calcic–plagioclase (Ca–feldspar).
antigorite A platy serpentine mineral formed at relatively high temperatures.
aquifer A saturated substratum, such as basalt or limestone, that holds more readily transmittable water than most kinds of substrata.
aquitard A practically impervious substratum that retards the flow of water to or from a saturated substratum.
asbestos Fibrous silicate minerals that have commercial applications. Chrysotile and several kinds of amphibole are asbestos minerals.
autochthonous Something that originated it the same place where it now resides.
available water Water available to plants through their roots.
available-water capacity (AWC) The amount of soil water that a soil wetted to field capacity (approximately 10 kPa, or 0.1 bar, suction) will release to plants before some cultivated plants wilt (about 1.5 MPa, or 15 bar, suction).

barren   An area with scant vegetative cover, or none at all.

basalt   Fine grained, dark-colored, mafic rock consisting mainly of calcic feldspar and clinopyroxene. Basalt lava is fluid and can flow long distances, although flows are attenuated by relatively rapid cooling in oceanic environments.

bog   A plant community of acidic wet areas. Decomposition rates are slow, favoring the development of peat, or undecomposed organic matter.

brucite   A mineral in which the basic unit is a sheet of magnesium octahedra: $Mg(OH)_2$

California Floristic Province   The portion of western North America that is characterized by a Mediterranean climate (cool rainy winters and hot dry summers). This includes most of California, except for the Modoc Plateau, eastern slope of the Sierra Nevada, and southern deserts; it also includes small amounts of adjacent southwestern Oregon and northwestern Baja California. The California Region is practically the same area.

cation   A positively charged ion.

cation exchange   The interchange of positively charged ions among solids or between solids and solutions surrounding the solids.

cation-exchange capacity   The capacity of solids (for example clay minerals and soil organic mater) to retain cations on sites from which they can be displaced by other cations that then occupy the sites. It is commonly designated as CEC.

chalcedony   A cryptocrystalline variety of quartz.

chelation   The binding of a metal atom to an organic molecule at more than one site. Chelates with multiple bonds to multivalent cations are more stable than those with single bonds.

chert   A sedimentary rock composed mainly of chalcedony.

chlorite   A class of 2:1 layer phyllosilicate minerals that are common in serpentinized rocks and soils. Those in serpentinite are ferromagnesian silicates, but aluminum may be a prominent ion in soil chlorites.

chromite   A spinel group mineral with various amounts of magnesium, iron, chromium, and aluminum in a framework of oxygen ions.

chronosequence   A sequence of related things (for example, soils) that differ from one another primarily because of their different ages.

chrysotile   A fibrous serpentine mineral.

Circum-Pacific margin   The edge of Pacific Ocean basin, from South America around past North America to Asia.

clay   Inorganic material with particle <2 mm nominal diameter.

clay mineral   Any mineral, generally a layer silicate, that occurs in very small particles that have nominal diameters <2 µm.

clinopyroxene   A pyroxene containing calcium. Clinopyroxenes are in the monoclinic crystal system, meaning that one axis is inclined, rather than perpendicular, to the other two axes.

colluvial, colluvium   A general term applied to any loose, heterogeneous, and incoherent mass of soil material or rock fragments deposited mainly by mass wasting, usually on a slope or at the base of a steep slope.

concretion   A discrete, more or less spherical, concentrically zoned mass of cemented soil or chemical deposit, such calcite, within a soil.

connate water   Water that has not been in contact with the atmosphere for a very long time, such as water entrapped in the interstices of sedimentary rocks when the sediments were deposited.

consolidated material A coherent mass of material that will not disintegrate in water.

conspecific Applied to individuals that belong to the same species.

continental crust The uppermost layer of the lithosphere that comprises continents. It is generally ~30–70 km thick.

continental margin The edge of a continent, adjacent to an ocean.

craton The more stable parts of continents that have been tectonically inactive for very long periods.

crust The outermost zone of the earth, above the mantle.

cumulate An accumulation of crystals that have settled down through magma before it solidified to form rock.

dehydration Loss of water.

diabase Dark-colored mafic igneous rock with crystalline grains that are smaller than those in gabbro. Chemically it is similar to the coarser-grained gabbro and the finer-grained basalt.

diapir A dome or fold in which the overlying rocks have been pushed upward by a core of (for example) relative light salt (specific gravity [SG] = 2.15) or serpentine (SG = 2.5)

dike (geology, petrography) A tabular body that cuts across the bedding or foliation of a rock.

dolomite A sedimentary rock composed of calcium and magnesium carbonate.

dunite A plutonic rock that is mostly olivine (olivine >90%).

ecotype A locally adapted population of a widespread species. Such species show minor variations in shape, form, and/or physiology that are related to habitat (such as serpentine habitat) and are genetically induced. Nevertheless they can still reproduce with members of other ecotypes of the same species.

edaphology The study of soil effects on plant growth and production.

eluviation The removal of material in solution or suspension from a layer of soil.

endemic A taxonomic class of plants, or other living organisms, that is restricted to limited geographic localities or occurs only on specific substrates, such as ultramafic rock or serpentine soil.

eolian Material transported by wind.

epipedon The uppermost layer of soil in a pedon.

Ericaceous Belonging to the heath family, Ericaceae.

eukaryote A cellular organism with a nucleus that is bound in a membrane.

evaporation The conversion of liquid to gas—for example, water to vapor.

evapotranspiration The combination of water evaporated from the surface of the earth and water transpired by plants.

feldspar A group of alumino–silicate minerals. The main subgroups of feldspars are K-feldspars containing potassium and plagioclase containing sodium and calcium.

fen An area of wet organic soil or organic material, typically alkaline in reaction, or sometimes neutral or very slightly acidic.

ferruginous (regolith) Reddish soil, or other regolith, with much ferric iron.

flora All the plant species that make up the vegetation of a given area.

footslope A comparatively gentle slope from the base of a steep mountain or hill.

gabbro Coarse-grained mafic intrusive rock composed principally of calcic feldspar and clinopyroxene, with or without olivine and orthopyroxene. It is commonly dark-colored, but it can be light-colored (leucogabbro).

geoecology The study of living organisms in their natural geosphere environments that include rock, soil, water, and air.

geosphere The earth, including the lithosphere (rock), pedosphere (soil), hydrosphere (water), and atmosphere (air).

glacial drift A general term for all kinds of materials deposited from a glacier by different processes.

glacial till Unsorted glacial drift.

glacier A large perennial mass of ice that moves slowly because of stresses from its own weight.

graminoids Grasses and similar plants, such as sedges and rushes.

harzburgite A plutonic rock that is composed mostly of olivine and orthopyroxene.

heme-Fe Iron held within a porphyrin ring.

herbivory Animal (for example, insects) feeding on plants.

horizon (soil) see soil horizon

humus Organic matter in soils that has decayed sufficiently that its origin is no longer evident visually.

hydration The addition of water.

hygroscopic (soil) water Water retained in air-dry soil.

hyperaccumulation (plants) The accumulation of a heavy element ($z > 20$) to concentrations greater than $1000\,\mu g/g$ (dry) in a plant.

illuviation The deposition of material that has been moved from one soil layer to another.

infiltration (soil) Passage of water from above ground into a soil.

ionic potential An indication of the polarizing power of an ion, a cation in particular, it is the ratio of the electrical charge of an ion to its ionic radius.

ionization potential The energy required to remove an electron from an atom or a molecule.

iron pellets (soil) Spherical nodules or concretions that contain enough magnetite or maghemite to be attracted to a magnet. They are common in some surface horizons of old serpentine soils.

kimberlite An alkali-rich peridotite consisting of olivine and phogophite (dark colored mica) that is found in diamond-bearing igneous pipes.

komatiite An ultramafic volcanic rock.

laterite Soil or regolith that is weakly cemented by iron oxides but hardens irreversibly upon drying. It is commonly referred to as plinthite.

lateritic soil Highly weathered soil in which aluminum and iron are concentrated by loss of other elements. Iron is highly concentrated in lateritic soils derived from the weathering and leaching of ultramafic parent materials.

lava A molten magma that flows across the surface of the lithosphere. The term is sometimes applied to the rock that solidifies from the magma; for example, komatiite or basalt lava.

leaching Transport of material within a solid, such as soil, by movement as a solute or suspension in water.

leucogabbro A light-colored gabbro in which plagioclase (feldspar) is the dominant mineral.

lherzolite A plutonic rock composed of olivine, orthopyroxenes, and clinopyroxenes. It has a chemical composition similar to that of the mantle.

ligand An anion or a neutral molecule that can combine with a metallic cation to form a complex.

lithosphere The solid earth; that is, not including the atmosphere.

lizardite A platy serpentine mineral formed at relatively low temperatures.

mafic rock Igneous rock composed chiefly of one or more ferromagnesian, dark-colored minerals.

magma Naturally occurring molten rock material generated within the earth, and capable of intrusion and extrusion, from which igneous rocks are derived by solidification.

magnetite A spinel group mineral with ferrous and ferric iron in a framework of oxygen ions. $FeFe_2O_4$

mantle A zone in the interior of the earth, below the crust and above the core. The depth to the core is 3480 km.

marsh A more or less permanently wet area of mineral soil, as opposed to organic soil (peat or muck).

mass wasting Downslope movement of rock or other detritus by gravity, either dry or in a fluid mass, rather than suspended in flowing water. Rock fall, creep, and debris flow are some the kinds of mass wasting.

meadow A treeless area with predominantly graminoids, and generally with some forbs.

mélange A heterogenous medley or mixture of rock materials.

mesic Applied to an environment that is neither extremely wet (hydric) nor dry (xeric). Also, see soil temperature regime.

metamorphism The mineralogical and structural adjustment of solid rocks to physical and chemical conditions which have been imposed at depth below the surface zones of weathering and cementation, and which differ from conditions under which these rocks originated.

metasomatism The process of practically simultaneous capillary solution and deposition by which a new mineral of partly or wholly different composition may grow in the body of an old mineral or mineral aggregate.

meteoric water Water from precipitation of moisture in the atmosphere, rather than water released from magma or rocks of the earth.

mineral A naturally occurring solid in which the atoms have a specific arrangement.

montane Term for that which is characteristic of high altitudes.

neoendemic A species of restricted distribution that has evolved in the location or habitat in which in now resides.

nickel laterite (geology) Lateritic soil developed in ultramafic materials in which nickel is concentrated by differential removal of other elements.

nodule (soil) A discrete, spherical to irregularly shaped mass of cemented soil or weathered rock fragment within a soil.

obduction The process of pushing heavier oceanic crust on top of continental crust by faulting or folding, or both, long continental margins.

ochre yellow (soil) Brownish to reddish yellow in soils with high iron oxyhydroxide contents.

octahedral sheet (mineralogy) A broad network of octahedra that are linked laterally in a sheet one octahedon thick; see octahedron

octahedron (mineralogy) An octahedron has eight sides. Those in phyllosilicates have oxygen and hydroxyl ions at there apices, with a cation in some or all octahedra of an octahedral sheet.

olistostrome A sedimentary deposit consisting of a chaotic mass of intimately mixed heterogenous materials that accumulated as a semi-fluid body by submarine gravity sliding or slumping of unconsolidated sediments.

ophiolite A group of mafic and ultramafic rocks in fossil oceanic crust that has been transported from ocean basins onto continental margins.

orogenic belt A linear region that has been subjected to folding and faulting during the formation of a mountain chain.

orthopyroxene A pyroxene lacking calcium. Orthopyroxenes are in the orthorhombic crystal system, meaning that there are three distinct axes and they are perpendicular to each other.

oxyanion A complex anion containing cations and oxygen; for example phosphate, PO_3^{3-}.

oxyhydroxide An oxide that contains both oxygen and hydroxyl ions; for example goethite, FeOOH.

paleoendemic A species that has been isolated from a once wider distribution.

parent material (soil) Loose or unconsolidated material from which a soil with A and B horizons has developed, or at least a soil with an A horizon.

pedology The study of the upper part of regolith, where changes in it are effected by meteoric water, by exchanges of gases between soil and aboveground atmosphere, and by biological activity.

pedon A three-dimensional body of soil with a ground surface of $1-10\,m^2$ and deep enough to reveal soil down to bedrock or other material from which the soil developed or rests upon.

peneplain A nearly featureless, gently undulating land surface of considerable area which presumably has been produced by subaerial erosion. The original peneplain surface may have been largely dissected, as in the Klamath Mountains where few undulating land surfaces remain, but there are many concordant mountain summits that represent a former peneplain.

peridotite A general term for a coarse-grained rock composed chiefly of olivine with or without other mafic minerals such as pyroxene and amphibole and containing little or no plagioclase. Peridotites encompass more specific rock terms such as dunite, harzburgite (or saxonite), lherzolite, and wehrlite. The upper mantle of the earth is believed to be peridotite.

petrography A discipline in which the concern is the description and classification of rocks.

photosynthesis The production of carbohydates by plants utilizing energy from solar radiation. It is a process in which carbon from carbon-dioxide is fixed as organic carbon.

phyllosilicate A layered mineral composed of silicon tetrahedral and cation octahedral sheets. Examples are mica, serpentine, talc, and clay minerals such as smectite and kaolinite.

physiognomy (vegetation) A physical description of vegetation, without regard to plant taxonomy.

physiography (geology) The physical description of landscapes.

plagioclase A group of triclinic feldspars. They are alumino–silicate minerals containing sodium and calcium in a series from albite, $NaAlSi_3O_8$, to anorthite, $CaAl_2Si_2O_8$.

plagiogranite Silicic plutonic rock with low potassium content.

plutonic Igneous rock that has cooled slowly from magma at appreciable depth within the Earth.

precipitation (meteorology) Water passed from the atmosphere to the ground or plants above ground as rain, snow, and condensation of water on plants.

prokaryote   A cellular organism lacking a membrane-bounded nucleus. Archaebacteria and eubacteria are prokaryotes.

protolith   An igneous or a sedimentary rock from which a given metamorphic rock was transformed by metamorphism.

pyroxene   A group of ferromagnesian minerals, some of which contain calcium, also. The major subgroups are orthopyroxenes, such as enstatite, and clinopyroxenes, such as diopside.

regolith   Disintegrated bedrock or gravel, sand, silt, clay, mud, and other materials at the surface of the earth that have not been consolidated to form rock; soils form within regolith.

rhizosphere   The area around roots that has special chemical properties and microbial populations related to the roots.

saline   Salty–saline soils and water are alkaline if the dominant anion is carbonate or bicarbonate, but they may be neutral or acid if the dominant anions are chlorine and sulfate.

saprolite   Bedrock that has been weathered to a soft mass without appreciable deformation.

sclerophyllous   (botany) Plants with stiff evergreen leaves.

serpentine   (mineralogy) A group of layered Mg-silicate minerals, including chrysotile, lizardite, and antigorite; (pedology, soil) Any soil with sand fractions that are dominated by Mg-silicate minerals; (popular ecology) Any rock that is dominated by minerals with high Mg-silicate concentrations.

serpentinite   A rock consisting mostly of serpentine minerals. $Mg_6Si_4O_{10}(OH)_8$

serpentization   The hydrothermal alteration of Mg-silicate minerals to produce serpentine minerals, which are specific kinds of Mg-silicate minerals.

siderophore   A chemical compound with affinity for iron, specifically Fe(III).

silica   Silicon dioxide, SiO_2. It can be crystalline, as in quartz; cyrptocrystalline, as in chalcedony; or amorphous, as in opal.

silica boxwork   A network of veins where silica has been deposited in weathered ultramafic rock.

silica–carbonate rock   Rock consisting of varying quantities of silica (chalcedony and opal) together with magnesium carbonates and stained rusty red by alteration of iron sulfide minerals. It is often the host rock for economic deposits of cinnabar (mercury) formed by hydrothermal alteration of serpentine. It has a cavernous appearance and forms resistant rock outcrops with sparse vegetation. The less thoroughly replaced material may retain some serpentine.

soil, soils   The nonlithic outermost layer (meters thick, or less) of our planet, where reactions involve water and gases from the atmosphere and the actions of living organisms. Engineers commonly think of the entire regolith as soil.

soil horizon   A layer of soil more or less parallel to the ground surface that is recognized as being different from adjacent layers in soil description.

soil temperature regime   (STR) The pattern and intensity of annual soil temperature cycles. There are nine STR classes in the US Soil Taxonomy: cryic, frigid, isofrigid, mesic, isomesic, thermic, isothermic, hyperthermic, and isohyperthermic. Also, there are three temperature classes for frozen soils: hypergelic, pergelic, and subgelic.

specific gravity The mass of a substance relative to the mass of water.

subduction The process of one crustal block descending beneath another by folding or faulting, or both.

substrate The material inhabited by a living organism or on which an organism is growing.

tabular (geology, petrography) A planar feature with much greater breadth than thickness.

talc A 2:1 layer magnesium silicate mineral. $Mg_3Si_4O_{10}(OH)_2$

talus Coarse colluvium composed of rock fragments, lacking fine earth (particles <2 mm).

terrain An extensive land feature, more general than a landform.

terrane A fault bounded package of rock types or strata that are genetically unrelated to those in adjoining packages.

tetrahedron (mineralogy) A tetrahedron has four sides. Those in phyllosilicates have oxygen ions at their apices, generally with a silicon ion in all tetrahedra of an tetrahedral sheet. Some ions of comparable size, such as aluminum, replace silicon ions in many phyllosilicate minerals.

transform fault A strike–slip fault characteristic of midoceanic ridges and along which the spreading ridges are offset.

transpiration The loss or water from photosynthesizing plants.

ultramafic rock An igneous rock composed chiefly of mafic minerals; for example, olivine, clinopyroxene, and orthopyroxene.

vascular plant A plant that has conductive tissues, such as xylem and phloem. Bryophytes are nonvascular plants.

vein (geology, petrography) A thin fissure, or crack, filling. Chrysotile is common in veins within ultramafic rocks.

weathering Weathering is the physical and chemical alteration of rock or regolith in contact with meteoric water or with air at atmospheric pressure.

wehrlite A plutonic rock that is composed mostly of olivine and clinopyroxene.

xeric A dry, as opposed to wet (hydric) or intermediate (mesic), environment.

PLANT INDEX

Abies 193
  amabilis 234, 354, 362
  concolor 194, 233–234, 277, 317, 339–341
  grandis 226, 234, 317, 353-354
  lasiocarpa 193, 228, 258, 353, 356, 362, 367-368, 371, 387
  magnifica 234
  magnifica var. *shastensis* 341
  procera 234
Acanthomintha obovata 202
Achillea millefolium 140, 240, 356, 366, 369, 371
Aconitum delfinifolium 387
Adenostoma fasciculatum 163–164, 196–197, 207, 217, 222, 225, 230, 268, 269, 276, 282–283, 286, 314, 321, 349
Adiantum
  aleuticum 364, 381
  pedatum 193, 362
Aegilops triuncialis 189, 242
Agave 264, 265
  shawii 243
Agropyron spicatum (see *Pseudoroegneria spicata*)
Aira caryophyllea 348, 366
alder
  green (see *Alnus crispa*)
  white (see *Alnus rhombifolia*)

Allium 191, 200–201, 321, 340, 357
  cratericola 219, 278, 285
  fibrillum 356
  hoffmannii 318
  hyalinum 287
  membranaceum 277
  sanbornii 277, 278
  sanbornii var. *tuolumnense* 285
Alnus
  crispa 235–236, 361, 386, 392, 395
  rhombifolia 235, 277
Alyssum 179
  bertolonii 136
  murale 96
Ambrosia dumosa 264
Amelanchier 194, 258, 337
  alnifolia 191–192, 226, 361, 363, 377
Anemone drummondii 371
Aquilegia eximia 201, 220, 301, 324
Arabidopsis thaliana 163
Arabis 200–201
  constancei 195, 278
  holboellii 364
  macdonaldiana 319
Arbutus menziesii 317, 325, 335–337, 366
Arctagrostis 393
Arctostaphylos 171, 191, 194, 214, 314
  alpina 385

Arctostaphylos (continued)
  bakeri 214–215, 327
  bakeri ssp. *bakeri* 327
  bakeri ssp. *sublaevis* 325
  canescens 316, 346
  glandulosa 230, 268, 325
  glauca 173, 207, 218–219, 222, 225, 230–231, 297, 299
  hookeri 214
  hookeri ssp. *montana* 328
  hookeri ssp. *ravenii* 293
  klamathensis 341
  manzanita 163, 218, 223, 327, 346
  mariposa 287
  nevadensis 173, 194, 226, 234, 238, 275, 277, 281, 316, 337–341, 354, 367-368, 370-371
  obispoensis 214, 304–305
  patula 281, 316, 337–339, 346
  pungens 301
  standfordiana 346
  uva–usi 226, 241, 258, 361-363, 376, 394
  viscida 9, 163–164, 195–197, 207, 218–219, 223, 225, 229–230, 277, 278, 280, 282–283, 286, 298, 314, 316, 321, 325–326, 334–339, 345–346, 349
Arenaria
  aculeata 356
  californica 277
  capillaris 357, 363
  congesta 340, 342
  obtusiloba 193, 216, 356, 370-371
Aristida hamulosa 287
Artemisia
  alaskana 173
  californica 218, 229, 268, 297
  norvegica 174
  tridentata 233, 237, 353, 355-356
Asclepias
  cordifolia 321
  solanoana 219, 321, 348
aspen (see *Populus*)
asphodel, bog (see *Narthecium californicum*)
Aspidotis densa 193, 201, 229, 243, 281, 326, 340, 348, 356-357, 360, 362-369, 371
Asteraceae 176

Astragalus 340
  clevelandii 201, 220
  gambelianus 141
Athyrium filix-femina 381
Atriplex
  barclayana 266
  canescens 264
  polycarpa 307
Avena 203
  barbata 229, 287, 328
  fatua 165, 198, 302

Balsamorhiza
  deltoidea 347
  sagitatta 370
Balsamroot (see *Balsamorhiza*)
barbed goatgrass (see *Aegilops triuncialis*)
bearberry (see *Arctostaphylos* uva-ursi)
Bebbia juncea 264
bedstraw (see *Galium*)
Berberis
  aquilinum 370
  nervosa 338
Berkheya coddii 134, 136
Betula 390
  glandulosa 239, 386-387, 392
  nana (see *B. glandulosa*)
  papyrifera 190, 193, 233, 235–236, 376-377, 386, 394-395
birch (see *Betula*)
bistort (see *Polygonum bistortoides*)
bladderpod (see *Isomeris arborea*)
Blechnum spicant 381
blueberry, bog (see *Vaccinium uliginosum*)
box–thorn (see *Lycium*)
Brahea armata 243
Brassicaceae 171, 176
brittlebush (see *Encilia farinosa*)
Brodiaea 199
  pallida 285
Bromus 285
  diandrus 276
  hordeaceus 165, 198–199, 203, 229, 240, 242, 276, 287, 302, 328, 347, 366
  madratensis 203, 229
  rubens 302, 347
  tectorum 167, 356
buckbrush (see *Ceanothus cuneatus*)
buffaloberry (see *Shepherdia*)

Bupleurum tiradiatum 394
burrobush (see *Ambrosia dumosa*)
Bursera 58, 192
   microphylla 264
   hindsiana 264

Calamagrostis
   canadensis 294
   koelerioides 270, 279
   ophiditis 198, 201, 328
   purpurascens 337, 395
Calliandra californica 265
Calocedrus decurrens 173, 194, 197, 225, 275, 278, 301, 316, 334–339, 345
Calochortus 191
   dunnii 269
   raichei 325
   superbus 277
   tiburonensis 328
Calycadenia multiglandulosa 328
Calycanthus occidentalis 201, 220, 321, 342
Calyptridium quadripetalum 219
Calystegia
   collina 187
   fulcrata 277
   stebbinsii 283
Camissonia benitensis 301
Campanula
   rotundifolia 190
   scabrella 342
candelillo (see *Pedilanthus macrocarpus*)
cardon (see *Pachycereus pringlei*)
Carex 222, 258
   bigelowii 192, 241, 393
   buxbaumii 342
   eburnea 173
   geyeri 228, 354-355
   gigas 341
   multicaulis 348
   rossii 327, 348, 355
   serratodens 201, 220–221, 301
   sitchensis 201, 243, 362
   spissa 268
Caryophyllaceae 176
Cassiope
   mertensiana 341, 362, 367, 379
   tetragona 192, 239–240, 387, 392-393
Castilleja 362
   affinis ssp. *neglecta* 328
   elmeri 371

   minneata 220
   minor 201
   rubicundula 202
Caulanthus 173
   amplexicaulus 171, 307
Ceanothus 171, 194, 214
   crassifolius 268, 269
   cuneatus 163, 197, 217, 225, 230–231, 276, 277, 278, 285–286, 298, 337–338, 345, 349
   ferrisiae 214, 225, 229
   integerrimus 281
   jepsonii 163, 196, 218, 223, 314, 319
   jepsonii ssp. *albiflorus* 321
   jepsonii ssp. *jepsonii* 327
   oliganthus var. *sorediatus* 298
   ophiochilus 214, 269
   prostratus 281
   pumilus 214, 281, 327
   roderickii 283
   thyrsiflorus 293
   velutinus 371
cedar
   incense (see *Calocedrus decurrens*)
   Port Orford (see *Chamaecyparis lawsoniana*)
   yellow (see *Chamaeciparis nootkatensis*)
cenizo (see *Atriplex canescens*)
Centaurea trichanthum 324
Cerastium
   arvense 193, 356
   fontanum 161, 163
Cercidium microphyllum 264
Cercocarpus ledifolius 234, 237, 341, 354, 356-357
Chaenactis thompsoni 193
Chamaebatia australis 269
Chamaecyparis
   lawsoniana 140-141, 194, 208, 221, 238, 316, 342
   nootkatensis 190, 234, 362-363, 367, 379
chamise (see *Adenostoma fasciculatum*)
Cheilanthes
   gracillima 371
   siliquosa 191, 193
Chimaphila umbellata 338
Chlorogalum
   grandiflorum 283, 285
   pomeridianum 241
   purpureum 304

cholla (see *Opuntia*)
Chrysolepis sempervirens 341
Chrysothamnus
  nauseosus 237
  nauseosus ssp. *albicaulis* 278
cinquefoil (see *Potentilla*)
Cirsium
  breweri 342
  fontinale 222
  fontinale var. *campylon* 297
  utahense 356
Clarkia 171
  franciscana 293, 296
Claytonia
  exigua 219
  nivalis 371
  saxosa 348
Cochlearia pyrenaica 161, 163
coffeeberry, thick-leaved (see *Frangula californica*)
Collinsia 191
Collomia 191
Comandra umbellata var. *californica* 346
copalquín (see *Pachychormus discolor*)
Cordylanthus 191, 281
  tenuis 327
Coreopsis stillmanii 285
Cornus canadensis 233, 376
creosotebush (see *Larrea divaricata*)
Crepis
  atrabarba 364
  pleurocarpa 340–342
crowberry (see *Empetrum nigrum*)
Cryptantha intermedia 277
Cupressus 197
  arizonica ssp. *stephensoni* 214, 216, 268
  bakeri 195
  forbesii 268
  lawsoniana 201
  macnabiana 172–173, 195–196, 218, 220, 229, 276, 278, 280, 321
  macrocarpa 293
  sargentii 163–164, 173, 195, 197, 201, 215, 219–220, 299, 320, 321, 325, 327
Cuscuta californica 136
Cymopterus terebinthinus 356
Cyperaceae 176
cypress, sargent (see *Cupressus sargentii*)

Cypripedium californicum 201, 325, 342

dactylillo (see *Agave*)
Dactylis glomeratus 184
Danthonia 202
  californica 348
Darlingtonia californica 201, 221, 238, 276, 325, 335
Daucus pusillus 366
deer grass (see *Muhlenbergia rigens*)
deerbrush (see *Ceanothus integerrimus*)
Delphinium uliginosum 201, 324
Deschampsia
  cespitosa 192, 240, 258, 348
  danthonioides 348
Dichelostemma pulchella 287
Distichlis spicata 324, 342
Douglas-fir (see *Pseudotsuga menziesii*) 379
Douglasia nivalis 371
Draba
  aureola 342
  densifolia 356
Drosera rotundifolia 221, 342
Dryas 240
  dummondii 192, 387
  octapetala 192, 241, 258, 393
Dulichium arundinaceum 324

Eleocharis 240
  macrostachys 240
  pauciflora 240
Elymus
  elymoides 198, 281, 319, 347, 356, 366, 371
  glaucus 198
  multisetus 240, 321
  triticoides 324
Empetrum nigrum 239–241, 379, 387, 393
Encilia farinosa 58, 192, 243, 264, 266
Epilobium
  angustifolium 235
  minutum 326
  nivium 319
Ephedra 264
Epipactis gigantean 201, 220
Ericacae 176
Ericameria ophitidis 346
Erigeron
  bloomeri 338

compositus 341
peregrinus 356
Eriodictyon californicum 320–321, 346
Eriogonum 192, 200–201, 264, 356, 362-363
  alpinum 342
  compositum 167
  dasyanthemum 321
  fasciculatum 237, 307
  flavum 237
  heracleoides 356
  kelloggii 319
  libertini 338, 348
  luteolum var. *canium* 328
  nudum 321
  ovalifolium 364
  pyrolaefolium 216
  siskiyouense 216, 342
  tripodum 277
  umbellatum 277, 338, 356, 370
  vimineum 321
  wrightii 173, 237,
  wrightii ssp. *subscaposum* 323
Eriophyllum
  confertiflorum 328
  lanatum 356, 371
  staechadifolium 293
Errazurizia benthamii 192, 264
Eryngium vaseyi 240
Erysimum tortulosum 371
Erythronium 191
  multiscapoideum 279
Eschscholzia californica 229, 240
Euphorbia
  magdaleneae 264
  misera 264
Euphorbiaceae 176

Fabaceae 176
fairyduster (see *Calliandra californica*)
Ferocactus 192, 265
Festuca 319
  altaica 192, 241, 376, 386-387, 392, 394
  californica 298, 327, 334–335
  idahoensis 194, 198, 224, 281, 334–335, 354, 356, 366
  pacifica 306
  rubra 161, 163–164
  viridula 371

Ficus petiolaris 243
fir
  grand (see *Abies grandis*)
  noble (see *Abies procera*)
  Pacific silver (see *Abies amabilis*)
  red (see *Abies magnifica*)
  subalpine (see *Abies lasiocarpa*)
fireweed (see *Epilobium angustifolium*; *Eriodictyon californicum*)
Fouquieria
  columnaris 243
  diqueti 58, 192, 264
Fragaria virginiana 371
Frangula californica 220–221
Frasera albicaulis 356
Fremontodendron 171
  californicum 173
  californicum ssp. *decumbens* 283
  californicum var. *napense*
Fritillaria 191
  agrestis 285
  pluriflora 202

Galium 281, 298
  yollabolliensis 347
Galvesia juncea 264
Garrya 194
  buxifolia 217, 337–338
  congdonii 196, 217–219, 314, 321, 349
  fremontii 281
Gastridium ventricosum 203
Gaultheria shallon 366
Gentiana calycosa 318
Geocaulon lividum 394
Githopsis pulchella 277
Grimmia 356
grouseberry (see *Vaccinium scoparium*)
Guillenia 171

Harmonia 170, 173
Hastingsia alba 201, 221
Haumaniastrum robertii 133
Helianthemum suffrutescens 283
Helianthus
  annuus 165
  bolanderi 277
  exilis 165, 187, 201
Hemizonia 171, 191
  congesta 198

Hemizonia (continued)
  congesta var. *congesta* 328
  congesta var. *luzulifolia* 297
hemlock
  mountain (see *Tsuga mertensiana*)
  western (see *Tsuga heterophylla*)
Hesperevax sparsiflora 198, 296
Hesperolinon 170, 191, 196, 219
  congestum 328
  micranthum 349
Heteromeles arbutifolia 196, 218, 223, 229–230, 239, 321, 325
Heuchera cylindrica 356
Hieracium 191
Holodiscus 325
  microphyllus 339
Hordeum 242
  brachyantherum 324
  murinum 307
Horkelia 356
Hulsea nana 341

Indian's dream (see *Aspidotis densa*)
Isomeris arborea 264
Ivesia
  gordonii 341
  tweedei 371

Jojoba (see *Simmondsia chinensis*)
Juncus
  balticus 222–223, 324
  parryi 341
  mexicanus 324
  xiphioides 222, 324
juniper
  low (see *Juniperus communis*)
  Rocky Mountain (see *Juniperus scopulorum*)
  western (see *Juniperus occidentalis*)
Juniperus
  californicus 173, 195
  communis 190–191, 226, 228, 235–236, 258, 316, 356, 361-363, 367-368, 371, 377, 379, 386, 394
  occidentalis 233
  scopulorum 193, 233, 243, 354, 366

Kalmia
  polifolia 221, 238
  polifolia var. *microphylla* 342
Koeleria macrantha 198, 366

Kruckeberg holly fern (see *Polystichum kruckbergii*)

Larrea divaricata 192, 264
Lasthenia 171, 231, 240
  californica 163, 167, 198–199, 231, 285, 296, 323
  glaberrima 202,
  glabrata 324
Lathyrus biflorus 318
laurel, bog (see *Kalmia polifolia*)
Layia 191
  discoidea 301
  galioides 323
  platyglossa 198
leather oak (see *Quercus durata*)
Ledum
  glandulosum 221, 227, 233, 238, 371
  groenlandicum 379
  palustre 392-394
Lentisco (see *Rhus lentii*)
Lepidium 356
Lesquerella occidentalis 341
Lessingia 136, 191
  micradenia var. *micradenia* 328
Leutkea pectinata 367
Lewisia
  leana 340–341
  rediviva 277, 285
Ligusticum grayi 356
Linanthus 171
  nuttallii 338
Lithocarpus densiflora 141, 194, 238, 316, 335–337
lodgepole pine (see *Pinus contorta*)
Lolium multiflorum 198, 202, 222, 240–242, 296, 328
Lomatium 191
  congdonii 285
  cuspidatum 193, 360, 371
  marginatum 195, 277
Lotus
  purshianus 203
  wrangelianus 141, 199, 349
lupine, miniature (see *Lupinus bicolor*)
Lupinus 192, 340, 362, 370, 387
  arbustus 364
  arcticus 241, 362, 393
  albifrons 287
  bicolor 258
  constancei 318

leucophyllus leucophyllus 356
   spectabilis 285
Lycium
   andersonii 264
Lythrum 222

Machaerocereus gumosus 192, 264, 265
Madia 191
Malosma laurina 268
mariola (see *Solanum hindsianum*)
manzanita
   bakeri (see *Arctostaphylos bakeri*)
   bigberry (see *Arctostaphylos glauca*)
   pinemat (see *Arctostaphylos nevadensis*)
   whiteleaf (see *Arctostaphylos viscida*)
Melastomataceae 176
Melica 337
   bulbosa 364
   harfordii 327
   torreyana 218, 298
Menyanthes trifoliata 342
Menziesii ferruginea 367
Microseris californica 198
Mimulus 171, 191
   guttatus 163–165, 168–170, 201, 222, 324
   layneae 279
   nudatus 165, 168–170, 201, 219, 324
   pardalis 168, 170
   torreyi 279
Minuartia 192, 362
   arctica 241, 243, 394
   californica 198
   decumbens 318
   douglasii 349
   nuttallii 340–341
   obtusiloba 258
   rubella 135, 190
Monardella 281
   hypoleuca ssp. *lanata* 270
   odoratissima 3430
mountain avens (see *Dryas*)
Muhlenbergia rigens 268
muskbush (see *Ceanothus jepsonii*)
Myrtaceae 176

Narthecium californicum 201, 221, 342
Nassella
   lepida 197
   pulchra 197–198, 229, 240–241, 287, 296, 305, 321

Navarretia 171, 191, 196
Nolina interrata 269

oak
   blue (see *Quercus douglasii*)
   huckleberry (see *Quercus vaccinifolia*)
   leather (see *Quercus durata*)
Opuntia 192, 265

Pachistema myrsenites 356, 362, 371
Pachycereus pringlei 192, 243, 265
Pachychormus discolor 58, 192, 243, 264
palo Adán (see *Fouquieria diqueti*)
palo verde (see *Cercidium*)
paper birch (see *Betula papyrifera*)
Parvisedum congdonii 277
Pedicularis 319
Pedilanthus macrocarpus 264
Pellea mucronata 326
Penstemon
   procerus 216, 356
   rupicola 341
Pentagramma triangularis 298
Perideridia 191, 201, 347
Phacelia 191, 196
   corymbosa 278, 340
   davidsonii 371
   fruticosus 371
   purpusii 277
Phlox, cespitose (see *Phlox pulvinata*)
Phlox
   diffusa 243, 277, 281, 338–342, 366, 368
   hirsuta 236
   pulvinata 258, 355
Phyllodoce
   empetriformis 341, 367-368
   grandulifera 367, 369
Picea 193
   breweriana 337
   engelmannii 229, 232, 258, 368, 371
   glauca 190, 193, 232–233, 376, 392, 394-395
   mariana 394
   sitchensis 190, 193, 317
Pickeringia montana 269, 280
pine
   foothill (see *Pinus sabiniana*)
   foxtail (see *Pinus balfouriana*)
   gray (see *Pinus sabiniana*)
   knobcone (see *Pinus attenuata*)

pine (*continued*)
  lodgepole (see *Pinus contorta*)
  sugar (see *Pinus lambertiana*)
  white (see *Pinus strobus*)
  whitebark (see *Pinus albicaulis*)
Pinguicula macroceras 221
Pinus
  albicaulis 193, 229, 234–235, 258, 338, 341, 353, 356, 362,264, 368, 371
  attenuata 141, 194–195, 197, 223, 268, 276, 305, 314, 334–335
  balfouriana 227, 338, 341
  brutia 176
  contorta 141, 167, 190, 194, 228, 258, 314, 335, 361-362, 367, 370-371, 379
  contorta ssp. *murryana* 341
  coulteri 223–225, 268, 301
  densiflora 176
  jeffreyi 140, 173, 194–195, 197, 224–225, 230, 233, 275, 301, 316, 334–345
  lambertiana 194, 227, 233, 335–340, 345
  monticola 194, 226, 228, 316, 335–341, 353, 370-371
  nigra 176
  pentaphylla 176
  ponderosa 193–194, 197, 207, 226, 228, 342, 353, 356, 363-364, 370-371
  radiata 293
  sabiniana 196, 219, 225, 229, 276, 277, 278, 285–286, 298, 320, 345
  strobus 234, 258
pitaya agria (see *Machaerocereus gummosus*)
pitcherplant, California (see *Darlingtonia californica*)
Plantago erecta 198, 225, 231, 240, 285, 293, 296
Poa
  curtifolia 193, 216, 360, 371
  pratensis 348
  pringlei 341
  secunda 198, 240, 299, 302, 348, 354
Polemonium
  chartaceum 342
  pulcherrimum 356
Polygala 281
Polygonum
  bistortoides 258, 356
  newberryi 371

Polypogon monspeliensis 222
Polystichum
  imbricans ssp. *imbricans* 348
  kruckebergii 360-361, 363, 375
  lemmonii 191, 195, 229, 339–340, 356, 363, 367, 371
  mohroides (see *P. lemmonii*)
  scopulinum 193, 201, 360
Populus
  balsamifera 235, 377
  tremuloides 235, 361, 377
Potentilla 258
  flabellifolia 356
  fruticosa 341, 395
Prunus
  emarginata 363, 371
  virginiana 363
Pseudoroegneria spicata 167, 173, 193, 226, 228, 236, 342, 354, 356, 362-364, 368-369, 371
Pseudotsuga menziesii 193–194, 207–208, 225–226, 228, 233, 278, 314, 325, 334–339, 345, 353, 356, 361-363, 367-368, 370-371
Ptelea crenulata 277
Purshia tridentata 228, 233, 370
Pyrola secunda 386

Quercus
  agrifolia 268, 327
  berberidifolia 163, 217, 268
  douglasii 236, 276, 278, 285
  durata 9, 163, 172–175, 197, 216, 218–219, 223–225, 229–230, 277, 278, 280, 297, 298, 314, 321, 325–326, 345
  garryana 236 345
  garryana var. *breweri* 345
  kelloggii 282–283
  sadleriana 316
  vaccinifolia 173, 194, 234, 238, 275, 281, 316, 335–337, 341
  wislizeni 282–283

Raillardiopsis scabrida 318
Rhamnus 194
  californica 140-141, 316, 321, 325, 335–339
  illicifolia 285
  rubra 277
  tomentella 201, 218, 222, 277, 321, 327

Rhododendron
  albiflorum 227, 361-362, 367
  occidentale 140, 201, 220, 316, 346
Rhus lentii 264
Ribes 356
Romanzoffia sitchensis 341
Rosa 361, 368
  acicularis 190, 235, 377
Rosaceae 176
Rubiaceae 176
Rubus
  chamaemorus 386
  leucodermis 243, 362
Rumex
  acetosa 190, 394
  acetosella 237

sagebrush (see *Artemisia*)
salidillo (see *Atriplex barclayana*)
Salix 192, 362, 385-387, 393
  breweri 201, 220, 301, 321, 323
  planifolia 239
  reticulata 239
Salvia
  apiana 268
  mellifera 238–239, 268, 297
  sonomensis 269
sandwort (see *Arenaria* and *Minuartia*)
Sanguisorba stipulata 387
Sanicula tracyi 318
Saxifraga
  brachialis 362
  mertensiana 341
  oppositifolia 190
  tricuspinata 241, 377
Schoepfia californica 266
Sebertia accuminata 135
sedge (see *Carex*)
Sedum
  albomarginatum 195
  divergens 356
  lanceolatum 338, 342, 354, 364, 387
  laxum ssp. *eastwoodiae* 319
  laxum ssp. *flavidum* 318
Selaginella
  densa var. *scopolorum* 356
  hansenii 287
Senecio 191, 277
  clevelandii 201, 324
  eurycephalus ssp. *lewisrosei* 278–279
  ganderi 270

  layaneae 277, 283
  pauperculus 371
Sequoia 171
serpentine morning glory (see *Calystegia collina*)
serpentine sunflower (see *Helianthus exilis*)
serviceberry (see *Amelanchier*)
Shepherdia canadensis 235, 361-362, 377
Sibbaldia procumbens 367
Sidalcea 171
  hirsuta 277
Silene
  acaulis 190, 192, 241, 367, 387, 393
  campanulata ssp. *campanulata* 319
  douglasii 356
Simmondsia chinensis 264
Sisyrinchium
  bellum 231, 223, 324
  idahoensis 342
Solanum hindsianum 265
Solidago guiradonis 301
Spiraea
  betulifolia 370
  beuvardiana 361
  densiflora 238
spruce 390
  brewer (see *Picea breweriana*)
  Engelmann (see *Picea engelmannii*)
  Sitka (see *Picea sitchensis*)
  white (see *Picea glauca*)
spurge (see *Euphorbia*)
Stachys albens 220
starflower (see *Trientalis latifolia*)
Stipa thrurberiana 354
Streptanthus 170–171, 173, 191, 196, 200–201
  albidus 170
  barbatus 348
  barbiger 135
  brachiatus 236
  breweri 219, 301
  glandulosus 170–171
  glandulosus var. *hoffmanii* 325
  morrisonii 219, 321
  niger 170, 328
  polygaloides 135, 277, 278, 285
Symphorocarpus 356
Syrax officinalis 218
subalpine fir (see *Abies lasiocarpa*)
sundew (see *Drosera rotundifolia*)
sweetbush (see *Bebbia juncea*)

Taxus brevifolia 363, 371
Tetracoccus dioicus 270
Thalaspi 163, 167
   alpestre 193, 356, 371
   goesingense 167
   montanum 134-136
thistle, Mount Hamilton (see *Cirsium fontinale*)
Thuja 193
Tofieldia glutinosa 221, 342
torote (see *Bursera*)
toyon (see *Heteromeles arbutifolia*)
Trichostema simulatum 277
Trientalis
   borealis 361
   latifolia 234, 347, 363
Trifolium fucatum 202
Triglochin maritima 342
Triteleia lutea var. *scabra* 287
Tsuga
   heterophylla 193, 208, 234, 367, 379
   mertensiana 190, 228, 234, 338, 341, 362, 367, 379

Umbellularia 171
   californica 196, 218, 236, 277, 278, 316, 325, 335

Vaccinium 194, 241
   alaskense 367
   caespitosum 361
   membranaceum 234, 361-363, 368
   myrtillus 227, 361
   ovalifolium 228, 367
   ovatum 317, 335
   parviflorum 316–317, 335
   scoparium 226, 234, 258, 354, 368
   uliginosum 190, 221, 233, 238–239, 342, 379, 393-394
   vitis-idaea 394
Verbena californica 285
Veronica copelandii 342
Viola
   cuneata 135
   douglasii 277
   pinetorum 340
Vulpia 347
   megalura 203
   microstachys 167–168, 184, 185, 199, 225, 232, 285, 296
   myuros 302

willow
   Brewer's (see *Salix breweri*)
   tea–leaf (see *Salix planifolia*)
Wyethia reticulata 283

Xerophyllum tenax 316, 337–340

yampa (see *Perideridia*)
yerba santa (see *Eriodictyon californicum*)
Yucca
   valida 243
   whipplei 305

GENERAL INDEX

absolane 65, 67
accreted terranes 257, 261, 354, 358, 375, 390
Agness 313, 315
Agua Blanca fault 259
Agua Tibia Wilderness 269
Ahklun Mountains 384
Alaska 4, 9, 55, 70, 263
Alaska Range 382, 385
Alaskan–type ultramafic body 49, 69, 267, 280–281, 363, 373, 375, 378, 381, 384
Aldrich Mountains 350, 353–354
Alfisols 57–58, 60, 63, 67, 71–73, 258, 267, 275, 277–278, 283–285, 287, 292, 297, 314, 317, 320, 333, 337, 339, 341–342, 353, 355
algae 5, 41, 80, 93–95, 97, 105, 114, 461
  essential elements 107
alkaline springs and alkali seeps 325, 342, 419
allochthonous terranes 257, 271, 274
alpine 258
alpine meadow 387
Alpine–type serpentinite 23
alpine tundra 382, 385, 386–387, 392
alteration, hydrothermal 23
aluminum 7, 14, 22, 51, 57, 62, 69, 75, 80, 83, 107, 359, 390, 427–429, 432–434, 437–441, 445
American Creek 395
American River 274, 282
amphibole 18. 23, 62, 127, 131, 267–268, 414, 429
amphibolite (including hornblendite) 264, 268, 277, 284, 293, 303, 306, 312, 384, 393
amphibolite faces 276, 323, 332
Andes Mountains 4
Andisols 55, 57, 70, 375
animals 6, 35, 41, 79–80, 84–86, 97, 461
  elemental toxicities 83–84, 128
  essential and beneficial elements 80–85, 197, 129–131
  biomass 84
anorthosite 17, 19, 258, 267
anthophyllite 276
antigorite 21–24, 76, 278, 285, 287, 362, 432–434
Antilles 4, 54, 62
ants 86
Appalachian Mountains 59
Appalachian Piedmont 58
aquifer 28, 418
Arctic 4, 141
Arctic Circle 390

Aridisols 58, 70, 77, 263–266
Arolik 385
arsenic 46
asbestos
  in dust 413–415
  fibers 6, 20, 23–24, 297, 300–301, 413–415
  minerals 6, 20, 23–24, 282, 284, 414
Asik Mountain
atmosphere 5, 6, 41, 46, 48, 70, 75, 80, 84, 93–94, 98, 105, 114, 126, 325
  dust 122, 124–125, 127, 129, 141
  wet deposition 114, 125, 127
Avan Hills 391

bacteria 37, 41, 80, 83, 86, 91, 93, 95–96, 105, 114, 123, 128–129, 132, 134, 460–461
  elemental toxicities 83–84
  essential and beneficial elements 80–85, 136
  symbiotic 97, 105, 114, 122
Baja California 4, 9, 50, 55, 58–58, 70, 259–270
Baldy Mountain 350, 356
barrens 7, 51, 140, 281, 317–318, 320–321, 326, 344, 348, 356, 382, 394, 418
basalt 17, 20, 27, 83, 264, 299, 310, 312, 322–323, 359, 385
Bear Mountain block 274, 282
Bear Mountain faults 274, 282–284
Beartooth Mountains 258
Bear Valley 322
Bear Wallow fault 343, 348
beetles 89–90
Bell Springs peneplain 329
Ben Bow Mine 258
beryllium 80
bicarbonate 24–25, 31, 325
Big Bend fault 274
Big Blue formation 307
Big Pine fault 291
Big Red Mountain 339–340
biomass 41, 80, 84–85, 92, 95–96, 136, 141
Blashke Islands 381
blueschist (including glaucophane) 263, 291, 294–295, 300, 303, 306, 312, 315, 328, 385
Border Ranges fault 375, 377

botanist 20
Boulder Peak 338
Bradford Mountain 326
Bridge River group (or assemblage) 359, 361
British Columbia 4, 9, 58, 358, 372, 382
bromine 46, 80
Brooks Range 4, 59, 70, 388, 390
brucite 18, 21–24, 32, 293–294, 297, 300, 302, 304, 431
bryophytes 141–142, 364, 381, 391, 462
bugs 89
Bureau of Land Management 301
Burro Mountain 302–303
butterflies 5, 87–89, 294

Cache Creek assemblage 359–360
cadmium 83
Calaveras fault 297
Calaveras River 283
calcic horizons 263–264
calcite 26, 31, 281
calcium 437
  in dust 280
  in mantle 81, 417
  in plants and other organisms 80–83, 85, 94–95, 112, 134
  in rocks and minerals 7, 13–14, 17, 22, 26, 67–69, 83, 94, 268, 276–277, 281, 342, 418–419
  in soils 54, 67–69, 110–114, 292, 324, 364
  in water 31–33, 343, 419
calcium carbonate (*see* carbonate)
calcium hydroxide 25–26, 31, 325
calcium/magnesium ratio 278
  soil 280, 282, 324, 326
California 4–6, 8, 49–52, 54–55, 57–67, 70, 74, 76–77, 85–88, 92, 94–95, 263
California Counties
  Alameda 295, 298
  Amador 274, 283
  Butte 271, 278
  Calaveras 283–284
  Colusa 310, 320–322
  Contra Costa 94, 122
  Del Norte 94, 308, 329, 334–335
  ElDorado 271, 276–277
  Fresno 271, 274, 286, 300, 302. 307
  Glenn 320
  Humboldt 317, 343

Kings 302
Lake 79, 114, 310, 322–323, 325
Los Angeles 415
Madera, 274
Marin 328
Mariposa 274, 285–286
Mendocino 59. 64, 305, 314, 319, 321
Merced 299
Monterey 302
Napa 79, 312, 323, 325–326, 420
Nevada 276, 280
Orange 267
Placer 61, 277
Plumas 271, 277–278
Riverside 267–269
San Benito 299–300, 307
San Diego 267–270
San Joaquin 55
San Luis Obispo 304, 306
San Mateo 79
Santa Barbara 305–306
San Bernardino 415
Santa Clara 79, 300
Santa Cruz 416
Shasta 329, 340, 422
Sierra 276, 278
Siskiyou 340
Sonoma 314, 323, 325–327
Stanislaus 55, 299–300
Tehama 320, 329, 343
Trinity 59, 60, 313, 317, 319, 340, 343
Tulare 271, 274
Tuolumne 284–285
Yolo 415–416
Yuba 278
Califonia Critical Areas Program 328
California Department of Health Services 415–416
California Floristic Province 420
Calmalli 58, 264–265
Cantwell fault 385
Canyon Mountain 354–355
Cape Newenham 384
carbon 80, 84
  fixation 80, 91, 141
  organic 37, 41–42, 46, 75, 124
carbon dioxide 22–23, 26, 80, 325
carbonate 22–23, 26, 32, 34, 54, 83, 142, 278, 291, 313, 353, 419
Cariboo Mountains 358
Caribou Mountain 392

Carolin Mine 362
Cedar Mountain 298–299
Cedars 313, 325
Cedros Island 259, 263–264
Cement Bluff 53
Central Metamorphic Belt (Klamath Mountains) 332
Chatham Straight 373
chert 20, 83, 293–294, 304, 312, 322–323, 332, 343, 359, 385, 394
Chetco River 310
Chilkat River 377
chlorine 80, 343, 419
chlorite (*see* clay minerals, chlorite) 433–434
Christian complex 390, 393
Chrome 320
Chrome Mountain 258
chromite 18, 50, 62, 64–65, 67, 77, 129, 157, 354, 368, 429
chromium 18, 22, 51, 67, 80, 129, 326, 429, 437–441
  in water 415–416
chrysotile 20–24, 49, 263, 278, 282, 284–285, 293–294, 297, 300, 302, 304, 335, 354, 432–434
Chugach Mountains 372–373, 375–377
Chulitna River 385
cinnabar 297, 300
Circum–Pacific margin 14, 17
clay 97
  particles 39, 42, 52–54
  soil clay 54, 57, 61–63, 68–72, 74, 76–77, 100–101, 275, 292, 313
clay minerals 41, 49, 53, 110, 127
  chlorite 18, 64–65, 74, 278, 333, 355, 445
  gibbsite 62, 75, 431, 445
  illite (and hydrous mica) 74, 445
  kaolinite 62, 275, 292, 313, 333, 431–434, 445
  serpentine 333, 355, 445
  smectite 62, 64–65, 67, 69, 72, 74–76, 275, 292, 296, 313–314, 333, 431–434
  vermiculite 74, 333
Cleveland National Forest 269–270
Cleveland Peninsula 378–379
Clinton Creek 388
Coast Range fault (zone) 312, 320–321, 323, 348

Coast Range ophiolite 295, 297, 302, 304
Coast Range thrust 299, 304, 312, 320
cobalt 7, 22, 64, 67, 84, 96, 132–133, 282, 326, 429, 437–441
Cold Fork fault 348
Columbia River basalt 350
Condrey Mountain 332
continental margin 14, 16, 17, 20–22, 24, 257, 312
copper 80, 83–84, 437–441
Cordilleran ice sheet 358
core of earth 7, 13–14
Costa Rica 4
Covello 319
Coyote Ridge 79, 296
cracking–clay (see soil)
craton 257
Cretaceous 261, 264, 291, 360–361, 364, 390
crust 7, 24, 131
  continental 13–14, 17, 23, 132, 417–419
  ocean 13–17, 20, 23, 161, 275, 291, 306, 310, 312, 320, 362, 375, 390, 417–419
Cuba 4, 74, 89, 96
Cuesta Ridge 304–305
cumulate 23
Cuyamaca Valley 288
Cuyamuca Peak 267, 269
cyanobacteria 41, 93, 95, 97, 114, 122
Cypress Island 366

dehydration 24, 141
Del Puerto Canyon 299
Denali fault 382–385
Devonian 332, 340, 385–386
diabase 13, 17, 20, 304, 312, 340, 386
Diablo Range 297, 300
diapir 16, 291, 300, 302, 313
dikes, sheeted 13, 15, 304, 310, 375
Dishna complex 390, 393
dolomite 23, 394
Dubakella Mountain 343
Duke Island 378
dunite 16, 18–19, 23, 276, 278–281, 285, 293, 299, 303, 315, 340, 360–361, 363, 366, 373, 375–377–378, 384–386, 391–395, 418, 431

earthworms 90–91, 461
East Bay Regional Parks District 296
Eastern Klamath Belt 332
eclogite 327
ecology 3
ecosystems 413
edaphologist 37, 38
edaphology 35
Edson Butte 315
Eel River 310
Eight Dollar Mountain 66, 336
Eklutna 376–377, 380
Elder Creek 320
electronegativity 439–431
electron orbits 437–439
Elsinor fault 257
endemic plants (see plants, endemic)
Endemic Species Act
endemism, endemic species 9, 85, 89, 135–136, 142, 420–421
Entisols 58–59, 70–71, 76–77, 263–266, 282, 301, 318, 337–338, 345, 363, 375
Environmental Protection Agency (U.S.) 414–415
Eocene 62, 78
eolian sand 293
erosion
  geologic 274
  soil (see soil, erosion)
evapotranspiration 27, 103–104, 463–466
Evergreen thrust fault 296

Fairweather Range 377–378
Feather River 276–278
Feather River complex 274, 275
feldspar 62, 69, 76, 127, 310
fen 333, 336, 342, 378
ferns 364
ferrihydrite 132
Fidalgo Island 364–366
Figueroa Mountain 306–307
fluorine 80, 437–438
fog 259, 288, 310, 330
Forest Hill Divide 276–277
Franciscan complex 54, 63, 107, 291, 293–295, 297, 299, 302–304, 306, 310, 312, 319, 321, 323, 327
Fraser River 358–359
Fraser River fault 358–359, 361–362

Frenzel Creek 320
fungi 5, 37, 79, 79, 91–94, 105, 128, 131, 136, 420, 462
   elemental toxicities 83–84
   essential and beneficial elements 80–85
   symbiotic 97, 114

gabbro 13, 17, 19–20, 69, 83, 258, 259, 261, 263, 266–270, 276–277, 279–280, 282, 294, 304, 312, 322–323, 332, 336, 340, 343, 353–354, 357, 361–362, 375, 377–378, 381, 384–386, 391–393, 419
garnet 393
Gelisols 59, 390, 392
geoecology 3–5, 417–419, 422
geoecosystems 5–6, 79, 141, 419–421
Geyser Peak 326
glacial drift (including till) 367, 369
glaciation 330, 358, 364, 372, 379, 328, 388, 420
Glacier Bay National Park 377–378
glaciers 377
glaucophane (see blueschist)
Glomales 92–93
Goat Mountain 310
goethite 64–65, 67, 72, 74–75, 128, 130, 132, 318, 429, 421, 445
Golden Gate National Park 293
Golden Mountain 378–379
Goodyears Creek fault 274
gophers 85
Gorda plate 15, 330
Grasshopper Mountain 363–364
graywacke 293–294, 310, 313, 335
Great Valley (California) 271, 274, 288, 308
Great Valley sequence 293, 295, 299, 302, 306, 312–313, 322–323, 325
greenschist facies 264, 385
greenstone 293–294, 313, 320, 322–323, 332, 361, 368, 394–395
Greenville fault 298–299
Guatay Mountain 269
Guatemala 4
gypsum 324, 445

Harbin Springs 323
Harrison Grade 327
harvestmen 85–86
harzburgite 16, 18–19, 32, 263, 276, 278–279, 285, 293, 303, 315, 320, 323, 340, 354, 361, 366, 384–386, 391–395, 418
Hayward fault 295–297
Healdsburg 326
hematite 65, 72, 74–75, 132, 318, 429, 431, 445
Histosols 55, 335, 345
Hog Mountain 287
Holocene 364
Hopland
Howell fen 60, 336
Hozameen fault 362
Hunters Creek bog 316
Hunters Point 293
Hurst complex 390, 393
hydration 47–49
hydrogen 22, 24, 83, 436–437
hydroxyl ion 26, 343, 431

Illinois River 335
Inceptisols 60, 258, 292, 314, 318, 325, 333, 337, 339, 341–342, 353, 355, 360–363, 368, 375–376, 379, 392, 394
inert gases 80
Ingalls complex 360, 367–371
Ingalls Mountain 359, 368
Ingalls Peak 359
insects 37, 41, 84–86, 89, 136
interior Alaska highland 388
interior Alaska lowland 388
Interior Plateau (British Columbia) 358–359
intrusion 17, 275
   layered 18
iodine 46, 80
Ione formation 62
ionic potential 438–439
iron 123, 131–133, 437–441
   pellets (nodules) 64–65, 130
   in plants and other organisms 80–84, 112, 127, 130–132
   in rocks and minerals 3, 7, 13–14 17–18, 22–23, 31, 49–51, 58–59, 67–68, 76, 81–83, 131, 418–419, 427–429
   in soils 49–51, 67–69, 71–72, 107, 123, 131–133, 268, 318, 359, 379, 390, 427, 445
   in water 415

Iron Mountain 258, 315, 369
iron oxides (and oxyhydroxides, see goethite, hematite, maghemite) 62, 65, 68, 75–77, 100, 123–124, 130, 275, 292, 314, 333, 418, 445
Iyikrok Mountain 391

jade 286
Jade Mountain 391
Jarbo Gap 278–279
Jasper Ridge 79
Jasper Ridge Biological Preserve 294
John Day 354
Josephine Mountain 335
Josephine ophiolite 333, 335
Juan de Fuca plate 15
Jurassic 261, 263, 274, 279, 285, 291, 294, 299, 323, 332, 343, 354, 359, 366, 390

Kanuti River 392
Kaweah River 287
Kenai Peninsula 372–373, 376
kingdoms of life 460–462
Kings River 286–287
Klamath Mountains 63–64, 71–73
Klamath National Forest 339
Klamath peneplain 329, 335–336, 338
Klamath River 310, 330, 338
Knik River 376
Knoxville 323
Kodiak Island 372–373, 375
komatiite 18
Koyukuk–Yukon basin 388, 390
Kuskokwim Mountains 388, 390

Laguna Mountain 269
landslide 61, 68, 73, 304, 313, 315, 343
La Perouse intrusion 377–378
Lassic Mountain 317
Lassics Botanical and Geological Area 318
lava 13, 18
Leech Lake Mountain 319
Lewis (acid) softness index 339–431
lherzolite 16, 18, 340
lichens 41, 91–94, 97, 105, 114, 377, 386–387, 390, 393
 crustose 281
 feathery 264, 267, 391
 leafy 386, 391–392, 394

limestone 274, 312, 332, 359, 394
lithic contact 291
lithosphere 5
litter (see plant detritus)
Little Red Mountain 318
Livengood ultramafic body 394
liverworts 41, 141–142
lizardite 19, 21–24, 49, 263, 284, 293–294, 297, 300, 302, 304, 336, 354, 362
loess 394–395
Loma Prieta 297
Los Pinos Mountain 267–268

Maclaren Glacier 386
Magdalena Island 259, 266–267
maghemite 65, 75, 431
Magitchlie Range 393
magma, basaltic 13, 17, 23, 258
magnesia 75
magnesite 50, 64, 74, 263, 281, 292, 299
magnesium 437–439
 in mantle 82, 417
 in plants and other organisms 80–83, 134, 112, 94–95
 in rocks 3, 6, 13, 17–18, 22–25, 49–51, 67–69, 82–83, 107, 133, 268, 418–419, 427–429
 in soils 49–51, 67–69, 83, 107, 133, 263, 280, 292, 342, 364, 427
 in water 31–33, 419
magnetism 65
magnetite 18, 21–24, 50, 63, 64, 68, 76–77, 293–294, 297, 300, 302, 304, 381
manganese 22, 67, 80, 437–441
 in plants and other organisms 80, 112, 127, 130–131
 in rocks and minerals 14, 22, 67, 130, 427–429
 in soils 130–131, 427
 in water 415–416
manganese oxide (and oxyhydroxide) 64, 123
mantle 7, 13–18, 22–23, 83, 417
Marmolejo Creek 304
Maryland 76
mass movement (see landslide) 306–307, 330
Mayacamas Mountains 325
McGinty Mountain 267

General Index 507

McLaughlin Reserve 79, 324
Mehrten formation 284
melange 20, 54, 261, 263, 274, 278, 285, 287, 291, 293–294, 297, 300, 302–303, 306, 312–313, 320, 332, 336, 361
Melones fault 274, 276–277, 284–286
Mentasta Pass 386
Merced River 286
mercury 83, 300, 326, 415
Mesozoic 287, 332, 343, 390
metamorphism
microorganisms 5, 35, 37, 41, 79, 84, 95–96
  elemental toxicities 83–84
  essential and beneficial slements 80–84, 95, 132
  nitrogen transformations 105, 122, 133
  symbiotic 105
Miocene 284, 300, 307, 329, 335
Misheguk Mountain 391
mites 37, 41
Moccasin Peak 284
Mokelumne River 283
Mollisols 61, 275, 284–286, 292–294, 296–300, 302–305, 314, 320, 322, 324, 326, 328, 333, 338–339, 341–342, 345, 353, 355, 363, 375–376, 387, 395
molybdenum 80, 95–96
Montana 18
Monumental Ridge 281
moss 80, 128, 141–142, 362–363, 367, 376–377, 387, 392, 394
  sphagnum 394
Mount Barnham 387
Mount Burnett 379
Mount Diablo 297
Mount Eddy 330, 340, 342
Mount Eklutna 376
Mount Lazaro 378
Mount Mazama 353
Mount Russell 385
Mount Saint Helena 323, 325
Mount Sidney Williams 360, 368
Mount Stuart 367
Mount Tamalpais
mycorrhiza 79, 91–93, 420

Nacimiento fault 291, 303
Nahlin 386–387

nematodes 35, 37, 89, 91, 461
neutron 436–437
Nevada (Esmeralda, Humboldt, Mineral, and Nye Counties) 58
Nevada City 279–280
New Almaden 297
New Caledonia 75
New Idria 300–302
nickel 7, 22, 51, 89, 96, 112, 133–134, 437–441
  in soil 282, 326, 364
nickel hyperaccumulation 89, 95–96, 134
nickel laterite (ferruginous lateritic soil) 55, 64, 67, 315, 317, 330, 335
Nickel Mountain 64, 335
nickeliferous phyllosilicate 64
nitrogen 37, 41–42, 46, 48, 79, 84, 92, 95, 114–123, 134
nitrogen–fixation 84, 95–96, 105, 114, 122–123, 128, 132, 134
nivation hollow 392
norite 69, 25, 258, 378
North American plate 15, 310, 330
Northern Cascade Mountains 358–359

Oakland Hills 296
obduction 14, 17, 310, 312
Occidental 327
oceans 132
octahedral sheets 432–434
octahedron 427–429
Okanagan highland 58
Oligocene 329
olistostrome 306, 323
olivine (including forsterite) 18–24, 31, 49–50, 54, 62, 64, 66–67, 69, 127, 130–131, 133–134, 263, 267–268, 276, 280, 285, 294, 297, 303, 312, 325, 354, 361–364, 366, 368, 378, 391, 393, 418, 439
Olivine Mountain 359, 363
ophiolite 8, 14–16, 20, 32, 257, 259, 269, 261, 263–267, 259, 261, 263, 266, 300, 304, 306, 317, 319, 322–323, 325–327, 333, 340, 366, 375, 385
ophiolite breccia 322
Oregon 4, 9, 55, 57, 61, 64–65, 67, 70, 93, 96, 101, 111, 122, 124
Oregon Counties
  Coos 60, 308, 315
  Curry 60, 71, 315–316, 334

Oregon Counties (*continued*)
  Douglas 64, 67, 308, 329, 334
  Josephine County 60, 96, 139, 142, 334
organic detritus 38–41, 46, 51, 60, 92, 96, 99, 125, 135, 140, 142
organic compounds 80–84, 97
organic matter
  see soil, organic matter
Orleans thrust fault 336
orogenic belt 4, 16–17, 23
orogeny 7

Otay Mountain 269
Oxisols 61–62, 277
oxygen 7, 22, 42, 49, 83–84, 101, 132–134, 427–434, 438

Pacific Ocean 288
Pacific plate 15, 303
Paleozoic 274, 284, 332, 390
Paleozoic–Triassic Belt 332, 334
palladium 258
paralithic contact 291
Parkfield 302
Paskenta 320
Pearsol Peak 334
pedologist 20, 37–38
pedology 35
Peninsular Ranges 259, 261, 267–270
peridotite 6, 16, 19–21, 28, 31–32, 39, 46, 50, 54, 59–60, 62–69, 73–75, 77, 100, 104, 124, 107, 130–133, 157, 261, 263–264, 285–287, 291, 294, 297–298, 302–304, 306, 310, 312–313, 318, 320, 322–323, 325, 332, 334–340, 342–344, 353–354, 360, 368, 373, 375, 377, 386, 392–393, 419, 431
permafrost 375, 388
Permian 357
Peshastin Creek 369–370
phosphorous 7, 41, 46, 51, 79–81, 92, 123, 124–126, 364, 417
Pine Hill 69, 282–283
Pioneer peridotite 361, 362
Pistol River 317
Pitka sequence 393
plagioclase 18–19, 267–268
plants 97–98
  biomass 85, 97, 136–141
  composition 46, 105
  communities 3, 8–9, 79, 257
  detritus (see *organic detritus*)
  drought tolerance 104–105
  endemic 264, 268–270
  nutrition 35, 40, 46, 79, 97–98, 105–136
  productivity 41–42, 85, 97–98, 136–139
  roots 35, 37, 40–41, 48, 79, 92–93, 96, 99, 102–104, 114
  succulent 263–265
  taxonomy 9
platinum 258
Pleistocene 358, 372, 382, 388
Pliocene 62, 330
podzolization 359, 361, 390
Point Sal 306
Pope Valley 324
potassium 7, 13–14, 51, 81, 83, 127, 417–418, 427–429
Precambrian 390
precipitation 27, 41–42, 49, 51, 54, 99, 138–141, 259
Presidio 293
Preston Peak 330, 332, 336
proton 436–437
protozoa 461
Puerto Nuevo 263–264
Puerto Rico 62
pyroxene (clinopyroxenes and orthopyroxenes) 13, 18–21, 23, 24, 31–32, 49, 50, 54, 74–75, 78, 80, 127, 130–134, 157, 263, 276, 279–281, 285, 294, 297, 303, 312, 325, 343, 354, 357, 361, 366, 368, 375, 418, 428–429, 439
  bronzite 258
  diopside 276
pyroxenite 19, 78, 80–81, 157, 298, 300, 346, 362–363, 372, 376–377, 381, 391, 393, 418

Quaternary 62, 330, 364

Rainbow Mountain 386
Rattlesnake Creek terrane 68, 73–74, 332
Red Flat 316–317
Red Hill 271, 277
Red Hills 285
Red Mountain 64, 271, 278, 287, 299, 304, 313, 318–319, 334–335, 338–340, 343, 376, 382

Red Ridge 322
Redwood Regional Park 296
regolith 37–38, 41, 75, 315
Research Natural Areas
　King Creek 268
　Organ Valley 268
revegetation 139–141
Riddle 313
Ring Mountain 328
Rocky Mountains 58, 257
rodingite 293, 336, 362, 368
Rogue River 310, 316, 330
roundworms (*see* nematodes)
Ruby anticline 389
Ruby Rock Provincial Park 360
Russian River 310

Sacramento River 310, 330, 340
Saddle Mountain 316
Saint Elias Mountains 372
Salinian block 291
Salt Creek fault 343
salts, soluble 445
San Andreas fault 291, 293, 300, 302
San Benito Island 259
San Benito Mountain 300–301
San Francisco Bay 288
San Juan Islands 358–359, 366
San Marcos Mountains 267, 269
San Miguel Mountain 269
San Simeon 304
Santa Ana Mountains 267–268
Santa Clara Valley 296–297
Santa Cruz Mountains 291
Santa Margarita Island 259, 266–267
saprolite 64–65, 315, 317
Sedgwick Range Reserve 307
Seiad Creek 337
selenium 46
serpentine
　colluvium 39
　endemic 4, 9
　etiology, definition 3, 6
　melange 54, 266
　minerals 6, 13, 16–18, 20–24, 26, 32, 49, 54, 263–264, 432–434
　syndrome 4–5, 7
serpentinite 6, 16, 28, 39, 46, 58, 63, 67–69, 74–77, 100, 104, 129, 133, 140, 264, 266, 274, 278–280, 285–286, 291, 293, 297–300, 302, 304, 306, 310, 313, 315, 317, 319–320, 323, 332, 336, 344, 385, 418
Serpentine Slide Research Natural Area 394
serpentinization 16, 31–32, 50
Sexton Mountain 334
Shuksan thrust 366
Shulaps ultramafic body 359, 361
siderite 74
Sierra de Placer 263
Sierra Nevada Metamorphic Belt
　Central Belt 274, 277, 279, 283–284
　Eastern Belt 274
　Western Belt 274, 282
Signal Buttes 316
silica (*see* silicon)
silica–carbonate 287, 297, 300, 325–326
silicate 13, 17, 31, 49, 133
　minerals 7, 18. 26, 31, 57, 62, 66, 69, 127, 129, 428–434
silicon (and silica)
　boxwork 64–65, 67, 271, 283, 315
　cementation 53, 263, 379, 445
　in rocks and minerals 14, 21–23, 25, 69, 81, 83, 131–132, 427–434
　in soil 49, 54, 66, 74–76, 271, 314, 427
　in water 26, 31–33, 343
silicon tetrahedra 49, 427–429
Siniktanneyak Mountain 391
Siskiyou Mountains 337, 339
Six Rivers National Forest 318
slope stability 68, 344
Smartsville complex 282
smectite (*see* clay minerals, smectite)
Smith River 335
Snake River 350
Snettisham Peninsula 381
Snow Camp Mountain 313, 316
sodium 7, 13–14, 83, 343, 419, 427–428
soil
　aggregation 68, 93, 99–100, 418
　aquic conditions 61, 70, 292, 333, 444, 447
　available–water capacity (AWC) 100–101, 104, 135–139
　cation–exchange capacity (CEC) 61, 63, 72, 77, 111, 445
　classification 8, 55–63, 442–459
　composition 46, 50–51
　cracking–clay 54, 74, 114, 292, 314
　definition 37–38
　development 38–51, 54, 55–67

soil (*continued*)
  drainage 74
  drainage classes 122–125, 128–132
  erosion, erodibility 7, 60, 62, 68, 72–74, 76, 99–100, 140, 300–302, 418
  exchangeable cations 57, 63, 70, 76–77, 107, 110, 127–128, 130, 134, 138, 141, 295, 305, 445
  families 449
  fertility 105–136, 267
  groups (great groups and subgroups) 448, 450–453
  hydric (or wet) 77, 345–346, 348
  living organisms 37, 41, 79–96
  net positive charge 62
  orders of Soil Taxonomy 448–449
  organic carbon (*see* soil, organic matter)
  organic matter (SOM) 41–42, 46, 51, 53, 63, 72, 75–76, 86, 96, 110, 359, 379, 390, 445
  reaction (pH) 62, 114, 129, 133, 138–139, 295, 305
  serpentine parent material 39, 55–78
  slickensides 57, 63, 77
  taxonomy (*see* soil classification)
  water 98–105
  water balance 102–104, 463
  water drainage 72, 74–75, 101–102
  water infiltration 99–100
  water retention 61, 71, 99
  water storage 27, 35, 42, 74, 100–101, 464
soil horizons 38–40, 51–53, 443–446
  albic 375, 379, 395
  argillic horizon 57–58, 61, 63, 72, 74, 77, 275, 294, 305, 313, 316, 320–322, 333, 337, 353, 370
  cambic horizon 59–61, 77, 275, 313, 333, 353, 360, 375, 390
  histic epipedon 313, 333, 375
  kandic horizon 61, 63, 72
  mollic epipedon 59, 61, 72, 74, 77, 275, 313, 333, 353, 375
  ochric epipedon 58, 275, 313, 333, 353
  oxic horizon 61–62, 67, 77–78
  spodic horizon 62, 375
  umbric epipedon 333
soil moisture regime (SMR) 70, 446–447
  aridic 70, 263
  perudic 70
  xeric 70, 275, 277, 292–294, 296–297, 299, 301, 303–304, 333, 335
  udic 70–71, 275, 292, 333, 335
soil series 449, 454–456
  Aldino 58
  Atravasada 300
  Bearvalley 61
  Beaughton 320, 454–455
  Blomidon 59
  Bramlet 344
  Cedar Camp 316
  Cerp 277
  Cerpone 278
  Chrome 58, 76
  Climara 63, 302, 454–455
  Copsey 61
  Cornut 57, 334, 336, 454–455
  Cuesta 305
  Culving 370
  Dann 318
  Deadfall 59
  Delpiedra 282–283, 285, 287
  Dingman 319–320
  Dubakella 57, 72, 277–278, 280–281, 320, 322, 326–327, 334–336, 344, 454–455
  Edfro 63
  Eightlar 334–335, 454–455
  Fancher 282, 287
  Flycatcher 72, 316
  Forbes 61, 67, 277
  Gasquet 63, 72
  Gravecreek 71, 454–455
  Greggo 316
  Grell 71, 353
  Guemes 366
  Henneke 61, 71, 285, 296, 298–300, 303, 305, 320–322, 324, 326–328, 345, 349, 454–455
  Hentine 300
  Hiltabidel 59, 318, 344
  Hungry 317
  Huse 325
  Hyampom 344
  Ishi Pishi 72
  Jorgensen 63
  Kang 61, 71
  Kingmont 58, 370
  Kinotrail 58
  Las Posas 267–270

Leesville 61, 322
Littlered 63, 72, 318
Madden 317
Marksbury 353
Maxwell 63, 320–321, 324
Mislatnah 316
Montara 61, 294–296, 302, 320–321, 324, 326–327, 454–455
Nipe 62
Obispo 294–295, 304–305
Okiota 322, 324, 326
Overholt 353–354
Pearsoll 60, 335, 454–455
Redflat 72, 316–317
Rescue 282
Rosario 62
Round Hill 59
Rustybutte 61, 317
Sebastian 61, 317
Serpen 370
Serpentano 60, 71–72, 316, 454–455
Shadeleaf 71
Snowcamp 71–72, 316
Stussi 353
Tangle 57, 72
Toadlake 57, 72, 454–455
Venado 63, 322
Walnett 57, 454–455
Weitchpec 60, 71, 454–455
Wildmad 281, 344
Woodleaf 278
soil temperature regime (STR) 70, 446
 cryic 70, 258, 333, 338–339, 341, 353, 360, 375
 frigid 70, 281, 292, 316–317, 333, 337–339, 341–342, 353, 360, 375
 hyperthermic 70, 263
 isohyperthermic 70, 292
 isomesic 70, 292–294, 317, 333
 isothernic 70
 mesic 70, 275, 277–278, 281, 292, 297, 299, 301, 316–317, 322, 326–327, 335–339, 341–342, 346
 thermic 70, 267, 275, 278, 282, 292, 294, 296–297, 301, 303–305, 321–322, 324, 326–327, 333, 346, 349
solifluction 392
Sonoran Desert 263
South Fork fault 343, 348

Sparta 357
spinel 429
Spodosols 55, 62–63, 360–361, 367–368, 375
Stanislaus River 284
Stanley Mountain 306–307
Stillwater complex 7, 18, 258
stone stripes 392
Stony Creek fault 323
Stonyford 320
Straight Creek fault 358–360
Strawberry Mountain 350
Strawberry Range 350, 356
Stuart Lake belt 360
subduction 7, 14–17, 24, 310, 323
sulfide 22, 48, 84, 123, 133, 427
sulfur 22, 41, 46, 80, 83–84, 123–127
Suzie Mountain

Table Mountain 284, 302
Takshanuk Mountains 377
talc 18, 21, 23–24, 31, 64, 276, 278, 286, 360, 385, 431–434
Talkeetna Mountains 385
talus 258, 281, 362, 366, 418
Tecate Peak 269
Tedoc Mountain 343, 348
termites 79, 85, 87
terrane
 Alisitos 259
 Baker
 Cochimí 259
 Eastern Hayfork 343
 Marble Mountain 336, 343
 Rattlesnake Creek 336, 343–348
 Smith River 348
 Wallowa 350
 Western Hayfork 343
 Wrangelia 364
Tertiary 282, 284, 300, 302, 312–313, 329, 335, 353–354, 378
tetrahedral sheets 432–434
tetrahedron 427–434
Thompson Peak 330
Tiburon Peninsula 328
timber site index 137–139, 367
Tom Martin Peak 338–339
Tonsina Island Arc complex 377
toxicity 80–84, 89, 90–93, 107, 110, 127–131, 133–136, 139, 141, 439–441

transition elements 80, 105, 107, 112, 123, 127–136, 415
Transverse Ranges 288, 291
travertine 31, 325, 419
tremolite 276, 278, 282, 286
trench 14–17, 24, 312, 335
Triassic 354, 357, 360
Trinity Alps 330
Trinity ultramafic sheet 340
troctolite 19, 378
Tulameen complex 359, 363–364
Tule River 287
Tuolumne River 284
Twin Sisters dunite 366
Twin Sisters Mountain 366–367

Ultisols 63, 275, 277, 314, 318, 333
Union Bay 378–379
Upper Chulitna fault 385

vanadium 80, 88, 123, 127–128, 429, 437–441
Van Duzen River 310
vegetation 8–9, 257
Vertisols 63, 292, 302, 314, 320–322, 324, 333, 345, 348
Viejas Mountain 267
Vinegar Hill 350, 353–354
Viscaíno Peninsula 58–59, 259
volcanic ash 353, 367–368

Walker Ridge 322
Washington 4, 9, 358

water (*see* soil, water)
  springs an seeps 31–32
  runoff 27–30, 32, 46–51, 103, 418
weathering 7, 46–50, 54, 57–59, 61–71, 75–77, 79, 83, 94, 97, 100–101, 124, 127–134, 139, 418
websterite 19, 377
Weimer fault 279
Wenatchee Mountains 359, 371
Western Jurassic Belt 332–334, 348
wetland 297, 318, 324, 419
wet meadow 342
wet soils (*see* soils, hydric)
wherlite 19, 263, 280–281, 377, 391, 393
Wilbur Springs 321–322
Willow Creek 385
Wimer formatiom 330, 335
Wolf Creek fault 274, 279
World Reference Base for Soil Resources 453, 457–459
Wrangle Gap 339

Yakobi Island 377–378
Yalakom fault 359
Yolla Bolly junction 348–349
Yolla Bolly Mountains 310
Yuba River 278, 280
Yuki 393
Yukon Territory 4, 9, 388

Zaca Peak 304
zinc 80, 83–84, 90, 96, 134, 437–441